Communications in Computer and Information Science 1215

Commenced Publication in 2007
Founding and Former Series Editors:
Simone Diniz Junqueira Barbosa, Phoebe Chen, Alfredo Cuzzocrea,
Xiaoyong Du, Orhun Kara, Ting Liu, Krishna M. Sivalingam,
Dominik Ślęzak, Takashi Washio, Xiaokang Yang, and Junsong Yuan

More information about this series at http://www.springer.com/series/7899

Le-Minh Nguyen · Xuan-Hieu Phan ·
Kôiti Hasida · Satoshi Tojo (Eds.)

Computational Linguistics

16th International Conference of the Pacific Association
for Computational Linguistics, PACLING 2019
Hanoi, Vietnam, October 11–13, 2019
Revised Selected Papers

Springer

Editors
Le-Minh Nguyen
Japan Advanced Institute of Science
and Technology
Ishikawa, Japan

Kôiti Hasida
Graduate School of Information Science
and Technology
The University of Tokyo
Tokyo, Japan

Xuan-Hieu Phan
University of Engineering and Technology
Hanoi, Vietnam

Satoshi Tojo
Japan Advanced Institute of Science
and Technology
Ishikawa, Japan

ISSN 1865-0929   ISSN 1865-0937 (electronic)
Communications in Computer and Information Science
ISBN 978-981-15-6167-2   ISBN 978-981-15-6168-9   (eBook)
https://doi.org/10.1007/978-981-15-6168-9

This Springer imprint is published by the registered company Springer Nature Singapore Pte Ltd.
The registered company address is: 152 Beach Road, #21-01/04 Gateway East, Singapore 189721, Singapore

Preface

This book constitutes the refereed proceedings of the 16th International Conference of the Pacific Association for Computational Linguistics (PACLING 2019), held in Hanoi, Vietnam, in October 2019. The 28 revised full papers and 14 short papers presented were carefully reviewed and selected from 70 submissions. The papers are organized into topical sections, including: Text Summarization; Relation and Word Embedding; Statistical Machine Translation; Text Classification; Web Analyzing; Question and Answering; Dialog Analyzing; Speech and Emotion Analyzing; Parsing and Segmentation; Information Extraction; Grammar Error; and Plagiarism Detection.

PACLING 2019 enjoyed many international participants, representing Vietnam, Japan, India, Canada, Taiwan, Thailand, Tunisia, and the UK. The conference is notable in that it included a session for discussing future directions and issues of computational linguistics in Vietnam. We would like to thank all the participants and the committee members for their contributions to PACLING 2019. Our gratitude also goes to the static members of FPT University for the local arrangements and hospitality. We would like to thank keynote speakers (Professor Kentaro Inui, Professor Danushka Bollegala, Dr. Kentaro Torisawa, and Professor Tomoko Matsui) for their interesting and great talks in PACLING 2019. PACLING 2019 was the 16th in the series of biannual meetings that started in 1989. The first two of these events were the Japan-Australia Joint Symposium on NLP held in Australia and then in Japan, followed by 13 PACLING conferences held not only in Australia and Japan, but also in Canada, Malaysia, Indonesia, Myanmar, and Vietnam.

February 2020

Le-Minh Nguyen
Xuan-Hieu Phan
Kôiti Hasida
Satoshi Tojo

Organization

General Chairs

Koiti Hasida   The University of Tokyo, Japan
Satoshi Tojo   JAIST, Japan
Nguyen Khac Thanh   FPT University, Vietnam

Program Chairs

Le Minh Nguyen   JAIST, Japan
Xuan Hieu Phan   VNU UET, Vietnam

Organizing Committee

Kim Anh Nguyen   FPT University, Vietnam
 (Co-chair)
Thi Minh Huyen Nguyen VNU HUS, Vietnam
 (Co-chair)
Hong Viet Le   FPT Corporation, Vietnam
Hung Quy Pham   FPT University, Vietnam
The Trung Tran   FPT University, Vietnam

Program Committee

Thomas Ahmad   University of Minho, Portugal
Kenji Araki   Hokkaido University, Japan
Vataya Chunwijitra   NECTEC, Thailand
Kohji Dohsaka   Akita Prefectural University, Japan
Alexander Gelbukh   Instituto Politécnico Nacional, Mexico
Choochart Haruechaiyasak NECTEC, Thailand
Koiti Hasida   AIST, Japan
Yoshihiko Hayashi   Waseda University, Japan
Kai Ishikawa   Data Science Research Laboratories - NEC
           Corporation, Japan
Hiroyuki Kameda   Tokyo University of Technology, Japan
Yoshinobu Kano   Shizuoka University, Japan
Vlado Keselj   Dalhousie University, Canada
Kiyoshi Kogure   Kanazawa Institute of Technology, Japan
Anh Cuong Le   Ton Duc Thang University, Vietnam
Huong Thanh Le   Hanoi University of Science and Technology, Vietnam
Phuong Le-Hong   Hanoi University of Science and Technology, Vietnam
Thang Ly   FPT University, Vietnam

Diego Molla	Macquarie University, Australia
Huy-Tien Nguyen	JAIST, Japan
Kiem-Hieu Nguyen	Hanoi University of Science and Technology, Vietnam
Kim Anh Nguyen	Institute of Big Data - Vingroup, Vietnam
Le-Minh Nguyen	JAIST, Japan
Ngan Nguyen	Vietnam National University, Vietnam
Thai Phuong Nguyen	Vietnam National University, Vietnam
Thi Minh Huyen Nguyen	Vietnam National University, Vietnam
Tri Thanh Nguyen	Vietnam National University, Vietnam
Truong-Son Nguyen	Vietnam National University, Vietnam
Tien Nguyen-Minh	Hung Yen Technical University, Vietnam
Minh Quang Nhat Pham	Alt Vietnam Co., Ltd.
Anh Phan	Le Quy Don Technical University, Vietnam
Xuan-Hieu Phan	Vietnam National University, Vietnam
Hiroaki Saito	Keio University, Japan
Kazutaka Shimada	Kyushu Institute of Technology, Japan
Akira Shimazu	JAIST, Japan
Kiyoaki Shirai	JAIST, Japan
Thepchai Supnithi	NECTEC, Thailand
Masami Suzuki	KDDI Research, Inc., Japan
Kumiko Tanaka-Ishii	The University of Tokyo, Japan
Satoshi Tojo	JAIST, Japan
Takenobu Tokunaga	TITECH, Japan
Oanh Tran	Vietnam National University, Vietnam
Vu Tran	JAIST, Japan
Hai-Long Trieu	AIST, Japan
Tran Van Khanh	JAIST, Japan
Vinh Van Nguyen	Vietnam National University, Vietnam
Chai Wutiwiwatchai	NECTEC, Thailand
Yang Xiang	University of Guelph, Canada
Ngo Xuan Bach	Posts and Telecommunications Institute of Technology, Vietnam

Contents

Text Summarization

A Submodular Approach for Reference Recommendation.   3
  Thanh-Binh Kieu, Son Bao Pham, Xuan-Hieu Phan,
  and Massimo Piccardi

Split First and Then Rephrase: Hierarchical Generation
for Sentence Simplification.   15
  Mengru Wang, Hiroaki Ozaki, Yuta Koreeda, and Kohsuke Yanai

Abstractive Text Summarization Using LSTMs with Rich Features   28
  Viet Nguyen Quoc, Huong Le Thanh, and Tuan Luu Minh

Relation and Word Embedding

SemSeq: A Regime for Training Widely-Applicable
Word-Sequence Encoders.   43
  Hiroaki Tsuyuki, Tetsuji Ogawa, Tetsunori Kobayashi,
  and Yoshihiko Hayashi

Learning to Compose Relational Embeddings in Knowledge Graphs   56
  Wenye Chen, Huda Hakami, and Danushka Bollegala

Context-Guided Self-supervised Relation Embeddings   67
  Huda Hakami and Danushka Bollegala

Evaluation of Embedded Vectors for Lexemes and Synsets Toward
Expansion of Japanese WordNet.   79
  Daiki Ko and Koichi Takeuchi

Neural Rasch Model: How Do Word Embeddings Adjust
Word Difficulty?.   88
  Yo Ehara

Machine Translation

Dynamic Fusion: Attentional Language Model for Neural
Machine Translation   99
  Michiki Kurosawa and Mamoru Komachi

Improving Context-Aware Neural Machine Translation
with Target-Side Context   112
  Hayahide Yamagishi and Mamoru Komachi

Learning to Evaluate Neural Language Models   123
  James O'Neill and Danushka Bollegala

Recommending the Workflow of Vietnamese Sign Language Translation
via a Comparison of Several Classification Algorithms   134
  *Luyl-Da Quach, Nghia Duong-Trung, Anh-Van Vu,
  and Chi-Ngon Nguyen*

Text Classification

Document Classification by Word Embeddings of BERT   145
  Hirotaka Tanaka, Hiroyuki Shinnou, Rui Cao, Jing Bai, and Wen Ma

Deep Domain Adaptation for Low-Resource Cross-Lingual Text
Classification Tasks   155
  Guan-Yuan Chen and Von-Wun Soo

Multi-task Learning for Aspect and Polarity Recognition
on Vietnamese Datasets   169
  *Dang Van Thin, Duc-Vu Nguyen, Kiet Van Nguyen,
  Ngan Luu-Thuy Nguyen, and Anh Hoang-Tu Nguyen*

Evaluating Classification Algorithms for Recognizing Figurative
Expressions in Japanese Literary Texts   181
  Mateusz Babieno, Rafal Rzepka, and Kenji Araki

Web Analysing

Model-Driven Web Page Segmentation for Non Visual Access   191
  *Judith Jeyafreeda Andrew, Stéphane Ferrari, Fabrice Maurel,
  Gaël Dias, and Emmanuel Giguet*

Update Frequency and Background Corpus Selection in Dynamic TF-IDF
Models for First Story Detection   206
  Fei Wang, Robert J. Ross, and John D. Kelleher

A Pilot Study on Argument Simplification in Stance-Based Opinions   218
  Pavithra Rajendran, Danushka Bollegala, and Simon Parsons

Automatic Approval of Online Comments
with Multiple-Encoder Networks   231
  Vu Dang

Question and Answering, Dialog Analyzing

Is the Simplest Chatbot Effective in English Writing Learning Assistance?. . . 245
Ryo Nagata, Tomoya Hashiguchi, and Driss Sadoun

Towards Task-Oriented Dialogue in Mixed Domains. 257
Tho Chi Luong and Phuong Le-Hong

Timing Prediction of Facilitating Utterance in Multi-party Conversation. 267
Tomonobu Sembokuya and Kazutaka Shimada

Evaluating Co-reference Chains Based Conversation History
in Conversational Question Answering. 280
Angrosh Mandya, Danushka Bollegala, and Frans Coenen

Speech and Emotion Analyzing

Multiple Linear Regression of Combined Pronunciation Ease
and Accuracy Index. 295
Katsunori Kotani and Takehiko Yoshimi

Rap Lyrics Generation Using Vowel GAN. 307
Tomoya Miyano and Hiroaki Saito

Emotion Recognition for Vietnamese Social Media Text 319
Vong Anh Ho, Duong Huynh-Cong Nguyen, Danh Hoang Nguyen,
Linh Thi-Van Pham, Duc-Vu Nguyen, Kiet Van Nguyen,
and Ngan Luu-Thuy Nguyen

Effects of Soft-Masking Function on Spectrogram-Based
Instrument - Vocal Separation. 334
Duc Chung Tran and M. K. A. Ahamed Khan

Parsing and Segmentation

Japanese Predicate Argument Structure Analysis with Pointer Networks. 347
Keigo Takahashi, Hikaru Omori, and Mamoru Komachi

An Experimental Study on Constituency Parsing for Vietnamese. 360
Luong Nguyen-Thi and Phuong Le-Hong

Antonyms-Synonyms Discrimination Based on Exploiting Rich
Vietnamese Features 374
Bui Van Tan, Nguyen Phuong Thai, Pham Van Lam,
and Dinh Khac Quy

Towards a UMLS-Integratable Vietnamese Medical Terminology 388
The Quyen Ngo, My Linh Ha, Thi Minh Huyen Nguyen,
Thi Mai Huong Hoang, and Viet Hung Nguyen

Vietnamese Word Segmentation with SVM: Ambiguity Reduction
and Suffix Capture   400
   Duc-Vu Nguyen, Dang Van Thin, Kiet Van Nguyen,
   and Ngan Luu-Thuy Nguyen

An Assessment of Substitute Words in the Context of Academic Writing
Proposed by Pre-trained and Specific Word Embedding Models   414
   Chooi Ling Goh and Yves Lepage

Effective Approach to Joint Training of POS Tagging and Dependency
Parsing Models.   428
   Xuan-Dung Doan, Tu-Anh Tran, and Le-Minh Nguyen

Information Extraction

Towards Computing Inferences from English News Headlines   439
   Elizabeth Jasmi George and Radhika Mamidi

Extraction of Food Product and Shop Names from Blog Articles Using
Named Entity Recognition   454
   Ryuya Ikeda and Kazuaki Ando

Transfer Learning for Information Extraction with Limited Data   469
   Minh-Tien Nguyen, Viet-Anh Phan, Le Thai Linh, Nguyen Hong Son,
   Le Tien Dung, Miku Hirano, and Hajime Hotta

Self-deprecating Humor Detection: A Machine Learning Approach   483
   Ashraf Kamal and Muhammad Abulaish

Grammar Error and Plagiarism Detection

Deep Learning Approach for Vietnamese Consonant Misspell Correction. . . .   497
   Ha Thanh Nguyen, Tran Binh Dang, and Le Minh Nguyen

Grammatical Error Correction for Vietnamese Using Machine Translation . . .   505
   Nghia Luan Pham, Tien Ha Nguyen, and Van Vinh Nguyen

Developing a Framework for a Thai Plagiarism Corpus   513
   Santipong Thaiprayoon, Pornpimon Palingoon,
   Kanokorn Trakultaweekoon, Supon Klaithin,
   Choochart Haruechaiyasak, Alisa Kongthon, Sumonmas Thatpitakkul,
   and Sawit Kasuriya

Author Index   523

Text Summarization

A Submodular Approach for Reference Recommendation

Thanh-Binh Kieu[1,2(✉)], Son Bao Pham[1,2], Xuan-Hieu Phan[1], and Massimo Piccardi[2]

[1] VNU University of Engineering and Technology,
Vietnam National University, Hanoi, Vietnam
{binhkt,sonpb,hieupx}@vnu.edu.vn
[2] University of Technology Sydney, Broadway, Ultimo, NSW 2007, Australia
binh.kieuthanh@student.uts.edu.au,
{SonBao.Pham,Massimo.Piccardi}@uts.edu.au

Abstract. Choosing appropriate references for a given topic is an important, yet challenging task. The pool of potential candidates is typically very large, in the order of tens of thousands, and growing by the day. For this reason, this paper proposes an approach for automatically providing a reference list for a given manuscript. The approach is based on an original submodular inference function which balances relevance, coverage and diversity in the reference list. Experiments are carried out using an ACL corpus as a source for the references and evaluated by MAP, MRR and precision-recall. The results show the remarkable comparative performance of the proposed approach.

Keywords: Reference recommendation · Submodular inference · Monotonic submodular functions

1 Introduction

Have you ever been overwhelmed by the large number of research papers in your research field? Are you wondering which ones you should read for your research, and afraid of missing some important, new ideas? Does it feel like the deluge is only getting worse?

To ease this challenge, this paper proposes a novel model, named **SubRef**, that can recommend an effective and adequate reference list for a user-provided query. The reference list is predicted based on a combination of relevance, coverage and diversity. Specifically, the relevance refers to the relevance of the recommended articles to the user query, which can be computed in terms of a similarity score between the query and the articles' contents; the coverage reflects the extent to which the recommended list is able to cover the query; and the enforced diversity across the recommended articles prevents redundancies and overlapping within the list. In addition, by its nature, the proposed model lends itself to accommodate specialized recommendations (for instance, for a given user profile: e.g., newcomer vs. expert).

© Springer Nature Singapore Pte Ltd. 2020
L.-M. Nguyen et al. (Eds.): PACLING 2019, CCIS 1215, pp. 3–14, 2020.
https://doi.org/10.1007/978-981-15-6168-9_1

The ability to provide reference recommendations that meet our three stated criteria stems from the powerful domain of submodular inference [1]. Submodular functions are capable of encapsulating many desirable properties of subsets while allowing for fast inference of good subsets with theoretical performance guarantees. As such, they are natural candidates for the selection of effective reference recommendations from large corpora of documents.

The main contributions of our paper are summarized as follows.

- We propose a novel model that recommends relevant, covering and diverse references which can help, amongst other, in the writing of scientific papers.
- We propose a class of effective submodular functions to be used for the provision of reference recommendations. Moreover, we provide a comparative performance evaluation.
- Thorough experiments are carried out using the ACL Anthology Reference Corpus to assess the effectiveness of the proposed model.

2   Related Work

Reference recommendation approaches could be divided into two main categories based on different styles of querying: global and inline. The global recommendation approaches predict references that are relevant to an entire manuscript. They typically employ the whole manuscript or significant parts thereof such as the title, abstract, author and venue as the query [2]. By contrast, inline, or contextual, reference recommendation approaches aim to suggest references for each possible citation placeholder in the text [3]. Although our approach can apply to both styles of reference recommendation, we only focus on global reference recommendation in the following.

From the perspective of the underlying technology, recommendation approaches can instead be divided into three main groups: collaborative filtering (CF) approaches, graph-based (GB) approaches and content-based filtering (CBF) approaches. Each group has its own rationale for basing the recommendations: CF approaches focus on the recommendations or ratings of other users whose profiles are similar to the user's query [4]. CBF computes the similarity score between keywords extracted from the user's query and from candidate papers [5]. In turn, GB methods construct a graph in which authors and papers (possibly including venues and other meta-information) are regarded as nodes, and the edges express paper-to-paper, author-to-author, and paper-to-author relationships [6]. Recommendations are eventually provided in terms of optimal graph traversals. In addition, various combinations of these three groups, generally referred to as *hybrid methods*, have also been proposed to improve the accuracy of the recommendation results and obtain better overall performance [7].

A submodular approach to reference recommendation has also been proposed in the literature [8]. To mollify computational issues, it introduces a streaming algorithm with a constant-factor approximation guarantee that uses only a limited amount of memory. Differently from [8], in this paper we propose a novel class of submodular functions of limited computational cost which is capable of providing effective recommendation lists out of large candidate pools.

3   Submodularity Background

3.1   Definitions

To clarify the connection between submodularity and reference recommendation, hereafter we recap the main properties of submodularity.

Definition 1 (Submodularity). A function $f : 2^V \rightarrow \mathbb{R}$ is *submodular* if for every $A, B \subseteq V$ it holds that

$$f(A \cap B) + f(A \cup B) \leq f(A) + f(B)$$

Equivalently, a function $f : 2^V \rightarrow \mathbb{R}$ is *submodular* if for every $A \subseteq B \subseteq V$ and $e \in V$ it holds that

$$\Delta(e \mid A) :=:= (f(A \cup e) - f(A)) \geq \Delta(e \mid B).$$

Intuitively, this definition says that the benefit achieved for adding element e to an existing summary is greater or equal if such a summary is smaller.

An important subclass of submodular functions is *monotone*, where enlarging the argument set cannot cause the function to decrease:

Definition 2 (Monotonicity). A function $f : 2^V \rightarrow \mathbb{R}$ is *monotone* if for every $A \subseteq B \subseteq V, f(A) \leq f(B)$.

The concept of submodularity fits reference selection in recommendation tasks well: in this case, V is the set of all the candidate references, e is an element in V, and A and B are two recommendation lists (i.e., subsets of V). Intuitively, there will be less "gain" for introducing another reference into a list if such a list is already substantial. Therefore, our aim becomes that of maximizing submodular functions, i.e. finding $\max_{S \subseteq V} f(S)$, subject to some constraints on S. The simplest type of constraint is the cardinality constraint, where we require that $|S| \leq k$, with k an acceptable number of references for a scientific publication, typically in the order of 20 to 30.

The problem of maximizing submodular functions is NP-hard and usually approximately solved via a simple, greedy algorithm which, however, enjoys theoretical guarantees for its worst-case approximation. The greedy algorithm is provided below.

The Greedy Algorithm (Algorithm 1). With S_0 the empty set, at iteration $i = 1 \ldots k$, add element $d \in V \setminus S_{i-1}$ maximizing the discrete derivative $\Delta(d \mid S_{i-1})$:

$$S_i = S_{i-1} \cup \{\mathrm{argmax}_d \Delta(d \mid S_{i-1})\} \tag{1}$$

A well-known result by Nemhauser et al. (1978) [1] proves that the greedy algorithm provides a good approximation to the optimal solution of the NP-hard optimization problem.

Algorithm 1. Greedy submodular function maximization

1: $S^* \leftarrow \emptyset$
2: $A \leftarrow V = \{d_1, d_2, ...\}$
3: **while** $A \neq \emptyset$ **and** $|S^*| < k$ **do**
4:    $z \leftarrow \mathrm{argmax}_{d \in A} F(S^* \cup d) - F(S^*)$
5:    $S^* \leftarrow S^* \cup \{z\}$
6:    $A \leftarrow A \setminus \{z\}$
7: **end while**

Theorem 1 [1]. Given a nonnegative monotone submodular function $f : 2^V \rightarrow \mathbb{R}_+$, let $S_1, S_2 \ldots S_k$ be the greedily selected sets defined in Eq. 1. Then, for any positive integer $l \leq k$,

$$f(S_l) \geq (1 - e^{-l/k})\mathrm{max}_{S:|S| \leq k} f(S) \tag{2}$$

In particular, for $l = k$, $f(S_k) \geq (1 - 1/e)\mathrm{max}_{S:|S| \leq k} f(S)$.

3.2 Submodular Functions Used in Document Summarization

Document summarization is a well-studied problem in the literature, with interesting analogies with reference recommendation. Document summarization is often solved at sentence level, whereas reference recommendation is solved at document level with each document as a reference. Hereafter, we describe the main submodular functions for document summarization that can form the basis for our proposal.

Lin and Bilmes [9] have been the first to frame document summarization as the maximization of a submodular function under a budget constraint. They used an objective consisting of a coverage term combined with a penalty for redundancy:

$$f_{MMR}(S) = \sum_{i \in V-S} \sum_{j \in S} w_{ij} - \lambda \sum_{i,j \in S:i \neq j} w_{ij}, \lambda \geq 0. \tag{3}$$

in which $S \subset V$ and w_{ij} is the similarity score between sentence i and sentence j. Please note that this scoring function is not guaranteed monotone: while the coverage term can be easily proved to be monotone, the penalty term may surpass it and make increments of (3) become negative. However, a monotone behavior can still be expected for reasonably small summaries.

In [10], the same authors proposed a class of monotone submodular functions that add up two terms, one for coverage and one for diversity. The summary score function is modeled as:

$$F(S) = F_c(S) + \lambda F_d(S) \tag{4}$$

where $F_c(S)$ presents the coverage and $F_d(S)$ presents the diversity. In turn:

$$F_c(S) = \sum_{i \in V} min\{C_i(S), \alpha C_i(S)\} \tag{5}$$

where $C_i(S)$ assesses the degree to which element i is covered by summary S, which "saturates" when $C_i(S)$ reaches a given fraction, α, of its largest possible value, $C_i(V)$; and:

$$F_d(S) = \sum_{i=1}^{K} \sqrt{\sum_{j \in P_i \cap S} r_j} \qquad (6)$$

where $P_i, i = 1, \ldots K$ is a partition of the original set, V, and $r_j \geq 0$ indicates the reward of putting i to the empty set. Since the square root grows less than linearly, this penalty favors selecting summary elements from different clusters.

Other studies have used a relatively similar submodular framework, but slightly different design and analysis. Among them, Dasgupta et al. [11] have formulated the objective function as a combination of submodular and non-submodular functions, in which the non-submodular functions use inter-sentence dissimilarities in other ways while targeting non-redundancy of the summary.

4   Submodular Reference Recommendation

Two very desirable properties of a good reference list are *relevance* and *non-redundancy*. Traditional recommendation systems usually measure relevance by ranking candidate recommendations from the most relevant to the least relevant based on a query. On the other hand, the submodular summarization approaches reviewed in the previous section provide summaries that are query-independent. For this reason, we propose extending the submodular scoring functions with an additional query-dependent *relevance* term, and we prove that the resulting function is capable of retaining submodularity.

4.1   Non-monotone Submodular Functions

The relevance term that we propose is defined as:

$$F_q(S) = \sum_{i \in S} sq_i \qquad (7)$$

where sq_i is the similarity score between the given query, q, and document i in the summary.

Theorem 2 (original). We prove that (7) is submodular by this simple argument: given a summary, S, the score increment provided by adding a new document, d, to it only depends on the document itself and not on the summary. Therefore, (7) satisfies the second non-strict inequality in Definition 1 with the equal sign. In addition, any convex combination of (7) with other submodular functions is submodular by construction.

Our first non-monotone submodular function considers two terms: relevance and non-redundancy. When adding a new element, k, to an existing summary, S, the increment of this function can be expressed as:

$$\Delta(k) = \lambda sq_k - (1 - \lambda) \max_{i \in S} w_{ik} \qquad (8)$$

The second non-monotone submodular function considers relevance, coverage and non-redundancy:

$$F(S,q) = \alpha F_q(S) + F_{cr}(S) \tag{9}$$

where $F_q(S)$ is as in (7), and:

$$F_{cr}(S) = \lambda \sum_{i \in V \backslash S} \sum_{j \in S} w_{ij} - (1 - \lambda) \sum_{i,j \in S} w_{ij} \tag{10}$$

As we said above, w_{ij} is the similarity score between document i and document j. In the simplest case, w_{ij} can be set to be the TFIDF cosine similarity [12], In alternative, the BM25 score [13] is able provide more sophisticated similarity measurements and rankings.

4.2   Monotone Submodular Functions

To also capture the information about shared authors and references, we also propose adding terms that account for the overlap between authors and references:

$$F(S) = \sum_{i=1}^{K} \sqrt{\sum_{j \in P_i \cap S} R(d_j)} \tag{11}$$

$$R(d_i) = \lambda F_q(d_i) + (1 - \lambda) F_{au}(d_i) \tag{12}$$

where:

$$F_{au}(d_i) = \sum_{Au(d_i) \cap Au(q) \neq \emptyset} \omega_1 + \sum_{\substack{d_j \in cite(d_i), \\ Au(d_j) \cap Au(q) \neq \emptyset}} \omega_2 \tag{13}$$

with $cite(d_i)$ being the reference list of document i, $Au(d_i)$ its set of authors, and ω_1, ω_2 the scores credited for sharing authors and references, respectively. With a preliminary evaluation, we have chosen $\omega_1 = 2.0$ and $\omega_2 = 1.0$. This function is submodular and also naturally monotone since it does not include any penalty.

5   Experiments

5.1   Corpus

In order to validate our model for reference recommendation, we use the ACL Anthology Reference Corpus (AAN) a set of 22,085 documents in the field of computational linguistics first presented in [14]. We ignore the papers which do not contain information about titles or abstracts in the data set, then we use the documents from 1965 to 2012 as the experimental data set. For evaluation purposes, we use all papers in 2012 containing ACL references for the test set. Table 1 summarizes the statistics of the test set.

For our retrieval experiments, we first construct a query by an article's title and abstract; then, we let our model provide the recommended references for this query and compare the results with the actual reference list of the article.

Table 1. Main statistics for the test set.

Year	Papers	Authors	Venues	References	Avg. refs
2012	1186	1657	34	11631	9.8

5.2 Evaluation Metrics

To assess the quality of our model's reference recommendations, we use the reference lists of the test papers as the ground truth. As the previous works, we have used the following metrics for evaluation:

Precision and Recall, which are the most commonly used metrics in the reference recommendation field (see survey [7]). These metrics are computed:

$$Precision = \frac{\sum_{d \in Q} |R(d) \cap T(d)|}{\sum_{d \in Q} |R(d)|} \tag{14}$$

$$Recall = \frac{\sum_{d \in Q} |R(d) \cap T(d)|}{\sum_{d \in Q} |T(d)|} \tag{15}$$

where Q is the test set of papers, $T(d)$ are the ground-truth references of paper d, and $R(d)$ are its recommended references.

In addition, the Average Precision (AvP) computes a precision-recall trade-off by calculating the precision at each point in the recommended list where a ground-truth article appears [15]. The AvP can be expressed as:

$$AvP(d) = \sum_{i \in [1, |R(d)|]} \frac{Precision(d)@N(i)}{|T(d)|} \tag{16}$$

where $Precision(d)$ is the precision for document d and $N(i)$ is the minimum number of top recommendations required to include reference i. In turn, the Mean Average Precision (MAP) averages the AP value over the entire test set:

$$MAP = \sum_{d \in Q} \frac{AvP(d)}{|Q|} \tag{17}$$

Eventually, the Mean Reciprocal Rank (MRR) is a measure for evaluating models that return an ordered list of recommendations over all the queries in the test set. The reciprocal rank of a recommendation is the inverse of the rank position of the first correct recommendation for a test paper. Therefore, The MRR is calculated as:

$$MRR = \frac{1}{|Q|} \sum_{d \in Q} \frac{1}{rank_d} \tag{18}$$

Table 2. SubRef-QFRv1 by Lambda

Lambda	No. correct	MRR	MAP	P@100	R@100
0.0	3674	0.4418	0.1228	0.0310	0.3159
0.1	3783	0.4433	0.1270	0.0319	0.3253
0.2	3947	0.4461	0.1318	0.0333	0.3394
0.3	4101	0.4495	0.1358	0.0346	0.3526
0.4	4265	0.4529	0.1412	0.0360	0.3666
0.5	4469	0.4569	0.1472	0.0377	0.3842
0.6	4629	0.4608	0.1523	0.0390	0.3980
0.7	4765	0.4649	0.1573	0.0402	0.4097
0.8	**4820**	0.4715	**0.1594**	**0.0406**	**0.4144**
0.9	4766	**0.4739**	0.1580	0.0402	0.4098
1.0	4585	0.4729	0.1516	0.0386	0.3942

5.3 Experimental Settings

For a comprehensive comparison, we have evaluated the proposed model in a number of variants together with various baselines and a state-of-the-art method:

ES-TFIDF: The similarity score based on TFIDF is a common method to search related documents. For every query in the test set, we use it to rank all the documents and select the top N.

ES-BM25: The fuller name of this method is Okapi BM25 and it is a bag-of-words retrieval function for measuring a set of documents based on the query terms appearing in each document. BM25 is actually a family of scoring functions with slightly different components and parameters. The newest version of the popular Elasticsearch (ES)[1] uses BM25 as its default similarity score. The authors state that there can be significant advantages in using BM25 over TFIDF as similarity measurement, at least in some cases [16]. In the experiments, the bias term b is set to 0.75 and bias term k_1 is set to 1.2 (the ES default). Table 6 shows that the results with ES-BM25 are much better than those with ES-TFIDF.

TopicCite: TopicCite is a state-of-the-art reference recommendation approach which leverages a joint model of feature regression and topic learning [2]. It first extracts several reference features from a citation network and subject-specific features, and then it integrates them with a topic model. Therefore, TopicCite is able to extract more relevant subject distributions from the citation data, providing a new perspective of topic discovery on the reference recommendation task. In [2], the authors carried out ample experiments on the AAN and DBLP corpora and evaluated performance by Precision, Recall and MRR.

---

[1] https://www.elastic.co/.

Table 3. SubRef-QFRv2 by Lambda

Lambda	No. correct	MRR	MAP	P@100	R@100
0.0	826	0.3697	0.0601	0.0070	0.0710
0.1	1170	0.3778	0.0629	0.0099	0.1006
0.2	1806	0.3922	0.0703	0.0152	0.1553
0.3	2490	0.4093	0.0811	0.0210	0.2141
0.4	3064	0.4264	0.0947	0.0258	0.2634
0.5	3526	0.4406	0.1084	0.0297	0.3031
0.6	3917	0.4508	0.1219	0.0330	0.3368
0.7	4194	0.4592	0.1329	0.0354	0.3606
0.8	4356	0.4672	0.1413	0.0367	0.3745
0.9	4489	0.4700	0.1475	0.0378	0.3859
1.0	**4585**	**0.4729**	**0.1516**	**0.0386**	**0.3942**

Table 4. SubRef-QAIv1 by Lambda

Lambda	No. correct	MRR	MAP	P@100	R@100
0.0	4265	0.4612	0.1281	0.0360	0.3667
0.1	4844	0.4885	0.1431	0.0408	0.4165
0.2	4913	**0.5073**	0.1521	0.0414	0.4224
0.3	4936	0.5036	0.1559	0.0416	0.4244
0.4	4938	0.4944	0.1550	0.0416	0.4246
0.5	4937	0.4772	0.1543	0.0416	0.4244
0.6	4936	0.4576	0.1497	0.0416	0.4244
0.7	4936	0.4418	0.1462	0.0416	0.4243
0.8	4936	0.4297	0.1445	0.0416	0.4244
0.9	4937	0.4277	0.1433	0.0378	0.4245
1.0	**5192**	0.4375	**0.1565**	**0.0438**	**0.4464**

SubRef-QFRv1: our model as per Eq. 8. As values for λ we have used range $(0.0, 1.0)$ in 0.1 steps. $\lambda = 0.0$ means that the scoring function only scores non-redundancy and $\lambda = 1.0$ means that function only scores query relevance, which makes it equivalent to the baseline methods.

SubRef-QFRv2: our model as per Eq. 9. We have used $\alpha = 5.0$ and λ as above.

SubRef-QAIv1: our model as per Eq. 11, with partition P obtained by clustering the venues. **SubRef-QAIv2:** our model as per Eq. 11, with partition P obtained by clustering the authors.

For all comparisons, we use budget $k = 100$. This value is rather large and is chosen to favor recall over precision, in the assumption that the average user

Table 5. SubRef-QAIv2 by Lambda

Lambda	No. correct	MRR	MAP	P@100	R@100
0.0	4585	0.4729	0.1516	0.0386	0.3942
0.1	4626	0.4754	0.1529	0.0390	0.3977
0.2	4671	0.4778	0.1543	0.0393	0.4016
0.3	4728	0.4809	0.1563	0.0399	0.4065
0.4	4797	0.4833	0.1591	0.0404	0.4124
0.5	4868	0.4919	0.1629	0.0410	0.4185
0.6	4946	0.5017	0.1687	0.0417	0.4252
0.7	5036	0.5164	0.1742	0.0425	0.4330
0.8	5131	**0.5179**	**0.1798**	0.0433	0.4411
0.9	**5192**	0.4949	0.1787	**0.0438**	**0.4464**
1.0	5190	0.4374	0.1564	0.0437	0.4463

Table 6. Performance comparison

Method	Budget	MRR	MAP	P@100	R@100
ES-TFIDF	100	0.2768	0.0640	0.0178	0.1814
ES-BM25	100	0.4729	0.1516	0.0386	0.3942
TopicCite	Unknown	**0.5713**	Unknown	Unknown	**0.5035**
SubRef-QFRv1	100	0.4739	0.1594	0.0406	0.4144
SubRef-QFRv2	100	0.4700	0.1475	0.0378	0.3860
SubRef-QAIv1	100	0.5073	0.1565	0.0438	0.4464
SubRef-QAIv2	100	**0.5179**	**0.1798**	0.0438	0.4464

would prefer a brief manual refinement of an abundant list over an unbounded search for false negatives.

5.4   Parameter Tuning

In this part, we examine the effect of two parameters, α and λ, in our non-monotone submodular functions. In the QFRv1 function, λ is the trade-off between relevance and non-redundancy. In the QFRv2 function, α is the trade-off between relevance and a coverage and non-redundancy combination, while λ trades off coverage and non-redundancy. In functions QAIv1 and QAIv2, λ is the trade-off between the relevance and the importance of the author clusters. As values, for α we have empirically chosen 5.0 based on preliminary trials. For λ, we report all results in range $[0.0, 1.0]$ in 0.1 steps. In turn, the trade-off between coverage and diversity in Eq. 11 is encapsulated by the square root term.

5.5   Performance Comparison

All results are shown in Tables 2, 3, 4, 5 and 6. It is evident that the submodular functions with suitably-chosen parameters significantly outperform the baseline methods *in all cases*. The state-of-the-art method TopicCite still outperforms our method in the available metrics. However, TopicCite is a fully trained method of significant computational cost, while our method *does not require any training* and runs in linear time. All experiments have been implemented on a high-performance computing system with a 12-core Intel Xeon(R) CPU E5-2697 v4 @ 2.30 GHz with an 8×16 GB DIMM ECC DDR4 @ 2400 MHz RAM.

6   Conclusion

In this paper, we have presented a novel approach for submodular inference of reference recommendations, an important information retrieval application. Our experiments have proven that submodular functions that had first been proposed for document summarization also give significant benefits in reference recommendation and are able to outperform relevant baselines. Although our results have not yet overcome state-of-the-art methods, there are many potential directions for further developing this work, including integrating a step of submodular inference into existing methods and training the parameters automatically against loss functions. We plan to incorporate these features in our approach in the near future.

References

1. Nemhauser, G.L., Wolsey, L.A., Fisher, M.L.: An analysis of approximations for maximizing submodular set functions-I. Math. Program. **14**(1), 265–294 (1978). https://doi.org/10.1007/BF01588971
2. Dai, T., Zhu, L., Wang, Y., Zhang, H., Cai, X., Zheng, Y.: Joint model feature regression and topic learning for global citation recommendation. IEEE Access **7**, 1706–1720 (2019). https://doi.org/10.1109/ACCESS.2018.2884981
3. He, Q., Pei, J., Kifer, D., Mitra, P., Giles, L.: Context-aware citation recommendation. In: Proceedings of the 19th International Conference on World Wide Web, WWW 2010, pp. 421–430. ACM, New York (2010). https://doi.org/10.1145/1772690.1772734
4. Goldberg, D., Nichols, D., Oki, B.M., Terry, D.: Using collaborative filtering to weave an information tapestry. Commun. ACM **35**(12), 61–70 (1992). https://doi.org/10.1145/138859.138867
5. Ding, Y., Zhang, G., Chambers, T., Song, M., Wang, X., Zhai, C.: Content-based citation analysis: the next generation of citation analysis. J. Assoc. Inf. Sci. Technol. **65**(9), 1820–1833 (2014). https://onlinelibrary.wiley.com/doi/abs/10.1002/asi.23256
6. Liang, Y., Li, Q., Qian, T.: Finding relevant papers based on citation relations. In: Wang, H., Li, S., Oyama, S., Hu, X., Qian, T. (eds.) WAIM 2011. LNCS, vol. 6897, pp. 403–414. Springer, Heidelberg (2011). https://doi.org/10.1007/978-3-642-23535-1_35

7. Bai, X., Wang, M., Lee, I., Yang, Z., Kong, X., Xia, F.: Scientific paper recommendation: a survey. IEEE Access **7**, 9324–9339 (2019)
8. Yu, Q., Xu, E.L., Cui, S.: Submodular maximization with multi-knapsack constraints and its applications in scientific literature recommendations. In: 2016 IEEE Global Conference on Signal and Information Processing (GlobalSIP), pp. 1295–1299, December 2016
9. Lin, H., Bilmes, J.: Multi-document summarization via budgeted maximization of submodular functions. In: Human Language Technologies, pp. 912–920. Association for Computational Linguistics, Los Angeles, June 2010. https://www.aclweb.org/anthology/N10-1134
10. Lin, H., Bilmes, J.: A class of submodular functions for document summarization. In: Human Language Technologies, pp. 510–520. Association for Computational Linguistics, Portland, June 2011. https://www.aclweb.org/anthology/P11-1052
11. Dasgupta, A., Kumar, R., Ravi, S.: Summarization through submodularity and dispersion. In: Proceedings of the 51st Annual Meeting of the Association for Computational Linguistics (Volume 1: Long Papers), pp. 1014–1022. Association for Computational Linguistics, Sofia, August 2013. https://www.aclweb.org/anthology/P13-1100
12. Salton, G., Buckley, C.: Term-weighting approaches in automatic text retrieval. Inf. Process. Manag. **24**(5), 513–523 (1988). https://doi.org/10.1016/0306-4573(88)90021-0
13. Robertson, S.E., Walker, S.: Some simple effective approximations to the 2-poisson model for probabilistic weighted retrieval. In: Croft, B.W., Rijsbergen, C.J. (eds.) SIGIR 1994, pp. 232–241. Springer, New York (1994). https://doi.org/10.1007/978-1-4471-2099-5_24. http://dl.acm.org/citation.cfm?id=188490.188561
14. Radev, D.R., Muthukrishnan, P., Qazvinian, V., Abu-Jbara, A.: The ACL anthology network corpus. Lang. Resour. Eval., 1–26, 2013. https://doi.org/10.1007/s10579-012-9211-2
15. Bethard, S., Jurafsky, D.: Who should I cite: learning literature search models from citation behavior. In: Proceedings of the 19th ACM International Conference on Information and Knowledge Management, CIKM 2010, pp. 609–618. ACM, New York (2010). http://doi.acm.org/10.1145/1871437.1871517
16. Robertson, S., Zaragoza, H.: The probabilistic relevance framework: BM25 and beyond. Found. Trends Inf. Retr. **3**(4), 333–389 (2009). https://doi.org/10.1561/1500000019

Split First and Then Rephrase: Hierarchical Generation for Sentence Simplification

Mengru Wang[(✉)], Hiroaki Ozaki, Yuta Koreeda, and Kohsuke Yanai

Hitachi, Ltd., Research & Development Group,
1-280 Higashi-koigakubo, Kokubunji-shi, Tokyo, Japan
{mengru.wang.uq,hiroaki.ozaki.yu,
yuta.koreeda.pb,kohsuke.yanai.cs}@hitachi.com

Abstract. Split-and-rephrase is a strategy known to be used when humans need to break down a complex sentence into a meaning preserving sequence of shorter ones. Recent work proposed to model split-and-rephrase as a supervised sequence generation problem. However, different from other types of sequence generations, the task of split-and-rephrase inevitably introduces overlaps across splits to compensate for the missing context caused by separating a sentence. Serving as the baseline of this task, the vanilla SEQ2SEQ model usually suffers from inappropriate duplication because of the lack of a mechanism to plan how the source sentence should be split into shorter units. This work demonstrates that the problem of inappropriate duplication can be tackled by explicitly modeling the hierarchy within split-and-rephrase: Our model first introduces a *separator* network capable of selecting semantic components from the source sentence to form a representation for each split. Then, a decoder generates each split on the basis of its representation. Analyses demonstrate that with the aid of the *separator*, a model can effectively learn attention to avoid duplication and detect clues for splitting a sentence. Experimental results on the WikiSplit corpus show that our model outperforms the non-hierarchical SEQ2SEQ model by 1.4 points in terms of duplication rate and by 0.3 points in terms of coverage rate.

Keywords: Text simplification · Hierarchical text generation · Split-and-rephrase

1 Introduction

The cumbersomeness of comprehending a long, complex sentence would be eased if it was expressed in multiple, shorter ones. Split-and-rephrase [1] is a strategy known to be used when humans need to break down a complex sentence into a meaning preserving sequence of shorter sentences to gain better understandability. Additionally, when performed as the pre-editing phase of many natural language processing tasks such as machine translation [2–5], parsing [6–9], and relation extraction [10], split-and-rephrase could boost system performance.

© Springer Nature Singapore Pte Ltd. 2020
L.-M. Nguyen et al. (Eds.): PACLING 2019, CCIS 1215, pp. 15–27, 2020.
https://doi.org/10.1007/978-981-15-6168-9_2

Table 1. The vanilla SEQ2SEQ model fails to generate the necessary overlap between the two splits ("she", instead it overly repeats words (underlined) which have already appeared in the preceding split. (At preprocessing step, we lowercased tokens for the ease of experiment.)

Source sentence	She was born in Los Angeles, California, grew up in Garden Grove, California, and attended California State University, Fullerton
After split-and-rephrase	She was born in Los Angeles, California, grew up in garden grove, California. /// she grew up in garden grove, California, and attended California State University, paratroopers

In the split-and-rephrase task [1], we are given a long, complex sentence and need to generate its rewrite separated into multiple shorter sentences while preserving meaning. Different from other text generation tasks, split-and-rephrase inevitably introduces overlaps across splits for the reason that information contained in the preceding split need to be referred in the subsequent ones. The range of the reference might vary from a single proper noun to an entire split. Therefore, to generate each split, the model is required to learn to continuously make reference to the same part of the source sentence. Due to the nature of the task of split-and-rephrase as discussed above, the vanilla SEQ2SEQ model could easily make inappropriate duplication: The subsequent split overly repeats words that have already appeared in the preceding one (Table 1). We attribute this unwanted behavior to the lack of a planning mechanism to facilitate the hierarchy within split-and-rephrase behavior. In the case of a SEQ2SEQ-alike editor, without a high-level plan for arranging semantic components in each split, the editor depends heavily on his "language model" to decide whether it is an appropriate time to end a split and then start another while verbalizing. Due to the unsettled range of each split, the problem of inappropriate duplication might arise.

In this work, we focus on the inappropriate duplication problem in the setting of split-and-rephrase. We propose a split-first-and-then-rephrase mechanism that is implemented with neural networks and trained in an end-to-end fashion. We first incorporate a *separator* network that is capable of selecting semantic components from the source sentence to form a representation for each split. Then, fed with one semantic representation at each time, the decoder outputs each token in the split until all the semantic components are verbalized. Our proposed model enables the decoder to focus more on the local context of each split and to be informed of information contained in the preceding splits that should be excluded from the current split. Thus, the problem of inappropriate duplication discussed above can be alleviated.

The rest of the paper is organized as follows. Section 2 briefly describes our proposed model. Section 3 reports the experimental results. Section 4, summarizes the related work. Finally, we conclude in Sect. 5.

2   Split-First-and-Then-Rephrase Model

Our proposed model draws on the intuition that just as a human editor first arranges semantic components of a complex sentence in sequential order and then conducts verbalization of each split to form a natural and understandable rewrite as a whole, we first derive the semantic representation of a complex sentence by an encoder and then apply a *separator* network to determine information each split should contain. At last, a decoder generates tokens within each split based on the *separator* outputs. We illustrate our proposed model in Fig. 1.

Notation. Let the source sentence $x = (x_1, ..., x_T)$ be a sequence of T tokens, and let the target splits $Y = (y_1, ..., y_L)$ contain L sentences that together express the same meaning as x. $y_s = (y_{s,1}, ..., y_{s,l_s})$ denotes the s-th split containing l_s tokens. The s-th ($s < L$) split ends with an end-of-split token '///', and the last split y_L ends with an end-of-sentence tag <EOS>.

Encoder. Encoder obtains semantic representations for each token and the source sentence. We first apply embedding function to each token in the source sentence:

$$u_t = \text{emb}(x_t)$$
$$= \text{onehot}(x_t) \cdot W^{emb} \tag{1}$$

where onehot denotes one-hot embedding function and W^{emb} denotes word embeddings.

We then apply a bi-directional [11] Long Short-Term Memory (LSTM) network [12] (denoted as BiLSTM$_{enc}$) to obtain contextual representations for each token.

$$h_t^{enc} = [\overrightarrow{h}_t^{enc}, \overleftarrow{h}_t^{enc}]$$
$$= \text{BiLSTM}_{enc}\left(u_t, \overrightarrow{h}_{t-1}^{enc}, \overleftarrow{h}_{t+1}^{enc}\right) \tag{2}$$
$$= \left[\overrightarrow{\text{LSTM}}_{enc}\left(u_t, \overrightarrow{h}_{t-1}^{enc}\right), \overleftarrow{\text{LSTM}}_{enc}\left(u_t, \overleftarrow{h}_{t+1}^{enc}\right)\right]$$

$(h_1^{enc}, ..., h_T^{enc})$ denotes a sequence of hidden states of BiLSTM$_{enc}$, and $[\cdot]$ denotes the concatenation of vectors.

Separator. The *separator* functions as a semantic-level decoder which decides the information that the current split should encapsulate. It consists of a LSTM (denoted as LSTM$_{sep}$) with an attention mechanism. For the first split, inputs to LSTM$_{sep}$ are the last hidden state of the encoder ($h_1^{split} = [\overrightarrow{h}_T^{enc}, \overleftarrow{h}_1^{enc}]$) and the embedding of a beginning-of-the-sentence tag <BOS>. For the subsequent splits ($s > 1$), inputs are the hidden state of the *separator* h'^{split}_{s-1} and the last hidden state of the decoder $h_{l_{s-1}}^{dec}$ at the preceding split (the $(s-1)$-th split) to inform the *separator* of information contained in preceding splits.

$$h_s^{split} = \text{LSTM}_{sep}\left(e_s^{split}, h'^{split}_{s-1}\right) \tag{3}$$

$$e_s^{split} = \begin{cases} \text{emb}(<\text{BOS}>), & s = 1 \\ h_{l_{s-1}}^{dec}, & s > 1 \end{cases} \tag{4}$$

We incorporate an attention mechanism that operates on the source sentence to obtain the local-context c_s^{split} for the s-th split. Following [13], we derive an attention weight $a_{s,t}^{split} \in \mathbb{R}$ over each h_t^{enc} conditioned on the *separator*'s output h_s^{split} at each split:

$$a_{s,t}^{split} = h_s^{split^T} \cdot h_t^{enc} \tag{5}$$

$$\bar{a}_{s,t}^{split} = \frac{\exp\left(a_{s,t}^{split}\right)}{\sum_{k=1}^{T} \exp\left(a_{s,k}^{split}\right)} \tag{6}$$

$$c_s^{split} = \sum_{t=1}^{T} \bar{a}_{s,t}^{split} h_t^{enc} \tag{7}$$

h_s^{split} and c_s^{split} are concatenated and then transformed back to the size of h_s^{split} by a linear function to be fed back to LSTM$_{sep}$ at the next split.

$$h'^{split}_s = \tanh\left(\boldsymbol{W}^{split}\left[c_s^{split}, h_s^{split}\right] + b^{split}\right) \tag{8}$$

where $\boldsymbol{W}^{split}$ and b^{split} are model parameters.

We use the gold target splits to explicitly guide the *separator* to learn information that should be included for forming the current split's representation, instead of anticipating our end-to-end model to intrinsically learn to do so. Concretely, we obtain representation $\tilde{h}_s^{gold} = \left[\overrightarrow{h}_{s,l_s}^{gold}, \overleftarrow{h}_{s,1}^{gold}\right]$ for the each gold split using the encoder,

$$\tilde{h}_{s,t}^{gold} = \text{BiLSTM}_{enc}\left(\text{emb}(\boldsymbol{y}_{s,t}), \overrightarrow{h}_{s,t-1}^{gold}, \overleftarrow{h}_{s,t+1}^{gold}\right) \tag{9}$$

and guide the *separator*'s output h'^{split}_s to resemble $\tilde{h}_s^{gold}$ by employing a semantic similarity loss (denoted as J_s^{split}).

$$J_s^{split} = \|\ h'^{split}_s - \tilde{h}_s^{gold}\ \|^2 \tag{10}$$

Decoder. After obtaining the semantic representation for the s-th split h_s^{split}, we generate tokens within the s-th split using a LSTM decoder with the attention mechanism in a similar manner to the *separator*. We first obtain the hidden state $h_{s,q}^{dec}$ of the LSTM (denoted as LSTM$_{dec}$)

$$h_{s,q}^{dec} = \text{LSTM}_{dec}\left(e_{s,q}^{dec}, h_{s,q-1}^{dec}\right) \tag{11}$$

$$e_{s,q}^{dec} = \begin{cases} \text{emb}(\text{<BOS>}), & q = 1 \\ \text{emb}(\hat{y}_{s,q-1}), & q > 1 \end{cases} \tag{12}$$

where $\hat{y}_{s,q-1}$ is the $(q-1)$-th predicted token in the s-th split. The hidden state of LSTM$_{dec}$ is initialized with the hidden state of the *separator* ($h_{s,1}^{dec} = h'^{split}_s$).

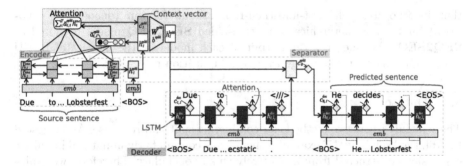

Fig. 1. Our proposed model which employs the split-first-and-then-rephrase mechanism

Following Eq. (5)–(7), we obtain a context vector $c_{s,q}^{word}$. $h_{s,q}^{dec}$ and $c_{s,d}^{word}$ are fed into a linear transformation followed by weight-tied transformation [14] to scores $p_{s,q} = (p_{s,q,1}, p_{s,q,2}, ...)$ for each token in vocabulary.

$$p_{s,q} = \left(\boldsymbol{W}^{dec} \left[h_{s,q}^{dec}, c_{s,q}^{word} \right] + b^{dec} \right) \cdot \boldsymbol{W}^{emb^T} \tag{13}$$

where $\boldsymbol{W}^{dec}$ and b^{dec} are model parameters. $p_{s,q}$ is then normalized over tokens to obtain probabilities $\bar{p}_{s,q} = (\bar{p}_{s,q,1}, \bar{p}_{s,q,2}, ...)$.

$$\bar{p}_{s,q,i} = \frac{\exp(p_{s,q,i})}{\sum_{k=1}^{T} \exp(p_{s,q,k})} \tag{14}$$

The loss of text generation (denoted as J_s^{dec}) is formulated as the sum of negative log likelihood of generating ground-truth tokens in $\boldsymbol{y}_s$.

At the inference time, a token with the highest probability is generated ($\hat{y}_{s,q} = \operatorname{argmax}_i \bar{p}_{s,q,i}$). The generation of a split is completed when the decoder predicts an end-of-split token '///' ($q = \hat{l}_s$) and the whole decoding process is terminated when the decoder predicts an end-of-sentence tag <EOS> ($q = \hat{l}_{\hat{L}}, s = \hat{L}$).

Training. The final loss for generating corresponding splits for the source sentence of interest is the weighted sum of two losses.

$$J = \sum_s^L \left(J_s^{dec} + \lambda J_s^{split} \right) \tag{15}$$

The model is then trained with mini-batch stochastic gradient decent by minimizing J. A ground-truth token is fed into $\hat{y}_{s,q-1}$ in Eq. (12) during training.

3   Experiments

We explore whether introducing hierarchy into split-and-rephrase can alleviate inappropriate duplication by making direct comparisons between a model that incorporates split-first-and-then-rephrase mechanism (denoted as **Proposal**) and a non-hierarchical attention-aided **SEQ2SEQ** model (denoted as **SEQ2SEQ**). We measure the output of each model by the metrics of BLEU scores [15], token-level coverage rate and duplication rate.

3.1 Dataset

There mainly exists two testbeds for conducting split-and-rephrase. As suggested by [16], the WebSplit [17] examples themselves contain fairly unnatural linguistic expression constructed from a relatively small vocabulary. Therefore, we evaluated our model on the alternative, namely the WikiSplit corpus, which is mined from publicly available Wikipedia[1] revision histories [16]. In the WikiSplit corpus, each example consists of an unsplit sentence and its rewrite containing 2 splits. All text in the WikiSplit corpus has been tokenized on punctuation. According to the original WikiSplit corpus [16], the training set consists of 989,944 pairs. 5000 examples are randomly extracted for tuning, validation and testing. All experiments below are performed on the same split of the orignal corpus.

3.2 Training Details

At preprocessing step, we lowercased tokens for the ease of experiment and then limit vocabularies to be the top 50K most frequent tokens. All out-of-vocabulary tokens are mapped to a special token UNK. We train each model with sentence lengths of up to 100 tokens.

We train our **Proposal** and the vanilla **SEQ2SEQ** in the same setting, such that the efficacy of the *separator* can be directly estimated. The dimension of word embeddings is 512 and hidden sizes of the encoder, the *separator* and the decoder are 256, 512, 512 respectively. Our encoder has 2 layers. Our decoder and *separator* have 1 layer. During training, we set the batch size to 10. At testing time, we decode with greedy search. We use Adam [18] with learning rate decay (decay rate $= 0.00005$ per epoch) as our optimization algorithm. The initial learning rate and betas are set by default as 0.001 and (0.9, 0.999) respectively. Early-stopping based on validation loss is used as a criterion for training. We perform batch normalization and clip gradients larger than 2. We use dropout on word embeddings and all hidden states with a rate of 0.3. All models are implemented in PyTorch[2].

3.3 Results

We first report models' generation performance. Table 2 shows the results of corpus-level BLEU, sentence-level BLEU, splits per source sentence, and tokens per split (micro-average). As shown in Table 2, both methods generate splits with

---

[1] Wikipedia is a registered trademark of the Wikimedia Foundation, Inc.

[2] https://pytorch.org/.

Table 2. Each model's generation performance on test set

	BLEU	sBLEU	Tokens per split	Splits per source sentence
Gold	–	–	18.69	2
SEQ2SEQ	68.97	66.79	17.65	1.998
Proposal	**69.26**	**67.72**	18.23	1.937

Table 3. Token-level stats

	Token-level coverage (%)	Token-level duplication (%)
Gold	95.99	–
SEQ2SEQ	87.75	4.66
Proposal	**88.07**	**3.28**

length close to the length of Gold split in the test set. Furthermore, our **Proposal** achieves 0.29 and 0.93 gains respectively in corpus-level and sentence-level BLEU score (p = 0.01, Mann-Whitney U test), which indicates the alleviation of inappropriate duplication.

Attention-Oriented Reference. We further investigate into each model's effectiveness in alleviating inappropriate duplication in terms of attention-oriented reference by measuring token-level coverage rate and duplication rate. Token-level duplication rate together with coverage rate can serve as an intuitive criterion for the reason that a model need make broad yet less repeated reference to tokens in the source sentence in order to both preserve semantic equivalency and in the meanwhile to minimize unnecessary duplication. To calculate coverage rate and duplication rate for each model, we extract the source token with the highest attention weight per generated token and then compare the extracted sequence of source-side tokens with the source sentence. Coverage rate for each model is derived as the recall of tokens in the source sentence with reference to the extracted sequence of source-side tokens according to attention weights and duplication rate is defined as the ratio of tokens extracted for multiple times among tokens within the source sentence. Additionally, we provide the gold coverage rate for the purpose of reference by comparing tokens in the splits with tokens in its corresponding source sentence. As shown in Table 3, compared to **SEQ2SEQ**, our **Proposal** achieves a lower token-level duplication rate of 3.28% and a higher coverage rate of 88.07%. These observations together demonstrate that introducing hierarchy into split-and-rephrase enables the attention mechanism to learn to effectively retrieve information for each split and thus to avoid duplication when unnecessary.

3.4 Segmentation Analysis

We conduct further analyses to better understand each model's ability to detect clues for splitting a sentence.

Table 4. Segmentation Agreement rates on test set

	Agreement rates
SEQ2SEQ	0.10
Proposal	**0.41**

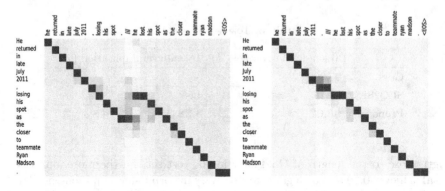

Fig. 2. Attention weights learnt by the decoders of **SEQ2SEQ** model (left) and our **Proposal** (right)

Segmentation Quality. To evaluate the model's capability of detecting clues for splitting a sentence, we measure segmentation agreement rate α of each model:

$$\alpha = \frac{\alpha_{Ref} \cap \alpha_{Pred}}{\alpha_{Ref} \cup \alpha_{Pred}}. \tag{16}$$

where α_{Ref} denotes the gold alignment between the corresponding clue token in the source sentence and the end-of-split token '///' in its splits. α_{Pred} denotes the attention-based alignment derived by each model. We run *fast align* [19] in the forward direction to obtain the gold alignment result. To obtain α_{Pred}, we extract the source token with the highest attention weight per target token.

Table 4 shows that our **Proposal** yields a large gain up to 0.31 points compared to **SEQ2SEQ**, which indicates that our model learns attention weights to detect the position for segmentation. Table 5 shows several sample generations. We observe an interesting case in the second example, which contains two "and". However, only the second "and" can serve as a clue for sentence segmentation. **SEQ2SEQ** incorrectly split the sentence at the first "and", whereas our **Proposal** detects the correct clue.

Decoder-Level Attention for Segmentation. To visualize how each model makes decisions for attention weights, Fig. 2 plots a heat map for the first example in Table 5. Figure 2 shows that at the prediction of '///', our **Proposal** pays heavy attention to "," in the source sentence, which in this case helps to avoid inappropriate duplication ("losing his spot"). However, in the case of **SEQ2SEQ**, the alignment between '///' and source-side tokens is incorrectly

Table 5. Sample generations by **SEQ2SEQ** model and our **Proposal**

(a) Reduced inappropriate duplication	
Sentence	He returned in late July 2011, losing his spot as the closer to teammate Ryan Madson
Gold	He returned in late July 2011. /// He lost his spot as the closer to teammate Ryan Madson
SEQ2SEQ	He returned in late july 2011, losing his spot. /// He lost his spot as the closer to teammate ryan madson
Proposal	He returned in late july 2011. /// He lost his spot as the closer to teammate ryan madson
(b) Better split position	
Sentence	Fantastic beings crawl or leap here and there, and satyrs appear from every side and surround the brigands
Gold	Fantastic beings crawl or leap here and there. /// Satyrs appear from every side and surround the brigands
SEQ2SEQ	Fantastic beings crawl or leap here. /// There, and satyrs appear from every side and surround the brigands
Proposal	Fantastic beings crawl or leap here and there. /// Satyrs appear from every side and surround the brigands
(c) Less information loss	
Sentence	According to Mnaseas, again cited in Athenaeus, Glaucus abducted Syme on a journey back from Asia, and had the island Syme named after her; according to Aeschrion of Samos, Glaucus was the lover of the semi-historical Hydne
Gold	According to Mnaseas, again cited in Athenaeus, Glaucus abducted Syme on a journey back from Asia, and had the island Syme named after her. /// And according to Aeschrion of Samos, Glaucus was the lover of the semi-historical Hydne
SEQ2SEQ	According to mnaseas, again cited in athenaeus, glaucus abducted syme on a journey back from asia. /// According to aeschrion of samo, glaucus was the lover of the semi-historical hydne
Proposal	According to mnaseas, again cited in athenaeus, glaucus abducted syme on a journey back from asia, and had the island syme named after her. /// According to aeschrion of samos, glaucus was the lover of the semi-historical hydne

learnt, which leaves the decoder repeatedly paying attention to a part of the semantic components in the preceding split in order to phrase the second split as a completed sentence. As shown in Table 5, our **Proposal** generates sentences with higher fluency and with less repeated words because of the improved decision-making power for splitting a sentence.

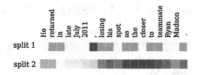

Fig. 3. Attention weights learnt by the *separator*

Separator-Level Attention for Segmentation. We further investigate into the question: How does incorporating the *separator* benefit the task? Fig. 3 plots the attention weights learnt by the *separator* in the case of the first example in Table 5. As shown in Fig. 3, we observe a tendency that at the first split our separator-level attention does not show any significant preference to the relevant part in the source sentence. However, the separator-level attention detects the position to split the source sentence by assigning a high weight to clues for splitting such as, ",", "and", and ";", which demonstrates the separator-level attention's capability of detecting clues for splitting a sentence. At the second split, the *separator* pays attention to the remaining part of the sentence thus to enable the decoder to focus on the local context of the current split. This behavior further supports the analysis for the first example in Table 5 discussed above in terms of reducing inappropriate duplication across splits and attention-aided clue-detection for splitting a sentence.

3.5   Error Analysis

A Potential Defect of Introducing Hierarchy into Split-and-Rephrase. Upon inspection of errors, a negative impact of our **Proposal** is identified. We observe a tendency that compared with the well-formed preceding split, errors more incline to occur in the subsequent split. Now that the quality of the preceding split is somehow deterministic to the generation of the subsequent ones due to the recurrent nature of the *separator*, this error tendency is quite self-explanatory. In the worst case, the scope of this unwanted impact may cover an entire split even if the discrepancy between the Gold and the generated preceding split is caused only by one token ("he") as shown in Table 6.

A General Challenge for Research on Split-and-Rephrase. We observe that models tend to output incorrect pronouns (Table 7) due to their incapacity of being conscious of deeper categorical information contained in each name entity. Incorrect pronouns are observed in the sample generations by both our **Proposal** and by **SEQ2SEQ**. Since none of the existing work (described in 4) employs well-designed learning paradigm such as multi-task or fine tuning to obtain categorical knowledge about name entity, we see it as a general challenge for the task of split-and-rephrase.

Table 6. A sample generation by our **Proposal** which reveals a tendency that errors more incline to occur in the second split when introducing hierarchy into the task of split-and-rephrase

Sentence	It was originally composed of 50 operators; however, this number has since increased to well over 90 full - time operators
Gold	It was originally composed of 50 operators. /// However, this number has since increased to well over 90 full - time operators
Proposal	It was originally composed of 50 stations. /// It is the most common in the

Table 7. A sample generation by our **Proposal** which reveals a general challenge for the task of split-and-rephrase that pronouns are usually incorrectly generated due to the lack of mechanism to learn about useful categorical knowledge of name entities

Sentence	The United States presidential election of 1876 was the 23rd quadrennial presidential election, held on Tuesday, November 7, 1876
Gold	The United States presidential election of 1876 was the 23rd quadrennial presidential electio. /// It was held on Tuesday, November 7, 1876
Proposal	The united states presidential election of 1876 was the 23rd quadrennial presidential election. /// He was held on tuesday, november 7, 1876

4   Related Work

4.1   Other Approaches to the Split-and-Rephrase Task

The task of split-and-rephrase was introduced by [1], along with the Web-Split corpus and several baseline models including the attention-aided vanilla SEQ2SEQ model. [17] reduced overlap in the data splits and established a stronger baseline by augmenting the SEQ2SEQ model with a copy mechanism. Following the vein of research, [16] introduced the WikiSplit, which contains sentence rewrites with richer vocabulary than the WebSplit. To release the SEQ2SEQ model from the pressure of memoring long source sentences, [20] adapted an architecture with augmented memory capacities. To the best of our knowledge, duplication problem in split-and-rephrase has not been investigated before. This is the first work to demonstrate that inappropriate duplication in split-and-rephrase can be alleviated by employing split-first-and-then-rephrase mechanism with a hierarchical text generation model. Comparison with the state-of-the-arts Copy512 model [17] is beyond the scope of our discussion. For the reason that the purpose of our work is to verify if the hierarchical generation architecture can alleviate inappropriate duplication that a non-hierarchical generation architecture suffers from, and we need to keep the elementary architecture simple for deeper analyses. The proposed *separator* can be incorporated into any of the non-hierarchical SEQ2SEQ model including the Copy512. We leave the comparison with other variants of the vanilla SEQ2SEQ model for future work.

4.2 Hierarchical Text Generation in Other Tasks

Early attempts in hierarchical text generation inspired our work. Such tasks include machine translation [21,22] and long documents generation [23]. [21,22] improved the translation quality of long sentences by first splitting a sentence into a sequence of short translatable clauses using coarsely made rules then encoding the sentence in a word-clause-sentence hierarchical structure. These works did not explore a model's capability of literally splitting a sentence. On the other hand, we proposed a RNN-based *separator* for the reason that a model's capability of learning to split serves as the fundamental element of split-and-rephrase task.

[23] explores a RNN model's capability of reconstructing documents consist of multiple sentences by an auto-encoder that hierarchically encodes and reconstructs a paragraph using its sub-units: Its words and sentences. Our work is significantly different than that of [23] in terms of task setting: In an attention-aided setting, reconstructing text is easier than conducting split-and-rephrase, which not only involves predicting the missing subject caused by segmentation, but also requires the RNN model to separate semantic components under the condition that no explicit clue (such as "." and the space between paragraphs in the case of [23]) can be seen.

5   Conclusion

This paper demonstrated that inappropriate duplication in split-and-rephrase can be alleviated by employing split-first-and-then-rephrase strategy with a hierarchical text generation model. We incorporate a *separator* network capable of arranging semantic components for each split. Based on this network, the decoder generates each split on the basis of its semantic components. Analyses demonstrate that with the aided of the *separator*, a model can effectively learn attention to avoid duplication and detect clues for splitting a sentence. Experimental results on the WikiSplit corpus show that our model outperforms the non-hierarchical SEQ2SEQ in terms of both token-level duplication rate and coverage rate, which indicates the alleviation of inappropriate duplication.

References

1. Narayan, S., Gardent, C., Cohen, S.B., Shimorina, A.: Split and rephrase. In: Proceedings of EMNLP, pp. 606–616 (2017)
2. Braud, C., Lacroix, O., Søgaard, A.: Cross-lingual and cross-domain discourse segmentation of entire documents. In: Proceedings of ACL, vol. 2, pp. 237–243 (2017)

3. Koehn, P., Knowles, R.: Six challenges for neural machine translation. In: Proceedings of the 1st Workshop on NMT, pp. 28–39 (2017)
4. Pouget-Abadie, J., Bahdanau, D., van Merrienboer, B., Cho, K., Bengio, Y.: Overcoming the curse of sentence length for neural machine translation using automatic segmentation. In: Proceedings of SSST-8, pp. 78–85 (2014)
5. Chandrasekar, R., Doran, C., Srinivas, B.: Motivations and methods for text simplification. In: Proceedings of COLING, vol. 2, pp. 1041–1044 (1996)
6. Tomita, M.: Efficient Parsing for Natural Language: A Fast Algorithm for Practical Systems. SECS, vol. 8. Springer, Boston (1986). https://doi.org/10.1007/978-1-4757-1885-0
7. Chandrasekar, R., Bangalore, S.: Automatic induction of rules for text simplification. Knowl.-Based Syst. **10**, 183–190 (1997)
8. McDonald, R., Nivre, J.: Analyzing and integrating dependency parsers. Comput. Linguist. **37**(1), 197–230 (2011)
9. Jelínek, T.: Improvements to dependency parsing using automatic simplification of data. In: Proceedings of LREC, pp. 73–77 (2014)
10. Zhang, Y., Zhong, V., Chen, D., Angeli, G., Manning, C.D.: Position-aware attention and supervised data improve slot filling. In: Proceedings of EMNLP, pp. 35–45 (2017)
11. Schuster, M., Paliwal, K.K.: Bidirectional recurrent neural networks. IEEE TSP **45**(11), 2673–2681 (1997)
12. Hochreiter, S., Schmidhuber, J.: Long short-term memory. Neural Comput. **9**(8), 1735–1780 (1997)
13. Bahdanau, D., Cho, K., Bengio, Y.: Neural machine translation by jointly learning to align and translate. In: Proceedings of ICLR (2015)
14. Press, O., Wolf, L.: Using the output embedding to improve language models. In: Proceedings of EACL, vol. 2, pp. 157–163 (2017)
15. Papineni, K., Roukos, S., Ward, T., Zhu, W.-J.: BLEU: a method for automatic evaluation of machine translation. In: Proceedings of ACL, pp. 311–318 (2002)
16. Botha, J.A., Faruqui, M., Alex, J., Baldridge, J., Das, D.: Learning to split and rephrase from Wikipedia edit history. In: Proceedings of EMNLP, pp. 732–737 (2018)
17. Aharoni, R., Goldberg, Y.: Split and rephrase: better evaluation and stronger baselines. In: Proceedings of ACL, vol. 2, pp. 719–724 (2018)
18. Kingma, D.P., Ba, J.: Adam: a method for stochastic optimization. In: Proceedings of ICLR (2015)
19. Dyer, C., Chahuneau, V., Smith, N.A.: A simple, fast, and effective reparameterization of IBM model 2. In: Proceedings of NAACL-HLT, pp. 644–648 (2013)
20. Vu, T., Hu, B., Munkhdalai, T., Yu, H.: Sentence simplification with memory-augmented neural networks. In: Proceedings of NAACL-HLT, pp. 79–85 (2018)
21. Zuo, S., Xu, Z.: A hierarchical neural network for sequence-to-sequences learning, vol. arXiv preprint arXiv:1811.09575 (2018)
22. Brown, P.F., Pietra, S.A.D., Pietra, V.J.D., Mercer, R.L., Mohanty, S.: Dividing and conquering long sentences in a translation system. In: Proceedings of the Workshop on Speech and Natural Language, pp. 267–271 (1992)
23. Li, J., Luong, T., Jurafsky, D.: A hierarchical neural autoencoder for paragraphs and documents. In: Proceedings of ACL and IJCNLP, vol. 1, pp. 1106–1115 (2015)

Abstractive Text Summarization Using LSTMs with Rich Features

Viet Nguyen Quoc[1], Huong Le Thanh[1(✉)], and Tuan Luu Minh[1,2]

[1] Hanoi University of Science and Technology, Hanoi, Vietnam
`nguyenquocviet1306@gmail.com,huonglt@soict.hust.edu.vn`
[2] National Economics University, Hanoi, Vietnam
`tuanlm@neu.edu.vn`

Abstract. Abstractive text summarization using sequence-to-sequence networks have been successful for short text. However, these models have shown their limitations in summarizing long text as they forget sentences in long distance. We propose an abstractive summarization model using rich features to overcome this weakness. The proposed system has been tested with two datasets: an English dataset (CNN/Daily Mail) and a Vietnamese dataset (Baomoi). Experimental results show that our model significantly outperforms recently proposed models on both datasets.

Keywords: Abstractive text summarization · Sequence to sequence · LSTM · Rich features

1 Introduction

Abstractive text summarization is the task of generating a condensed text that captures the main content of the original one. It is not done by selecting the most important sentences from the input document as in extractive summarization. Instead, it rewrites and compresses the original text, similar to how human does when summarizing a document.

Most recent research on abstractive summarization is based on a sequence to sequence (seq2seq) network, as it can generate new text from the original one. The input of the original version of seq2seq is often short since the character of seq2seq is "short term memory". That means, it often processes the recent sentences, but forgets sentences in a longer distance. As a result, most summarization models using seq2seq networks tend to ignore the first part of the long input text. This is a challenge for summarization systems whose processing target is news articles because, in such type of text, the important information often situates at the beginning of the text.

In this paper, we propose a model for abstractive summarization that can take into account the whole input article and generate a multi-sentence summary. We propose a new seq2seq network by using sentence position and term frequency as features. Our system is evaluated with two datasets, an English dataset (CNN/DailyMail) and a Vietnamese dataset (BaoMoi).

© Springer Nature Singapore Pte Ltd. 2020
L.-M. Nguyen et al. (Eds.): PACLING 2019, CCIS 1215, pp. 28–40, 2020.
https://doi.org/10.1007/978-981-15-6168-9_3

Experimental results showed that our system provides better performance compared to existing ones.

The remainder of this paper is organized as follows: Sect. 2 introduces related works on abstractive summarization, using seq2seq networks. Section 3 describes in detail the baseline model and our proposal to improve this model. Section 4 discusses all issues involving our experiments and evaluation. Finally, Sect. 5 concludes the paper and gives some insight into future work.

2   Related Work

Recent work on abstractive text summarization often uses seq2seq networks, since they are quite suitable and promising in solving this task. Rush et al. [1] applied a Convolutional Neural Network (CNN) seq2seq model with an attention-based encoder for the abstractive summarization task. They used CNN for the encoder, and a context-sensitive attentional feed-forward neural network to generate the summary. The abstractive summarization dataset DUC2004 was used to evaluate their system. The weakness of Rush et al.'s system is that it can only generate headlines (approximately 75 bytes) with many grammatical errors.

Narayan et al. [13] used Topic Sensitive Embeddings and Multi-hop Attention to fix long-range dependencies problems and achieving good results for single document summarization on a dataset of BBC articles accompanying with single sentence summaries whose length is limited to 90 tokens.

Nallapati et al. [2] used attentional encoder-decoder Recurrent Neural Network (RNN) to create a system that generates longer summaries by capturing the hierarchical document structure with hierarchical attention. In their experiments, they created a large training dataset named CNN/DailyMail, consisting of original texts and their multi-sentence summaries. By testing the system with the new dataset, they establish performance benchmarks for further research.

There are two main weaknesses in early works on abstractive text summarization. First, a word can be generated several times in the output (e.g., "I'm Vietnamese Vietnamese Vietnamese"). Second, numbers and private names, which are considered as out-of-vocabulary (OOV) words by the system, cannot be recovered correctly in the output. For example, if the input is "Peter go to school", the summary will be "<UNK> go to school" or "John goes to school". Nallapati et al. [2] 's system encountered the first problem when summarizing long texts. As for the second problem, Nallapati et al. [2] used modeling rare/unseen words that used switching generator-pointer to overcome.

Gu et al. [3] proposed a CopyNet network based on pointer-networks [4] to deal with OOV words. This network is modified by See et al. [5] to create a more powerful system that can solve the OOV problem and word repeat errors by using a pointer-generator network [3] and a distraction mechanism [6]. The limitation of their system is that it does not consider the full article as the input. Instead, it cuts off the last part of the article to guarantee that the important information of the input article will be included in the summary. This process reduces the generality of the system.

The next section will introduce our proposed approach to deal with the problems mentioned above.

3 Proposed Model

In this section, we first present Nallapati et al. [2]'s model, which is used as the baseline in our research. Then we briefly introduce two mechanisms used in [5] to solve the weaknesses in Nallapati et al. [2]'s system. Finally, we propose our method to resolve the weakness in See et al. [5]'s system to enhance its generality and its accuracy.

3.1 Baseline Model

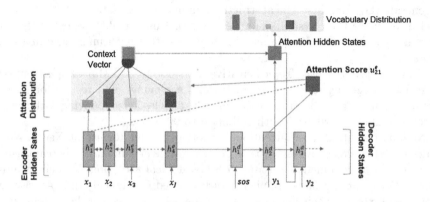

Fig. 1. General model. SOS represents the start of a sequence, respectively.

Nallapati et al. [2]'s model is a seq2seq with attention architecture which use a bidirectional Long Short Term Memory (LSTM) for the encoder and a unidirectional LSTM for the decoder. As shown in Fig. 1, the encoder reads a sequence of words $x = (x_1, x_2, \ldots, x_J)$ from the article, and transforms it to encoder hidden states $h^e = (h^e_1, h^e_2, \ldots, h^e_J)$. We denote target summary $y = (y_1, y_2, \ldots, y_T)$, on each step t, the decoder receives as input the previous word y_{t-1} of the target summary and uses it to update the decoder hidden state h^d_t.

At each decoding step t for generating output word y_t, an attention score u^e_{tj} is computed based on the encoder hidden state h^e_j and the decoder hidden state h^d_t as in [7]:

$$u^e_{tj} = \vartheta^T \tanh(W_{\text{align}}(h^e_j \oplus h^d_t) + b_{\text{align}}) \tag{1}$$

In which ϑ, W_{align} and b_{align} are learnable parameters.

At each step t, the attention distribution a_{tj}^e over the source words $u_{t_1}^e, u_{t_2}^e, \ldots, u_{t_J}^e$ is calculated as follows:

$$a_{tj}^e = \frac{\exp(u_{tj}^e)}{\sum_{k=1}^{J} \exp(u_{tk}^e)} \tag{2}$$

The weighted sum of the encoder hidden states is:

$$c_t^e = \sum_{j=1}^{J} a_{tj}^e h_j^e \tag{3}$$

With the current decoder hidden state h_t^d, we calculate vocabulary distribution as follows:

$$P_{vocab,t} = softmax(W_{d2v}(W_c[c_t^e, h_t^d] + b_c) + b_{d2v}) \tag{4}$$

with W_{d2v}, W_c, b_c and b_{d2v} are learnable parameters.

Copying from Source Article. Nallapati et al. [2] have pointed out a weakness in the LSTM network in general and in his model in particular, that is, the model cannot recover correctly OOV words (e.g numbers, private names) in the output. See et al. [5] solved this problem by allowing the model to occasionally copy words directly from the source instead of generating a new word based on their attention weights. This is done by a *pointer-generator network* - a specially designed seq2seq attentional model that can generate the summary by copying words in the article or generating words from a fixed vocabulary at the same time. The generation probability of a word $p_{gen,t}$ at time t is:

$$p_{gen,t} = \sigma(W_{s,c}c_t^e + W_{s,h}h_t^d + W_{s,y}y_t + b_s) \tag{5}$$

in which $W_{s,c}$, $W_{s,h}$, $W_{s,y}$ and b_s are learnable parameters.

The final distribution that can deal with OOV words is computed as:

$$P(y_t) = p_{gen,t}P_g(y_t) + (1 - p_{gen,t})P_c(y_t) \tag{6}$$

in which:

$P_g(y_t)$ is the vocabulary distribution corresponding to the "generator" function of *the pointer-generator network*.

$$P_g(y_t) = \begin{cases} P_{vocab,t}(y_t) \text{ if } y_t \in V \\ 0 \quad\quad\quad otherwise \end{cases} \tag{7}$$

with V is the vocabulary.

$P_c(y_t)$ represents the attention distribution corresponding to the "copy" function of *the pointer-generator network*.

$$P_c(y_t) = \begin{cases} \sum_{j:x_j=y_t} a_{tj}^e \quad y_t \in V_1 \\ 0 \quad\quad\quad otherwise \end{cases} \tag{8}$$

with V_1 is the word sequence of the input text.

Coverage Mechanism. The coverage model was first proposed by Tu et al. [8] for the NMT task, then See et al. [5] applied this mechanism for abstract summarization to overcome word repeat errors. In each decoder step t, they calculate the coverage vector cov_t^e as the sum of attention distributions of the previous decoding steps:

$$cov_t^e = \sum_j^{t-1} a_{tj}^e \tag{9}$$

The coverage vector cov_t^e is used to calculate attention score as:

$$u_{tj}^e = \vartheta^T \tanh(W_{align}(h_j^e \oplus h_t^d \oplus cov_t^e) + b_{align}) \tag{10}$$

Besides, See et al. [5] defined a coverage loss to penalize repeatedly attending to the same locations when generating multi-sentence summaries.

3.2   Our Proposed Model

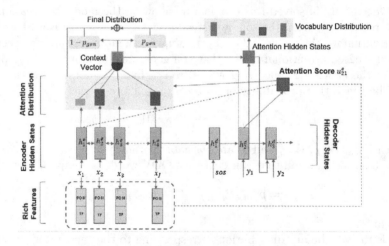

Fig. 2. Our proposed model using rich features

Rich Features. According to Pascanu et al. [9], a weakness of the models developed based on RNN is a vanishing gradient problem. That means, when the input is too long, the first part of the text will be forgotten. The LSTM model does not completely solve this problem. Since the main content of articles is often located at the beginning, See et al. [5] deal with this problem by using only the first part of the article to put into the model. However, that solution reduces the generality of the system since not all textual types put important content in the first part of the text. To deal with this problem, we add information about sentence position (POSI) as a feature in the network, to enhance the weight of the first sentences without cutting off the input text. The sequence of input words $x = (x_1, x_2, \ldots, x_J)$ now re-expressed as:

$$x = (x_{1_1}, x_{2_1}, \ldots, x_{J_k}) \tag{11}$$

where x_{j_k} means word x_j in the kth sequence. From there, we can easily create $x_{j_{position}}$ as:

$$x_{j_{position}} = k \tag{12}$$

in which k denotes the sentence position of x_j in the article.

The output word y_t is generated based on the attention distributions of all input words in the encoder side and previous output words. Since we now include the entire article without cutting off the last part, when the array's size grows, the attention distribution of each word will decrease, thus the effect of each attention will be reduced. To fix this, we use term frequency (TF) to help the model focus on important words. The frequency of each word is calculated as follows:

$$tf(x_j, x) = \frac{f(x_j, x)}{\max\{f(x_i, x)|i = 1 \to J\}} \tag{13}$$

where $f(x_j, x)$ is the number of occurrences of x_j in the article, $\max\{f(x_i, x)|i = 1 \to J\}$ is the highest number of occurrences of any word in the article. Based on this, we denote frequency of x_j as $x_{j_{TF}}$ as follows:

$$x_{j_{TF}} = tf(x_j, x) \tag{14}$$

By using **Rich Features**, we improve the formula to compute the attention score in [5] by a new formula as shown below:

$$u_{tj}^e = \frac{\vartheta^T \tanh(W_{align}(h_j^e \oplus h_t^d \oplus u_t^e) + b_{align})x_{j_{TF}}}{x_{j_{position}}} \tag{15}$$

The value of u_{tj}^e is inversely proportional to the sentence position. Therefore, sentences at the end of the article will have less effect than sentences at the beginning of the article. Also, the word that has high term frequency will have a high attention score.

Our proposed model is shown in Fig. 2. Sentence positions and term frequencies are represented as two vectors that have equal lengths with the input article. In our experiments, this length is set to 550 (words) with Vietnamese articles and 800 (words) with English articles. These two vectors are concatenated together with word vectors as input to the encoder.

At each decoder step, information of the sentence positions and the term frequencies are used to compute the attention score as in (15). Then the attention score is used to calculate the attention distribution a_{tj}^e as in (2). Because of this, the attention distribution of the first part of the text will be higher than that of the last part.

4   Experiments and Results

4.1   Dataset

Our experiments were carried out with two datasets: CNN/Daily Mail dataset for English and Baomoi for Vietnamese. The purpose of using the first dataset is to compared results with recent works in abstract summarization. Experiments with the second dataset to evaluate our proposed method on another language and to guaranty the generality of our approach.

The CNN/Daily Mail dataset for abstractive summarization is first established by Nallapati et al. [2]. This dataset contains 287,113 training samples, 13,368 validation samples, and 11,490 test ones. Each sample consists of a news article from CNN or Daily Mail websites accompanying with its summary. The average length of the original articles and the summaries are 781 words and 3.75 sentences (or 56 words), respectively. This dataset has two versions: anonymized and non-anonymized. In the anonymized version, each named entity, e.g., IBM, was replaced by its unique identifier, e.g., @entity10. We use the unanonymized version of the data since it more like the nature of the text summarization task.

There is no available corpus for the Vietnamese text summarization task. Therefore, we have to create a corpus by ourselves. This is done by gathering articles from a Vietnamese online newspaper (http://baomoi.com). The structure of each article consists of 3 parts: headline, abstract, and the article's body. The headline is the first sentence of the article. The abstract is the first paragraph of the article, after the headline. The remaining part is the article's body. The abstract is more likely the key information of the article, which guides readers on whether to continue reading the article or not, rather than a complete summary. In other words, the abstract sometimes can lack important information from the article. However, since we cannot find any better source, Baomoi is still our best choice to be used as the summarization corpus at the moment. We take the article part and the abstract part to serve as the original text and its summary. The average length of the original text and its summary are 503 words and 45 words, respectively.

Length's distributions of abstracts and articles in the Baomoi dataset are shown in Fig. 3 and Fig. 4 below. The final dataset consists of 1,187,000 news articles, in which 951,000 samples are used for training, 117,000 samples for validation, and 119,000 ones for testing.

4.2   Processing Data

For the Vietnamese dataset, we cleaned the data by removing words that have no meaning such as the newspaper's address dantri.vn, the author's name at the end of the article, etc. since they do not contribute to the article's content. The articles that are too short (less than 50 characters) were also removed. The tool UETSegment[1] and Stanford CoreNLP[2] were used to tokenize Vietnamese and English text, respectively.

---

[1] Available at http://github.com/phongnt570/UETsegmenter.
[2] Available at http://stanfordnlp.github.io/CoreNLP/.

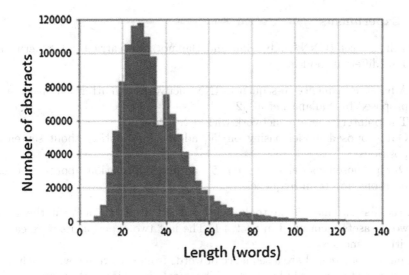

Fig. 3. The distribution of the length of the abstracts

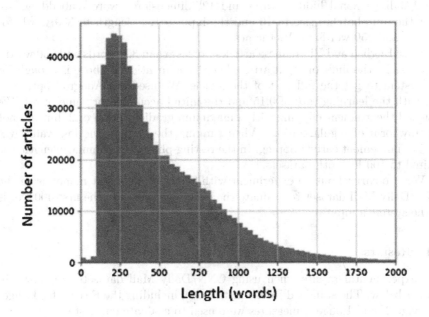

Fig. 4. The distribution of the length of the articles

4.3   Experiments

For each dataset (CNN/Daily Mail and Baomoi), we carried out experiments with four different models:

(i) A baseline sequence-to-sequence RNNs network with attention mechanism, proposed by Nallapati et al. [2]
(ii) The pointer generator network with coverage of [5]
(iii) Our proposed model basing on [5], adding information about sentences' positions
(iv) Our proposed model basing on [5], adding information about sentences' positions and term frequencies.

With the first two models, we took the source code from [5][3], and run them with the two datasets mentioned in Sect. 4.1. The last two experiments were carried out with our models.

Inputs of our model are a sequence of words from the article, with each word being represented as a one-hot vector. The vocabulary size in these experiments is 50,000 for both English and Vietnamese data. For all experiments, our model has 256-dimensional hidden states and 128-dimensional word embedding. We limit the mini-batches size to 16 and the input article length to 800 words for English and 550 words for Vietnamese.

Since English and Vietnamese articles are less than 800 words and 550 words, respectively, the limit of input article lengths as mentioned above is enough for the system to get the full text of the article. We used the Adagrad optimizer [10] with the learning rate of 0.15 and the initial accumulator value of 0.1. We assigned the gradient clipping with a maximum gradient norm of 2, but did not use any form of regularization. When tunning the system, the loss value was used to implement early stopping. In the testing phase, the summary length was limited to 100 for both datasets.

We also carried out an experiment with See et al. [5]'s best model using the CNN/Daily Mail dataset to evaluate the effect of using only the first 400 words as the system's input.

4.4   Results

Our experimental results when using CNN/Daily Mail datasets are shown in Table 1 below. The standard Rouge metric [11], including the F-score for Rouge-1, Rouge-2, and Rouge-L measures were used to evaluate our systems.

---

[3] Available at https://github.com/abisee/pointer-generator.

Table 1. Results on the CNN/DailyMail dataset – (*) is See et al. [5]'s model

	ROUGE		
	1	2	L
Seq2seq + attn baseline	27.21	10.09	24.48
Pointer-Gen + coverage(*)	29.71	12.13	28.05
Our model: (*) + POSI	31.16	12.66	28.61
Our model: (*) + POSI + TF	**31.89**	**13.01**	**29.97**

When repeated experiments in [5] using the first 400 words of articles as the input, we got a 35.87% Rouge-1 score. However, when using the full article as the input, the Rouge-1 score reduce to 29.71%. This is because when feeding a long text to the model, the first part of the text is "forgot" by the system. Unfortunately, the first part of an article usually keeps the main content of the article. However, summarizing articles in this way will reduce the generality of the system, as in other cases, important information may not locate at the first 400 words of the document.

As can be seen from Table 1, when using the full text of articles as the input, both of our proposed models outperform the systems of Nallapati et al. [2] and See et al. [5] in all the three Rouge scores. It indicates that the position is important information in generating a summary. The experimental results also show that term frequencies are a good indicator for summarization tasks using deep learning techniques. When information about sentence positions and term frequencies are added to the model, the Rouge-1 score is significantly improved with 2.18% Rouge-1 higher than that of See et al. [5]'s system.

Table 2 shows our experimental results with the Baomoi dataset.

The results in Table 2 also pointed out that both of our systems achieve higher Rouge scores than that of the other two systems. Our best model obtained 2.25% Rouge-1 higher than that of See et al. [5] and 3.91% Rouge-1 higher than the baseline.

Table 3 show an output of See et al. [5]'s model and our best model, using the full article in the CNN/Daily Mail dataset as the input.

As can be seen from Table 3, the summary of our proposed system is more informative than that of [5], and it does not have the problem of repeated words.

Table 2. Results on the BaoMoi dataset – (*) is See et al. [5]'s model

	ROUGE		
	1	2	L
Seq2seq + attn baseline	26.68	9.34	16.49
Pointer-Gen + coverage(*)	28.34	11.06	18.55
Our model: (*) + POSI	29.47	11.31	18.85
Our model: (*) + POSI + TF	**30.59**	**11.53**	**19.45**

Table 3. Outputs of See et al. [5]'s model and our best model, using an input from the CNN/Daily Mail dataset

---

REFERENCE SUMMARY: Mary Todd Lowrance, teacher at Moises e Molina high school, turned herself into dallas independent school district police on Thursday morning.

Dallas isd police said she had been in a relationship with student, who is older than 17 years old, for a couple of months.

She confided in coworker who alerted authorities and police eventually got arrest warrant.

Lowrance was booked into county jail on $ 5,000 bond and has been released from the Dallas county jail, according to county records.

She has been on leave for several weeks while investigators worked on the case, police said.

Pointer-Generator, Coverage (See et al. [5]):

Towrance Lowrance was accused of a male school on a $ 5,000 bond.

Lowrance has been on leave for several weeks while investigators worked on the case.

The student has been offered counseling warrant.

Pointer-Generator, Coverage, Rich Features (our system): Mary Todd Lowrance, 49, arrested after being accused of having an improper relationship with a male student older than 17 years old.

Miller said the teacher had been in a relationship with the student for a couple of months.

Officers learned of the alleged relationship after Lowrance disclosed details to a coworker who then alerted a Dallas isd officer.

Experimental results with a Baomoi's article are shown in Table 4 below.

The main information of the article in Table 4 is "1,000 kg *of dirty oil from a processing factory in Ma Diem, Giao Chau has been delivered to an illegal factory in Tuc_Mac, Qingdao to process moon cakes. On August 31, a working group of the food safety department, the police and the town government coordinated in investigating this case*". The reference summary contains most of the above information. The summary generated by [5] does not contain the key point "1,000 kg *of dirty oil* ", and only provides half of the necessary information. Also, although the output of [5] is short and lack of the main information, the phrase "*the incident was uncovered on August 30*" is repeated twice. Meanwhile, the summary generated by our system provides more information than that of [5] and does not contains redundance phrases. Besides, our system's output is easier to understand without grammatical errors for both of the English dataset and the Vietnamese dataset.

Table 4. Outputs of See et al. [5]'s model and our best model, using an input from the Baomoi dataset

---

REFERENCE SUMMARY: một xưởng sản_xuất phi_pháp tại __Tức_Mặc__, Thanh_đảo, trung_quốc vừa bị phát_hiện sử_dụng 1.000 kg dầu bẩn để chế_biến bánh_trung_thu.

An illegal factory in Tuc_ Mac, Qingdao, China has just been found out of using 1,000 kg of dirty oil to process moon cakes.

Pointer-Generator, Coverage (See et al. [5]):

Sự_việc bị phanh_phui vào ngày 30/8 vừa_qua. Cơ_quan an_toàn thực_phẩm, phòng công_thương, công_an và chính_quyền thị_xã đã phối_hợp điều_tra sự_việc bị phanh_phui vào ngày 30/8 vừa_qua.

The incident was uncovered on August 30. Food safety agencies, industrial and commercial bureaus, police and town authorities coordinated to investigate the uncovered incident on August 30

Pointer-Generator, Coverage, Rich Features (our system):

1.000 kg dầu bẩn vừa được chuyển tới bởi một xe_tải chở hàng. Theo nguồn_tin thân_cậy, số dầu này có xuất_xứ từ một xưởng chế_biến tại mã_điểm, giao_châu. ngày 31/8, một tổ công_tác gồm phòng an_toàn thực_phẩm, công_an và chính_quyền thị_xã đã phối_hợp điều_tra sự_việc này.

1,000 kg of dirty oil has just been delivered by a freight truck. According to reliable sources, this oil comes from a processing factory in Ma Diem, Giao Chau. On August 31, a working group of the food safety department, the police and the town government coordinated in investigating this case

---

5 Conclusions

In this paper, we present our work on abstractive summarization, using the pointer generator network with a coverage mechanism, combining with information about sentence positions and term frequencies. Our proposed approach solves the problem of recurrent neural networks that focus on the last part of the input text. Our experimental results with both of the English dataset and the Vietnamese dataset indicate that our approach is language-independent, as our system provides higher Rouge scores than previous systems of Nallapati et al. [2] and See et al. [5]. It also proves that our proposed features (sentence position and term frequency) are important in abstractive summarization tasks using the seq2seq network.

Nevertheless, there are still rooms for future works. Since the quality of abstracts in the Baomoi dataset are not good, it is necessary to build a summarization dataset with better quality. Also, since the structure of a document plays an important role in understanding that document, we will investigate methods to learn that structure, in order to provide a better summary for long texts. Hierarchical Attention Networks [12] is a candidate for this purpose since it is good in capturing document structures.

References

1. Rush, A.M., Chopra, S., Weston, J.: A neural attention model for abstractive sentence summarization. arXiv preprint arXiv:1509.00685 (2015)
2. Nallapati, R., Zhou, B., dos Santos, C., Gulcehre, C., Xing, B.: Abstractive text summarization using sequence-to-sequence RNNs and beyond. arXiv preprint
   . arXiv:1602.06023 (2016)
3. J. Gu, Z. Lu, Li, H., Li, V.O.K.: Incorporating copying mechanism in sequence-to-sequence learning. arXiv preprint arXiv:1603.06393 (2016)
4. Vinyals, O., Fortunato, M., Jaitly, N.: Pointer networks. In: Advances in Neural Information Processing Systems, pp. 2692–2700 (2015)
5. See, A., Liu, P.J., Manning, C.D.: Get to the point: summarization with pointer-generator networks. arXiv preprint arXiv:1704.04368 (2017)
6. Chen, Q., Zhu, X., Ling, Z., Wei, S., Jiang, H.: Distraction-based neural networks for modeling document. In: International Joint Conference on Artificial Intelligence (2016)
7. Luong, T., Pham, H., Manning, C.D.: Effective approaches to attention-based neural machine translation, pp. 1412–1421 (2015)
8. Tu, Z., Lu, Z., Liu, Y., Liu, X., Li, H.: Modeling coverage for neural machine translation (2016)
9. Pascanu, R., Mikolov, T., Benigo, Y.: On the difficulty of training recurrent neural networks. In: ICML 2013 Proceedings of the 30th International Conference on International Conference on Machine Learning, vol. 28 (2013)
10. Duchi, J., Hazan, E., Singer, Y.: Adaptive subgradient methods for online learning and stochastic optimization. J. Mach. Learn. Res. **12**, 2121–2159 (2011)
11. Lin, C.Y.: Rouge: a package for automatic evaluation of summaries. In: Text Summarization Branches Out: ACL Workshop (2004)
12. Yang, Z., Yang, D., Dyer, C., He, X., Smola, A., Hovy, E.: Hierarchical attention networks for document classification. In: Proceedings of the 2016 Conference of the North American Chapter of the Association for Computational Linguistics: Human Language Technologies, pp. 1480–1489 (2016)
13. Narayan, S., Lapata, M., Cohne, S.B.: Don't give me the details, just the summary! topic-aware convolutional neural networks for extreme summarization. In: EMNLP (2018)

Relation and Word Embedding

SemSeq: A Regime for Training Widely-Applicable Word-Sequence Encoders

Hiroaki Tsuyuki, Tetsuji Ogawa, Tetsunori Kobayashi,
and Yoshihiko Hayashi[✉]

Faculty of Science and Engineering, Waseda University, Tokyo, Japan
tsuyuki@pcl.cs.waseda.ac.jp,
{ogawa,koba}@waseda.jp, yshk.hayashi@aoni.waseda.jp

Abstract. A sentence encoder that can be readily employed in many applications or effectively fine-tuned to a specific task/domain is highly demanded. Such a sentence encoding technique would achieve a broader range of applications if it can deal with almost arbitrary word-sequences. This paper proposes a training regime for enabling encoders that can effectively deal with word-sequences of various kinds, including complete sentences, as well as incomplete sentences and phrases. The proposed training regime can be distinguished from existing methods in that it first extracts word-sequences of an arbitrary length from an unlabeled corpus of ordered or unordered sentences. An encoding model is then trained to predict the adjacency between these word-sequences. Herein an unordered sentence indicates an individual sentence without neighboring contextual sentences. In some NLP tasks, such as sentence classification, the semantic contents of an isolated sentence have to be properly encoded. Further, by employing rather unconstrained word-sequences extracted from a large corpus, without heavily relying on complete sentences, it is expected that linguistic expressions of various kinds are employed in the training. This property contributes to enhancing the applicability of the resulting word-sequence/sentence encoders. The experimental results obtained from supervised evaluation tasks demonstrated that the trained encoder achieved performance comparable to existing encoders while exhibiting superior performance in unsupervised evaluation tasks that involve incomplete sentences and phrases.

Keywords: Sentence encoding · Unsupervised representation learning · Semantic tasks

1 Introduction

In recent NLP research, one focus of research interest has been shifted from context-independent word embeddings [2,14,15] to context-dependent word embeddings [6,17] and sentence embeddings. Sentence encoders that can properly encode the meaning of a sentence have been actively studied [16]. Such a

© Springer Nature Singapore Pte Ltd. 2020
L.-M. Nguyen et al. (Eds.): PACLING 2019, CCIS 1215, pp. 43–55, 2020.
https://doi.org/10.1007/978-981-15-6168-9_4

sentence encoder should be readily applicable to many tasks/domains (universality), or effectively transferable to a given target task/domain (transferability). Besides, an encoder that could capture adequate semantic representation from a word-sequence of even an incomplete sentence could achieve a broader range of applications.

The research effort toward a universal sentence encoder, with respect to the learning strategy, can be divided into: (a) supervised learning with a labeled corpus of a carefully chosen target task/domain (for instance, natural language inference (NLI) [5]), and (b) unsupervised learning with a large-scale unlabeled corpus of general domain. The efficacy of the former methods is still unclear, as there could be a more suitable NLP task than NLI. Moreover, there may exist a trade-off between the applicability of the resulting encoder and its performance in a specific target task/domain other than, for example, NLI. Therefore, the recent sentence encoding trend is in the latter approach, often followed by a domain/task-dependent fine-tuning process.

The latter approach that employs an unlabeled corpus in the training is characterized by the usage of language modeling as the training task. The research in [8–11,13,18,21] achieved some level of universality by training a sentence encoder with an unlabelled corpus of sequentially ordered sentences. Among them, the work presented in [9,18] has demonstrated the importance and effectiveness of fine-tuning with a target task corpus. In particular, Howard et al. [9] proposes language modeling as a fine-tuning method for domain adaptation and emphasizes its efficacy. Although some research argues that language modeling-based methods tend to exhibit lower performances [3], the recent trend, as chiefly demonstrated by BERT [6], emphasizes the importance of language modeling as a pre-training task, provided a large-scale corpus is readily available.

In summary, the unsupervised training methods are becoming popular, and many of the existing methods train a sentence encoding model by relying on a corpus of *ordered sentences*. BERT [6] also adopts the next sentence prediction as one of the training tasks. These methods generally exhibit good performances, but a training method applicable to *unordered sentences* would be highly desirable, as it enables an encoding model to be refined in certain types of NLP tasks, in which contextually independent sentences have to be processed. Here, an unordered sentence means an individual sentence without neighboring contextual sentences.

Our goal in the present study is to train a word-sequence encoder, which can also be used as a sentence encoder, from unlabeled and even unordered sentences. If unordered sentences can be effectively used in the training, an encoder could be further fine-tuned with a target task corpus that consists of a set of unordered sentences, as argued in [9]. To achieve this goal, our proposed training regime adopts word-sequences with an arbitrary length as the training data. By using rather unconstrained word-sequences extracted from a large corpus, without heavily relying on complete sentences, it is expected that linguistic expressions of various kinds are employed in the training. A word-sequence encoder is then trained by predicting the continuity between two word-sequences. To implement the proposed training regime efficiently, we cast the prediction problem to a selection problem, where a correct sequence should be distinguished from negative samples provided. The models obtained through our proposed training regime are referred to as **S**entence **E**ncoding Models by word-**SEQ**uences (**SemSeq**).

The experimental results obtained from supervised evaluation tasks demonstrated that the trained encoder achieved performance comparable to existing encoders while exhibiting superior performance in unsupervised evaluation tasks that involve incomplete sentences and phrases.

2   Related Work

This section reviews relevant work in sentence encoding and highlights our proposed method. The approaches for realizing an effective sentence encoder can be classified by the way it is trained: supervised and unsupervised. Unsupervised approaches can be classified in several ways depending on a classification aspect.

Supervised Approaches. With this approach, a sentence encoder is trained by using a specific semantic task, insisting that the choice of the task is crucially important. Conneau et al. [5] argue that a sentence encoder that has been trained through natural language inference (NLI) task generalizes well to many different semantic tasks. Their encoding model, referred to as InferSent, adopts BiLSTM with a max-pooling layer, and classification layers on top of it. The series of work proposed by Socher et al. as represented by [20], can also be classified as a supervised approach, where recursive neural tensor networks are trained towards a sentiment classification task. The trend in sentence encoding, however, has shifted to unsupervised approaches given the situation where large-scale corpora are becoming available whereas a labeled corpus is still expensive to build.

BoW-Based Unsupervised Approaches. Probably the most straight-forward unsupervised approach to sentence encoding is BoW approaches. Although the BOW model does not consider word order, it is adequate for capturing the rough meaning of sentences such as the topic [1]. Hill et al. [8] proposed FastSent model, which uses a BOW encoder to reconstruct surrounding sentences. They confirmed that a BOW encoder can capture richer sentence representation on semantic similarity task than RNN encoders. Similarly, Kenter et al. [10] proposed the Siamese Continuous Bag of Words (Siamese CBOW) model which optimizes the embeddings of the words in a sentence by predicting its surrounding sentences. Tang et al. [21] also improved sentence representations on semantic similarity tasks by using an ensemble of two views, in which, one view encodes the input sentence with BOW, and the other view encodes with RNN.

Language Modeling-Based Unsupervised Approaches. As unlabeled corpora are readily available, language modeling-based unsupervised approaches whose objective function is somehow associated with language modeling have been proposed. Among them, Howard et al. [9,18] argue the importance of language modeling as ways of domain-independent pre-training and domain-dependent fine-tuning; ELMo [17] combined forward and backward language modeling and successfully produced context-dependent word embeddings. BERT [6] extended the framework of conventional language modeling (prediction of next word) and employed masked language modeling as one of the training tasks. The other training task is next sentence prediction, which tries to properly capture the meaning of a sentence by considering the relationship between adjacent sentences.

Training with Next Sentence Prediction. SkipThought [11] and its successor QuickThought [13] extended the Skip-gram model [14] to sentence-level. The SkipThought model adopts an encoder-decoder model to reconstruct adjacent sentences of an input sentence. The QuickThought framework successfully improved the sentence representations, while reducing the computational cost by changing from sentence reconstruction to sentence selection. As depicted in Fig. 1 (a), given the input sentence "My son likes playing tennis", the model is trained to distinguish between true succeeding sentence ("Playing abroad is his dream".) from competing negative samples ("It's delicious" and others). As described in the following section, we adopt the QuickThought framework. Note however that both of these frameworks require a corpus of ordered sentences in the training, which may restrict their applicability. That is, these models cannot be trained with a corpus of unordered independent sentences.

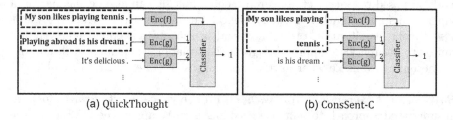

(a) QuickThought   (b) ConsSent-C

Fig. 1. QuickThought approach and ConsSent-C approach.

Training by a Corpus of Unordered Sentences. A sentence encoder would achieve a broader range of applications if it could be trained by a corpus of unordered sentences. The use of unordered sentences is present in two recent papers [3,19]. In the fake sentence detection tasks proposed in [19], an original sentence has to be distinguished from its altered ones (fake sentences) generated through some corrupting operations. On the other hand, Brahma [3] devised a set of six methods (referred to as ConsSent-X), in which two of them (ConsSent-C and ConsSent-N) are closely related to our methods.

- The ConsSent-C method, illustrated in Fig. 1 (b), generates two sub-sentences S_i^1 and S_i^2 by partitioning a sentence S_i at a random point. Given S_i^1, the ConsSent-C model is trained to choose the correct S_i^2 from a set of negative samples extracted from $S_j^2 (j \neq i)$.

- The ConsSent-N method also partitions a sentence S_i into two sub-sentences. S_i^1 is constructed by choosing a random sub-sentence of S_i^1, and S_i^2 is formed by $S_i \setminus S_i^1$. The negative samples are constructed by pairing S_i^1 with $S_j^2 (j \in \{i+1, k\})$.

As detailed in the next section, our models are similar to ConsSent models in that both use word-sequences rather than sentences as the training data. However, unlike ConsSent models, ours relax the constraint in word-sequence extraction: our models do not constrain a gold word-sequence pair to coincide with the original sentence.

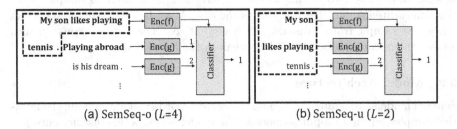

(a) SemSeq-o (L=4)   (b) SemSeq-u (L=2)

Fig. 2. Proposed approaches. L is the number of words in a word-sequence.

3   Proposed Approach

Our proposed training regime for sentence encoders is primarily characterized by the ways of word-sequence extraction from the training corpus, which can consist of unordered sentences. The model structure itself is adopted from the QuickThought model [13].

3.1   Word-Sequence Extraction

Figure 2 shows our proposed SemSeq approaches (Fig. 2 (a, b)) compared with related approaches (Fig. 1 (a, b)). In the figure, gold word-sequence pairs are presented in bold font. Each model is trained to choose the *context word-sequence* (sentences/sequences indexed by #1) against the negative samples (indexed by #2). For example, in Fig. 1(a), the sentence "Playing abroad is his dream" (#1) should be chosen as the succeeding sentence of "My son likes playing tennis", instead of "It's delicious" (#2).

Using Ordered Sentences (SemSeq-o). (Figure 2 (a)). Given a training corpus $T = \{w_1, w_2, \cdots, w_N\}$, this method successively extracts sequences of L-words, yielding N/L word-sequences. The remainder words are discarded. Note that a word-sequence may cross a sentence boundary as displayed in the figure. Negative samples are taken from nearby contexts. More precisely, they are randomly chosen from the same paragraph most of the time. If a negative sample is extracted from a nearby context, it could be more difficult to distinguish it from the gold one. This difficulty may contribute to the better training of an encoder.

Using Unordered Sentences (SemSeq-u). (Figure 2 (b)). This method extracts a pair of contiguous word-sequences from a sentence whose length is longer than $2L$. The sentences whose length is shorter than $2L$ are simply discarded. To increase the linguistic varieties, we do not always sample from the beginning of a sentence. Instead, we start to extract a sample from the n-th word in the sentence and continue to extract L-word sequences. The remainder words are discarded. Note that a word-sequence never crosses a sentence boundary, as this method only considers unordered sentences even in a context that consists of ordered sentences. Negative samples are randomly taken from the whole training corpus. This means that a negative sample can be largely irrelevant to the gold sample. In the Fig. 2 (d) example, the negative sample "tennis" is sampled from the same sentence, but it is not always the case. With larger L, we often extract a negative sample from a distant and irrelevant sentence in the corpus. In such a case, the negative sample could be weak in the adversarial training sense.

3.2   Model Architecture

SemSeq models are trained to choose the correct word-sequence among the negative samples, given an input sequence. The model employs two distinct encoders f and g that share the same architecture but with different parameters. Following QuickThought [13], the encoder f is used to encode an input word-sequence, and g is used to encode each of the corresponding candidate context word-sequences, including a correct one as well as negative ones.

Note that a SemSeq-trained model achieved by our proposed training regime should be distinguished from the model trained with existing methods (in particular ConsSent-C) in two aspects: (1) Our methods do not constrain a word-sequence (or sub-sentence) to be extracted from a complete sentence, rather sample unconstrained word-sequences while not considering sentence boundaries; (2) A model is trained to predict the succeeding context as well as the preceding context.

Given a training corpus T, we denote an input word-sequence as $s \in T$, and the corresponding gold sample that appears before or after of s by c. The set of negative samples that are supplied in the minibatch is denoted as $S_{ng}(s)$. Then the probability of $\hat{c}$ being chosen as the preceding/succeeding word-sequence from the set of candidates $S_{cand}(s) = \{c\} \cup S_{ng}(s)$ is formulated as follows.

$$p(\hat{c}|s, S_{cand}(s)) = \frac{\exp[f(s) \cdot g(\hat{c})]}{\sum_{c_j \in S_{cand}(s)} \exp[f(s) \cdot g(c_j)]}. \tag{1}$$

The overall objective function to be maximized is then stated as:

$$\sum_{s \in T} \log p(c|s, S_{cand}(s)). \tag{2}$$

As formulated in Eq. (1) and (2), SemSeq models learn to maximize the inner product of the encoding of s achieved by f and that of $c \in S_{cand}(s)$ given by g. In the testing time, the representation of a sentence is obtained by concatenating the vectors produced by f and g.

4   Experiments

4.1   Training Data

We train the sentence encoding models by using the UMBC corpus[1] [7]. This dataset maintains 133M sentences (3.3B words) that were crawled from 100M WebPages. The average number of words per sentence is 24.9. We shuffle the sentences to create an unordered sentences corpus, which amounted to 0.4B words ($L = 25$).

4.2   Models and the Training Setup

Figure 3 anatomizes our encoding model that employs a single layer BiGRU-Max model with 2048 hidden dimensions. We devise the encoders by modifying the original implementation of QuickThought[2] provided by the authors. As illustrated in the figure, LSTM in BiLSTM-Max [5] was replaced by GRU.

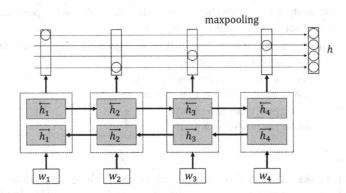

Fig. 3. The anatomy of the proposed encoder: BiLSTM-Max model.

At each time step, we concatenate hidden states of a forward GRU $\overrightarrow{h_t}$ and a backward GRU $\overleftarrow{h_t}$ that read the sentence in two opposite directions. We apply max pooling, selecting the maximum value over each dimension of the hidden states, to get a sentence representation. Conneau et al. [5] argued that the encoder with max-pooling performed better than any other encoders such as the self-attentive model [12] and the hierarchical convolution model [22].

We use a batch size of 400 and the learning rate of 5e−4 with the Adam optimizer. We use the 300-dimensional fastText vectors[3] [2] trained on Common Crawl 600B dataset for representing words. These vectors are frozen during the training.

---

[1] http://swoogle.umbc.edu/umbc_corpus.tar.gz.

[2] https://github.com/lajanugen/S2V.

[3] https://dl.fbaipublicfiles.com/fasttext/vectors-english/crawl-300d-2M.vec.zip.

4.3 Evaluation Setup

We evaluate the resulting sentence representations by using them as *features* in several downstream NLP tasks, which means that we did not perform any fine-tuning on the trained encoders. We make use of SentEval toolkit [4] that allows efficient and fair comparisons. The downstream tasks employed in the evaluation are divided into supervised and unsupervised tasks.

Table 1. Supervised evaluation tasks provided by SentEval.

Name	Examples	Label
MR	"Too slow for a younger crowd, too shallow for an older one"	neg
CR	"We tried it out christmas night and it worked great"	pos
SUBJ	"A movie that doesn't aim too high, but doesn't need to"	subj
MPQA	"Don't want"; "would like to tell"	(neg, pos)
STS	"Not a bad journey at all"	pos
TREC	"What are the twin cities?"	LOC: city
MSRP	"The study is being published today in the journal Science" "Their findings were published today in Science"	paraphrase
SICK-E	"A man is typing on a machine used for stenography" "The man isn't operating a stenograph"	contradiction
SICK-R	"A man is singing a song and playing the guitar" "A man is opening a package that contains headphones"	1.6
STSB	"Two zebras play in an open field" "Two zebras are playing in a field"	4.2

Supervised Tasks. Table 1 displays the supervised evaluation tasks used in the present work. The dataset for these tasks are provided by SentEval: MR, CR, STS (sentiment), SUBJ (subjectivity), MPQA (opinion polarity), and MSRP (paraphrase) are binary classification tasks; TREC (question type) and SICK-E (natural language inference) are multi-class classification tasks; SICK-R and STSB are semantic textual similarity prediction tasks which require similarity score regression.

Furthermore, these tasks are also classified in terms of the number of input sentences: MR, CR, SUBJ, MPQA, STS, and TREC take a single sentence, whereas MSRP, SICK-E, SICK-R, and STSB require a pair of sentences. For the sentence-pair classification tasks (MSRP, SICK-E, and SICK-R), we first obtained sentence vectors u and v for the paired sentences by using the encoders, and then created vector h by concatenating the offset vector $|u - v|$ and the element-wise multiplication vector $u * v$, as formulated in Eq. 3.

$$h = (|u - v|, u * v) \tag{3}$$

In these supervised tasks, we utilize logistic regression classifiers whose batch size was 64. We adopt Adam optimizer and terminate the training process if the validation loss did not improve in five successive epochs. We use accuracy as the evaluation metrics. We apply 10-fold cross-validation on the tasks (MR, CR, SUBJ, and MPQA) for which train/validation/test data divisions are not provided.

For the semantic textual similarity tasks, we cast the regression task as a classification task whose target class is defined by the integers ranging from one to five. To be more specific, the similarity score r of a sentence-pair is calculated by using Eq. 4, which converts the predicted class probabilities to an integrated score. The resulting scores are finally compared with the ground-truth similarities by using Pearson correlation coefficients.

$$r = \sum_{n \in \{1...5\}} p(n|s, s') * n \tag{4}$$

Table 2. Unsupervised evaluation task provided by STS14.

Name	Examples	Score
deft-news	"Tskhinvali is the capital of Georgia" "Paris is the capital of France"	0.4
deft-forum	"could be a lot of things" "Could be a simple mistake"	2.2
OnWN	"Create code or computer programs" "Create code, write a computer program"	4.6
tweet-news	"The tricks of moving from wall street to tech # dealbook" "Dealbook: moving from wall street to the tech sector proves tricky"	4.2
images	"A passenger train waiting in a station" "A passenger train sits in the station"	4.8
headlines	"Stocks rise in early trading" "US stocks ease in choppy trading"	2.0

Unsupervised Tasks. We also compare our models with QuickThought in an unsupervised task setting, which relies on the SemEval-2014 Task 10 (semantic textual similarity task) (STS14) dataset. This dataset is divided into six parts as listed in Table 2: deft-news (news articles), deft-forum (forum discussions), OnWN (OntoNotes and WordNet sense definition mappings), tweet-news (news title and tweet comments), images (image descriptions), and headlines (news headlines). Among these dataset divisions, OnWN and tweet-news provide incomplete sentences and phrases as exemplified in Table 2. We predict the similarity between the paired sentences by using the cosine of the corresponding sentence vectors that are created by the encoders. Same as the SICK-R and STSB tasks in the supervised evaluation, we compare the predicted similarities with ground-truth similarities by using Pearson correlation coefficients.

4.4 Word-Sequence Length Impact

As pre-experiments, we examined the best number of words in a word-sequence and compared the models trained from ordered sentences with the same number of words. Figure 4 displays the average performance on the supervised and unsupervised evaluation tasks with respect to the sequence length L, suggesting that the best number of words for a word-sequence is 25. Accordingly, SemSeq models were trained on the condition $L = 25$, which coincides with the average sentence length in the UMBC corpus. The rest of this paper thus presents the results obtained with this condition.

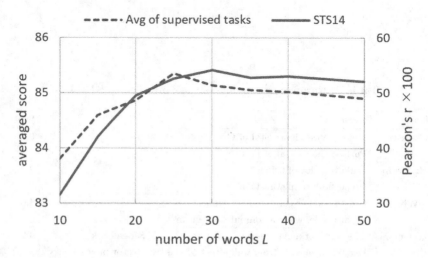

Fig. 4. Impact of word-sequence lengths: averaged performance on the supervised tasks and the unsupervised task (STS14).

5   Results and Discussion

5.1 Supervised Tasks

Table 3 summarizes the results on the supervised evaluation tasks. The SemSeq-o model achieved performance comparable to that of QuickThought and even better performance than the ConsSent-C and N models. The comparison with QuickThought may imply the power of the classification layer with logistic regression, which tries to adapt to the differences in sentence representations. In sum, these results suggest that the SemSeq-o model is advantageous compared to other models, yet potentially applicable to task-oriented fine-tuning.

On the other hand, the SemSeq-u model exhibited slightly poorer performance than the model trained from ordered sentences, which could be attributed to the random negative word-sequence sampling. Better sampling, such as adversarial sampling, could improve the performance.

5.2 Unsupervised Tasks

Table 4 summarizes the results. The most notable result displayed in the table is that SemSeq-u largely outperformed the QuickThought model in the dataset divisions, such as deft-forum, OnWN, and tweet-news. These results imply that our model is more robust in tasks that require handling partial and/or casual fragmentary phrases.

Another important observation from this table is that the SemSeq-u model performed better than the SemSeq-o model irrespective of the dataset divisions. The reason for this outcome may attribute to the nature of a STS task: fine-grained syntactic and/or word-order information may not be necessarily required to assess sentence similarities. Thus the SemSeq-u model behaved more robust than SemSeq-o model that had been trained to capture more fine-grained syntactic/word-order information. Yet another possibility is that we do not need any classification layer in these semantic similarity tasks: the similarity scores are directly affected by the sentence vectors produced by an encoder.

Table 3. Results on the supervised tasks. The figures are accuracies, except SICK-R that was evaluated using Pearson correlations $r \times 100$. The results for language model (LM), ConsSent-C and ConsSent-N were adopted from [3], while the rest were measured with our implementations.

Model	MR	CR	SUBJ	MPQA	SST	TREC	MSRP	SICK-E	SICK-R	STSB	Avg
LM	72.1	72.0	87.8	88.1	77.4	75.0	75.4	77.7	70.3	54.4	75.0
ConsSent-C	80.1	83.7	93.6	89.5	83.1	90.0	75.9	86.0	83.2	74.4	84.0
ConsSent-N	80.1	84.2	93.8	89.5	83.4	90.8	**77.3**	**86.1**	83.8	75.8	84.4
QuickThought	**82.8**	**86.9**	**95.3**	89.4	**85.5**	89.6	75.0	83.8	**87.1**	**79.3**	**85.5**
SemSeq-o	82.7	86.0	94.7	**89.6**	85.4	**91.4**	76.1	83.2	87.0	77.4	85.4
SemSeq-u	82.5	86.4	94.8	89.3	84.2	90.4	75.6	82.6	86.5	76.0	84.8

Table 4. Results on the unsupervised task STS14 (pearson's $r \times 100$). L is the number of words in a word-sequence.

Model	deft-news	deft-forum	OnWN	tweet-news	images	headlines	Overall
QuickThought	**71.2**	29.9	53.5	63.7	**72.5**	**63.2**	59.0
SemSeq-o	46.4	26.8	57.9	64.8	61.5	58.4	52.6
SemSeq-u	63.0	**35.6**	**62.6**	**68.8**	64.4	61.8	**59.4**

6 Conclusion

This paper proposed a regime for training a sentence encoder from an unlabeled corpus with ordered and unordered sentences. Unlike existing work that relies on a corpus of ordered sentences, the proposed method (SemSeq) uses word-sequences extracted from a training corpus as the training data. The use of unordered sentences as training data will allow us to fine-tune a sentence encoder toward a target task that requires the processing of independent sentences. We experimentally demonstrated that our proposed encoder exhibited comparable performance to QuickThought while outperforming it on the unsupervised textual similarity tasks involving incomplete sentences, such as word definitions and Tweets. This good trait that extends the applicability of the sentence encoder comes from the proposed regime, which can consume almost arbitrary word-sequences rather than complete sentences. For future work, we will fine-tune a sentence encoder trained with the proposed method using a target task corpus, which will allow us to compare the results with that of language modeling-based ULMFiT [9].

Acknowledgment. The present work was partially supported by JSPS KAKENHI Grants number 17H01831.

References

1. Arora, S., Liang, Y., Ma, T.: A simple but tough-to-beat baseline for sentence embeddings. In: ICLR, pp. 385–399 (2017)
2. Bojanowski, P., Grave, E., Joulin, A., Mikolov, T.: Enriching word vectors with subword information. In: TACL, pp. 135–146 (2017)
3. Brahma, S.: Unsupervised learning of sentence representations using sequence consistency. arxiv:1808.04217 (2018)
4. Conneau, A., Kiela, D.: SentEval: an evaluation toolkit for universal sentence representations. In: LREC (2018)
5. Conneau, A., Kiela, D., Schwenk, H., Barrault, L., Bordes, A.: Supervised learning of universal sentence representations from natural language inference data. In: EMNLP, pp. 670–680 (2017)
6. Devlin, J., Chang, M., Lee, K., Toutanova, K.: BERT: pre-training of deep bidirectional transformers for language understanding. In: NAACL, pp. 4171–4186 (2018)
7. Han, L., Kashyap, A.L., Finin, T., Mayfield, J., Weese, J.: Umbc_ebiquity-core: semantic textual similarity systems. In: ACL, pp. 44–52 (2013)
8. Hill, F., Korhonen, A.: Learning distributed representations of sentences from unlabelled data. In: NAACL, pp. 1367–1377 (2016)
9. Howard, J., Ruder, S.: Universal language model fine-tuning for text classification. In: ACL, pp. 328–339 (2018)
10. Kenter, T., Borisov, A., de Rijke, M.: Siamese CBOW: optimizing word embeddings for sentence representations. In: ACL, pp. 941–951 (2016)
11. Kiros, R., et al.: Skip-thought vectors. In: NIPS, pp. 3294–3302 (2015)
12. Lin, Z., et al.: A structured self-attentive sentence embedding. In: ICLR, pp. 1–15 (2017)
13. Logeswaran, L., Lee, H.: An efficient framework for learning sentence representations. In: ICLR (2018)
14. Mikolov, T., Yih, W., Zweig, G.: Linguistic regularities in continuous space word representations. In: NAACL, pp. 746–751 (2013)

15. Pennington, J., Socher, R., Manning, C.D.: Glove: global vectors for word representation. In: EMNLP, pp. 1532–1543 (2014)
16. Perone, C.S., Silveira, R., Paula, T.S.: Evaluation of sentence embeddings in downstream and linguistic probing tasks. arXiv:1806.06259 (2018)
17. Peters, M.E., et al.: Deep contextualized word representations. In: NAACL, pp. 2227–2237 (2018)
18. Radford, A., Narasimhan, K., Salimans, T., Sutskever, I.: Improving language understanding by generative pre-training (2018). https://s3-us-west-2.amazonaws.com/openai-assets/research-covers/language-unsupervised/language_understanding_paper.pdf
19. Ranjan, V., Kwon, H., Balasubramanian, N., Hoai, M.: Fake sentence detection as a training task for sentence encoding. arXiv:1808.03840v2 (2018)
20. Socher, R., et al.: Recursive deep models for semantic compositionality over a sentiment treebank. In: EMNLP, pp. 1631–1642 (2013)
21. Tang, S., de Sa, V.R.: Improving sentence representations with multi-view frameworks. arxiv:1810.01064 (2018)
22. Zhao, H., Lu, Z., Poupart, P.: Self-adaptive hierarchical sentence model, pp. 4069–4076 (2015)

Learning to Compose Relational Embeddings in Knowledge Graphs

Wenye Chen[1], Huda Hakami[1,2]([✉]), and Danushka Bollegala[1]

[1] Department of Computer Science,
University of Liverpool, Liverpool, UK
W.Chen29@student.liverpool.ac.uk
{h.a.hakami,danushka}@liverpool.ac.uk
[2] Department of Computer Science,
Taif Univeristy, Taif, Saudi Arabia
hoda.h@tu.edu.sa

Abstract. Knowledge Graph Embedding methods learn low-dimensional representations for entities and relations in knowledge graphs, which can be used to infer previously unknown relations between pairs of entities in the knowledge graph. This is particularly useful for expanding otherwise sparse knowledge graphs. However, the relation types that can be predicted using knowledge graph embeddings are confined to the set of relations that already exists in the KG. Often the set of relations that exist between two entities are not independent, and it is possible to predict what other relations are likely to exist between two entities by composing the embeddings of the relations in which each entity participates. We introduce *relation composition* as the task of inferring embeddings for unseen relations by combining existing relations in a knowledge graph. Specifically, we propose a supervised method to compose relational embeddings for novel relations using pre-trained relation embeddings for existing relations. Our experimental results on a previously proposed benchmark dataset for relation composition ranking and triple classification show that the proposed supervised relation composition method outperforms several unsupervised relation composition methods.

Keywords: Knowledge graphs · Knowledge graph embeddings · Relation composition · Novel relation types

1 Introduction

Knowledge graphs (KGs) such as Freebase [1] organise the knowledge that we have about entities and the relations that exist between entities in the form of labelled graphs, where entities are denoted by the vertices and the relations are denoted by the edges that connect the corresponding entities. A KG can be represented using a set of relational tuples of the form (h, R, t), where the relation $R \in \mathcal{R}$ exists between the (head) entity $h \in \mathcal{E}$ and the (tail) entity $t \in \mathcal{E}$ such

© Springer Nature Singapore Pte Ltd. 2020
L.-M. Nguyen et al. (Eds.): PACLING 2019, CCIS 1215, pp. 56–66, 2020.
https://doi.org/10.1007/978-981-15-6168-9_5

that the direction of the relation is from h to t. Here, $\mathcal{E}$ and $\mathcal{R}$ respectively denote the sets of entities and relations in the KG. For example, the relational tuple (Donald Trump, `president_of`, US) indicates that the `president_of` relation holds between Donald Trump and US. Despite the best efforts to create complete and large-scale KGs, most KGs remain incomplete and does not represent all the relations that exist between entities. In particular, new entities are constantly being generated and new relations are formed between new as well as existing entities. Therefore, it is unrealistic to assume that a real-world KG would be complete at any given time point.

Knowledge graph embedding (KGE) methods [2,4,14,17,18,24,25,28] learn representations (also known as *embeddings*) for the entities and relations in a given KG. The learnt KGEs can be used for *link prediction*, which is the task of predicting whether a particular relation exists between two given entities in the KG. Specifically, given KGEs for entities and relations, in link prediction we predict the R that is most likely to exist between h and t according to some scoring formula. For example, in translational embeddings (TransE), h, t, R are all embedded into the same d-dimensional vector space respectively by $\boldsymbol{h} \in \mathbb{R}^d$, $\boldsymbol{t} \in \mathbb{R}^d$ and $\boldsymbol{R} \in \mathbb{R}^d$ and the tuple (h, R, t) is scored by $\|\boldsymbol{h} + \boldsymbol{R} - \boldsymbol{t}\|_2$. Once KGEs for all the entities and relations in a given KG are learnt, for two entities h' and t' that are not connected by a relation, TransE finds the relation $R' \in \mathcal{R}$ that minimises $\|\boldsymbol{h}' + \boldsymbol{R}' - \boldsymbol{t}'\|$. However, the relation types that can be predicted using KGEs are confined to $\mathcal{R}$, the set of relation types that *already exists* in the KG. In other words, we *cannot* predict novel relation types using the pre-trained KGEs alone.

On the other hand, the relations that exist in a KG are often closely related [22]. For example, given the embeddings for the relations `country_of_film` and `currency_of_country` relations, we can compose the embedding for a previously unseen relation such as `currency_of_film_budget` because entities are shared across many tuples such as (Movie, `country_of_film`, Country), (Country, `currency_of_country`, Currency), where Movie, Country, and Currency can be replaced respectively by valid instances such as The Italian Job, UK and GBP.

In this paper, we propose a method for composing relation embeddings for novel (unseen in the KG) relation types by composing the embeddings for existing relation types. Our problem setting differs from that of KGE in two important ways. First, we do not learn relation embeddings from scratch for a given KG, but instead use pre-trained KGEs and learn a composition function to predict the embeddings for the relations that currently do not exist in the KG. In our experiments, we use the state-of-the-art matrix embeddings produced by the Relational Walk (RelWalk) method [2] as the relation embeddings. Second, the composition functions we learn are *universal* [19] in the sense that they are not parametrised by the entities or relations in the KG, thereby making the composition function independent from a particular KG. This is attractive because, theoretically the learnt composition function can be used to compose *any* relation type, not limited to the relations that exist in the KG used for training.

2　Background

2.1　Knowledge Graph Embedding Methods

Various methods have been proposed in the literature for representing entities and relations from a given KG. A popular choice for representing entities is to use vectors, whereas relations have been represented by vectors, matrices or tensors. For example, TransE [4], TransH [26], TransD [8], TransG [27], TransR [14], lppTransD [29], DistMult [28], HolE [17] and ComplEx [24] represent relations by vectors, whereas Structured Embeddings [4], TranSparse [9], STransE [16], RESCAL [18], RelWalk [2] use matrices and Neural Tensor Network (NTN) [21] uses three dimensional tensors. ComplEx [24] introduced complex embedding vectors for KGEs to capture the asymmetry in semantic relations. [5] obtained state-of-the-art performance for KGE by imposing non-negativity and entailment constraints to ComplEx.

Given that the number of entities in a KG is significantly larger than that of its relations, from a space complexity point-of-view, it is desirable to represent entities using vectors. On the other hand, relations are often asymmetric and directional, which cannot be modelled using vector addition or multiplication. For example, the is_father_of relation requires the head entity to be the father of the tail entity whereas, the is_son_of relation requires the reverse. Matrices are a natural choice for representing relations because matrix multiplication in general is non-commutative and this property can be used to encode the directionality of a relation. On the other hand, it is desirable to embed both entities and relations in the same vector space for carrying out linear algebraic operations for the purpose of learning KGEs. This requires a $\mathcal{O}(d^2)$ space for storing relation embeddings relative to the $\mathcal{O}(d)$ required for the entity embeddings. This requirement can be infeasible for large KGs such as Freebase, which covers over 39 million topics[1]. Therefore, much prior work on KGE has opted to represent both entities as well as relations using vectors.

As already stated in the introduction, we emphasise that we *do not* propose a method for learning KGEs in this paper. Instead, given pre-trained entity and relation embeddings, our goal is to compose relation embeddings for the relations that currently do not exist in the KG. Our proposed method is agnostic to the algorithm used to learn the input KGEs. In this regard, it can be used to compose relation embeddings using KGEs produced by any KGE learning method. As a concrete example, we use the state-of-the-art relation embeddings produced by RelWalk [2] in our experiments. RelWalk represents relations using matrices and is further detailed in Subsect. 2.2.

2.2　Relational Walk

In this section, we briefly describe RelWalk, the method that produces the relation embeddings we use in this paper. For a detailed overview refer [2].

---

[1] https://developers.google.com/freebase/guide/basic_concepts.

RelWalk assumes that the task of generating a relational triple (h, R, t) in a given KG to be a two-step process. First, given the current knowledge vector at time k, $\boldsymbol{c} = \boldsymbol{c}_k$ and the relation R, the probability of an entity h satisfying the first argument of R to be given by (1).

$$p(h \mid R, \boldsymbol{c}) = \frac{1}{Z_c} \exp\left(\boldsymbol{h}^\top \mathbf{R}_1 \boldsymbol{c}\right). \tag{1}$$

Here, $\mathbf{R}_1 \in \mathbb{R}^{d \times d}$ is a relation-specific orthogonal matrix that evaluates the appropriateness of h for the first argument of R and Z_c is a normalisation coefficient such that $\sum_{h \in \mathcal{E}} p(h \mid R, \boldsymbol{c}) = 1$. After generating h, the state of our random walker changes to $\boldsymbol{c}' = \boldsymbol{c}_{k+1}$, and the second argument of R is generated with probability given by (2).

$$p(t \mid R, \boldsymbol{c}') = \frac{1}{Z_{c'}} \exp\left(\boldsymbol{t}^\top \mathbf{R}_2 \boldsymbol{c}'\right). \tag{2}$$

Here, $\mathbf{R}_2 \in \mathbb{R}^{d \times d}$ is a relation-specific orthogonal matrix that evaluates the appropriateness of t as the second argument of R and $Z_{c'}$ is a normalisation coefficient such that $\sum_{t \in \mathcal{E}} p(t \mid R, \boldsymbol{c}') = 1$. In RelWalk, we consider ($\mathbf{R}_1$ and $\mathbf{R}_2$) to collectively represent the embedding of R. Bollegala et al. [2] proved Lemma 1 for a slow random walk over the KG.

Lemma 1 (Concentration Lemma). *If the entity embedding vectors satisfy the Bayesian prior $\boldsymbol{v} = s\hat{\boldsymbol{v}}$, where $\hat{\boldsymbol{v}}$ is from the spherical Gaussian distribution, and s is a scalar random variable, which is always bounded by a constant κ, then the entire ensemble of entity embeddings satisfies that*

$$\Pr_{c \sim \mathcal{C}}[(1 - \epsilon_z)Z \leq Z_c \leq (1 + \epsilon_z)Z] \geq 1 - \delta, \tag{3}$$

for $\epsilon_z = O(1/\sqrt{n})$, and $\delta = \exp(-\Omega(\log^2 n))$, where $n \geq d$ is the number of words and Z_c is the partition function for c given by $\sum_{h \in \mathcal{E}} \exp\left(\boldsymbol{h}^\top \mathbf{R}_1 \boldsymbol{c}\right)$.

Under the conditions required to satisfy Lemma 1, Bollegala et al. [2] proved the following theorem.

Theorem 1. *Suppose that the entity embeddings satisfy* (1). *Then, we have*

$$\log p(h, t \mid R) = \frac{\left\|\mathbf{R}_1{}^\top \boldsymbol{h} + \mathbf{R}_2{}^\top \boldsymbol{t}\right\|_2^2}{2d} - 2 \log Z \pm \epsilon. \tag{4}$$

for $\epsilon = O(1/\sqrt{n}) + \tilde{O}(1/d)$, where

$$Z = Z_c = Z_{c'}. \tag{5}$$

Theorem 1 states that the log-likelihood of observing R between h and t is related to the squared ℓ_2 norm of $\mathbf{R}_1{}^\top \boldsymbol{h} + \mathbf{R}_2{}^\top \boldsymbol{t}$. This provides an objective function for learning KGEs. Specifically, we can randomly initialise entity and relation embeddings and iterate such that above ℓ_2 norm is approximately equal to the empirical probabilities $p(h, t \mid R)$, estimated from a KG.

2.3 Inference in Knowledge Graphs

Lao et al. [12] used the Path Ranking Algorithm (PRA) [11] for predicting relations between two entities in a KG. The number of paths connecting two entities in a large KG can be large and it is difficult to systematically enumerate them all. Instead, PRA performs random walks and selects paths that cover most of the entities in a given training set. Next, the likelihood of a path is computed as the product of the transition probabilities when moving from one vertex to another. Neelakantan et al. [15] used PRA to find paths connecting entity pairs and ran a recurrent neural net (RNN) to combine the vector embeddings of relations to compose an embedding for the relation between two entities. They learnt separate RNNs for each relation type, but also proposed a zero-shot [13] version, where they used pre-trained KGEs and learn a single composition function. However, the performance of their zero-shot model was significantly worse than the relation-specific model.

Guu et al. [6] considered path queries in a knowledge graph connecting two entities and proposed a composition method that multiplies the relation embedding matrices corresponding to the relations along the connecting path. They considered relation composition under the TransE model [3], where relational embedding vectors are added, and under the bilinear-diagonal model [28], where relations are represented using diagonal matrices. These composition operators can be seen as *unsupervised* in the sense that there are no learnable parameters in the composition function. In our experiments, we use both matrix addition and multiplication as unsupervised baseline methods for comparisons. On the other hand, our proposed method is a supervised relation composition method and we consider relations represented by orthogonal matrices, which are not diagonal in general.

3 Relation Composition

Let us assume that the two relations R_A and R_B jointly implies a third relation R_C. In this paper, we use the notation $R_A \wedge R_B \Rightarrow R_C$ to express this fact. Moreover, let us assume that the relational embeddings produced by RelWalk for R_A and R_B to be respectively $(\mathbf{R}_1^A, \mathbf{R}_2^A)$ and $(\mathbf{R}_1^B, \mathbf{R}_2^B)$. For simplify the explanation, let us assume all relation embedding matrices are in $\mathbb{R}^{d \times d}$. We model the problem of composing a relation embedding $(\hat{\mathbf{R}}_1^C, \hat{\mathbf{R}}_2^C)$ for r_C as learning two joint compositional operators (ϕ_1, ϕ_2) such that:

$$\phi_1 : \mathbf{R}_1^A, \mathbf{R}_2^A, \mathbf{R}_1^B, \mathbf{R}_2^B \longrightarrow \hat{\mathbf{R}}_1^C \tag{6}$$

$$\phi_2 : \mathbf{R}_1^A, \mathbf{R}_2^A, \mathbf{R}_1^B, \mathbf{R}_2^B \longrightarrow \hat{\mathbf{R}}_2^C \tag{7}$$

3.1 Unsupervised Relation Composition

When the compositional operators ϕ_1, ϕ_2 do not have learnable parameters we call them *unsupervised*. In the case of matrix relation embeddings, we consider the following unsupervised operators.

Addition

$$\mathbf{R}_1^A + \mathbf{R}_1^B = \hat{\mathbf{R}}_1^C \tag{8}$$

$$\mathbf{R}_2^A + \mathbf{R}_2^B = \hat{\mathbf{R}}_2^C \tag{9}$$

Matrix Product

$$\mathbf{R}_1^A \mathbf{R}_1^B = \hat{\mathbf{R}}_1^C \tag{10}$$

$$\mathbf{R}_2^A \mathbf{R}_2^B = \hat{\mathbf{R}}_2^C \tag{11}$$

Hadamard Product

$$\mathbf{R}_1^A \odot \mathbf{R}_1^B = \hat{\mathbf{R}}_1^C \tag{12}$$

$$\mathbf{R}_2^A \odot \mathbf{R}_2^B = \hat{\mathbf{R}}_2^C \tag{13}$$

Here, $\odot$ denotes the Hadamard (elementwise) product of two matrices. Unlike the matrix product, both addition and Hadamard product are commutative.

3.2 Supervised Relation Composition

The unsupervised compositional operators described in Subsect. 3.1 are not guaranteed to correctly predict the embeddings because they cannot be tuned to the relations in a given KG. Moreover, each unsupervised operator considers either one of $\mathbf{R}_1$ or $\mathbf{R}_2$, and do not model their possible interactions. Therefore, we propose to learn two *supervised* relation composition operators with shared parameters. The parameter sharing enables the two operators to learn a consistent relation embedding.

Different models can be used to express ϕ_1 and ϕ_2. In this paper, we use feed-forward neural nets, which are universal approximators [7] for this purpose. We first linearise the input $d \times d$ matrix relation embeddings to d^2-dimensional vector embeddings via a linearisation operator $\mathfrak{L}$. We then concatenate the four linearised relational embeddings $\mathfrak{L}(\mathbf{R}_1^A), \mathfrak{L}(\mathbf{R}_2^A), \mathfrak{L}(\mathbf{R}_1^B), \mathfrak{L}(\mathbf{R}_2^B)$ and feed it to the neural net. The weight and bias for the first layer are respectively $\mathbf{W} \in \mathbb{R}^{4d^2 \times m}$ and $\boldsymbol{b} \in \mathbb{R}^m$, where m is the number of neurones in the hidden layer. A nonlinear activation function, f is applied at the hidden layer. In our experiments, we used tanh as the activation function. The weight and bias for the output layer, respectively $\mathbf{U} \in \mathbb{R}^{m \times 2d^2}$ and $\boldsymbol{b}' \in \mathbb{R}^{2d^2}$, are chosen such that by appropriately splitting the output into two parts and applying the inverse mapping of the linearisation, we can predict $\hat{\mathbf{R}}_1^C$ and $\hat{\mathbf{R}}_2^C$. Denoting the concatenation by $\oplus$ and inverse linearisation by $\mathfrak{L}^{-1}$, we can write the predicted embeddings for r_C as follows:

$$x = \mathcal{L}(\mathbf{R}_1^A) \oplus \mathcal{L}(\mathbf{R}_2^A) \oplus \mathcal{L}(\mathbf{R}_1^B) \oplus \mathcal{L}(\mathbf{R}_2^B) \tag{14}$$

$$h = f(\mathbf{W}x + b) \tag{15}$$

$$y = \mathbf{U}h + b' \tag{16}$$

$$\hat{\mathbf{R}}_1^C = \mathcal{L}^{-1} y_{:d^2} \tag{17}$$

$$\hat{\mathbf{R}}_2^C = \mathcal{L}^{-1} y_{d^2:} \tag{18}$$

Using a training set of relational tuples $\{(R_A, R_B, R_C)\}$, where $R_A \wedge R_B \Rightarrow R_C$ and their RelWalk embeddings, using Adam [10], we find the network parameters that minimise the squared Frobenius norm given in (19).

$$L(\mathbf{W}, \mathbf{U}, b, b') = \left\| \mathbf{R}_1^C - \hat{\mathbf{R}}_1^C \right\|_2^2 + \left\| \mathbf{R}_2^C - \hat{\mathbf{R}}_2^C \right\|_2^2 \tag{19}$$

4 Experiments

4.1 Datasets

We use the FB15k-237 dataset[2] created by Toutanova and Chen [23] for training KGEs using RelWalk. This dataset was created by removing the reverse relations between train and test portions in the original FB15k dataset and is considered as a more appropriate benchmark dataset for evaluating KGEs. FB15k-237 dataset contains 237 relation types for 14541 entities. To preserve the asymmetry property for relations, we consider that each relation $R^<$ in the relation set has its inverse $R^>$, so that for each triple $(h, R^<, t)$ in the KG we regard $(t, R^>, h)$ is also in the KG. Thus as a total we have 474 relation types to be learnt (we call this extended version as FB15k-474). The train, test and validation parts of this dataset contains respectively 544230, 40932 and 35070 tuples. Following the recommendations by the authors, RelWalk is trained using 100 minibatches for 1000 epochs until convergence. Negative sampling rate is set to 50 and we learn KGEs of dimensionalities $d = 20, 50$ and 100. We consider two tasks to evaluate relational composition operators namely, relation composition (Subsect. 4.2) and triple classification task (Subsect. 4.3). The matrix relational embeddings produced by RelWalk are used in the subsequent experiments described in the paper when learning supervised compositional operators.

To evaluate the ability of a relation composition operator, we use the dataset created by Takahashi et al. [22] from FB15-23k as follows. For a relation R, they define the *content set* $\mathcal{C}(R)$ as the set of (h, t) pairs such that (h, R, t) is a fact in the KG. Likewise, they define $\mathcal{C}(R_A \wedge R_B)$ as the set of (h, t) pairs such that $(h, R_A \to R_B, t)$ is a path in the KG. Next, $R_A \wedge R_B \Rightarrow R_C$ is considered as a compositional constraint if their content sets are similar; that is, if $|\mathcal{C}(R_A \wedge R_B) \cap \mathcal{C}(R_C)| \geq 50$ and the Jaccard similarity between $\mathcal{C}(R_A \wedge R_B)$ and $\mathcal{C}(R_C)$ is greater than 0.4. They obtained 154 compositional constraints of the form $R_A \wedge R_B \Rightarrow R_C$ after this filtering process. We name this dataset as the Relation Composition (**RC**) dataset in the remainder of the paper.

---

[2] https://www.microsoft.com/en-us/download/details.aspx?id=52312.

Table 1. Performance of relation composition ranking task.

Method	d = 20			d = 50			d = 100		
	MR	MRR	Hits@10	MR	MRR	Hits@10	MR	MRR	Hits@10
Addition	238	0.010	0.012	250	0.008	0.019	247	0.007	0.000
Matrix product	225	0.018	0.032	233	0.012	0.025	231	0.010	0.019
Hadamard product	215	0.020	0.051	192	0.037	0.051	209	0.016	0.032
Supervised (proposed)	**75**	**0.412**	**0.581**	**64**	**0.390**	**0.729**	**49**	**0.308**	**0.703**

We perform 5-fold cross validation on the RC dataset to train a supervised relation composition operator using our proposed method described in Subsect. 3.2. Using a separate validation dataset, we set the initial learning rate for Adam to 5E–4 and minibatch size to 25. We apply dropout with rate 0.5 and ℓ_2 regularisation with coefficient 1E–10 to avoid overfitting during training. For $d = 20$ dimensional embeddings, we use a single hidden layer of 300 neurones, whereas for $d = 50$ and 100 we used two hidden layers, where each has 600 neurones. In all settings training converged after 25k epochs. The source code implementation of the proposed method, datasets and FB15K-474 KGEs are publicly released[3]. We consider two tasks to evaluate relational composition operators namely, relation composition ranking and triple classification that requires involving entity embeddings.

4.2 Relation Composition Ranking

Let us assume that the composition of the two relations R_A and R_B is the relation R_C. Moreover, let us denote the pre-trained RelWalk embeddings for a relation R_x to be $\mathbf{R}_1^x$ and $\mathbf{R}_2^x$, where $x \in \{A, B, C\}$. We will denote the composed embedding for R_C by $\hat{\mathbf{R}}_1^C$ and $\hat{\mathbf{R}}_2^C$.

Relation composition ranking task aims to measure the similarity between a composed embedding for an unseen relation and all other relation embeddings. Following Takahashi et al. [22], we rank the test relations R_L by its similarity to $\hat{R}_C$, the composed version of R_C using the distance function, $d(R_L, \hat{R}_C)$, given by (20).

$$d(R_L, \hat{R}_C) = \left\|\mathbf{R}_1^L - \hat{\mathbf{R}}_1^C\right\|_F + \left\|\mathbf{R}_2^L - \hat{\mathbf{R}}_2^C\right\|_F \tag{20}$$

If the R_C is ranked higher for $\hat{R}_C$, then it is considered better. We consider the 474 relation types in FB15K-474 as candidates (i.e., R_L). We use Mean Rank (MR), Mean Reciprocal Rank (MRR) and Hits@10 to measure the accuracy of the composition.

Table 1 presents the average performance of relation compositions using 5-folds cross validation on **RC** compositional constraints. We consider the 474 relation types in FB15K-474 for this evaluation. Lower MR indicates better performance. As can be observed, the supervised relation composition achieves

---

[3] https://github.com/Huda-Hakami/Relation-Composition-for-Knowledge-Graphs.

Table 2. Results of triple classification.

Method	d = 20	d = 50	d = 100
Addition	68.9	70.44	69.45
Matrix product	67.6	65.24	75.71
Hadamard product	58.44	63.01	70.94
Supervised (proposed)	**77.55**	**77.73**	**77.62**

the best results for MR, MRR and Hits@10 with significant improvements over the unsupervised compositional operators. Hadmard product is the best among unsupervised relation compositional operators. However, the unsupervised operators collectively perform as the random baseline, which picks a relation uniformly at random from the candidate set of relations.

4.3   Triple Classification

To evaluate the effectiveness of the learnt operator for generating composed relation embeddings, we consider the triple classification task using the composed embeddings for R_C. Triple classification task is originally proposed by Socher et al. [20], and aims to predict whether a triple (h, R, t) is a valid triple or not given entity and relation embeddings and a scoring function that map the embeddings to a confidence score. Specifically, in this paper, we use the embeddings learnt by RelWalk for the entities and the relations in FB15k-474 and the joint probability $p(h, R, t)$ given by Theorem 1 to determine whether a relation R exists between two given entities h and t. We need positive and negative triples for classification. The negative triples are generated by randomly corrupting entities of the positive examples. For example, for a test triple (h, R, t), we consider (h, R, t') as a negative example where t' is sampled from all entities that appear in the corresponding argument in the entire KG.

We perform 5 folds cross-validation on **RC** compositional constraints. Once the proposed supervised relation composition is learnt using a training set, we perform triple classification for those triples in FB15K-474 testing set that are connected by the relation types in the test compositional constraints of **RC**. We evaluate the performance using the accuracy which is the percentage of the correctly classified test triples. We use the validation set to find a threshold θ for each test relation such that if $p(h, R, t) > \theta$, the relation (h, R, t) holds, otherwise we consider it as a negative triple.

The performance of the supervised and unsupervised relational compositional operators for triple classification is shown in Table 2. Across the relational compositional operators and for different dimensionalities, the proposed supervised relational composition method achieves the best accuracy for this task. Despite increasing the dimensionality of relation embeddings from 20 to 100 leading to a complex model with a large number of parameters to be tuned using a small set of compositional constraints as in **RC**, the trained operator shows better performance in all cases.

5   Conclusion

In this paper, we addressed the problem of composing pre-trained relation embeddings in KGs. Given a set of compositional constraints over relations in the form $R_A \wedge R_B \Rightarrow R_C$, our proposed method learns a supervised operator that maps the relation embeddings of two relations to a new relation embedding. The learnt operator can be used to infer relation embeddings for rare or unseen relation types. Evaluating the predicted relation embeddings for triple classification task indicates the effectiveness of the proposed relation composition method.

Acknowledgement. We would like to thank Ran Tian for sharing the relation composition benchmark dataset.

References

1. Bollacker, K., Evans, C., Paritosh, P., Sturge, T., Taylor, J.: Freebase: a collaboratively created graph database for structuring human knowledge. In: Proceedings of SIGMOD, pp. 1247–1250 (2008)
2. Bollegala, D., Hakami, H., Yoshida, Y., Kawarabayashi, K.i.: Relwalk - a latent variable model approach to knowledge graph embedding (2019). https://openreview.net/forum?id=SkxbDsR9Ym
3. Bordes, A., Usunier, N., Garcia-Durán, A., Weston, J., Yakhenko, O.: Translating embeddings for modeling multi-relational data. In: Proceedings of NIPS (2013)
4. Bordes, A., Weston, J., Collobert, R., Bengio, Y.: Learning structured embeddings of knowledge bases. In: Proceedings of AAAI (2011)
5. Ding, B., Wang, Q., Wang, B., Guo, L.: Improving knowledge graph embedding using simple constraints. In: Proceedings of ACL, pp. 110–121 (2018)
6. Guu, K., Miller, J., Liang, P.: Traversing knowledge graphs in vector space. In: Proceedings of EMNLP, pp. 318–327 (2015)
7. Hornik, K.: Multilayer feedforward networks are universal approximators. Neural Netw. **2**, 359–366 (1989)
8. Ji, G., He, S., Xu, L., Liu, K., Zhao, J.: Knowledge graph embedding via dynamic mapping matrix. In: Proceedings of ACL, pp. 687–696 (2015)
9. Ji, G., Liu, K., He, S., Zhao, J.: Knowledge graph completion with adaptive sparse transfer matrix. In: Proceedings of AAAI, pp. 985–991 (2016)
10. Kingma, D.P., Ba, J.L.: Adam: a method for stochastic optimization. In: Proceedings of ICLR (2015)
11. Lao, N., Cohen, W.W.: Relational retrieval using a combination of path-constrained random walks. Mach. Learn. **81**(1), 53–67 (2010). https://doi.org/10.1007/s10994-010-5205-8
12. Lao, N., Mitchell, T., Cohen, W.W.: Random walk inference and learning in a large scale knowledge base. In: Proceedings of EMNLP, pp. 529–539 (2011)
13. Larochelle, H., Erhan, D., Bengio, Y.: Zero-data learning of new tasks. In: Proceedings of AAAI, pp. 646–651 (2008)
14. Lin, Y., Liu, Z., Sun, M., Liu, Y., Zhu, X.: Learning entity and relation embeddings for knowledge graph completion. In: Proceedings of AAAI, pp. 2181–2187 (2015)
15. Neelakantan, A., Roth, B., McCallum, A.: Compositional vector space models for knowledge base completion. In: Proceedings of ACL, pp. 156–166 (2015)

16. Nguyen, D.Q., Sirts, K., Qu, L., Johnson, M.: Stranse: a novel embedding model of entities and relationships in knowledge bases. In: Proceedings of NAACL-HLT, pp. 460–466 (2016)
17. Nickel, M., Rosasco, L., Poggio, T.: Holographic embeddings of knowledge graphs. In: Proceedings of AAAI (2016)
18. Nickel, M., Tresp, V., Kriegel, H.P.: A three-way model for collective learning on multi-relational data. In: Proceedings of ICML, pp. 809–816 (2011)
19. Riedel, S., Yao, L., McCallum, A., Marlin, B.M.: Relation extraction with matrix factorization and universal schemas. In: Proceedings of NAACL, pp. 74–84 (2013)
20. Socher, R., Chen, D., Manning, C.D., Ng, A.: Reasoning with neural tensor networks for knowledge base completion. In: Advances in Neural Information Processing Systems, pp. 926–934 (2013)
21. Socher, R., Chen, D., Manning, C.D., Ng, A.Y.: Reasoning with neural tensor networks for knowledge base completion. In: Proceedings of NIPS (2013)
22. Takahashi, R., Tian, R., Inui, K.: Interpretable and compositional relation learning by joint training with an autoencoder. In: Proceedings of ACL, pp. 2148–2159 (2018)
23. Toutanova, K., Chen, D.: Observed versus latent features for knowledge base and text inference. In: Proceedings of 3rd Workshop on Continuous Vector Space Models and their Compositionality, pp. 57–66 (2015)
24. Trouillon, T., Welbl, J., Riedel, S., Gaussier, É., Bouchard, G.: Complex embeddings for simple link prediction. In: Proceedings of ICML (2016)
25. Wang, Q., Mao, Z., Wang, B., Guo, L.: Knowledge graph embedding: a survey of approaches and applications. IEEE Trans. Knowl. Data Eng. **29**(12), 2724–2743 (2017). https://doi.org/10.1109/TKDE.2017.2754499
26. Wang, Z., Zhang, J., Feng, J., Chen, Z.: Knowledge graph embedding by translating on hyperplanes. In: Proceedings of AAAI, pp. 1112–1119 (2014)
27. Xiao, H., Huang, M., Zhu, X.: TransG: a generative model for knowledge graph embedding. In: Proceedings of ACL, pp. 2316–2325 (2016)
28. Yang, B., Yih, W.T., He, X., Gao, J., Deng, L.: Embedding entities and relations for learning and inference in knowledge bases. In: ICLR (2015)
29. Yoon, H.G., Song, H.J., Park, S.B., Park, S.Y.: A translation-based knowledge graph embedding preserving logical property of relations. In: Proceedings of NAACL, pp. 907–916 (2016)

Context-Guided Self-supervised Relation Embeddings

Huda Hakami[1,2($\boxtimes$)] and Danushka Bollegala[1]

[1] Department of Computer Science, University of Liverpool, Liverpool, UK
{h.a.hakami,danushka}@liverpool.ac.uk
[2] Department of Computer Science, Taif Univeristy, Taif, Saudi Arabia
hoda.h@tu.edu.sa

Abstract. A semantic relation between two given words a and b can be represented using two complementary sources of information: (a) the semantic representations of a and b (expressed as word *embeddings*) and, (b) the contextual information obtained from the co-occurrence contexts of the two words (expressed in the form of lexico-syntactic patterns). Pattern-based approach suffers from sparsity while methods rely only on word embeddings for the related pairs lack of relational information. Prior works on relation embeddings have pre-dominantly focused on either one type of those two resources exclusively, except for a notable few exceptions. In this paper, we proposed a self-supervised context-guided Relation Embedding method (CGRE) using the two sources of information. We evaluate the learnt method to create relation representations for word-pairs that do not co-occur. Experimental results on SemEval-2012 task2 dataset show that the proposed operator outperforms other methods in representing relations for unobserved word-pairs.

Keywords: Relation embeddings · Relational patterns · Compositional approach · Learning relation representations

1 Introduction

Representing relations between words benefits various Natural Language Processing (NLP) tasks such as relational information retrieval [1,2], statistical machine translation [3], question answering [4] and textual entailment [5]. For example, given a premise sentence P, *a man ate an apple*, and a hypothesis H, *a man ate fruit*, a model that can infer the existence of *is-a* relation between *fruit* and *apple* would correctly predict that H entails P.

We consider the problem of creating a semantic representation r for the relation r that holds between two given words a and b. This problem has been approached from a *compositional* direction, wherein given the pre-trained word embeddings for a and b, respectively denoted by d-dimensional real vectors $\boldsymbol{a}, \boldsymbol{b} \in \mathbb{R}^m$, the goal is to learn a relation embedding function, $f(\boldsymbol{a}, \boldsymbol{b}; \boldsymbol{\theta})$,

© Springer Nature Singapore Pte Ltd. 2020
L.-M. Nguyen et al. (Eds.): PACLING 2019, CCIS 1215, pp. 67–78, 2020.
https://doi.org/10.1007/978-981-15-6168-9_6

parametrised by θ [6,7]. Unsupervised solutions to this problem have been proposed such as ones that use a fixed operator such as the vector offset [8], or supervised approaches that implement f as a multi-layer feed forward neural network [5,7,9].

On the other hand, the contexts in which a and b co-occur provide useful clues regarding the relations that exist between the two related words. We call this approach *pattern-based* relation representation because it requires contextual patterns in which the two related words co-occur in a text corpus [10,11]. The main drawback of the pattern-based approach is data sparseness. Because not every related word-pair co-occurs within a co-occurrence window even in a larger corpus, pattern-based relation representation methods fail to represent relations between words that never co-occur.

Despite the above-mentioned limitations, we argue that word embeddings and co-occurrence contexts collectively provide complementary information for the purpose of learning relational embeddings. For example, Bollegala et al. [1] observed a *duality* between word-pair and pattern-based approaches for representing relations where they refer to the former as an *intentional* definition of relation representation and the latter an *extensional* definition of relation representation. Bollegala et al. [1] used this duality to propose a sequential co-clustering algorithm for discovering relations from a corpus. Riedel et al. [12] further developed this line of research and proposed a *universal schema* for representing relations, which was then used to produce relation embeddings via matrix decomposition. However, despite these prior work on relation embeddings, the two types of information sources are often used independently.

Our focus in this paper is to propose a relation representation method that uses the contextual information from a text corpus to generalise the learnt operator for unobserved word-pairs. For this purpose, we propose Context-Guided Relation Embeddings (CGRE) to represent relations between words. Specifically, CGRE are learnt using the word embeddings of related word-pairs along with the co-occurrence contexts of the word-pairs extracted from a corpus. Our experimental results on the SemEval-2012 task 2 benchmark dataset show the ability of the learnt operator to generalise to unobserved word-pairs outperforming previously proposed relational operators.

2   Related Work

As already described in the previous section, two main approaches can be identified in the literature for representing a semantic relation between two words: *pattern-based* and *compositional* methods. The pattern-based approach uses lexical patterns in which the two words of interest co-occur, while the compositional approach, on the other hand, attempts to represent the relation between two words from their word embeddings.

2.1 Pattern-Based Approach for Relations

An unstructured text corpus forms important resource to extract information for numerous NLP tasks such as relation extraction where the task is to identify the relation that holds between two named entities. Lexical patterns in which two related words co-occur within a corpus provide useful insights into the semantic relations that exist between those two words. Patterns for hypernym relation (i.e., is-a) have been studied extensively since it plays a vital role in building ontologies covering entities. For instance, Hearst's [13] patterns such as *is a, is a kind of* and *such as* have been used to identify the hypernym relation between words. Lexical patterns for other relations have also been studied such as Mernonymy [14] and Causial [15].

Turney [16] introduced *latent relation hypothesis* which states that word-pairs that co-occur in similar patterns tend to have similar semantic relations, and proposed Latent Relational Analysis (LRA) to represent the relation between two words using a vector. Specifically, in LRA we first create a pair-pattern matrix where the elements correspond to the number of times a pair co-occurs with a pattern [10]. Next, dimensionality reduction techniques such as Singular Value Decomposition (SVD) is applied to this pair-pattern matrix to smooth the co-occurrence data and produce low-dimensional vector representations. When evaluated on a benchmark dataset containing word analogy questions collected from scholastic aptitude tests (SATs), LRA obtains an accuracy of 56.7%, while the average high school student's accuracy on SAT word analogy questions has been 57%. Along similar lines, Jameel et al. [17] extend Global Vectors [18] for learning word embedding to learn word-pair vectors considering the 3-ways co-occurrences between the two words and the context words in which they co-occur. We collectively refer to these methods as pattern-based representations because is strongly assume the availability of contexts that links the related words in texts.

Despite the success of the pattern-based methods for representing relations, it suffers from data sparsity. To represent the relation between two words, such methods require the two words to co-occur in a specified context for a pattern to be extracted. However, not every related words co-occur even in a large corpus. For example, in Fig. 1 we show the co-occurrence distribution of the word-pairs in SemEval-2012 Task 2 dataset, where we count the number of sentences in Wikipedia containing both words in each word-pair. In the SemEval 2012 Task 2, dataset there are 3,307 related word-pairs, out of which 490 word-pairs never co-occur in any sentence in Wikipedia corpus, resulting in a highly sparse co-occurrence distribution as shown in Fig. 1. Therefore, the pattern-based approach fails to handle such unobserved but related words.

Our proposed method differs from these existing pattern-based approaches in two important ways. First, we do not require the two words to co-occur within the same sentences in a corpus to be able to represent the relation between them. Second, the parametrised operator we learn generalises in the sense that it can be applied to any new word-pair or relation type, not limited to the words and relations that exist in the training data.

2.2 Compositional Approach for Relations

Prior works on word embedding learning have found that relations between words could be represented by the difference of the corresponding word embeddings (from here onwards we call it PairDiff) [18,19]. The most famous example is $king - man \approx queen - woman$. We call such approaches to represent relations between words as *compositional* because the relation representation is composed using the semantic representations of the two constituent words of the related pairs without requiring co-occurring patterns. The compositional approach for relations overcomes the sparseness issues in the pattern-based methods as it relaxes the assumption that related pairs have to co-occur in the same context.

Since Mikolov et al. findings in 2013 [19], a renewed interest in exploring the relations in the semantic spaces of words has been sparked. Several recent works have targeted to evaluate different combination methods that can be applied on word embeddings to generate word-pair embeddings [7,20,21]. Hakami and Bollegala [20] investigated several unsupervised operators, such as vector concatenation, addition, difference and elementwise multiplication, that map the embeddings of two related words to a vector representing the relation between them.

On the other hand, recent researches have raised concerns on claims about word embedding's ability to represent relations via PairDiff [22–24]. Given an analogy prediction problem in the form a is to b as c is to d, in some cases even by ignoring c it is possible to correctly predict d using the fact that d is similar to a and b individually. Roller et al. [25] showed that Hearst's patterns are more valuable for hypernym detection tasks than distributional word embeddings. Vylomova et al. [26] also showed the limitations of PaiDiff by applying it for representing semantic relations outside those in the Google dataset, which were used initially for the evaluation of PairDiff. These findings suggest that to represent a diverse set of relation types, we must combine the strengths in the pattern-based and compositional approaches, which is a motivation for our current work.

2.3 Hybrid Approaches for Relations

As described earlier, pattern-based and compositional approaches have complementary properties when it comes to representing relations. Hybrid approaches try to balance between the data sparsity in the pattern-based methods and the lack of relational information in the compositional approaches. However, few recent studies have been devoted to incorporate the two types of information to improve the relation representations.

Zilah et al. [27] measure the relational similarities between word-pairs by combining heterogeneous models including distributional word embeddings and lexical patterns. In their work, the compositional method based on PairDiff reported encouraging results for many relation types in SemEval-2012 task 2 dataset. More recently, Washio and Kato [28] proposed Neural Latent Relational Analysis (NLRA) an unsupervised relational operator that is learnt to make the compositional and pattern representations similar using a negative sampling training objective. They also found that NLRA can be used for the purpose of predicting missing dependency paths between word-pairs that don't co-occur in a corpus [9].

3   Method

Our main goal is to represent relations between words accurately. We propose to learn a parametrised operator for relations that maps a word-pair to a relation embedding considering two sources of information: (a) word embeddings of related words, and (b) the contexts in which two related words co-occur. We want the learnt operator to overcome the sparseness problem in pattern-based relation representations. Motivated by this, our objective is to create relation representations for word-pairs that do not co-occur or belong to unseen relations.

Given a set of related word-pairs along with their relation labels $\mathcal{D} = \{(a_i, b_i, r_i)\}_i^N$, pre-trained word embeddings that represent the semantics of words, and a text corpus, we propose a method for learning m-dimensional relation embeddings $r_{(c,d)} \in \mathbb{R}^m$ for an unseen word-pair (c, d). Relation labels for word-pairs can be the manually annotated gold labels provided in the relational dataset such as DiffVec [26], Google [19], and BATS [29], or can be pseudo labels generated from word-pair features as described in Sect. 3.1. Following the prior work [5,7,9,28], a word-pair (a, b) is fed to a deep multilayer neural network with a nonlinearity activation for the hidden layers. The input layer of the network is the concatenation of embeddings a and b and their difference, $(a; b; b - a)$. As described in Hakami and Bollegala [7], the output of the last layer of the neural network that is given by $f(a, b, \theta_f)$ is considered as a representation for a word-pair, and is passed to a fully connected softmax layer and the overall network is

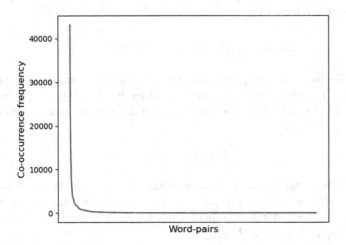

Fig. 1. Co-occurrence frequency for word-pairs in SemEval-2012 task2 from Wikipedia corpus.

trained to predict the relation label for the given pair. For this purpose, we use ℓ_2 regularised cross-entropy loss defined in (1) as the training objective.

$$\mathcal{J}_C = - \sum_{(a,b,r)\in\mathcal{D}} \log \; p(r|f(a,b,\theta_f)) \tag{1}$$

Here, θ_f collectively denotes the parameters of the network.

As we see, $\mathcal{J}_C$ given in (1) does not consider the co-occurrence contexts. Therefore, we consider a relation representation, $g(\mathcal{P}(a,b),\theta_g)$, that encodes a set of contextual co-occurrences between a and b according to (2).

$$g(\mathcal{P}(a,b),\theta_g) = \sum_{p\in\mathcal{P}(a,b)} w(a,p,b)\boldsymbol{h}(a,p,b,\theta_h) \tag{2}$$

Here, $p \in \mathcal{P}(a,b)$ is a lexical pattern that co-occurs with a and b. We use LSTMs [30] to map a sequences of words to a fixed-length vector $\boldsymbol{h}(a,p,b,\theta_h)$. It is worth noting that other models such as simple averaging or convolutional neural network can also be used to encode contextual patterns, but it is not the focus now to compare all alternative embeddings for patterns; rather, we aim to show the effectiveness of regularising the compositional method (i.e., $f(a,b,\theta_f)$) with contextual information. To incorporate the representativeness of a pattern of a relation, we assign a weight $w(a,b,p)$ given by (3).

$$w(a,p,b) = \frac{c(a,p,b)}{\sum_{t\in\mathcal{P}(a,b)} c(a,t,b)} \tag{3}$$

Here, c denotes the number of co-occurrences between p and (a,b).

Because the pattern-based and compositional-based methods represent the same semantic relation we require them to be close in the ℓ_2 space, captured by the constraint given by (4).

$$\mathcal{J}_{Patt} = \frac{1}{2}\|f(a,b,\theta_f) - g(\mathcal{P}(a,b),\theta_g)\|_2^2 \tag{4}$$

We would like to learn word pair embeddings that simultaneously minimise both (1) and (4). Therefore, we formulate the objective function of the proposed Context-Guided Relation Embeddings (i.e., CGRE) as a linear combination of (1) and (4) as follows:

$$\mathcal{J} = \mathcal{J}_C + \lambda\mathcal{J}_{Patt} \tag{5}$$

Here, $\lambda \in \mathbb{R}$ is a regularisation coefficient that determines the influence of the contextual patterns of the word-pairs for the learnt relational operator. After learning CGRE, we generate representations for a given word-pair (a,b) by concatenating $f_{\theta_f}(\boldsymbol{a},\boldsymbol{b})$ and $f_{\theta_f}(\boldsymbol{b},\boldsymbol{a})$.

3.1 Pseudo Relation Labels

To train CGRE, we require a dataset containing word-pairs annotated with relation labels. However, the cost of annotating word-pairs with relation labels can be high for specialised domains such as biomedical [31]. To make our proposed method self-supervised, we induce pseudo labels for word-pairs via clustering. Specifically, we cluster the PairDiff vectors of the training word pairs using k-means clustering algorithm with different k number of clusters. Because the ground truth class labels are given in DiffVec training data, we evaluate the quality of the generated clusters using the V-measure [32], which is an entropy-based measure for a harmonic mean between homogeneity and completeness of the clusters. V-measure scores are between 0 (imperfect) and 1 (perfect). We examine k from 10 to 80, in steps of 10. Consistent with Vylomova et al. findings [26], we find that $k = 50$ clusters to perform well with a V-measure of 0.416.

4 Experiments

4.1 Datasets

Measuring the Degrees of Prototypicality. We evaluate the relation embeddings on measuring degrees of relational similarity task using SemEval-2012 Task 2 dataset. The task is to rank word-pairs in a relation according to their degrees of prototypicality (i.e., the extent to which they exhibit the relation). The dataset has 79 relation types in total, and it is split into two sets of 69 test relations and ten train relations. Following the standard practice, we report performance on the test set and use the train set for setting hyperparameters.

Training Data. We use the DiffVec dataset [26] that contains $12,458$ triples (a, b, r), where words a and b are connected by an asymmetric relation r out of 36 fine-grained relation types. We use the word-pairs set $\mathcal{D}$ of the training relations and their reverse pairs to obtain relational patterns. Word-pairs in DiffVec that also appear in SemEval test data are excluded from the training set. Following Turney [33], we extract the context of one to five words in between the two related words considering the order in which they appear in the specified context ($\mathcal{P}(a, b)$ consists of all patterns in which a occurs before b). We filter patterns with out of vocabulary words. To reduce noise, we filter out the patterns that occur between less than ten distinct word-pairs in the corpus. As a result, we obtain $5,017$ contextual patterns and the number of training triples (a, b, p) after removing out-of-vocabulary words is $158,920$. For reproducibility, the code, training data and pre-trained embeddings are publicly available[1]. We use pre-trained 300-dimensional GloVe embedding for representing words[2]. To extract co-occurrence contexts, we use the English Wikipedia corpus, which consists of ca. 337M sentences.

---

[1] https://github.com/Huda-Hakami/Context-Guided-Relation-Embeddings.
[2] http://nlp.stanford.edu/data/glove.6B.zip.

4.2 Comparison Methods

We compare the proposed method with unsupervised compositional operators PairDiff and Concatenation (Concat) for the given pre-trained word embeddings. We also compare against the supervised Multi-class Neural Network Penulti-mate Layer (MnnPL) method proposed by Hakami and Bollegala [7]. Specifically, MnnPL learns a relation classifier using a relation labelled word-pairs and does not use contextual patterns (corresponds to $\lambda = 0$).

We also compare the proposed CGRE with NLRA using the contextual patterns provided by the original authors [28]. Because we are interested in relation representation methods that can generalise to word-pairs that *do not* co-occur in the corpus, we re-train NLRA using the same training data that we used for our proposed method such that NLRA doe not observe the word-pairs in SemEval dataset. LRA requires all word-pairs to be represented using lexical patterns extracted from the co-occurrence contexts. Because we strictly focus on evaluating relation representations for word-pairs without using their contextual patterns, LRA is excluded from the evaluations. Following Washio and Kato [28], we also assess the performance of each learnt relation representation method when it is combined with PairDiff by averaging the scores of a learnt method and the PairDiff score for each target word-pair.

4.3 Implementation Details

For a given word-pair (a, b), we compose their embeddings a and b using a multi-layer feedforward neural networks with three hidden layers followed by the batch normalisation and a *tanh* nonlinearity function. All the word vectors were first normalised to unit ℓ_2 length before feeding them to the neural net. The size of the hidden layers is set to 300. We did not update the input word embeddings during training to preserve their distributional regularity. A unidirectional LSTM

Table 1. Average MaxDiff accuracy and Spearman correlation for the 69 test relations in SemEval 2012 Task 2. Best results are in bold.

Method	MaxDiff	Correlation
PairDiff	43.48	0.31
Concat	41.67	0.29
NLRA	42.32	0.29
NLRA+PairDiff	44.35	0.33
MnnPL($\lambda = 0$)	43.75	0.31
MnnPL+PairDiff	45.42	0.35
CGRE-Gold	44.87	0.34
CGRE-Gold+PairDiff	**45.92**	**0.37**
CGRE-Proxy	44.34	0.34
CGRE-Proxy+PairDiff	45.49	0.36

with a 300-dimensional hidden state is used to encode the contextual patterns. AdaGrad [34] with mini-batch size 100 is used to learn the parameters of the proposed operator. All parameters are initialised by uniformly sampling from $[-1, +1]$, and the initial learning rate is set to 0.1. The best model was selected by early stopping using the MaxDiff accuracy on the SemEval train set.

4.4 Experimental Results

Table 1 shows the macro-averaged MaxDiff accuracy and Spearman correlations for the 69 test relations in the SemEval2012 Task 2 dataset. Our proposed method (GCRE) achieved the best results on both evaluation metrics when combined with PairDiff. CGRE trained using pseudo labels (CGRE-Proxy) can

Table 2. Average MaxDiff (top) and Spearman correlation (bottom) for each major relation in the test set of SemEval 2012-task2. The values between parentheses indicate the performance of a method combined with PairDiff. Best results for each relation are in bold.

Relation	MaxDiff			
	PairDiff	MnnPL	CGRE-Gold	CGRE-Proxy
CLASS-INCLUSION	48.50	**52.00** (51.60)	51.40 (51.67)	50.45 (49.35)
PART-WHOLE	43.50	41.33 (43.36)	39.61 (42.80)	43.35 (**44.38**)
SIMILAR	41.26	36.20 (41.15)	40.02 (40.82)	**41.68** (41.10)
CONTRAST	33.72	38.57 (38.73)	**40.21** (38.44)	36.39 (36.67)
ATTRIBUTE	46.32	44.84 (47.23)	46.19 (**47.97**)	45.44 (47.83)
NON-ATTRIBUTE	39.11	42.45 (41.82)	42.41 (42.79)	**43.00** (41.85)
CASE RELATIONS	46.49	49.53 (49.57)	**52.04** (51.67)	49.46 (50.21)
CAUSE-PURPOSE	44.43	44.17 (46.89)	47.57 (**48.59**)	47.74 (48.17)
SPACE-TIME	49.48	45.53 (48.50)	48.62 (**50.21**)	45.36 (49.79)
REFERENCE	41.92	45.94 (**47.84**)	41.32 (44.74)	41.52 (45.74)
Relation	Correlation			
	PairDiff	MnnPL	CGRE-Gold	CGRE-Proxy
CLASS-INCLUSION	0.375	0.519 (**0.537**)	0.533 (0.516)	0.515 (0.462)
PART-WHOLE	0.287	0.245 (0.288)	0.228 (0.292)	0.314 (**0.321**)
SIMILAR	0.252	0.186 (0.260)	0.245 (**0.286**)	0.280 (0.282)
CONTRAST	0.113	0.160 (0.202)	0.209 (**0.226**)	0.157 (0.171)
ATTRIBUTE	0.410	0.351 (0.409)	0.396 (**0.444**)	0.387 (0.437)
NON-ATTRIBUTE	0.209	0.264 (0.265)	0.287 (0.279)	**0.313** (0.274)
CASE RELATIONS	0.383	0.425 (0.467)	**0.475** (0.466)	0.419 (0.445)
CAUSE-PURPOSE	0.343	0.332 (0.384)	0.422 (**0.436**)	0.400 (0.404)
SPACE-TIME	0.422	0.373 (0.433)	0.432 (**0.455**)	0.385 (0.437)
REFERENCE	0.303	0.323 (**0.377**)	0.212 (0.323)	0.295 (0.375)

successfully reach the performance of CGRE trained using the gold labels in the DiffVec dataset (CGRE-Gold). This is encouraging because it shows that GCRE can be trained in a self-supervised manner, without requiring manually labelled data. Overall, for all the methods, adding the relational similarity scores from PairDiff improves the performance of ranking the word-pairs, which confirm the complementary properties between the two approaches when it comes to representing relations. As seen in Table 1, NLRA performs poorly when it is trained on DiffVec using patterns extracted for the word-pairs in DiffVec and tested on SemEval[3]. This shows that NLRA is unable to generalise well to the relations in the SemEval dataset, not present in the DiffVec dataset.

To evaluate the performance for different relation types, we breakdown the results for the ten major relations in the 69 SemEval test set as presented in Table 2. By incorporating contextual patterns when training CGRE, we obtain better performance in 8 out of the 10 test relations in terms of MaxDiff and Spearman correlation. These improvements are statistically significant according to a paired t-test ($p < 0.01$). MnnPL reports the best accuracy and correlation for CLASS-INCLUSION and REFERENCE relations (either without or with the addition of PairDiff).

5   Conclusion

We consider the problem of representing relations between words. Specifically, we proposed a method that uses the contextual patterns in a corpus to improve the compositional relation representation using word embeddings of the related word-pairs. For this purpose, we proposed a parametrised relational operator using the contexts where two words co-occur in a corpus and require that pattern-based representation to be similar to a compositional-based representation computed using the corresponding word embeddings. Experiments on measuring degrees of relational similarity between word pairs show that we can overcome the sparsity problem of the pattern-based approaches for relations.

References

1. Bollegala, D.T., Matsuo, Y., Ishizuka, M.: Relational duality: Unsupervised extraction of semantic relations between entities on the web. In: Proceedings of the 19th International Conference on World Wide Web, pp. 151–160. ACM (2010)
2. Cafarella, M.J., Banko, M., Etzioni, O.: Relational web search. In: WWW Conference (2006)
3. Nakov, P.: Improved statistical machine translation using monolingual paraphrases. In: Proceedings of ECAI 2008: 18th European Conference on Artificial Intelligence, 21–25 July 2008, Patras, Greece: Including Prestigious Applications of Intelligent Systems (PAIS 2008), vol. 178, p. 338. IOS Press (2008)

---

[3] The accuracy of NLRA when its trained on pattern extracted using word pairs in the entire SemEval dataset is 45.28%, which is similar to the result reported in the original paper.

4. Yang, S., Zou, L., Wang, Z., Yan, J., Wen, J.-R.: Efficiently answering technical questions-a knowledge graph approach, pp. 3111–3118. In: AAAI (2017)
5. Joshi, M., Choi, E., Levy, O., Weld, D.S., Zettlemoyer, L.: pair2vec: compositional word-pair embeddings for cross-sentence inference. arXiv preprint arXiv:1810.08854 (2018)
6. Hakami, H., Hayashi, K., Bollegala, D.: Why does pairdiff work? - A mathematical analysis of bilinear relational compositional operators for analogy detection. In: Proceedings of the 27th International Conference on Computational Linguistics (COLING) (2018)
7. Hakami, H., Bollegala, D.: Learning relation representations from word representations (2018)
8. Mikolov, T., Yih, W.-T., Zweig, G.: Linguistic regularities in continuous space word representations. In: Proceedings of HLT-NAACL, pp. 746–751 (2013)
9. Washio, K., Kato, T.: Filling missing paths: modeling co-occurrences of word pairs and dependency paths for recognizing lexical semantic relations. arXiv preprint arXiv:1809.03411 (2018)
10. Turney, P.D.: Measuring semantic similarity by latent relational analysis. arXiv preprint cs/0508053 (2005)
11. Snow, R., Jurafsky, D., Ng, A.Y.: Learning syntactic patterns for automatic hypernym discovery. In: Advances in Neural Information Processing Systems, pp. 1297–1304 (2005)
12. Riedel, S., Yao, L., McCallum, A., Marlin, B.M.: Relation extraction with matrix factorization and universal schemas. In: Proceedings of the 2013 Conference of the North American Chapter of the Association for Computational Linguistics: Human Language Technologies, pp. 74–84 (2013)
13. Hearst, M.A.: Automatic acquisition of hyponyms from large text corpora. In: Proceedings of the 14th Conference on Computational Linguistics, vol. 2, pp. 539–545. Association for Computational Linguistics (1992)
14. Girju, R., Badulescu, A., Moldovan, D.: Learning semantic constraints for the automatic discovery of part-whole relations. In: Proceedings of the 2003 Conference of the North American Chapter of the Association for Computational Linguistics on Human Language Technology, vol. 1, pp. 1–8. Association for Computational Linguistics (2003)
15. Marshman, E.: The cause-effect relation in a biopharmaceutical corpus: English knowledge patterns. In: Terminology and Knowledge Engineering, pp. 89–94 (2002)
16. Bigham, J., Littman, M.L., Shnayder, V., Turney, P.D.: Combining independent modules to solve multiple-choice synonym and analogy problems. In: Proceedings of the International Conference on Recent Advances in Natural Language Processing, pp. 482–489 (2003)
17. Jameel, S., Bouraoui, Z., Schockaert, S.: Unsupervised learning of distributional relation vectors. In: Proceedings of the 56th Annual Meeting of the Association for Computational Linguistics (Volume 1: Long Papers), vol. 1, pp. 23–33 (2018)
18. Pennington, J., Socher, R., Manning, C.: Glove: global vectors for word representation. In: Proceedings of the 2014 Conference on Empirical Methods in Natural Language Processing (EMNLP), pp. 1532–1543 (2014)
19. Mikolov, T., Chen, K., Corrado, G., Dean, J.: Efficient estimation of word representations in vector space. arXiv preprint arXiv:1301.3781 (2013)
20. Hakami, H., Bollegala, D.: Compositional approaches for representing relations between words: a comparative study. Knowl.-Based Syst. **136**, 172–182 (2017)

21. Gábor, K., Zargayouna, H., Tellier, I., Buscaldi, D., Charnois, T.: Exploring vector spaces for semantic relations. In: Proceedings of the 2017 Conference on Empirical Methods in Natural Language Processing, pp. 1814–1823 (2017)

22. Levy, O., Remus, S., Biemann, C., Dagan, I.: Do supervised distributional methods really learn lexical inference relations? In: Proceedings of the 2015 Conference of the North American Chapter of the Association for Computational Linguistics: Human Language Technologies, pp. 970–976 (2015)

23. Linzen, T.: Issues in evaluating semantic spaces using word analogies. arXiv preprint arXiv:1606.07736 (2016)

24. Rogers, A., Drozd, A., Li, B.: The (too many) problems of analogical reasoning with word vectors. In: Proceedings of the 6th Joint Conference on Lexical and Computational Semantics (* SEM 2017), pp. 135–148 (2017)

25. Roller, S., Kiela, D., Nickel, M.: Hearst patterns revisited: automatic hypernym detection from large text corpora. arXiv preprint arXiv:1806.03191 (2018)

26. Vylomova, E., Rimell, L., Cohn, T., Baldwin, T.: Take and took, gaggle and goose, book and read: evaluating the utility of vector differences for lexical relation learning. arXiv preprint arXiv:1509.01692 (2015)

27. Zhila, A., Yih, W.-T., Meek, C., Zweig, G., Mikolov, T.: Combining heterogeneous models for measuring relational similarity. In: Proceedings of the 2013 Conference of the North American Chapter of the Association for Computational Linguistics: Human Language Technologies, pp. 1000–1009 (2013)

28. Washio, K., Kato, T.: Neural latent relational analysis to capture lexical semantic relations in a vector space. arXiv preprint arXiv:1809.03401 (2018)

29. Gladkova, A., Drozd, A., Matsuoka, S.: Analogy-based detection of morphological and semantic relations with word embeddings: what works and what doesn't. In: Proceedings of the NAACL Student Research Workshop, pp. 8–15 (2016)

30. Hochreiter, S., Schmidhuber, J.: Long short-term memory. Neural Comput. **9**(8), 1735–1780 (1997)

31. Patel, P., Davey, D., Panchal, V., Pathak, P.: Annotation of a large clinical entity corpus. In: Proceedings of the 2018 Conference on Empirical Methods in Natural Language Processing, pp. 2033–2042 (2018)

32. Rosenberg, A., Hirschberg, J.: V-measure: a conditional entropy-based external cluster evaluation measure. In: Proceedings of the 2007 Joint Conference on Empirical Methods in Natural Language Processing and Computational Natural Language Learning (EMNLP-CoNLL) (2007)

33. Turney, P.D.: The latent relation mapping engine: algorithm and experiments. J. Artif. Intell. Res. **33**, 615–655 (2008)

34. Duchi, J., Hazan, E., Singer, Y.: Adaptive subgradient methods for online learning and stochastic optimization. J. Mach. Learn. Res. **12**(Jul), 2121–2159 (2011)

Evaluation of Embedded Vectors for Lexemes and Synsets Toward Expansion of Japanese WordNet

Daiki Ko[✉] and Koichi Takeuchi

Graduate School of Natural Science and Technology, Okayama University,
Okayama, Okayama 7008530, Japan
pk7y3fkx@s.okayama-u.ac.jp, takeuc-k@okayama-u.ac.jp

Abstract. In this paper, we discuss the possibility to expand Japanese WordNet using AutoExtend that can produce embedded vectors based on dictionary structure. Recently several kinds of NLP tasks showed that the distributed representations for words are effective, however, the word-embedded vectors constructed based on contexts of surrounded words would be difficult to discriminate meanings of a word because every vector is produced for a word. On the other hand, AutoExtend that can produce embedded vectors for meanings and concepts as well as words taking into account thesaurus structure of dictionary, has been proposed and applied into English WordNet. Thus, in this paper, we apply AutoExtend into a Japanese dictionary i.e., Japanese WordNet to construct embedded vectors for lexems and synsets as well as words taking into account thesaurus structure of Japanese WordNet. The experimental results show that embedded vectors constructed by AutoExtend can be helpful to find corresponding meanings for unregistered words in the dictionary.

Keywords: Synsets · Japanese WordNet · AutoExtend

1 Introduction

Lexical resources are useful not only for humanities but also various tasks in natural language processing: For example, WordNet is applied to a question answering system Watson [8] as well as word-sense disambiguation task [10]; FrameNet[1] and VerbNet[2] in PropBank [11] are used as lexical resources in sematic role labeling task [12].

The performance of these tasks depends heavily on the number of registered words in a lexical resource, thus developing a method to update and expand existing lexical resources is inevitable for improving the performance of these tasks.

---

[1] https://framenet.icsi.berkeley.edu/fndrupal/.

[2] http://verbs.colorado.edu/~mpalmer/projects/verbnet.html.

© Springer Nature Singapore Pte Ltd. 2020
L.-M. Nguyen et al. (Eds.): PACLING 2019, CCIS 1215, pp. 79–87, 2020.
https://doi.org/10.1007/978-981-15-6168-9_7

Previous studies proposed methods that can expend WordNet in another language using various types of language resources; e.g., using parallel corpora corpus [4] and using syntagmatic information [3]. In recent studies, embedded vectors for words such as word2vec [9] and fasttext [7], which is a technique to construct abstracted word meaning using raw texts, are revealed to be effective in various kinds of NLP tasks. These two methods of extracting embedded vectors, however, assign an embedded vector even for a polysemous word, so that the embedded vector has mixed several meanings of a word. For adding new entries to WordNet, embedded vectors with mixed meanings of a word might be difficult to be directly applied to adding unregistered words to WordNet.

On the other hand, AutoExtend [13] that can produce embedded vectors for meanings and concepts as well as words taking into account thesaurus structure of dictionary, has been proposed and applied into English WordNet. Thus, in this paper, we apply AutoExtend into a Japanese thesaurus dictionary i.e., Japanese WordNet (JWN) [6] to construct embedded vectors for lexemes and synsets. The experimental results show that embedded vectors constructed by AutoExtend can be helpful to find corresponding meanings for unregistered words in the dictionary.

2   Basic Approach to Construct Embedded Vectors Taking into Account Thesaurus Structure

Thesaurus-based dictionary has a hierarchical structure of concepts and polysemous words are categorized into their corresponding concepts in the concept tree structure, that is, several meanings of a word can be expressed in which concepts the word is assigned to. Thus the thesaurus-based dictionary is understandable for humans and useful in NLP, then several kinds of thesaurus-based dictionaries are constructed in Japanese, e.g., JWN, Predicate Thesaurus[3], and Lexeed [5]. To see the possibility to develop a method to add new entries to thesaurus-based dictionary, we focus on AutoExtend and clarify the possibility how AutoExtend can be applied to constructing embedded vectors for lexemes and synsets in non-English environment. In this section, we describe the basic formalisms defined in AutoExtend and show the framework has a potential to be applied to any kind of hierarchical structured dictionary.

2.1   AutoExtend

The neural network architecture of AutoExtend consists of embedded vectors among words, lexemes, and synsets. The input layer and the final output layer of the neural network are embedded vectors for words, and the hidden layer represents the gloss vectors, i.e., vectors for definitions of synsets [13]. The lexeme and synset embedded vectors are the same dimension for the word embedded vectors. This characteristic is convenient to be applied to new registration to the

---

[3] http://pth.cl.cs.okayama-u.ac.jp/.

dictionary because word embedded vectors can be directly compared with the lexeme/synset embedded vectors using conventional evaluation metrics between vectors.

In the basic idea of the network architecture, AutoExtend converts the word embedded vectors into synset embedded vectors and converts the synset embedded vectors into word embedded vectors through the lexeme embedded vectors. The formulas to realize this approach are as follows. In the lexical structure assumed in AutoExtend, a word has several lexemes, and a synset has a set of some lexemes. Thus the vector $l^{(i,j)}$ is regarded as a lexeme vector in a word $w^{(i)}$; besides the several lexeme vectors will belong to a synset $s^{(j)}$.

$$w^{(i)} = \sum_j l^{(i,j)} \tag{1}$$

$$s^{(j)} = \sum_i l^{(i,j)} \tag{2}$$

Here, We define $E^{(i,j)}$ as an identity matrix that converts word embedded vectors to lexeme embedded vectors:

$$l^{(i,j)} = E^{(i,j)} w^{(i)}. \tag{3}$$

Using Eq. 2 and Eq. 3, the $s^{(j)}$ is derived as

$$s^{(j)} = \sum_i E^{(i,j)} w^{(i)}. \tag{4}$$

After Eq. 4 is applied to the all words and synsets, the following formula is derived:

$$S = E \otimes W. \tag{5}$$

In Eq. 5, S is the matrix of synset embedded vectors and W is the matrix of word embedded vectors, and E is the rank four tensor containing $E^{(i,j)}$. From the view of the network structure of AutoExtend, Eq. 5 indicates that the E is an encoder of the input vector W.

In the same way, we assume $\overline{l}^{(i,j)}$ to be a lexeme vector and $\overline{w}^{(i)}$ to be a word vector at the decoder side of AutoExtend; where several lexeme vectors are assumed to come from a synset vector $s^{(j)}$. The formulas of these relations are defined below:

$$s^{(j)} = \sum_i \overline{l}^{(i,j)} \tag{6}$$

$$\overline{w}^{(i)} = \sum_j \overline{l}^{(i,j)}. \tag{7}$$

Here, let assume $D^{(j,i)}$ to be an identity matrix that converts a synset embedded vector to a lexeme embedded vector:

$$\overline{l}^{(i,j)} = D^{(j,i)} s^{(j)}. \tag{8}$$

Using Eq. 7 and Eq. 8, $\overline{w}^{(i)}$ can be derived as

$$\overline{w}^{(i)} = \sum_j D^{(j,i)} s^{(j)}. \tag{9}$$

When Eq. 9 is applied to the all words and synsets, the following formula is derived:

$$\overline{W} = D \otimes S. \tag{10}$$

In Eq. 10, $\overline{W}$ is the matrix of word embedded vectors and D is the rank four tensor containing $D^{(j,i)}$. From the view of the network structure of AutoExtend, Eq. 10 indicates that D is a decoder for outputting word vectors. In the learning shceme, AutoExtend reconstructs W with minimizing D and E:

$$\underset{E,D}{\mathrm{argmin}} \, \|D \otimes E \otimes W - W\|. \tag{11}$$

After the minimization, we obtain estimated lexeme vector, synset vector and word vector. Since the above formalism is based on a framework of autoencoder, this approach can be applicable in wider range of thesaurus-based dictionary.

3   Applying AutoExtend-Based Approach to JWN

We extracted embedded vectors for the synsets of JWN with AutoExtend. The method of AutoExtend requires initial embedded vectors for words and a thesaurus-based dictionary with descriptions of explaining concepts, i.e., synsets. As an input for AutoExtend, we use nwjc2vec [1] that are made by applying fasttext to large-scale Web corpus containing 100 million words [2]. As for the embedded vectors of descriptions of concepts, i.e., gloss vectors, we use English gloss texts in JWN because English gloss texts are described much better than the translated Japanese glosses. The embedded vector for an English gloss is constructed by calculating average vectors of embedded word vectors of English. Besides, we use optional information such as synset-relations, the hyponymy-relations and the similarity relations in calculating embedding vectors.

4   Experiments

4.1   Experimental Setup

To evaluate embedded vectors constructed by the proposed approach, we assume a simple task to find a nearest synset for unregistered words. Where, we defined the unregistered word indicates that the word does not exist in JWN, but exist in embedded vector calculated on large text corpus. Figure 1 depicts the area of unregistered words comparing with registered words in JWN.

We conduct ranking evaluations comparing embedded vectors for unregistered words with the other embedded vectors for synsets. The metrics of calculating similarity between vectors of words and synsets is cosine similarity.

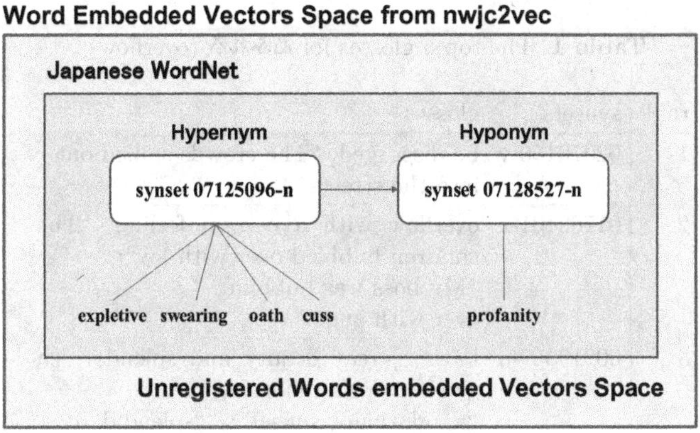

Word Embedded Vectors Space from nwjc2vec

Fig. 1. Area of unregistered words comparing with registered words in JWN

If the score of cosine similarity is close to 1, that is, the two vectors are similar, the synset will be in higher rank. After the ranking lists are obtained, we conduct a qualitative analysis of how appropriate the ranked synsets correspond to the meanings of the unregistered word.

As evaluation targets, we use seven unregistered words selected from 1.5 million unregistered words of JWN. The seven words are あふれる (overflow), 押さえつける (restrain), 蹴破る (break), ふっかける (rip off), 詰め替える (refill), とける (melt), 行き詰まる (go nowhere). If the highly ranked synsets are appropriate to the meaning of target unregistered word, the proposed approach can be considered as useful.

4.2 Experimental Results

Tables 1, 2 and 3 show the results of raking evaluation for the three words. Table 4 shows the synsets for the remaining words because of short space. All in all, these tables show that all the obtained synsets correspond to some meaning of their target unregistered words. Thus, the proposed approach of extracting embedded vectors for synsets using AutoExtend can capture thesaurus structure of JWN. Besides, this result indicates that we can obtain appropriate embedded vectors for synsets by giving lexical resource's structure even though we use embedded vectors of English gloss texts. This indicates that we can find near synsets for an unregistered word of JWN using embedded vectors taking into account thesaurus structure and word embedding calculated on large-scale Web corpus.

5 Discussions

From the results of Sect. 4.2, we found that extracting synsets embedded vectors with AutoExtend is useful for new words registration with JWN. If we can

Table 1. The top 5 glosses for あふれる (overflow)

rank	synsets	glosses
1	02070150-v	be disgorged; "The crowds spilled out into the streets"
2	01763101-v	overflow with a certain feeling; "The children bubbled over with joy"; "My boss was bubbling over with anger"
3	00219389-a	having great beauty and splendor; "a glorious spring morning"; "a glorious sunset"; "splendid costumes"; "a kind of splendiferous native simplicity"
4	00219705-a	dazzlingly beautiful; "a gorgeous Victorian gown"
5	01457369-a	pleasantly full and mellow; "a rich tenor voice"

Table 2. The top 5 glosses for 押さえつける (restrain)

rank	synsets	glosses
1	00462092-v	to put down by force or authority; "suppress a nascent uprising"; "stamp down on littering"; "conquer one's desires"
2	01568630-v	conceal or hide; "smother a yawn"; "muffle one's anger"; "strangle a yawn"
3	02424128-v	put down by force or intimidation; "The government quashes any attempt of an uprising"; "China keeps down her dissidents very efficiently"; "The rich landowners subjugated the peasants working the land"
4	00178898-v	rip off violently and forcefully; "The passing bus tore off her side mirror"
5	01872645-v	press against forcefully without moving; "she pushed against the wall with all her strength"

Table 3. The top 5 glosses for 蹴破る (break)

rank	synsets	glosses
1	01238640-v	deliver a sharp blow or push :"He knocked the glass clear across the room"
2	01240308-v	collide violently with an obstacle; "I ran into the telephone pole"
3	02570684-v	enter someone's (virtual or real)property in an unauthorized manner, usually with the intent to steal or commit a violent act; "Someone broke in while I was on vacation"; "They broke into my car and stole my radio!"; "who broke into my account last night?"
4	02185373-v	make light, repeated taps on a surface; "he was tapping his fingers on the table impatiently"
5	01239862-v	cause to come or go down; "The policeman downed the heavily armed suspect"; "The mugger knocked down the old lady after she refused to hand over her wallet"

Table 4. The top 5 glosses for remaining four words

rank	ふっかける rip off synsets	詰め替える refill synsets	とける melt synsets	行き詰まる go nowhere synsets
1	02319050-v	00453803-v	00444629-v	14015361-n
2	02259829-v	00140751-v	00397192-v	07366145-n
3	02214190-v	01270199-v	00396997-v	14478684-n
4	00775156-v	00161987-v	00446329-v	01972131-v
5	02018524-v	03704376-n	00419137-v	14408086-n

expand the number of words in lexical resources like JWN, we can improve the performance of NLP tasks using dictionaries.

We have still two issues as follows: First, we have not evaluated all unregistered words. Second the proposed approach requires the extra manual steps to decide how many synsets we should assign an unregistered word to.

For the second issue, we conducted additional experiments without manual steps, that is, excluding randomly 100 words as pseudo unregistered words in JWN. Then, two types of synset vectors are estimated: The first one is made by AutoExtend, and the second one is by averaging word vectors of nwjc2vec belonging to the synset. Top 10 lists of similar synsets for the pseudo unregistered words are made by comparing with the cosine similarity between the vector of pseudo unregistered words and their synset vectors. We evaluated recall rates of how the correct synsets are contained in the top 10 list. The experimental results show that the recall rates of applying AutoExtend and averaging vectors to synset vectors are 0.312 and 0.268, respectively. Thus, AutoExtend is also promising for finding synsets of the 100 pseudo unregistered words.

The prospects for the future about evaluation problems are as follows: to investigate whether all synsets embedded vectors are useful, we need to extract the similar synsets in the top 10 to the random sampling of unregistered words. After that human annotators evaluate them and give scores on how many synsets are similar to unregistered words; and then we investigate whether AutoExtend is useful for new words registration with lexical resources by the accuracy of the selected synsets. In terms of deciding how many synsets we should assign an unregistered word to, we can get the score of the cosine similarity when calculating cosine similarity between a word and a synset. Thus, we will decide automatically the number of synsets which should link an unregistered word to referring to this score. For example, the threshold value is 0.6. When the cosine similarity between an unregistered word and a synset is larger than 0.6, we link the unregistered word to the synset.

6   Conclusions

In this paper, we extracted synset embedded vectors by applying AutoExtend to JWN. The experimental results show that the synset embedded vectors are useful for new words registration with JWN. We also revealed that the relationships of Japanese unregistered words and synsets in JWN is usually correct using English glosses. In the future we should evaluate our approach using more examples as well as develop an automatic method to add unregistered words, especially for predicates, to several synsets.

Acknowledgment. A part of the research reported in this paper is supported by JSPS KAKENHI (JP19K00552) and the NINJAL project "Development of and Research with a parsed corpus of Japanese" by JSPS KAKENHI (JP15H03210).

References

1. Asahara, M.: NWJC2Vec: word embedding dataset from 'NINJAL Web Japanese Corpus'. Terminol. Int. J. Theor. Appl. Issues Spec. Commun. **24**(2), 7–25 (2018)
2. Asahara, M., Maekawa, K., Imada, M., Kato, S., Konishi, H.: Archiving and analysing techniques of the ultra-large-scale web-based corpus project of NINJAL, Japan. Alexandria **26**(1–2), 129–148 (2014)
3. Bentivogli, L., Pianta, E.: Extending wordnet with syntagmatic information. In: Proceedings of The Second Global WordNet Conference, pp. 47–53 (2004)

4. Fišer, D.: Leveraging parallel corpora and existing wordnets for automatic construction of the slovene wordnet. In: Vetulani, Z., Uszkoreit, H. (eds.) LTC 2007. LNCS (LNAI), vol. 5603, pp. 359–368. Springer, Heidelberg (2009). https://doi.org/10.1007/978-3-642-04235-5_31

5. Fujita, S., Tanaka, T., Bond, F., Nakaiwa, H.: An implemented description of Japanese: the Lexeed dictionary and the Hinoki treebank. In: COLING/ACL06 Interactive Presentation Sessions, pp. 65–68 (2006)

6. Isahara, H., Bond, F., Uchimoto, K., Utiyama, M., Kanzaki, K.: Development of the Japanese WordNet. In: Proceedings of the 6th International Conference on Language Resources and Evaluation, pp. 2420–2423 (2008)

7. Joulin, A., Grave, E., Bojanowski, P., Douze, M., Jégou, H., Mikolov, T.: FastText.zip: compressing text classification models. arXiv preprint arXiv:1612.03651 (2016)

8. Lally, A., Prager, J.M., et al.: Question analysis: how Watson reads a clue. IBM J. Res. Dev. 56(34), 2:1–2:14 (2012)

9. Mikolov, T., Chen, K., Corrado, G., Dean, J.: Efficient estimation of word representations in vector space. CoRR (2013). http://arxiv.org/abs/1301.3781

10. Navigli, R.: Word sense disambiguation: a survey. ACM Comput. Surv. (CSUR) 41(2), 1–69 (2009)

11. Palmer, M., Gildea, D., Kingsbury, P.: The proposition bank: an annotated corpus of semantic roles. Comput. Linguist. 31(1), 71–105 (2005)

12. Palmer, M., Gildea, D., Xue, N.: Semantic Role Labeling. Morgan & Claypool Publishers, San Rafael (2010)

13. Rothe, S., Schütze, H.: AutoExtend: extending word embeddings to embeddings for synsets and lexemes. In: Proceedings of the Association for Computational Linguistics (2015)

Neural Rasch Model: How Do Word Embeddings Adjust Word Difficulty?

Yo Ehara[✉]

Shizuoka Institute of Science and Technology (SIST), 2200-2, Toyosawa, Fukuroi,
Shizuoka 437-8555, Japan
i@yoehara.com
http://yoehara.com/

Abstract. The Rasch model is a probabilistic model for analyzing the psychological test response data such as second language tests; it is especially useful to automatically obtain a common scale of difficulty for various types of test questions by fitting the model to the response data of the test takers obtained from various methods such as reading comprehension questions and vocabulary questions. Because a test-taker can answer only some hundreds of vocabulary questions from tens of thousands of words in second language vocabulary, there exists a strong need to estimate the difficulty of the words that were not used during the test. For this purpose, the word frequency in a large corpus was previously used as a major clue. Although recent advancements in natural language processing enable us to obtain considerable semantic information about a word in the form of word embeddings, the manner in which such embeddings can be utilized while adjusting the word difficulty estimation has not been considerably investigated. Herein, we investigate how to effectively leverage word embeddings for adjusting the word difficulty estimates. We propose a novel neural model to fit the test response data. Further, we use the trained weights of our neural model to estimate the difficulty of the words that were not tested. The quantitative and qualitative experimental results denote that our model effectively leverages word embeddings to adjust simple frequency-based word difficulty estimates.

Keywords: Second language vocabulary · Word embeddings · Item response theory

1 Introduction

Comparing the difficulties of various kinds of language test questions is an important task in second-language education. A typical application is computer adaptive testing in which the abilities of the test takers are measured based on the

This work was supported by JST, ACT-I Grant Number JPMJPR18U8, Japan. We used the ABCI infrastructure by AIST for computational resources.

© Springer Nature Singapore Pte Ltd. 2020
L.-M. Nguyen et al. (Eds.): PACLING 2019, CCIS 1215, pp. 88–96, 2020.
https://doi.org/10.1007/978-981-15-6168-9_8

adaptively selected questions. Moreover, in educational natural language processing (NLP), the vocabulary test questions and reading comprehension questions must be of comparable difficulty. This is essential for building a system that can automatically select reading materials based on quick-response vocabulary tests of the data learners.

A naïve method to compare the difficulty of various types of test questions is to use the number of learners who inaccurately answered each question in the test as its difficulty, as previously used in [19,28]. However, this method inherently assumes that all learners have the same ability, which is not realistic. Practically, the questions that are incorrectly answered by the highly ability learners should be classified as more difficult than those that are incorrectly answered by the low ability learners.

To incorporate learner ability when estimating the difficulty of a test question, educational psychology employs item response theory (IRT) models. Herein, we focus on a simple IRT model called the Rasch model [23], which considers the test response data of the learners. These data record which learners accurately or inaccurately answered which test questions. Given these data, the model jointly estimates the difficulty d of each test question and the ability θ of each learner as parameters. The simple $\theta - d$ formula models the likelihood of the learner responding accurately to the question; a larger $\theta - d$ means that the learner will more likely provide an accurate answer.

IRT models can estimate the difficulty of various types of questions if we can distinguish the accurate and inaccurate responses in a binary manner. Using these difficulty values, we can not only compare various types of questions but also estimate how high a learner's ability should be for him/her to accurately respond to the question. In this way, the difficulty parameters estimated in the IRT models have several direct applications that adaptively support language learners [2].

Unfortunately, difficulty parameters are costly to measure. In typical cases, a meaningful difficulty parameter for *each* word in an IRT model [2] requires approximately 100 test takers, each needing approximately 30 min to complete a 100-word vocabulary test [6]. In contrast, language learners typically need to learn approximately 10,000 words to understand a university-level text in a second language [17]. We must also consider the fatigue of test takers; intuitively, one can understand that test takers answer randomly when asked to complete too many questions [17]. Therefore, testing all words is an impractical proposition. Instead, estimating the difficulty parameters of most of the words from small response data of vocabulary tests is more practical.

In vocabulary testing, the difficulty parameter of a word in the IRT models is known to correlate well with its word frequency in a large corpus. While the semantic features of the words may be useful for adjusting the estimated difficulty parameters of words, to the best of our knowledge, no previous study has seriously applied semantic features to difficulty-parameter estimation, presumably because obtaining high-quality semantic features was prohibitively expensive or difficult before the emergence of *word embeddings*.

Herein, we investigate the following research questions to understand the effectiveness of the effective word embeddings for estimating the difficulty parameters of words in the Rasch model. We propose a novel neural model to fit the test response data. We subsequently use the trained weights of our neural model to estimate the difficulty of the words that were not tested. The purpose of this study is to answer the following research questions:

1. Can word embeddings in IRT models be effectively used for estimating the word-difficulty parameters of words that are not in the vocabulary tests?
2. How effective are the word embeddings for estimating the word difficulty parameters when compared with the word frequencies in large corpora?

2   Related Work

Complex Word Identification. In NLP, the task of identifying complex words in running sentences is known as the *complex word identification* (CWI) task. Previous studies have identified two major types of shared tasks [19,28]. CWI, which typically is used as a pre-process of *text simplification*, identifies the complex words that need to be simplified.

Text simplification typically assumes a type or group of people who can read only simplified texts, such as beginning language learners. Hence, a fully-personalized, accurate identification of complex words that cannot be understood by a given learner is of less focus in CWI. However, there exist a few studies that addressed fully-personalized CWI [6–8,15,27]: these studies employ machine-learning-based binary classifiers that are trained using each learner's vocabulary test result. Then, for a given learner, the classifier makes a binary prediction as to whether the learner knows a given word of interest. This approach is more flexible than vocabulary-size-based approach because a classifier can use the word list or the vocabulary size as the classifier feature. In NLP/CL studies, this approach has been proposed to create personalized reading support systems [9,27] to select good translators in crowdsourcing [6] and to create personalized lexical simplification systems [15]. However, to our knowledge, no previous study reports how effective word embeddings in estimating word difficulty.

Vocabulary Testing Studies. Vocabulary test studies for modeling second language vocabulary usually assume that all learners memorize words in the precise order of a fixed word list, such as the word frequency list in a large balanced corpus. By estimating the known elements in this list, this method then tries to estimate the entire set of words known by the learner [13,17,25].

In these approaches, the word difficulties are expressed as rankings in the list, and the learner's language ability is summarized as the vocabulary size. For example, in [18], they simply sampled 100 words from a 20,000-word list, compiled 100 questions into a vocabulary test of these words, and calculated the vocabulary size as 20× the number of accurately answered questions.

Obviously, the effectiveness of this approach depends on the corpus from which the word list is sourced. The British National Corpus (BNC) [3] and the Contemporary Corpus of American English [4] are two widely used corpora [16, 18]. Although previous studies have acknowledged that learners do not always memorize second language words by a word list [14], vocabulary size is an intuitively understood concept. Hence, word difficulty and learners' language ability are often assessed by vocabulary size in second-language education studies.

3   Previous Models

3.1   Rasch Model

The Rasch model is one of the most basic cognitive models for understanding educational test data. Let us suppose that I questions $\{v_1, \ldots, v_I\}$ are to be answered by J learners (i.e., test-takers) $\{l_1, \ldots, l_J\}$. Then, we introduce the one-hot representation of the questions and learners. Let $\phi_r(v_i)$ be the I-dimensional one-hot vector, i.e., the i-th element in the vector is 1 and all the other elements in the vector are 0. Let $\phi_l(l_j)$ be the J-dimensional one-hot vector, i.e., the j-th element in the vector is 1 and all the other elements in the vector are 0. Then, the Rasch model is written as follows.

$$P(y_{i,j} = 1 | \mathbf{w}_l, \mathbf{w}_r) = \sigma(\mathbf{w}_l^\top \phi_l(l_j) - \mathbf{w}_r^\top \phi_r(v_i)) \tag{1}$$

In Eq. 1, σ denotes the logistic sigmoid function, i.e., $\sigma(x) := \frac{1}{1+\exp(-x)}$ and $y_{i,j}$ denotes the *test response data*. $y_{i,j} = 1$ if learner l_j answers correctly to question v_i, and $y_{i,j} = 0$ otherwise. $\mathbf{w}_l$ and $\mathbf{w}_r$ are J- and I-dimensional parameter vectors, respectively, to be fitted to the test response data.

Equation 1 can be intuitively explained as follows: first, σ is a monotonously increasing function with $\lim_{x \to \infty} \sigma(x) = 1$ and $\lim_{x \to -\infty} \sigma(x) = 0$. σ is simply used to make the range of $\mathbf{w}_l^\top \phi_l(l_j) - \mathbf{w}_r^\top \phi_r(v_i)$ to $(0, 1)$ and is interpreted as a probability. Once fitted to the test response data $y_{i,j}$, $\mathbf{w}_l^\top \phi_l(l_j)$ denotes the *ability* of learner l_j, and $\mathbf{w}_r^\top \phi_r(v_i)$ denotes the *difficulty* of question v_i.

According to Eq. 1, the larger the learner l_j's ability, he/she answers correctly to question v_i in higher probability, and the above probability approaches 1. On the contrary, the larger the question v_i's difficulty, learner l_j answers correctly to question v_i in lower probability, and the probability diminishes toward 0. In this way, the value $\mathbf{w}_l^\top \phi_l(l_j) - \mathbf{w}_r^\top \phi_r(v_i)$ determines the probability that learner l_j correctly answers question v_i, and σ simply converts this value to a probability.

The logistic sigmoid function in Eq. 1 is a binary version of the *softmax* function, which is widely used in modeling of neural network layers. Hence, the Rasch model can be described by neural-network diagrams, which are familiar to recent NLP researchers, as in Fig. 1.

Throughout this paper, when we use neural-network diagrams, **Linear** denotes a linear layer, **ReLU** denotes the Rectified Linear Unit [12] function, **BN** denotes the Batch Normalization technique [11], **Dropout** denotes the dropout technique [26], and **tanh** denotes the hyperbolic tangent function.

As illustrated in Fig. 1, the model inputs are the one-hot vectors $\phi_l(l_j)$ and $\phi_r(v_i)$. When training the model, the prediction Fig. 1 is also set to the test response data $y_{i,j}$. In neural-network diagrams, the parameters to be fitted to the test response data, namely $\mathbf{w}_l, \mathbf{w}_r$, are implicitly expressed as the weights of the two **Linear** layers.

3.2   Logistic Regression Models

From Eq. 1, we notice that features about each question can be introduced into the model. In this case, Eq. 1 can be simply regarded as a logistic regression model as shown in Fig. 2. This extension was previously used [6,8,9] in NLP.

Herein, we solely focus on vocabulary tests and assume that each question is a question about a word in second language vocabulary; hence, we interchangeably write a question or a word as v_i. We focus on two types of word features. One is corpus-based features written as $\phi_c(v_i)$, and another is word-embeddings based features written as $\phi_e(v_i)$. To use these features in the logistic regression models, we simply concatenate both types of features vectors for word v_i as illustrated in Fig. 2.

While logistic regression models leverage word features to estimate word difficulty, there are several pitfalls. First, because they are single-layer models from the viewpoint of neural networks, they cannot take combination effects between features into account. Second, word frequency and word embeddings are very different types of features weighted only by linear weights. Therefore, the model cannot be easily interpreted by analyzing the resulting weights: for instance, a heavily weighted feature may take a very small value.

4   Proposed Model

To account for the pitfalls described in the previous section, we propose a neural Rasch model as illustrated in Fig. 3, in which the corpus-based features and word embedding features are fed into the model separately. Unlike the logistic regression model, this model accounts for the effects of combining the corpus-based and word-embedding features.

By separately treating the two types of models, we can interpret the importance of each feature type. The corpus-based features provide a rough estimate of the word's difficulty. The word-embedding based features adjust its estimate. To adjust, the difficulty and its adjustment must have the same scale; the adjustment can be both positive or negative. Hence, we used hyperbolic tangent functions, ranging $(-1, 1)$, in the two ends of both feature types in our model. Finally, the importance of the estimates are determined from the weights of the the upper-most softmax function as illustrated in Fig. 3.

5   Experiments

We made the dataset for the experiment by using a crowdsourcing service called "Lancers" [6]. As the service is Japan-based, most of its workers are Japanese native speakers who are learning English. We asked 100 learners to answer 55 questions in the Vocabulary Size Test [18]. A dataset built under a similar setting was later publicly available [5].

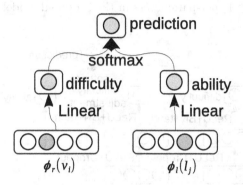

Fig. 1. Rasch model illustrated using neural network diagram [23].

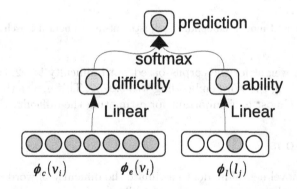

Fig. 2. Logistic regression model used in [8] illustrated using neural network diagram.

As for the features, we used [3] and [4] for corpus features, and GLoVe [22] for word embedding features. We used the 100-dimensional GLoVe word embedding vectors trained over 1 Giga word of the Wikipedia corpus, which is openly available in the **gensim** toolkit [24].

For evaluation, we split the test response data into the training/validation set and the test set; in the $100 \times 55 = 5,500$ responses, we first randomly selected 550 for the test set. In the remaining 4,950 data disjoint from the test, we randomly selected its 25%, i.e., 1,237, as the validation set, and the remaining 75%, i.e., 3,713 as the training set. To implement and optimize our model, we used the **PyTorch** [20] toolkit. To implement the logistic regression model, we used **scikit-learn** [21], which uses LIBLINEAR [10] internally. Its hyper-parameter C was chosen from the range of 0.01 to 100 in a log uniform manner by 200 trials using the **optuna** [1] hyper-parameter tuning toolkit.

The accuracies of the logistic regression model and the proposed model were 0.745 and 0.6981, respectively. The accuracies differ because the hyper-parameters are tuned in the logistic regression using the validation set while we used only the default hyper-parameters in the proposed model.

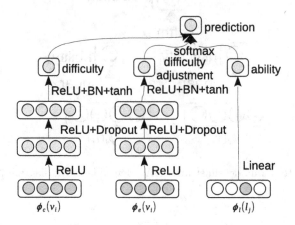

Fig. 3. Proposed model: embedding-adjustment-aware neural Rasch model.

In Fig. 3, the weight for the corpus-based word difficulty is -2.0331 and the adjustment by using the word embedding was 0.9732. This means that the word embeddings are almost half important for estimating the difficulty of words.

6   Conclusions

We proposed a novel neural model to estimate the difficulty of words. To answer the research questions on how word embeddings adjust word difficulty in the Rasch models, our model can compare the weight for corpus-based features and word-embedding based features. In experiments, the semantic adjustment by using word embedding when estimating word difficulties amounts to approximately a half of the corpus-based word difficulty.

References

1. Akiba, T., Sano, S., Yanase, T., Ohta, T., Koyama, M.: Optuna: a next-generation hyperparameter optimization framework. In: Proceedings of KDD, pp. 2623–2631 (2019)

2. Baker, F.B., Kim, S.H.: Item Response Theory: Parameter Estimation Techniques. Marcel Dekker, New York (2004)
3. BNC Consortium: The British National Corpus
4. Davies, M.: The 385+ million word Corpus of Contemporary American English (1990–2008+): design, architecture, and linguistic insights. Int. J. Corpus Linguist. **14**(2), 159–190 (2009)
5. Ehara, Y.: Building an English vocabulary knowledge dataset of Japanese English-as-a-second-language learners using crowdsourcing. In: Proceedings of LREC (2018)
6. Ehara, Y., Baba, Y., Utiyama, M., Sumita, E.: Assessing translation ability through vocabulary ability assessment. In: Proceedings of IJCAI, pp. 3712–3718 (2016)
7. Ehara, Y., Miyao, Y., Oiwa, H., Sato, I., Nakagawa, H.: Formalizing word sampling for vocabulary prediction as graph-based active learning. In: Proceedings of EMNLP (2014)
8. Ehara, Y., Sato, I., Oiwa, H., Nakagawa, H.: Mining words in the minds of second language learners: learner-specific word difficulty. In: Proceedings of COLING (2012)
9. Ehara, Y., Shimizu, N., Ninomiya, T., Nakagawa, H.: Personalized reading support for second-language web documents. ACM TIST **4**(2) (2013). https://doi.org/10.1145/2438653.2438666. Article No. 31
10. Fan, R.E., Chang, K.W., Hsieh, C.J., Wang, X.R., Lin, C.J.: LIBLINEAR: a library for large linear classification. JMLR **9**, 1871–1874 (2008)
11. Ioffe, S., Szegedy, C.: Batch normalization: accelerating deep network training by reducing internal covariate shift. In: Proceedings of ICML (2015)
12. Krizhevsky, A., Sutskever, I., Hinton, G.E.: ImageNet classification with deep convolutional neural networks. In: Proceedings of NIPS, pp. 1097–1105 (2012)
13. Laufer, B.: What percentage of text-lexis is essential for comprehension. In: Special Language: From Humans Thinking to Thinking Machines, pp. 316–323 (1989)
14. Laufer, B., Ravenhorst-Kalovski, G.C.: Lexical threshold revisited: lexical text coverage, learners' vocabulary size and reading comprehension. Read. Foreign Lang. **22**(1), 15–30 (2010)
15. Lee, J., Yeung, C.Y.: Personalizing lexical simplification. In: Proceedings of COLING, August 2018
16. Meara, P.M., Alcoy, J.C.O.: Words as species: an alternative approach to estimating productive vocabulary size. Read. Foreign Lang. **22**(1), 222–236 (2010)
17. Nation, P.: How large a vocabulary is needed for reading and listening? Can. Mod. Lang. Rev. **63**(1), 59–82 (2006)
18. Nation, P., Beglar, D.: A vocabulary size test. Lang. Teach. **31**(7), 9–13 (2007)
19. Paetzold, G., Specia, L.: Collecting and exploring everyday language for predicting psycholinguistic properties of words. In: Proceedings of COLING (2016)
20. Paszke, A., et al.: Automatic differentiation in PyTorch (2017)
21. Pedregosa, F., et al.: Scikit-learn: machine learning in Python. JMLR **12**, 2825–2830 (2011)
22. Pennington, J., Socher, R., Manning, C.: GloVe: global vectors for word representation. In: Proceedings of EMNLP, pp. 1532–1543 (2014)
23. Rasch, G.: Probabilistic Models for Some Intelligence and Attainment Tests. Danish Institute for Educational Research, Copenhagen (1960)
24. Řehůřek, R., Sojka, P.: Software framework for topic modelling with large corpora. In: Proceedings of LREC, pp. 45–50 (2010)
25. Schmitt, N., Jiang, X., Grabe, W.: The percentage of words known in a text and reading comprehension. Mod. Lang. J. **95**(1), 26–43 (2011)

26. Srivastava, N., Hinton, G., Krizhevsky, A., Sutskever, I., Salakhutdinov, R.: Dropout: a simple way to prevent neural networks from overfitting. JMLR **15**(1), 1929–1958 (2014)
27. Yeung, C.Y., Lee, J.: Personalized text retrieval for learners of Chinese as a foreign language. In: Proceedings of COLING, pp. 3448–3455 (2018)
28. Yimam, S.M., et al.: A report on the complex word identification shared task 2018. In: Proceedings of BEA (2018)

Machine Translation

Dynamic Fusion: Attentional Language Model for Neural Machine Translation

Michiki Kurosawa[✉] and Mamoru Komachi

Tokyo Metropolitan University, 6-6 Asahigaoka, Hino, Tokyo 191-0065, Japan
kurosawa-michiki@ed.tmu.ac.jp, komachi@tmu.ac.jp

Abstract. Neural Machine Translation (NMT) can be used to generate fluent output. As such, language models have been investigated for incorporation with NMT. In prior investigations, two models have been used: a translation model and a language model. The translation model's predictions are weighted by the language model with a hand-crafted ratio in advance. However, these approaches fail to adopt the language model weighting with regard to the translation history. In another line of approach, language model prediction is incorporated into the translation model by jointly considering source and target information. However, this line of approach is limited because it largely ignores the adequacy of the translation output.

Accordingly, this work employs two mechanisms, the translation model and the language model, with an attentive architecture to the language model as an auxiliary element of the translation model. Compared with previous work in English–Japanese machine translation using a language model, the experimental results obtained with the proposed Dynamic Fusion mechanism improve BLEU and Rank-based Intuitive Bilingual Evaluation Scores (RIBES) scores. Additionally, in the analyses of the attention and predictivity of the language model, the Dynamic Fusion mechanism allows predictive language modeling that conforms to the appropriate grammatical structure.

Keywords: Language model · Neural machine translation · Attention mechanism

1 Introduction

With the introduction of deep neural networks to applications in machine translation, more fluent outputs have been achieved with neural machine translation (NMT) than with statistical machine translation [17]. However, a fluent NMT output requires a large parallel corpus, which is difficult to prepare. Therefore, several studies have attempted to improve fluency in NMT without the use of a large parallel corpus.

To overcome the data-acquisition bottleneck, the use of a monolingual corpus has been explored. A monolingual corpus can be collected relatively easily, and has been known to contribute to improved statistical machine translation [2]. Various attempts to employ a monolingual corpus have involved the

© Springer Nature Singapore Pte Ltd. 2020
L.-M. Nguyen et al. (Eds.): PACLING 2019, CCIS 1215, pp. 99–111, 2020.
https://doi.org/10.1007/978-981-15-6168-9_9

following: pre-training of a translation model [12], initialization of distributed word representation [4, 11], and construction of a pseudo-parallel corpus by back-translation [14].

Here, we focus on a language modeling approach [3, 16]. Although recent efforts in NMT tend to output fluent sentences, it is difficult to reflect the linguistic properties of the target language, as only the source information is taken into consideration when performing translation [13]. Additionally, language models are useful in that they contain target information that results in fluent output and can make predictions even if they do not know the source sentence. In previous works utilizing a language model for NMT, both the language model and the conventional translation model have been prepared, wherein the final translation is performed by weighting both models. In the Shallow Fusion mechanism [3], the output of the translation and language models are weighted at a fixed ratio. In the Cold Fusion mechanism [15], a gate function is created to dynamically determine the weight of the language model considering the translation model. In the Simple Fusion mechanism [16], outputs of both models are treated equally, whereas normalization steps vary.

In this research, we propose a "Dynamic Fusion" mechanism that predicts output words by attending to the language model. We hypothesize that each model should make predictions according to only the information available to the model itself; the information available to the translation model should not be referenced before prediction. In the proposed mechanism, a translation model is fused with a language model through the incorporation of word-prediction probability according to the attention. However, the models retain predictions independent of one another. Based on the weight of the attention, we analyze the predictivity of the language model and its influence on translation.

The main contributions of this paper are as follows:

- We propose an attentional language model that effectively introduces a language model to NMT.
- We show that fluent and adequate output can be achieved with a language model in English–Japanese translation.
- We show that Dynamic Fusion significantly improves translation accuracy in a realistic setting.
- Dynamic Fusion's ability to improve translation is analyzed with respect to the weight of the attention.

2   Previous Works

2.1   Shallow Fusion

Gulcehre et al. [3] proposed Shallow Fusion, which translates a source sentence according to the predictions of both a translation model and a language model. In this mechanism, a monolingual corpus is used to learn the language model in advance. The translation model is improved through the introduction of the knowledge of the target language.

In Shallow Fusion, a target word $\hat{y}$ is predicted as follows:

$$\hat{y} = \underset{y}{\operatorname{argmax}} \log P_{\mathrm{TM}}(\mathbf{y}|\mathbf{x}) + \lambda \log P_{\mathrm{LM}}(\mathbf{y}) \tag{1}$$

where $\mathbf{x}$ is an input of the source language, $P_{\mathrm{TM}}(\mathbf{y}|\mathbf{x})$ is the word-prediction probability according to the translation model, and $P_{\mathrm{LM}}(\mathbf{y})$ is the word prediction probability according to the language model. Here, λ is a manually-determined hyper-parameter that determines the rate at which the language model is considered.

2.2 Cold Fusion

In addition to Shallow Fusion, Gulcehre et al. [3] proposed Deep Fusion as a mechanism that could simultaneously learn a translation model and a language model. Sriram et al. [15] extended Deep Fusion to Cold Fusion to pass information on a translation model for the prediction of a language model.

In this mechanism, a gating function is introduced that dynamically determines the weight, taking into consideration both a translation model and a language model. Therein, the language model predicts target words by using information from the translation model. Accuracy and fluency are improved through the joint learning of both models.

In Cold Fusion, a target word $\hat{y}$ is predicted as follows:

$$h_{\mathrm{LM}} = W_{\mathrm{LM}} S_{\mathrm{LM}}(\mathbf{y}) \tag{2}$$
$$g = W_{\mathrm{gate}}[S_{\mathrm{TM}}(\mathbf{y}|\mathbf{x}); h_{\mathrm{LM}}] \tag{3}$$
$$h' = [S_{\mathrm{TM}}(\mathbf{y}|\mathbf{x}); g \cdot h_{\mathrm{LM}}] \tag{4}$$
$$S_{\mathrm{cold}} = W_{\mathrm{output}} h' \tag{5}$$
$$\hat{y} = \underset{y}{\operatorname{argmax}} \ \operatorname{softmax}(S_{\mathrm{cold}}) \tag{6}$$

where both $S_{\mathrm{TM}}(\mathbf{y}|\mathbf{x})$ and $S_{\mathrm{LM}}(\mathbf{y})$ are word-prediction logits[1] with the translation model and the language model, respectively; g is a function that determines the rate at which the language model is considered; W_{LM} ($|h| \times |V|$), W_{gate} ($2|h| \times |h|$), and W_{output} ($2|h| \times |V|$) are the weights of the neural networks; and $[a; b]$ is the concatenation of vectors a and b.

2.3 Simple Fusion

Stahlberg et al. [16] proposed Simple Fusion, which simplifies Cold Fusion. Unlike Cold Fusion, Simple Fusion does not use a translation model to predict words output by a language model.

---

[1] A logit is a probability projection layer without softmax.

For Simple Fusion, two similar methods were proposed: PostNorm (7) and PreNorm (8). In PostNorm and PreNorm, a target word $\hat{y}$ is predicted as follows:

$$\hat{y} = \underset{y}{\text{argmax}} \ \text{softmax}(\text{softmax}(S_{\text{TM}}(\mathbf{y}|\mathbf{x})) \cdot P_{\text{LM}}(\mathbf{y})) \tag{7}$$

$$\hat{y} = \underset{y}{\text{argmax}} \ \text{softmax}(S_{\text{TM}}(\mathbf{y}|\mathbf{x}) + \log P_{\text{LM}}(\mathbf{y})) \tag{8}$$

where $S_{\text{TM}}(\mathbf{y}|\mathbf{x})$ denotes the word prediction logits with the translation model and $P_{\text{LM}}(\mathbf{y})$ denotes the word prediction probability according to the language model.

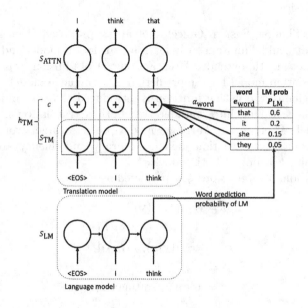

Fig. 1. Dynamic Fusion mechanism.

In PostNorm, the output probability of the language model is multiplied by the output probability of the translation model, wherein both models are treated according to the same scale.

In PreNorm, the log probability of the language model and the unnormalized prediction of the translation model are summed, wherein the language and translation models are treated with different scales.

Though the Simple Fusion model is relatively simple, it achieves a higher BLEU score compared to other methods that utilize language models.

3   Dynamic Fusion

An attentional language model called "Dynamic Fusion," is proposed in this paper. In the Shallow Fusion and Simple Fusion mechanisms, information from the language model is considered with fixed weights. However, translation requires that source information be retained, such that the consideration ratios should be adjusted from token to token. Thus, both models should not be mixed with fixed weights. The Cold Fusion mechanism dynamically determines the weights of mix-in; however, the Cold Fusion mechanism passes information from the translation model to the language model before prediction, and the language model thus does not make its own prediction.

Furthermore, in the previous research, it was necessary to make the vocabularies of the translation model and language model identical because the final softmax operation is performed in the word vocabulary dimension. However, since the proposed mechanism mixes a language model as an attention, the vocabularies of the translation model and language model do not have to be completely consistent, and different word-segmentation strategies and subword units can be used. Therefore, the proposed mechanism allows the use of a language model prepared in advance.

In the proposed mechanism, the language model serves as auxiliary information for prediction. Thus, the language model is utilized independently of the translation model. Unlike Cold Fusion, this method uses a language model's prediction score multiplied by word attention.

First, the word-prediction probability of the language model $P_{\mathrm{LM}}(y)$ is represented as follows:

$$P_{\mathrm{LM}}(\mathbf{y}; y = \mathrm{word}) = \mathrm{softmax}(S_{\mathrm{LM}}(\mathbf{y})) \tag{9}$$

Next, hidden layers of the translation model attending to the language model h_{TM} are represented as follows:

$$\alpha_{\mathrm{word}} = \frac{\exp(e_{\mathrm{word}}^{\mathrm{T}} S_{\mathrm{TM}}(\mathbf{y}|\mathbf{x}))}{\sum_{\mathrm{word} \in V} \exp(e_{\mathrm{word}}^{\mathrm{T}} S_{\mathrm{TM}}(\mathbf{y}|\mathbf{x})))} \tag{10}$$

$$c_{\mathrm{word}} = \alpha_{\mathrm{word}} e_{\mathrm{word}} \tag{11}$$

$$c_{\mathrm{LM}} = \sum_{\mathrm{word}} c_{\mathrm{word}} \cdot P_{\mathrm{LM}}(\mathbf{y}; y = \mathrm{word}) \tag{12}$$

$$h_{\mathrm{TM}} = [S_{\mathrm{TM}}(\mathbf{y}|\mathbf{x}); c_{\mathrm{LM}}] \tag{13}$$

$$S_{\mathrm{ATTN}} = W h_{\mathrm{TM}} \tag{14}$$

where e_{word} is the embedding of a word, c_{word} is the conventional word attention for each word, c_{LM} is the word attention's hidden state for the proposed Dynamic Fusion, and W $(2|h| \times V)$ is a weight matrix of neural networks. In Eq. (12), c_{LM} considers the language model by multiplying $P_{\mathrm{LM}}(\mathbf{y}; y = \mathrm{word})$ with a word attention. In this mechanism, the prediction of the language model only has access to the target information up to the word currently being predicted. Additionally, the language model and translation model can be made independent by using the conventional attention mechanism.

Finally, a target word $\hat{y}$ is predicted as follows:

$$\hat{y} = \underset{y}{\mathrm{argmax}} \; \mathrm{softmax}(S_{\mathrm{ATTN}}) \tag{15}$$

A diagram of this mechanism is shown in Fig. 1, wherein the language model is used for the translation mechanism by considering the attention obtained from both the translation model and language model.

The training procedure of the proposed mechanism follows that of Simple Fusion and is performed as follows:

1. A language model is trained with a monolingual corpus.
2. The translation model and word attention to the language model are learned by fixing the parameters of the language model.

Table 1. Corpus details.

	# sentences	# maximumtoken
Language model (monolingual)	1,909,981	60
Train (parallel)	827,188	60
Dev (parallel)	1,790	,
Test (parallel)	1,812	

Table 2. Experimental setting.

	Setting
Pre training epoch	15 epoch
Maximum training epoch	100 epoch
Optimization	AdaGrad
Training rate	0.01
Embed size	512
Hidden size	512
Batch size	128
Vocabulary size (w/o BPE)	30,000
# BPE operation	16,000

4 Experiment

Here, the conventional attentional NMT [1,6] and Simple Fusion models (PostNorm, PreNorm) were prepared as baseline methods for comparison with the proposed Dynamic Fusion model. We performed English-to-Japanese and Japanese-to-English translation. Using this, the translation performance of the proposed model was evaluated by taking the average of two runs with BLEU [10] and Rank-based Intuitive Bilingual Evaluation Score (RIBES) [5]. In addition, a significant difference test was performed using Travatar[2] with 10,000 bootstrap resampling.

The experiment uses two types of corpora: one for a translation model and the other for a language model. Thus, training data of the Asian Scientific Paper Excerpt Corpus (ASPEC) [9] are divided into two parts: a parallel corpus and a monolingual corpus. The parallel corpus, for translation, is composed of one million sentences with a high confidence of sentence alignment from the training data. The monolingual corpus, for language models, is composed of two million sentences from the target side of the training data that are not used in the parallel corpus. Japanese sentences were tokenized by the morphological analyzer MeCab[3] (IPADic), and English sentences were preprocessed by Moses[4]

---

[2] http://www.phontron.com/travatar/evaluation.html.
[3] https://github.com/taku910/mecab.
[4] http://www.statmt.org/moses/.

(tokenizer, truecaser). We used development and evaluation set on the official partitioning of ASPEC as summarized in Table 1[5]. Vocabulary is determined using only the parallel corpus. For example, words existing only in the monolingual corpus are treated as unknown words at testing, even if they frequently appear in the monolingual corpus to train the language model. Additionally, experiments have been conducted with and without Byte Pair Encoding (BPE) [7]. BPE was performed on the source side and target side separately.

The in-house implementation [8] of the NMT model proposed by Bahdanau et al. [1] and Luong et al. [6] is used as the baseline model; all the other methods were created based on this baseline. For comparison, settings are unified in all experiments (Table 2). In the pre-training process, only the language model is learned; the baseline performs no pre-training, as it does not have access to the language model.

Table 3. Results of English-Japanese translation. (Average of 2 runs.)

| Vocabulary | TM | w/o BPE | | w/ BPE | | w/ BPE | |
| | LM | w/o BPE | | w/ BPE | | w/o BPE | |
		BLEU	RIBES	BLEU	RIBES	BLEU	RIBES
Baseline		31.28	80.78	32.35	81.17	32.35	81.17
PostNorm		31.01	80.77	32.43	80.97	N/A	N/A
PreNorm		31.61	80.78	32.69	81.24	N/A	N/A
Dynamic Fusion		**31.84***	**81.13***	**33.22***	**81.54***	**33.05***	**81.40***

Table 4. Results of Japanese–English translation. (Average of 2 runs.)

| Vocabulary | TM | w/o BPE | | w/ BPE | | w/ BPE | |
| | LM | w/o BPE | | w/ BPE | | w/o BPE | |
		BLEU	RIBES	BLEU	RIBES	BLEU	RIBES
Baseline		22.64	73.57	22.80	73.54	22.80	73.54
PostNorm		21.49	73.13	22.10	72.88	N/A	N/A
PreNorm		22.38	73.65	22.71	73.36	N/A	N/A
Dynamic Fusion		**22.78**	**73.74**	**23.45**	**74.01**	**23.08**	**73.73**

5 Discussion

5.1 Quantitative Analysis

The BLEU and RIBES scores results are listed in Table 3 (English–Japanese) and Table 4 (Japanese–English). In both scores, we observed similar tendencies with and without BPE. Compared with the baseline model, Dynamic

---

[5] We exclude sentences whose number of tokens with more than 60 tokens in training.

Fusion yielded improved results in terms of BLEU and RIBES scores. However, between the baseline model and Simple Fusion, Simple Fusion was equal or worse except PRENORM on English–Japanese translation. Compared with PRENORM, Dynamic Fusion has improved BLEU and RIBES scores. Accordingly, the improvement of the proposed method is notable, and the use of attention yields better scores.

In the English–Japanese translation, it was also confirmed that BLEU and RIBES were improved by using a language model. RIBES was improved for the translation with Dynamic Fusion, suggesting that the proposed approach outputs adequate sentences.

The proposed method has statistically significant differences ($p < 0.05$) in BLEU and RIBES scores compared to the baseline except English–Japanese translation without BPE. There was no significant difference between Simple Fusion and the proposed method on Japanese–English translation.

In addition, we conducted additional experiments in a more realistic setting. We experimented with the translation model in which BPE was performed, whereas the language model was trained on a raw corpus without BPE[6]. It was found that the translation scores were improved as compared to the baseline model with BPE.

5.2   Qualitative Analysis

Examples of the output of each model are given in Tables 5 and 6.

In Table 5, compared with the baseline, the fluency of PRENORM and Dynamic Fusion resulted in improved translation. Additionally, it can be seen that the attentional language model provides a more natural translation of the inanimate subject in the source sentence. Unlike in English, inanimate subjects are not often used in Japanese. Thus, literal translations of an inanimate subject sounds unnatural to native Japanese speakers. However, POSTNORM translates "線量 (dose)" into "用量 (capacity)", which reduces adequacy.

PRENORM in Table 6 appears as a plain and fluent output. However, neither of the Simple Fusion models can correctly translate the source sentence in comparison with the baseline. In contrast, with Dynamic Fusion, the content of the source sentence is translated more accurately than in the reference translation; thus, without loss of adequacy, Dynamic Fusion maintains the same level of fluency.

This shows that the use of a language model contributes to the improvement of output fluency. Additionally, Dynamic Fusion maintains relatively superior adequacy.

---

[6] We did not perform an experiment with Simple Fusion because Simple Fusion requires the vocabularies of both the language model and translation model to be identical.

In Japanese–English translation, not only our proposed method but also other language models can cope with voice changes and inversion such as in Table 7. The use of active voice in Japanese where its counterpart is using passive voice is a common way of writing in Japanese papers [18], and this example shows an improvement using a language model.

5.3　Influence of Language Model

Table 8 shows an example wherein the language model compensates for the adequacy. In general, if there is a spelling error exists in the source sentence, a proper translation may not be performed owing to the unknown word. In this example, the word "temperature" is misspelled as "temperture." Thus, the baseline model translates the relevant part but ignores the misspelled word. However, PRENORM and Dynamic Fusion complemented the corresponding part appropriately thanks to the language model. The proposed method was able to translate without losing adequacy. This result is attributed to the language model's ability to predict a fluent sentence.

Table 5. Example of fluency improvement by language model.

Model	Sentence (Output)
Source	responding to these changes DERS can compute new dose rate .
Reference	DERS は これら の 変化 に 対応 して 新た な 線量 率 を 計算 できる 。
Baseline	これら の 変化 に 対応 する 応答 は，新しい 線量 率 を 計算 できる 。
	(Responses corresponding to these changes can calculate new dose rates.)
Simple Fusion (POSTNORM)	これら の 変化 に 対応 する 応答 は 新しい 用量 率 を 計算 できる 。
	(Responses corresponding to these changes can calculate new capacity rates.)
Simple Fusion (PRENORM)	これら の 変化 に 対応 すると，新しい 線量 率 を 計算 できる 。
	(In response to these changes, new dose rates can be calculated.)
Dynamic Fusion	これら の 変化 に 対応 する こと により，新しい 線量 率 を 計算 できる 。
	(By responding to these changes, new dose rates can be calculated.)

Table 6. Example of adequacy decline in Simple Fusion.

Table 6. Example of adequacy decline in Simple Fusion.

Model	Sentence (Output)
Source	the magnetic field is given in the direction of a right angle or a parallel (reverse to the flow) to the tube axis .
Reference	磁場 は 管 軸 に 直角 か 平行 逆 方向 に 加え た 。
Baseline	磁場 は 右 角 または 平行 (流れ) の 方向 に 与え られ，管 軸 に 平行 である 。
	(The magnetic field is given in the right angle or parallel (flow) direction and parallel to the tube axis.)
Simple Fusion (POSTNORM)	磁場 は 右 角度 または 平行 (流れ に 逆 に 逆) 方向 に 与え られ た 。
	(The magnetic field was applied at right angle or parallel (opposite to opposite to the flow) direction.)
Simple Fusion (PRENORM)	磁場 は 右 角 または 平行 (流れ に 逆 方向) の 方向 に 与え られ た 。
	(The magnetic field was applied in the right angle or parallel (opposite to the flow) direction.)
Dynamic Fusion	磁場 は，管 軸 に 直角 または 平行 (流れ に 逆 方向) の 方向 に 与え られる 。
	(The magnetic field is given in a direction perpendicular or parallel (reverse to the flow) to the tube axis.)

Table 7. Examples robust to changes in state.

Model	Sentence (Output)
Source	変形 が 対 密度 分布 に 影響 し て いる こと が 分かった 。
Reference	it was found that the deformation gave effects to the pairing density distribution .
Baseline	it was found that deformation was affected by the pair density distribution .
Simple Fusion (POSTNORM)	it was found that deformation affects the logarithmic density distribution .
Simple Fusion (PRENORM)	it was found that deformation affected the pair density distribution .
Dynamic Fusion	it was found that the deformation affected the pair density distribution .

5.4　Influence of Dynamic Fusion

Fluency. Excerpts from the output of Dynamic Fusion and word attention (top 5 words) are presented in Table 9.

Except for the first token[7], the word attention includes the most likely outputs. For example, if "**start bracket** (「) " is present in the sentence, there is a tendency to try to close it with "**end bracket** (」)". Additionally, it is not desirable to close brackets with "発電 (**power generation**)"; therefore, it predicts that the subsequent word is "所 (plant)". This indicates that the attentional language model can improve fluency while maintaining the source information.

Regarding attention weights, there are cases in which only certain words have highly skewed attention weights, among other cases in which multiple words have uniform attention weights. The latter occurs when there are many translation options, such as the generation of function words on the target side. This topic requires further investigation.

Adequacy. In contrast, it is extremely rare for Dynamic Fusion itself to return an adequate translation at the expense of fluency. Even if a particular word has a significantly higher weight than other words, the prediction of the translation model may likely be used for the output if it changes the meaning of the source sentence. In fact, the example in Table 9 contains many tokens in which the output of the language model is not considered, including at the beginning of the sentence.

One of the reasons for this is considered to be the difference in contributions between the translation model and the language model. We decomposed the transformation weight matrix in Eq. (12) into the translation model and the language model matrices, and we calculated the Frobenius norm for each matrix. The result reveals that the translation model contributes about twice as much as the language model.

---

[7] The language model cannot predict that the first token correctly because it starts with <BOS>.

Table 8. Comparison of adequacy by language model.

Model	Sentence (Output)
Source	this paper explains the application of chemical processes utilizing supercritical phase where a liquid does not make phase change irrespective of temperture or pressure .
Reference	流体 が 温度・圧力 に かかわら ず 相変化 しない 状態 である 超臨界相 を 利用 した 化学 プロセス の 応用 について 解説 した 。
Baseline	液体 が 相変化 を 持た ない 超臨界相 を 利用 した 化学 プロセス の 応用 について 解説 した 。 (The application of chemical processes using supercritical phase in which the liquid has no phase change is described.)
Simple Fusion (POSTNORM)	液体 が 相変化 を 起こす こと なく，圧力 や 圧力 に 関係 なく 相変化 を 生じる 化学 プロセス の 適用 について 解説 した 。 (The application of the chemical process which causes the phase change regardless of the pressure and the pressure without the liquid causing the phase change is described.)
Simple Fusion (PRENORM)	液体 が 相変化 を 起こさ ない 超臨界相 を 利用 した 化学 プロセス の 応用 について，温度 や 圧力 に 関係 なく 解説 した 。 (The application of chemical processes using supercritical phase in which liquid does not cause phase change is described regardless of temperature and pressure.)
Dynamic Fusion	液体 が 温度 や 圧力 に 関係 なく 相変化 を 起こさ ない 超臨界相 を 利用 した 化学 プロセス の 応用 について 解説 した 。 (We have described the application of chemical processes that use a supercritical phase in which the liquid does not undergo a phase change regardless of temperature and pressure.)

Table 9. Dynamic Fusion output and attention example (excerpt).

モデル	出力
Source	details of dose rate of ” Fugen Power Plant ” can be calculated by using <unk> software .
Reference	<unk> ソフトウエア を 用いて 「 ふげん 発電所 」 の 線量 率 を 詳細 に 計算 できる 。
Dynamic Fusion	「 ふげん 発電所 」 の 線量 率 の 詳細 を，<unk> ソフトウェア を 用いて 計算 できる 。 (The details of the dose rate of ”Fugen power plant”, can be calculated by using the <unk> software.)

Dynamic Fusion (excerpt)	「		ふ (Fu)		げん (gen)		発電 (Power)		所 (Plant)		」		の (of)	
Word attention (Top5 word) and weights	本	9.9e-1	この	5.5e-1	」	9.9e-1	」	1.0	所	9.9e-1	」	1.0	について	7.7e-1
	標記	8.7e-5	その	3.5e-1	ね	3.2e-6	号	2.7e-8	機	1.3e-4	発電	3.2e-12	の	1.7e-1
	この	4.2e-5	日本	7.0e-2	げん	2.0e-9	げん	1.4e-11	」	1.2e-6	の	1.7e-18	における	4.5e-2
	また	8.5e-6	1	2.7e-2	出	1.1e-10	<unk>	1.1e-12	設備	7.7e-11	<unk>	7.6e-19	で	6.4e-3
	これら	1.5e-6	高	4.7e-3	り	3.6e-11	・	1.8e-14	装置	2.6e-12	用	6.3e-19	と	3.2e-3

Role of Language Model. Currently, most existing language models do not utilize the source information. Accordingly, to eliminate noise in the language model's fluent prediction, language models should make predictions independently of translation models and thus be used in tandem with attention from translation models. However, language models are useful in that they have target information that results in fluent output; they can thus make a prediction even if they do not know the source sentence.

Ultimately, the role of the language model in the proposed mechanism is to augment the target information in order for the translation model to improve the fluency of the output sentence. Consequently, the fusion mechanism takes translation options from the language model only when it improves fluency and does not harm adequacy. It can be regarded as a regularization method to help disambiguate stylistic subtleness such as in the successful example in Table 5.

6 Conclusion

We proposed Dynamic Fusion for machine translation. For NMT, experimental results demonstrated the necessity of using an attention mechanism in conjunction with a language model. Rather than combining the language model and translation model with a fixed weight, an attention mechanism was utilized with the language model to improve fluency without reducing adequacy. This further improved the BLEU scores and RIBES.

The proposed mechanism fuses the existing language and translation models by utilizing an attention mechanism at a static ratio. In the future, we would like to consider a mechanism that can dynamically weight the mix-in ratio, as in Cold Fusion.

References

1. Bahdanau, D., Cho, K., Bengio, Y.: Neural machine translation by jointly learning to align and translate. In: Proceedings of ICLR (2015)
2. Brants, T., Popat, A.C., Xu, P., Och, F.J., Dean, J.: Large language models in machine translation. In: Proceedings of EMNLP-CoNLL, pp. 858–867 (2007)
3. Gulcehre, C., et al.: On using monolingual corpora in neural machine translation. arXiv (2015)
4. Hirasawa, T., Yamagishi, H., Matsumura, Y., Komachi, M.: Multimodal machine translation with embedding prediction. In: Proceedings of NAACL, pp. 86–91, June 2019
5. Isozaki, H., Hirao, T., Duh, K., Sudoh, K., Tsukada, H.: Automatic evaluation of translation quality for distant language pairs. In: Proceedings of EMNLP, pp. 944–952 (2010)
6. Luong, T., Pham, H., Manning, C.D.: Effective approaches to attention-based neural machine translation. In: Proceedings of EMNLP, pp. 1412–1421 (2015)
7. Luong, T., Sutskever, I., Le, Q., Vinyals, O., Zaremba, W.: Addressing the rare word problem in neural machine translation. In: Proceedings of ACL, pp. 11–19 (2015)
8. Matsumura, Y., Komachi, M.: Tokyo Metropolitan University neural machine translation system for WAT 2017. In: Proceedings of WAT, pp. 160–166 (2017)
9. Nakazawa, T., et al.: ASPEC: Asian scientific paper excerpt corpus. In: Proceedings of LREC, pp. 2204–2208 (2016)
10. Papineni, K., Roukos, S., Ward, T., Zhu, W.J.: BLEU: a method for automatic evaluation of machine translation. In: Proceedings of ACL, pp. 311–318 (2002)
11. Qi, Y., Sachan, D., Felix, M., Padmanabhan, S., Neubig, G.: When and why are pre-trained word embeddings useful for neural machine translation? In: Proceedings of NAACL, pp. 529–535 (2018). https://doi.org/10.18653/v1/N18-2084
12. Ramachandran, P., Liu, P., Le, Q.: Unsupervised pretraining for sequence to sequence learning. In: Proceedings of EMNLP, pp. 383–391 (2017). https://doi.org/10.18653/v1/D17-1039
13. Sennrich, R., Haddow, B.: Linguistic input features improve neural machine translation. In: Proceedings of WMT, pp. 83–91 (2016). https://doi.org/10.18653/v1/W16-2209
14. Sennrich, R., Haddow, B., Birch, A.: Improving neural machine translation models with monolingual data. In: Proceedings of ACL, pp. 86–96 (2016). https://doi.org/10.18653/v1/P16-1009
15. Sriram, A., Jun, H., Satheesh, S., Coates, A.: Cold fusion: training Seq2Seq models together with language models. arXiv (2017)
16. Stahlberg, F., Cross, J., Stoyanov, V.: Simple fusion: return of the language model. In: Proceedings of WMT, pp. 204–211 (2018)

17. Tu, Z., Lu, Z., Liu, Y., Liu, X., Li, H.: Modeling coverage for neural machine translation. In: Proceedings of ACL, pp. 76–85 (2016). https://doi.org/10.18653/v1/P16-1008
18. Yamagishi, H., Kanouchi, S., Sato, T., Komachi, M.: Improving Japanese-to-English neural machine translation by voice prediction. In: Proceedings of IJCNLP, pp. 277–282, November 2017

Improving Context-Aware Neural Machine Translation with Target-Side Context

Hayahide Yamagishi[✉] and Mamoru Komachi

Tokyo Metropolitan University, 6-6 Asahigaoka, Hino, Tokyo 191-0065, Japan
yamagishi-hayahide@ed.tmu.ac.jp, komachi@tmu.ac.jp

Abstract. In recent years, several studies on neural machine translation (NMT) have attempted to use document-level context by using a multi-encoder and two attention mechanisms to read the current and previous sentences to incorporate the context of the previous sentences. These studies concluded that the target-side context is less useful than the source-side context. However, we considered that the reason why the target-side context is less useful lies in the architecture used to model these contexts.

Therefore, in this study, we investigate how the target-side context can improve context-aware neural machine translation. We propose a weight sharing method wherein NMT saves decoder states and calculates an attention vector using the saved states when translating a current sentence. Our experiments show that the target-side context is also useful if we plug it into NMT as the decoder state when translating a previous sentence.

Keywords: Neural machine translation · Document · Context · Weight sharing

1 Introduction

Neural machine translation (NMT; Sutskever et al. [1], Bahdanau et al. [2], Vaswani et al. [3]) has become popular in recent years because it can handle larger contexts compared to conventional machine translation systems. However, most of the NMTs do not employ document-level contexts due to lack of an efficient mechanism, similar to other machine translation systems.

Recently, a few studies have attempted to expand the notion of a sentence-level context in NMT to that of a document-level context[1]. It is reported that the information of one or more previous sentences improves the scores of automatic and human evaluations.

Context-aware NMT systems typically have two encoders: one is for a current sentence and the other is for a previous sentence. For instance, Bawden et al. [4] showed that encoding a previous target sentence does not improve the performance in an English–French task even though encoding a previous source sentence works well. Other studies that utilized a multi-encoder (Jean et al. [5], Voita et al. [6], Zhang et al. [7]) did not use a previous target sentence. Thus, there are a few works on handling the

---

[1] Hereinafter, "document-level context" is simply referred to as a "context".

© Springer Nature Singapore Pte Ltd. 2020
L.-M. Nguyen et al. (Eds.): PACLING 2019, CCIS 1215, pp. 112–122, 2020.
https://doi.org/10.1007/978-981-15-6168-9_10

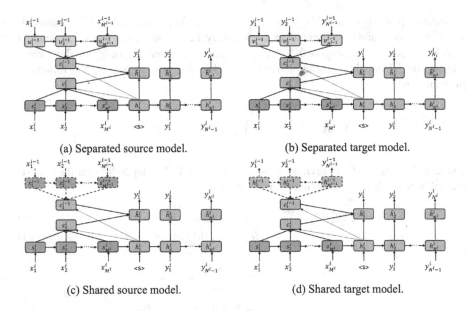

(a) Separated source model.   (b) Separated target model.

(c) Shared source model.   (d) Shared target model.

Fig. 1. Proposed methods: dashed line represents the weight sharing with the encoders or decoders.

target-side context. Moreover, these previous works mainly used language pairs that belonged to the same language family. In distant language pairs, the information of discourse structures in the target-side document might be useful because distant languages might have different discourse structures.

Therefore, this study investigates how the target-side context can be used in context-aware NMT. We hypothesize that the source-side contexts should be incorporated into an encoder and the target-side contexts should be incorporated into a decoder. To validate this hypothesis, we propose a weight sharing method, in which NMT saves the decoder states and calculates an attention vector using the saved states when translating a current sentence. We find that target-side contexts are also useful if they are inserted into the NMT as the decoder states. This method can obtain competitive or even better results compared to a baseline model using source-side features.

The main findings of this study are as follows:

- The target-side context is as important as the source-side context.
- The effectiveness of source-side context depends on language pairs.
- Weight sharing between current and context states is effective for context-aware NMT.

2   Model Architecture

Figure 1 presents our methods. We build context-aware NMT based on the multi-encoder model proposed by Bawden et al. [4]. A parallel document D consisting of

L sentence pairs, is denoted by $D = (X^1, Y^1), ..., (X^i, Y^i), ..., (X^L, Y^L)$, where X and Y are source and target sentences, respectively. Each sentence, X^i or Y^i, is denoted as $X^i = x_1^i, ..., x_m^i, ..., x_{M^i}^i$ or $Y^i = y_1^i, ..., y_n^i, ..., y_{N^i}^i$, where x_m^i or y_n^i are the tokens, and M^i or N^i are the sentence lengths. The objective is to maximize the following probabilities:

$$p(Y^i|X^i, Z^{i-1}) = \prod_{n=1}^{N^i} p(y_n^i|y_{<n}^i, X^i, Z^{i-1}) \tag{1}$$

where Z^{i-1} represents a previous sentence, X^{i-1} or Y^{i-1}, depending on the experimental settings. Each p is calculated as follows:

$$p(y_n^i|y_{<n}^i, X^i, Z^{i-1}) = \text{softmax}(W_o \tilde{h}_n^i) \tag{2}$$

$$\tilde{h}_n^i = W_h[h_n^i; c_n^i; c_n^{i-1}] \tag{3}$$

$$c_n^i = \sum_{m=1}^{M^i} \alpha_{n,m}^i s_m^i \tag{4}$$

$$\alpha_{n,m}^i = \text{softmax}(s_m^i \cdot h_n^i) \tag{5}$$

where s_m^i, h_n^i, and c_n^i represents encoder states, decoder states, and attention, respectively. $W_o \in \mathbb{R}^{V \times H}$ and $W_h \in \mathbb{R}^{H \times 3H}$ represents weights. We calculate the encoder state s_m^i and the decoder state h_n^i as follows:

$$s_m^i = \text{LSTM}_{enc}(W_x x_m^i, s_{m-1}^i) \tag{6}$$

$$h_n^i = \text{LSTM}_{dec}(W_y y_n^i, h_{n-1}^i) \tag{7}$$

where $W_x \in \mathbb{R}^{E \times V}$ and $W_y \in \mathbb{R}^{E \times V}$ represents word embeddings of source- and target sides, respectively. We use the dot product of encoder states and hidden states as an attention score $\alpha_{n,m}^i$, proposed by Luong et al. [8].

The multi-encoder model has an additional attention, c_n^{i-1}, which is for using the information of a previous sentence.

$$c_n^{i-1} = \sum_{t=1}^{|Z^{i-1}|} \beta_{n,t}^{i-1} z_t^{i-1} \tag{8}$$

$$\beta_{n,t}^{i-1} = \text{softmax}(z_t^{i-1} \cdot h_n^i) \tag{9}$$

We experiment using two methods, *separated model* and *shared model*. The separated model represents the conventional multi-encoder model, and the shared model is our proposed method. The difference between the two methods is the calculation of z_t^{i-1}.

2.1 Separated Model

Context-aware NMT saves and encodes a source or target sentence in a context encoder when translating a current sentence. Previous works on multi-encoder models have an

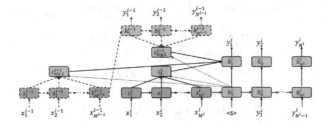

Fig. 2. Shared mix model.

additional encoder, referred to as a context encoder. Each context encoder u^{i-1} or v^{i-1} reads a previous source-side or target-side sentence as context, respectively.

$$u_t^{i-1} = \mathrm{LSTM_{src\_enc}}(W_x x_t^{i-1}, u_{t-1}^{i-1}) \tag{10}$$

$$v_t^{i-1} = \mathrm{LSTM_{trg\_enc}}(W_y y_t^{i-1}, v_{t-1}^{i-1}) \tag{11}$$

We refer to this architecture as a *separated model* in this paper. In the separated model, the weights of a context encoder are different from those of a current encoder which encodes a current source sentence. If u_t^{i-1} is used as z_t^{i-1}, we call this model *separated source* model; otherwise, we call this model *separated target* model.

2.2   Shared Model

A *shared model* saves the hidden states of an encoder or decoder and then calculates c_n^{i-1} using these states when translating a current sentence. The strength of this model is that the target-side context can be incorporated into a decoder instead of an encoder. Moreover, the shared model does not require much additional parameters and extra computational times because this model simply loads the saved hidden states. Thus, we can see these models as examples of weight sharing between a current encoder or decoder and a context encoder. The *shared source* model uses s_t^{i-1} as z_t^{i-1}, and the *shared target* model uses h_t^{i-1} as z_t^{i-1}.

2.3   Shared Mix Model

We propose a *shared mix* model, which incorporates the source- and target-side contexts. Figure 2 presents the shared mix model. The attention vector of the shared mix model c^{i-1} is calculated as $c^{i-1} = c_{source}^{i-1} + c_{target}^{i-1}$, where c_{source}^{i-1} and c_{target}^{i-1} are the context attentions calculated by the Eq. (8). The reason for calculating the sum of two attention is to arrange the same number of parameters as the other shared models. Other architectures are the same as the other shared models.

Table 1. Number of sentences in each dataset.

Corpus	Train	Dev	Test
TED De–En	203,998	888	1,305
TED Zh–En	226,196	879	1,297
TED Ja–En	194,170	871	1,285
Recipe Ja–En	108,990	3,303	2,804

Table 2. BLEU scores of our context-aware NMT in each language pair. Each score is the average of three runs. "$*$" represents the statistically significant results against the baseline at $p < 0.05$ in all the runs.

Experiment	Baseline	Separated		Shared		
		Source	Target	Source	Target	Mix
TED De–En	26.55	26.29 ± .37	26.52 ± .12	*27.20 ± .11	***27.34** ± .11	27.18 ± .21
TED En–De	21.26	21.04 ± .64	20.77 ± .10	21.63 ± .27	**21.83** ± .30	21.50 ± .29
TED Zh–En	12.54	12.52 ± .33	12.63 ± .24	*13.36 ± .41	***13.52** ± .10	*13.23 ± .09
TED Zh–Zh	8.97	8.94 ± .11	8.71 ± .06	9.45 ± .22	***9.58** ± .13	9.42 ± .19
TED Ja–En	5.84	*6.64 ± .26	*6.37 ± .12	*6.95 ± .07	***6.96** ± .18	*6.81 ± .16
TED En–Ja	8.40	8.58 ± .12	8.26 ± .00	8.51 ± .31	8.59 ± .08	**8.66** ± .14
Recipe Ja–En	25.34	*26.51 ± .09	*26.69 ± .15	*26.90 ± .17	***26.92** ± .10	*26.78 ± .11
Recipe En–Ja	20.81	*21.87 ± .12	*21.45 ± .14	***22.02** ± .20	*21.97 ± .09	*21.81 ± .15

3   Experiments

3.1   Data

We mainly use the IWSLT2017 German–English, Chinese–English, and Japanese–English datasets from TED [9] for experiments. We consider each talk of TED as a document, which includes sentences that cannot be translated using only sentence-level information. Japanese and Chinese sentences are segmented by the MeCab[2] (dictionary: IPADic 2.7.0) and jieba[3], respectively. English and German sentences are segmented by `tokenizer.perl` included in Moses[4]. The documents that include sentences consisting of more than 100 words are eliminated from the training corpus. We evaluate our methods on the 2014 test set. The statistics of preprocessed corpora are shown in Table 1. Byte pair encoding [10] is used separately for source and target languages for subword segmentation. The number of merge operations is 32,000.

Moreover, we use the Recipe Corpus[5], which consists of Japanese–English user-posted recipes, to investigate the influences in the different domains. The procedures of data preprocessing are the same as those for the TED corpus, except for the number of merge operations (8,000).

---

[2] http://taku910.github.io/mecab/.

[3] https://github.com/fxsjy/jieba.

[4] http://www.statmt.org/moses/.

[5] http://lotus.kuee.kyoto-u.ac.jp/WAT/recipe-corpus/.

3.2 Settings

The baseline system of this experiment is our implementation of RNN-based NMT. The encoder is two-layer bi-LSTM, and the decoder is two-layer uni-LSTM. The dimensions of hidden states and embeddings are set to be 512. We use dropout with $p = 0.2$. The optimizer is AdaGrad with initial learning rate $= 0.01$. Each batch consists of up to 128 documents. These settings are the same in the baseline and all context-aware models. Dot global attention is used for calculating context attention c^{i-1}. We set $c^0 = 0$ because the first sentences in documents do not have any previous contexts.

The context-aware models are pretrained with the baseline system. Each model is trained for 30 epochs; then, the best model is selected with a development set. The results are evaluated using BLEU [11]. We calculate the statistical significance between the baseline and our methods by the bootstrap resampling toolkit in Travatar [12]. Experiments are performed three times with different random seeds.

3.3 Results

Table 2 shows the results. The shared target model improves the performances in all language pairs. In the experiments on several language pairs, the separated target model used in Bawden et al. [4] also improves performances compared to the baseline. However, improvement is less compared to the shared target model. Therefore, these results show that the target-side context should be introduced from a decoder.

4 Discussion

4.1 Weight Sharing

We expected that there would be no differences between the results of the shared source and separated source models because both models can introduce source-side context into the encoder. However, the results obtained for the language pairs used in this study show that the shared source model also improves the BLEU scores with fewer parameters. Dabre et al. [13] found that translation performances could be boosted even if the weights of stacked layers were shared. Our shared models can be seen as an instance of weight sharing for stacking sentence-level RNNs in chronological order. Shared models can also be seen as an instance of multitask learning that shares the same weights for encoder–decoders of neighboring sentences such as skip-thought [14]. Thus, it is possible that weight sharing leads to a more efficient model space by regularization, rather than by learning discourse structures.

4.2 Language Dependency

The tendencies of the scores vary depending on language pairs. The result of the TED English–German task shows that the source-side context decreases the performance. Müller et al. [15] obtained similar results in other datasets using the concatenation method proposed in Tiedemann et al. [16]. However, in the Japanese–English and English–Japanese tasks, the importance of the source-side context is equivalent to that of the target-side one. The reason is that Japanese requires contexts more than English because Japanese is a pro-drop language, which allows for the omission of agents and object arguments when they are pragmatically or syntactically inferable. Comparing the result of the TED and Recipe corpora, the difference of corpus domains does not affect such tendencies. In the Chinese–English task, where they have more similar word order, the importance of target-side context is equivalent or even better compared to that of the source-side one. Therefore, these results imply that the necessity of the source-side context depends on language pairs, while the target-side context is generally important.

The shared mix model obtains competitive results compared to the shared source model, in most of the language pairs. Therefore, either of contexts helps the improvement without both side information if we choose the source- or target-side context depending on the language pairs.

4.3   Output Examples

We analyze the output examples in terms of the phrase coherence. We select the Recipe Japanese–English task because Japanese is a pro-drop language that needs context due to many omissions but it is difficult to draw any definitive conclusions on the TED Japanese–English task as the BLEU score is too low to analyze. Table 3 shows the examples. When the model translates the previous sentence, this model does not use the context information because this is the first sentence of a document. The examples written in each lower row are the result using the information of the upper sentence as a context.

Looking at the result of the baseline, "長ねぎ" (naga negi, *Japanese leek*) is translated into "Japanese leek" in the previous sentence, even though this is translated into "leek" in the current sentence. This phenomenon can be commonly seen in the results of separated models. If we independently evaluate these sentences, these sentences will be rated with high fluency and adequacy. BLEU scores are also high because reference sentences also follow this translation. However, these sentences have low coherence because the same noun phrase in the Japanese sentence is translated into different phrases.

On the contrary, "長ねぎ" is translated into "Japanese leek" in both sentences in the experiments of shared target model and shared mix model, which use target-side context. Our models improve phrase coherence using weight sharing.

4.4   Convergence of Training

Fig. 3 plots the BLEU scores on development sets. Shared models and separated source model seem to be stable. However, the separated target model is unstable and does not lead to an improvement. This is due to the exposure bias problem [17] in the context encoder as well as the decoder. At the test phase, the separated target model has to read the low-quality sentence with well-trained encoder if the learning speed of encoding is faster than that of decoding. Thus, the separated target model should fill the gap between the learning speed of the context encoder and decoder.

Table 3. The output examples in Recipe Japanese–English experiments. Each upper sentence represents a previous sentence, and each lower sentence represents a current sentence. Each sequence may comprise several sentences because each sentence in Recipe corpus corresponds to "one step" of cooking.

Experiment	Sentences
Input	わかめ は よく 洗って 塩 を 落とし、 10分 ほど 水 に 浸けて おいて から ざく 切り に する。 長ねぎ は 小口切り に する。 熱した 鍋 に ごま油 を ひき、 わかめ と 長ねぎ を 入れて 30秒 ほど 軽 く 炒める。
Reference	wash the wakame well to remove the salt , put into a bowl of water for 10 minutes and drain . cut into large pieces . slice the **Japanese leek** . heat a pan and pour the sesame oil . stir fry the wakame and **leek** for 30 seconds .
Baseline	wash the wakame seaweed well and remove the salt . soak in water for 10 minutes , then roughly chop . cut the **Japanese leek** into small pieces . heat sesame oil in a heated pot , add the wakame and **leek** , and lightly sauté for about 30 seconds .
Separated Source	wash the wakame well , remove the salt , soak in water for about 10 minutes , then roughly chop . cut the **Japanese leek** into small pieces . heat sesame oil in a heated pot and add the wakame and **leek** . stir-fry for about 30 seconds .
Shared Source	wash the wakame well, remove the salt , soak in water for about 10 minutes , then roughly chop . cut the **Japanese leek** into small pieces . heat sesame oil in a heated pot and add the wakame and **leek** . stir-fry for about 30 seconds .
Separated Target	wash the wakame well , soak in water for about 10 minutes . cut into small pieces. cut the **Japanese leek** into small pieces . heat the sesame oil in a frying pan , add the wakame and **leek** , and stir-fry for about 30 seconds .
Shared Target	wash the wakame well , remove the salt , soak in water for about 10 minutes , then roughly chop . chop the **Japanese leek** into small pieces . heat sesame oil in a heated pan , add the wakame and **Japanese leek** , and lightly stir-fry for about 30 seconds .
Shared Mix	wash the wakame well , remove the salt , soak in water for about 10 minutes , then roughly chop . chop the **Japanese leek** into small pieces . heat sesame oil in a heated pan , add the wakame and **Japanese leek** , and stir-fry for about 30 seconds .

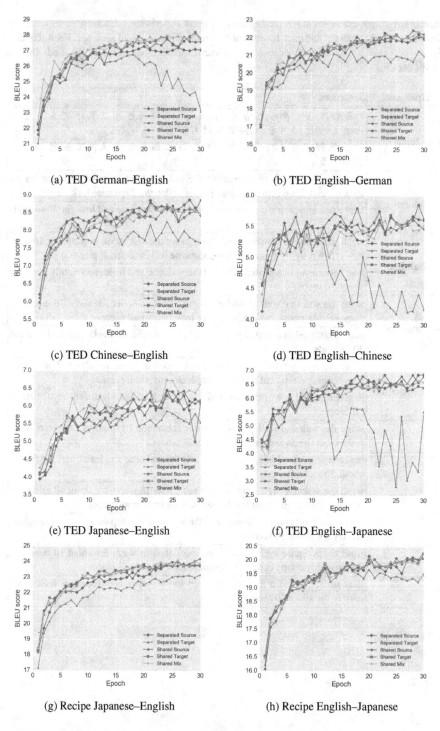

(a) TED German–English

(b) TED English–German

(c) TED Chinese–English

(d) TED English–Chinese

(e) TED Japanese–English

(f) TED English–Japanese

(g) Recipe Japanese–English

(h) Recipe English–Japanese

Fig. 3. The graph of BLEU scores using each development set. BLEU score is calculated at the end of each epoch.

5 Related Works

Wang et al. [18], Maruf et al. [19], and Tu et al. [20] incorporated the information of previous sentences by using a hierarchical encoder, a memory network and cache mechanism respectively. Although they used several sentences as contexts, the former two works found that the information of distant sentences in a document does not improve translation quality. Our investigation is focused on a previous sentence.

Tiedemann et al. [16] used the concatenation of a previous sentence and a current sentence as an input or output sentence to incorporate source-side and target-side contexts in conventional NMT. Müller et al. [15] evaluated the performance of existing context-aware NMT in the English–German task in terms of pronoun translation. They concluded that generating concatenated sentence is more effective than inputting concatenated sentence. Our results of the shared target model support their results.

Voita et al. [6] and Zhang et al. [7] proposed Transformer-based context-aware NMT. The former suggested that self-attention solves anaphora resolution. The latter performed fine-tuning with small document-level data to adapt a single-sentence NMT trained with large data to context-aware NMT. However, they did not investigate the influence of the target-side contexts.

6 Conclusion

We reported how context-aware neural machine translation effectively employs target-side contexts. We proposed a weight sharing to model the target-side context in a decoder. This method achieves high performances in several language pairs, even though it does not require much additional parameters. In the future, we will analyze whether this model can handle longer contexts.

References

1. Sutskever, I., Vinyals, O., Le, Q.V.: Sequence to sequence learning with neural networks. In: Advances in Neural Information Processing Systems, vol. 27, pp. 3104–3112 (2014)
2. Bahdanau, D., Cho, K., Bengio, Y.: Neural machine translation by jointly learning to align and translate. In: Proceedings of the International Conference on Learning Representations (2015)
3. Vaswani, A., et al.: Attention is all you need. In: Advances in Neural Information Processing Systems, vol. 30, pp. 5998–6008 (2017)
4. Bawden, R., Sennrich, R., Birch, A., Haddow, B.: Evaluating discourse phenomena in neural machine translation. In: Proceedings of the 2018 Conference of the North American Chapter of the Association for Computational Linguistics: Human Language Technologies (Long Papers), vol. 1, pp. 1304–1313 (2018)
5. Jean, S., Lauly, S., Firat, O., Cho, K.: Does neural machine translation benefit from larger context? CoRR, vol. abs/1704.05135 (2017)
6. Voita, E., Serdyukov, P., Sennrich, R., Titov, I.: Context-aware neural machine translation learns anaphora resolution. In: Proceedings of the 56th Annual Meeting of the Association for Computational Linguistics (Volume 1: Long Papers), pp. 1264–1274 (2018)

7. Zhang, J., et al.: Improving the transformer translation model with document-level context. In: Proceedings of the 2018 Conference on Empirical Methods in Natural Language Processing, pp. 533–542 (2018)
8. Luong, T., Pham, H., Manning, C.D.: Effective approaches to attention-based neural machine translation. In: Proceedings of the 2015 Conference on Empirical Methods in Natural Language Processing, pp. 1412–1421 (2015)
9. Cettolo, M., Girardi, C., Federico, M.: WIT[3]: web inventory of transcribed and translated talks. In: Proceedings of the 16th Conference of the European Association for Machine Translation, pp. 261–268, May 2012
10. Sennrich, R., Haddow, B., Birch, A.: Neural machine translation of rare words with subword units. In: Proceedings of the 54th Annual Meeting of the Association for Computational Linguistics (Volume 1: Long Papers), pp. 1715–1725 (2016)
11. Papineni, K., Roukos, S., Ward, T., Zhu, W.-J.: BLEU: a method for automatic evaluation of machine translation. In: Proceedings of the 40th Annual Meeting on Association for Computational Linguistics, pp. 311–318 (2002)
12. Neubig, G.: Travatar: a forest-to-string machine translation engine based on tree transducers. In: Proceedings of the 51st Annual Meeting of the Association for Computational Linguistics: System Demonstrations, pp. 91–96 (2013)
13. Dabre, R., Fujita, A.: Recurrent stacking of layers for compact neural machine translation models. In: Proceedings of the AAAI Conference on Artificial Intelligence, vol. 33, pp. 6292–6299 (2019)
14. Kiros, R., et al.: Skip-thought vectors. In: Advances in Neural Information Processing Systems, vol. 28, pp. 3294–3302 (2015)
15. Müller, M., Rios, A., Voita, E., Sennrich, R.: A large-scale test set for the evaluation of context-aware pronoun translation in neural machine translation. In: Proceedings of the Third Conference on Machine Translation: Research Papers, pp. 61–72 (2018)
16. Tiedemann, J., Scherrer, Y.: Neural machine translation with extended context. In: Proceedings of the Third Workshop on Discourse in Machine Translation, pp. 82–92 (2017)
17. Ranzato, M., Chopra, S., Auli, M., Zaremba, W.: Sequence level training with recurrent neural networks. In: Proceedings of the International Conference on Learning Representations (2016)
18. Wang, L., Tu, Z., Way, A., Liu, Q.: Exploiting cross-sentence context for neural machine translation. In: Proceedings of the 2017 Conference on Empirical Methods in Natural Language Processing, pp. 2826–2831 (2017)
19. Maruf, S., Haffari, G.: Document context neural machine translation with memory networks. In: Proceedings of the 56th Annual Meeting of the Association for Computational Linguistics (Volume 1: Long Papers), pp. 1275–1284 (2018)
20. Tu, Z., Liu, Y., Shi, S., Zhang, T.: Learning to remember translation history with a continuous cache. Trans. Assoc. Comput. Linguist. **6**, 407–420 (2018)

Learning to Evaluate Neural Language Models

James O'Neill[✉] and Danushka Bollegala[✉]

Department of Computer Science, University of Liverpool, Liverpool L69 3BX, UK
{james.o-neill,danushka}@liverpool.ac.uk

Abstract. Evaluating the performance of neural network-based text generators and density estimators is challenging since no one measure perfectly evaluates language quality. Perplexity has been a mainstay metric for neural language models trained by maximizing the conditional log-likelihood. We argue perplexity alone is a naive measure since it does not explicitly take into account the semantic similarity between generated and target sentences. Instead, it relies on measuring the cross-entropy between the targets and predictions on the word-level, while ignoring alternative incorrect predictions that may be semantically similar and globally coherent, thus ignoring quality of neighbouring tokens that may be good candidates. This is particularly important when learning from smaller corpora where co-occurrences are even more sparse. Thus, this paper proposes the use of a pretrained model-based evaluation that assesses semantic and syntactic similarity between predicted sequences and target sequences. We argue that this is an improvement over perplexity which does not distinguish between incorrect predictions that vary in semantic distance to the target words. We find that models that outperform other models using perplexity as an evaluation metric on Penn-Treebank and WikiText-2, do not necessarily perform better on measures that evaluate using semantic similarity.

Keywords: Language models · Semantic evaluation · Neural networks

1 Introduction

Language modelling is an important task in natural language processing and is partly responsible for many of the recent successess of transferable pretrained models that require very little fine-tuning to perform well on a host of various downstream tasks [6,8,18,19]. However, language models are currently evaluated using perplexity, which is directly proportional to the log-likelihood. Although log-likelihood is efficient in that it only evaluates the loss of each token and the corresponding predicted probability for that token, it fails to account for incorrect predictions that may be synonymous with the token, or share some semantic equivalence between predicted sentences and target sentences as a whole. Other conditional language modelling tasks use n-gram overlap measures, such

© Springer Nature Singapore Pte Ltd. 2020
L.-M. Nguyen et al. (Eds.): PACLING 2019, CCIS 1215, pp. 123–133, 2020.
https://doi.org/10.1007/978-981-15-6168-9_11

as BLEU in machine translation [17] and CIDEr [22] for image captioning. Similarly, these too do not account for global coherence and semantic similarity.

We argue that an evaluation measure that takes semantic and syntactic similarities into account results in a measure that reflects human judgements better than perplexity given that model-based evaluators are learned from human annotations of sentence similarity. This also results in generated text that allows for more diverse predictions since it is not restricted to word-level cross-entropy loss.

Contribution: This paper proposes to transfer pretrained models from semantic textual similarity tasks which can be used for evaluating text generation models. Pretraining on semantic textual similarity tasks allows us to optimize for generated sentences that are semantically similar to the ground truth tokens. This becomes particularly important for words that lie in the long-tail of the unigram distribution that have less context to disambiguate from. In such cases, the perplexity is likely to increase given that alternative incorrect predictions are ignored. Concretely, an evaluation measure that accounts for incorrectly predicted alternatives to the target token are is critically important for less frequent tokens. Moreover, we are indirectly leveraging the readily available annotations of semantic similarity between sentences that we argue should be considered for evaluating text generation models. We also include a simpler non-parametric embedding evaluation measures such as average cosine similarity and Word Mover's Distance (WMD) [10] between the embeddings corresponding to the predicted and target sequences. These model-free embedding similarity measures are relatively fast to compute, and evaluates embedding similarities at the word-level.

2 Related Work

2.1 Sentence Representations

We briefly introduce recent research on learning universal sentence representations before introducing the corresponding models that can be used for evaluation.

SkipThought: These representations are widely used and have shown good performance for semantic relatedness, paraphrase classification, question classification, sentiment analysis, among many other pairwise sentence tasks, as shown when used as input to simple linear classifiers [9].

QuickThought

InferSent: Infersent [5] use RNNs with supervision from natural language inference sentence-pairs provided in the Stanford Natural Language Inference (SNLI) [2] dataset. They showed that this *InferSent* model generalized well to other tasks, and in some cases outperforms the unsupervised SkipThought model. We include this in our analysis as to compare how this supervised pre-trained model compares to SkipThought in evaluation.

Weighted Meta-Embeddings: Meta-embeddings with a weighted combination of multiple pretrained word-embeddings are used in either an unsupervised way prior to a task, or learned directly for a supervised task on multiple intrinsic and downstream tasks [15]. In the latter case, a reconstruction loss for an autoencoder is used as an auxiliary loss. Similarly, pretrained sentence meta-embeddings can be obtained when used in a Siamese network for the sentence-based pairwise tasks e.g semantic relatedness, paraphrasing etc.

2.2 Learning the Evaluation Measure

Prior work [11] has argued that current n-gram overlap based evaluation measures are biased and insufficiently correlate with human judgements of the generated response quality. They propose to use an evaluation model that learns to predict the manually annotated scores using a proposed human response dataset. They found that the predictions from the evaluation model had better correlations with human judgements than an n-gram overlap measures such as BLEU.

Novikova et al. [16] have also considered alternative measures for evaluating natural language generation systems. They include a comprehensive evaluation using word-based metrics (tf-idf frequency), n-gram overlap metrics (BLEU, ROUGE, CIDEr), semantic similarity between Latent Semantic Analysis (LSA) word representations with WordNet measures and grammar based metrics (readability, grammaticality). Importantly, they report that metric performance is both data and system-specific, but find that automated metrics reliably work at the system-level. This also reinforces the idea of considering a range of different word-level and sentence-level evaluation metrics, including automated model-based measures.

The difference in our work is that we are proposing pretrained neural network model-based evaluation that not only includes models that are trained on semantic similarity tasks (such as their use of LSA) but other tasks such as paraphrasing and natural language inference, leading to universal sentence representations. We also propose faster model-free alternatives such as WMD and Average Cosine Similarity between word embeddings. Additionally, the model-based scores obtained when computing similarity between sentence representations indirectly takes into account the readily available human scores that were used for supervision, thus, avoiding extra manual annotations.

2.3 Evaluating Language Models

Chen et al. [4] compare perplexity and word error rate for n-gram models on language modelling and surprisingly find a strong Pearson correlation between both quantities. They suggest that a linear combination of both may lead to a stronger measure. This is one of the first papers that draw attention to alternatives to perplexity and its relation to other measures. Similarly, we analyze perplexity with respect to other measures for neural language models instead of n-gram language models

Marvin and Linzen [12] focus on evaluating syntax in language models by proposing a dataset that assesses the grammaticality of predictions. They pair minimally different grammatical and ungrammatical sentence pairs and train an LSTM language model to analyze whether the LSTM performs better on grammatical sentences as expected. They conclude that although multi-task learning with a syntactic objective can improve performance, there still remains a considerable gap compared to human performance.

Chaganty et al. [3] propose control variates to debias metrics such as BLEU with the help of human judgements, as a tradeoff between improved correlation and labor.

3   Methodology

3.1   Language Models

Recurrent Neural Network Language Model. For a sequence pair (X, Y) where $X = \{x_1, ..x_t, ..x_T\}$ is the input sequence and $Y = \{y_1, ..y_t, ..y_T\}$ is the target sequence, for $t \in T$ timesteps, x_t is passed to a parametric model along with the previous hidden state vector h_{t-1} to produce an output from its last layer $\tilde{h} = f_\theta(x_t, h_{t-1})$. Here, θ are the parameters of the sequential parametric model (e.g RNN), $h_t = f_\theta(h_{t-1}, y_{t-1})$ is an encoded hidden state vector. The probability $P_\theta(y_t|h_t)$ is then computed using a linear projection, leading to a prediction $\hat{y} = \arg\max\phi(h \cdot W + b)$, where ϕ is a softmax function that normalizes the output to calibrate $P_\theta(y_t|h_t)$, W are the decoder weights and the predicted token is retrieved via the arg max operator. The log probability $P(Y|X)$ is maximized as shown in Eq. 1, to directly minimize the perplexity in Eq. 2.

For unconditional generation (i.e language modelling) $X = \{\langle start \rangle \cup Y_{1:t-1}\}$, which is the case this paper focuses on. Although, this also applies for conditional generation such as machine translation, question answering and other such sequence-to-sequence problems.

$$\log P(Y_1^t|X) = \sum_{t=1}^{T} \log P(y_t|y_1^{t-1}, X)$$

$$P(y_t|y_1^{t-1}, X; \theta) = \log P(y_t|h_t; \theta) \tag{1}$$

$$2^{-\frac{1}{T}\log P(x_1,..,x_T)} = \sqrt[T]{\prod_{i=1}^{T} \frac{1}{P(w_i|w_1..w_{i-1})}} \tag{2}$$

The standard way to train the aforementioned RNN language model is to maximize the word-level log likelihood by minimizing the cross entropy loss $\mathcal{L}_{ce}(\theta)$ shown in Eq. 3.

$$\mathcal{L}_{ce}(\theta) = -\sum_{t=1}^{T} y_t \log \left(p_\theta(y_t|y_{t-1}, ..y_0) \right) \tag{3}$$

In our experiments, we use a standard 2-hidden layer LSTM sequence model [20] and a 2-hidden layer GRU network [14] with an embedding dimension and hidden state dimension size of $|e| = |h| = 400$. Xavier uniform initialization is used with ($\mu = 0, \sigma = 0.1$) and gradients are clipped at 0.5 threshold if exceeded at each update, for a batch size $|X_s| = 56$.

Table 1. Language Modelling Evaluation on Penn Treebank (Validation/Test Scores scaled [0, 100]), omitting perplexity

Model	BLEU2 ↑	BLEU3	BLEU4	Cosine ↑	WMD	InferSent ↑	SkipThought	Transformer	Perplexity ↓
GRU	24.12/25.98	11.54/12.76	7.71/7.83	69.63/68.20	78.31/79.53	80.52/82.59	87.10/91.23	83.02/81.48	91.45/86.12
GRU-SS	26.40/26.11	12.79/13.14	8.56/9.51	72.59/74.19	80.93/80.16	88.48/90.10	92.14/93.72	85.09/82.02	87.21/82.08
LSTM	26.09/26.20	12.56/12.24	8.34/8.91	68.91/70.08	81.04/80.23	82.01/84.91	84.02/84.33	82.77/82.03	89.32/83.27
LSTM-SS	27.90/28.63	13.15/13.87	8.80/9.04	73.59/72.48	81.43/82.78	84.42/85.06	90.28/91.70	86.10/86.24	84.71/81.11
Transformer	31.49/30.29	14.08/14.82	9.12/9.35	72.09/71.16	79.20/76.25	90.63/91.52	88.02/89.50	84.10/84.55	82.07/79.83

Transformer-Based Language Model. We consider the decoder of a multi-layer multi-attention head Transformer language model [19] which uses language model attention heads [19][1] as another evaluation based model. We include this in the analysis for completeness, as the Transformer-based models have recently shown improved performance when compared to recurrent-based models in language modelling and conditional generation tasks such as machine translation [19].

Instead of recurrent connections, Transformers use self-attention over the input context tokens, layer normalization (LayerNorm) and positional embeddings. The output of each attenttion head is concatenated and fed to batch-normalized feedforward layer to generate an output probability distribution for the next target token $t \in T$.

Assume that we have a sequence of vectors $x_1, ..., x_n$ where each vector $x_i \in \mathbb{R}^d$. We define $E \in \mathbb{R}^{n \times d}$ to be a matrix representing the sequence. We define parameters $W_e \in \mathbb{R}^{d \times l}, W_p \in \mathbb{R}^{d \times l}$ and $W_o \in \mathbb{R}^{d \times o}$. Z is defined in Eq. 4 where EC is an $n \times o$ matrix of new embeddings. $EW_e W_p^T E^T$ is an $n \times n$ matrix representing the inner products in a new l-dimensional space. softmax$(EW_e W_p^T E^T)$ is matrix where each row entry is positive and sums to 1.

$$Z = \text{softmax}(EW_e W_p^T E^T)EW_o \tag{4}$$

In our experiments we set, $d = 512$, $o = 64$, and the parameters $W_e^j, W_p^j \in \mathbb{R}^{d \times l}$, $W_o^j \in \mathbb{R}^o$ for $j = 1, ..., 8$. Therefore, for the multi-head attention expressed in Eq. 5 we use $\mathbb{Z}^j \in \mathbb{R}^{n \times 64}$ and once outputs are concatenated to form $Z^{n \times 512}$.

$$\mathbb{Z}^j = \text{softmax}(EW_e^j(W_p^j)^T E^T)EW_o^j$$
$$\mathbb{Z} = \text{concat}(Z^1, ... Z^8) \tag{5}$$
$$\mathbb{Z}' = \text{Feedforward}(\text{LayerNorm}(Z + E))$$

---

[1] we follow the hugging face implementation available here: https://github.com/huggingface/pytorch-openai-transformer-lm.

We use 8-hidden layers with embedding dimension and hidden state dimensions respectively $|e| = |h| = 512$ with 8 attention heads. An adaptive softmax [7] is used to normalize the output to a probability distribution which we denote as $\phi_{as}(\mathbb{Z}')$, splitting based on term frequency into top $1/10$ of words $w \in \mathcal{V}$, $1/10$–$4/10$ for the second bin and the remaining $4/10$–1 for the third bin. Lastly, input, hidden layer and attention dropout is set to 0.1.

Table 2. Language modelling evaluation on WikiText-2 using model-based (InferSent & SkipThought) and model-free evaluation measures

Model	BLEU2 ↑	BLEU3	BLEU4	Cosine ↑	WMD	InferSent ↑	SkipThought	Transformer	Perplexity ↓
GRU	16.24/16.63	7.02/7.71	4.09/4.68	65.30/63.08	74.95/75.49	76.91/78.02	81.27/79.43	80.22/82.76	148.62/132.90
GRU-SS	16.90/17.07	7.56/7.98	4.35/4.85	67.01/65.48	75.22/75.86	83.78/85.93	84.11/86.03	81.27/79.43	141.31/127.80
LSTM	16.58/16.84	7.31/7.76	4.09/4.68	64.15/61.59	76.71/78.03	80.02/76.37	81.90/79.68	81.26/81.52	150.02/137.11
LSTM-SS	17.82/17.49	8.18/8.02	4.09/4.68	69.42/71.09	77.22/80.11	82.01/83.17	83.20/85.41	83.15/85.79	141.66/135.19
Transformer	17.49/17.83	7.25/7.70	3.59/3.85	69.24/72.47	76.95/78.20	82.12/88.28	80.43/84.69	87.35/85.71	120.31/117.98

3.2   Pretrained Models for Evaluation

InferSent Evaluation. Conneau et al. [5] use the scoring function in Eq. 6 between two encoded sentence pairs (h_1, h_2) where $h_1, h_2 \in \mathbb{R}^d$, correspond to the two sentences $(\mathcal{S}_1, \mathcal{S}_2)$. This scoring function showed to be useful for universal representations, not only natural language inference. We also use the same scoring function for the pretrained InferSent model.

$$\phi([h_1, h_2, |h_1 - h_2|, h_1 \cdot h_2] \cdot W + b) \tag{6}$$

The model is a Bidirectional-GRU (or BiLSTM) with max (or mean) pooling, as in Eq. 7 where g represents the pooling function and e; is the embedding corresponding to word x_i.

$$h = g_{\text{max-pool}}([\overrightarrow{\text{GRU}}(e_1, .., e_T), \overleftarrow{\text{GRU}}(e_1, .., e_T)]) \tag{7}$$

We also use the self-attentive variation in Eq. 8, where the max-pooling operation g is replaced with self-attention that produces a weighted average where $g_{\text{avg}}(\cdot)$ sum the weights to 1 $\forall t \in T$. Hence, attention focuses on the hidden states of important tokens prior to using the scoring function.

$$h = \sum_{t=1}^{T} g_{\text{avg}}(\tanh(W h_t + b)) h_t \tag{8}$$

SkipThought. The original Skipthought [9] paper includes a comprehensive evaluation of CNN-RNN, RNN-RNN and LSTM-LSTM and GRU-GRU encoder-decoder architectures. In our evaluation, we use a Bidirectional GRU encoder decoder architecture which is called bi-skip. SkipThought vectors provide a good unsupervised baseline for model-based text generation evaluation, as they have shown competitive performance against supervised models on the aforementioned tasks of semantic relatedness and paraphrase detection [9].

Transformer Evaluation. We also use the original Transformer model [21] for a non-recurrent neural network model-based evaluation, which can also be compared to SkipThought, an RNN-based model that is also trained using unsupervised(-self) learning. To compute scores we use the inner product between the hidden representations of the K most upper layers of the decoder, and compute mean over all K dot products, followed by a Sigmoid non-linearity to normalize $\tilde{s} \in [0, 1]$.

$$\tilde{s}(s_1, s_2) = \sigma\left(\frac{1}{N}\sum_{i=1}^{K}\langle h_1^i, h_2^i\rangle\right) \tag{9}$$

3.3 Model-Free Embedding Evaluation

Word Mover's Distance Sentence Similarity. We also include the (WMD) [10] for measuring semantic similarity between ℓ_2 normalized embeddings associated with predicted and target words. We also include the average cosine similarity between sentences, which can be considered as an approximation to the optimal transport in WMD. Model-free in this sense means that we do not require a pretrained network for evaluation, only word-level vectors corresponding to predictions and the corresponding targets. Word-level embedding similarities offer a faster alternative to model-based sentence-level evaluation, hence we include it for our experiments.

Average Cosine Similarity. Alternatively, a faster method to evaluate $(\hat{Y}, Y)$ is to compute the average cosine similarity which can be achieved by one of two ways. The first is a straightforward similarity between adjacent word pairs $cos(Y_i, \hat{Y}_i)$ $\forall t \in T$ as shown in Eq. 10.

$$s(E_Y, E_Y) = \sigma\left(\frac{1}{|Y|}\sum_{t=1}^{T} cos(Y_t, \hat{Y}_t)\right) \tag{10}$$

We also considered decayed k-pairwise cosine similarity where k is a sliding window span that compares embeddings corresponding to n-gram groupings with a decay factor $\gamma \in [0, 1]$ that depends on the distance such that $\gamma_{(i,j)} = d(Y_i, Y_j)/k$ $\forall i, j \in T$.

$$\tilde{s}(Y, \hat{Y}) = \frac{1}{Tk}\sum_{t=1}^{T}\sum_{j=i-k}^{t+k} \gamma_{(i,j)} cos(Y_i, \hat{Y}_j)$$
$$s.t, \quad t \leq i \leq T - k \tag{11}$$

Dealing with Skewed Scores. Since we are optimizing to exactly predict the target tokens in language modelling, the predicted sequences often result in high embedding similarities by the end of training. Therefore, the differences in [0, 1] normalized scores can appear to be minuscule. To address this we allocate exponentially scaled scores $\tilde{y}$, shown in Eq. 12, so that smaller changes of highly similar sentences are distinct, yet still bounded in $[0, 1]$, analogous to perplexity in that it exponentiates the cross entropy loss.

$$\tilde{s} = 1 - (1 - s)2^s = 1 + 2^s s - 2^s \tag{12}$$

We also note that for the average cosine similarity between predicted and target embeddings (denoted as Cosine in Table 1 and Table 2), we pass the similarity as input to a logistic unit $\sigma(\cdot)$ in Eq. 13 in order for Cosine scores ($s \in [-1, 1]$) to be comparable with others in $[0, 1]$ (Fig. 1).

$$\tilde{s}(s_1, s_2) = \frac{1}{(1 + \exp(-s/\tau + b))} \tag{13}$$

In our experiments $\tau = 0.1$ and the bias is manually set $b = 8$, as shown in Fig. 2 e.g $\tilde{s} = 0.8$ corresponds to an average cosine similarity of $s = 0.82$.

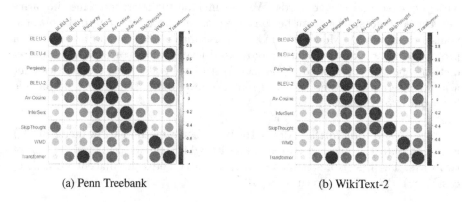

(a) Penn Treebank   (b) WikiText-2

Fig. 1. Correlation matrix of validation and test evaluation metrics

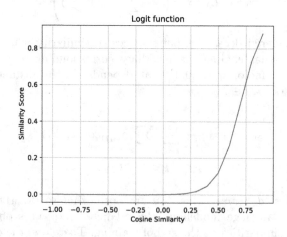

Fig. 2. Scaled similarity score $s \rightarrow \tilde{s}$ for average Cosine Similarity (COS) Evaluation

4   Results

Table 1 and Table 2 show the results evaluated on both n-gram overlap based measures, word-level (WMD, Average Cosine Similarity) and sentence-level embedding based evaluation measures (Infersent, SkipThought and Transformer), and lastly perplexity. We consider n-gram overlap to be a strict measure of text generation quality, compared to the embedding-based measures that account for word-level or sentence-level semantic similarity.

The model-based measures that use unsupervised training (SkipThought, transformer) tend to produce high similarity measures between $(\hat{Y}, Y)$ on average, when compared to model-based measures that use supervision (InferSent trained on NLI data).

Interestingly, using scheduled sampling [1] with LSTMs (LSTM-SS) leads to improvements over LSTM without SS, this is similarly found for GRU-SS as well. This is somewhat surprising considering this has been used in the context of language modelling where ground truth tokens are given at test time, unlike text generation, where SS can help mitigate compounding errors. We view SS as inducing noisy samples for language modelling that empirically shows to improve generalization, similar to how multiplicative Gaussian noise on input embeddings achieves superior performance over models without input embedding dropout [13].

We also find that the similarity between $\hat{Y}$ and Y is generally lower across each model for pretrained evaluators that use supervised learning. Concretely, it is more difficult to score well on semantic similarity with InferSent in comparison to SkipThought and Transformer *evaluators*. This can be explained by the fact that the supervision provided by humans in InferSent can be more difficult for a neural language model to learn from scratch without any labels.

From these findings, we posit that standard conditional log-likelihood training is limited in accounting for semantically similar sentences since cross-entropy is evaluated at each target token which does not consider good alternative predictions. Moreover, averaging word-level losses does ignores the global coherence of each sentence or paragraph.

Lastly, we note that scaling functions allow for useful comparisons across each measure and lead to better separation between relatively low performing and high performing models on sentence-level semantic similarity metrics, such as the two presented in Eq. 12 and Eq. 13. Perplexity itself can be considered an exponential scaling of CE loss and is often used to compare language models, as small CE loss alone make it difficult to assess the performance differences between each model.

5   Conclusion

This paper proposed to reconsider how we evaluate neural language models by advocating the use of sentence-based similarity measures between generated sequences and predicted sequences using pretrained pairwise learned models.

These models can be trained on supervised tasks such as natural language inference and semantic textual similarity (InferSent) or trained in an unsuperised
fashion (SkipThought and Transformer). We argue that this approach can be
used in conjunction with perplexity and word-overlap measures, or as an alternative for evaluating text generation systems, along with other model-free embedding similarity measures used and defacto metrics such as perplexity and BLEU.
We conclude that neural language models should be evaluated with a variety of
different metrics since there is not a clear correlation between them.

References

1. Bengio, S., Vinyals, O., Jaitly, N., Shazeer, N.: Scheduled sampling for sequence
   prediction with recurrent neural networks. In: Advances in Neural Information
   Processing Systems, pp. 1171–1179 (2015)
2. Bowman, S.R., Angeli, G., Potts, C., Manning, C.D.: A large annotated corpus for
   learning natural language inference. arXiv preprint arXiv:1508.05326 (2015)
3. Chaganty, A.T., Mussman, S., Liang, P.: The price of debiasing automatic metrics
   in natural language evaluation. arXiv preprint arXiv:1807.02202 (2018)
4. Chen, S.F., Beeferman, D., Rosenfeld, R.: Evaluation metrics for language models
   (1998)
5. s Conneau, A., Kiela, D., Schwenk, H., Barrault, L., Bordes, A.: Supervised learning
   of universal sentence representations from natural language inference data. In:
   Proceedings of the 2017 Conference on Empirical Methods in Natural Language
   Processing, pp. 670–680. Association for Computational Linguistics, Copenhagen,
   Denmark, September 2017. https://www.aclweb.org/anthology/D17-1070
6. Devlin, J., Chang, M.W., Lee, K., Toutanova, K.: Bert: pre-training of deep bidirectional transformers for language understanding. arXiv preprint arXiv:1810.04805
   (2018)
7. Grave, E., Joulin, A., Cissé, M., Jégou, H., et al.: Efficient softmax approximation for GPUs. In: Proceedings of the 34th International Conference on Machine
   Learning-Volume 70, pp. 1302–1310. JMLR. org (2017)
8. Howard, J., Ruder, S.: Universal language model fine-tuning for text classification.
   arXiv preprint arXiv:1801.06146 (2018)
9. Kiros, R., et al.: Skip-thought vectors. In: Advances in neural information processing systems, pp. 3294–3302 (2015)
10. Kusner, M., Sun, Y., Kolkin, N., Weinberger, K.: From word embeddings to doc-
    ument distances. In: International Conference on Machine Learning, pp. 957–966
    (2015)
11. Lowe, R., Noseworthy, M., Serban, I.V., Angelard-Gontier, N., Bengio, Y., Pineau,
    J.: Towards an automatic Turing test: learning to evaluate dialogue responses.
    arXiv preprint arXiv:1708.07149 (2017)
12. Marvin, R., Linzen, T.: Targeted syntactic evaluation of language models. arXiv
    preprint arXiv:1808.09031 (2018)
13. Merity, S., Keskar, N.S., Socher, R.: Regularizing and optimizing LSTM language
    models. arXiv preprint arXiv:1708.02182 (2017)
14. Miyamoto, Y., Cho, K.: Gated word-character recurrent language model. arXiv
    preprint arXiv:1606.01700 (2016)
15. Neill, J.O., Bollegala, D.: Semi-supervised multi-task word embeddings. arXiv
    preprint arXiv:1809.05886 (2018)

16. Novikova, J., Dušek, O., Curry, A.C., Rieser, V.: Why we need new evaluation metrics for NLG. arXiv preprint arXiv:1707.06875 (2017)
17. Papineni, K., Roukos, S., Ward, T., Zhu, W.J.: Bleu: a method for automatic evaluation of machine translation. In: Proceedings of the 40th Annual Meeting on Association for Computational Linguistics, pp. 311–318. Association for Computational Linguistics (2002)
18. Peters, M.E., Ammar, W., Bhagavatula, C., Power, R.: Semi-supervised sequence tagging with bidirectional language models. arXiv preprint arXiv:1705.00108 (2017)
19. Radford, A., Narasimhan, K., Salimans, T., Sutskever, I.: Improving language understanding by generative pre-training (2018). https://s3-us-west-2.amazonaws.com/openai-assets/research-covers/languageunsupervised/languageunderstanding paper.pdf
20. Sundermeyer, M., Schlüter, R., Ney, H.: LSTM neural networks for language modeling. In: Thirteenth Annual Conference of the International Speech Communication Association (2012)
21. Vaswani, A., et al.: Attention is all you need. In: Advances in Neural Information Processing Systems, pp. 5998–6008 (2017)
22. Vedantam, R., Lawrence Zitnick, C., Parikh, D.: Cider: consensus-based image description evaluation. In: Proceedings of the IEEE Conference on Computer vision and Pattern Recognition, pp. 4566–4575 (2015)

Recommending the Workflow of Vietnamese Sign Language Translation via a Comparison of Several Classification Algorithms

Luyl-Da Quach[1] , Nghia Duong-Trung[1] , Anh-Van Vu[2],
and Chi-Ngon Nguyen[3](✉)

[1] FPT University, Can Tho City, Vietnam
luyldaquach@gmail.com, duong-trung@ismll.de
[2] Korea Advanced Institute of Science and Technology, Daejeon, South Korea
vuanhvan@kaist.ac.kr
[3] Can Tho University, Can Tho City, Vietnam
ncngon@ctu.edu.vn

Abstract. The Vietnamese deaf community communicates via a special language called Vietnamese Sign Language (VSL). Three-dimensional space and hand gesture are primarily used to convey meanings that allow deaf people to communicate among themselves and with non-deaf people around them. It maintains syntax, grammar, and vocabulary which is completely different from regular spoken and/or written Vietnamese. The normal procedure of transformation from spoken and/or written language (SWL) to VSL consists of consecutive steps: (i) Vietnamese word tokenization, (ii) machine translation into written sign language sentences, and (iii) conversion of these written sign language sentences into visual sign gesture. In this procedure, the second step gets the most attention due to the completion of the conveyed message. The basic challenge is that sign language, in general, has limited vocabulary compared to spoken/written language. If the machine translation is poorly performed, the complete message might not be successfully communicated, or in some cases, the conveyed message has a different meaning from the original. Consequently, the high accuracy translation should be maintained in any circumstances. This research is the efforts of evaluating an effective classification algorithm that the authors recommend to be integrated into the workflow of VSL translation. We believe that this is the first showing a quantitative comparison of several classification algorithms used in the workflow. The experimental results show that the translation accuracy rate is 96.7% which strongly support our recommendation.

Keywords: Machine translation · Vietnamese Sign Language · Example-Based Machine Translation · Linguistics

© Springer Nature Singapore Pte Ltd. 2020
L.-M. Nguyen et al. (Eds.): PACLING 2019, CCIS 1215, pp. 134–141, 2020.
https://doi.org/10.1007/978-981-15-6168-9_12

1 Introduction

A recent statistics revealed that approximately 5% of the world population is the deaf and severe hearing people [18]. Consequently, they combine hand gesture, head movements, and body posture to exchange their feelings and/or ideas. According to census data [22], 6.7 million Vietnamese people, e.g. equivalent to 7.8% of the country population, were affecting by a disability of which the deaf contributes to a large part. Deaf people are facing many difficulties when communicating with other normal people and there are limited resources that are written in their language. VSL is now considered a minority language which coexists with majority languages. Although the noticeable achievement of VSL in Vietnam, the development of its sign utterance will require years to become independent. The conversion from text to VSL and its application attract many researchers [4, 14–16, 24] thanks to the modern advancement of linguistic research [21]. The deaf has their rights to access information and services as for the normal people.

As a result, the workflow of VSL translation should be seriously taken by the community. The Vietnamese government has provided regulation and protocol to assist the deaf by combining signers into several national TV channels. Real people are hired to express sign language during the news. Vietnamese Television channel VTV2 has news programs supporting VSL. That is useful because it is possible to express the intonation, to directly communicate to the deaf. However, that takes a lot of time in compiling the content accordingly. To perform the compilation of contents, it requires a team consisting of at least three people: one person will be in charge of reducing the content of the text, one person will be responsible for compiling the content, and one person will be express the sign language on television. However, the automation procedure should replace the second person due to a large amount of text to be compiled. It helps reduce the loss of compilation and the ease of handling big textual resources as well.

2 Vietnamese Sign Language

VSL has components consisting of lexicons, phonemes, morphemes, grammar, and sentences [8]. However, before expressing by the visual gesture of the signers, the written sentence needs to be converted into an intermediate structure which is quite similar to Vietnamese grammar [1], see Fig. 1.

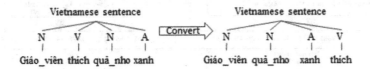

Fig. 1. The conversion of "a teacher likes a green apple".

3   Materials and Methodology

3.1   Workflow of Sign Language Translation

The workflow of sign language translation includes three principal steps.

Word tokenization Vietnamese language processing using a tool in project VLSP of Ho Tu Bao *et al.* [2], include VnTokenizer tool [9] to analyze and separate words, JvnTagger [2] to tag word segmentations with labels of word form.

Machine translation consists of learning the similarity between the Vietnamese text and the structure of the SL, collecting the dataset of conversion syntax, and determining algorithms to use machine translation. The correlation between Vietnamese and its SL is indicated by these algorithms. During the conversion from Vietnamese texts tp VSL structure, sentences with different content might have the same grammatical structure.

Conversion to visual gesture Using avatar 3D JASigning and HamNoSys signal system [7] of ViSiCast project [11] has built the dataset expression of Vietnamese sciences with 3.873 Signing Gesture Markup Language (SiGML) codes [14].

3.2   Dataset Collection for Syntax Conversion

The Vietnamese text pre-processing consists of word separation and tagging. While the former is done by using VnTokenizer tool [9], the later is conducted by utilizing JvnTagger tool [2] Then the indexing process is performed by numbering the words' types according to their position in the sentence because the words' arrangement makes a difference in meaning. The final result of indexing process has been evaluated by the Center for Research and Education of the Deaf and Hard of Hearing (CED) which is based in Ho Chi Minh city to transform to the structures of sign language. The linguistic experts labeled word form for both Vietnamese sentences and VSL structures, see Fig. 2.

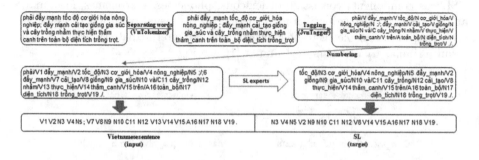

Fig. 2. The illustration of the process of building the training dataset.

3.3   Machine Translation Based on Examples

Nagao introduced the example-based machine translation (EBMT) algorithm as a Japanese-English mechanical translation framework [13], but soon it is effectively applicable to use in a wide variety of languages. The EBMT algorithm has several advantages such as any language system can be used, as long as there is a rich set of examples. Vocabulary and grammar are not required before the source and the target languages. The machine translation process of EBMT algorithms includes (1) Matching input data matching with existing data, (2) Identifying the data corresponding to the input data, and (3) recombining the output data that corresponds to the input data.

In regular text mining practice, the angle of similarity [3] is utilized to estimate the relationship between the text's length and its content. We denote σ as a triangle's edge. Then the other two edges are the size of two evaluated sentences. An empty string is denoted by ϕ. Given sentence x and y, a triangle with 3 edges, e.g. $\sigma(x, \phi), \sigma(y, \phi), \sigma(x, y)$, is formed. Hence, the angle θ_{xy} between x and y is calculated by half-sine formula as follows.

$$\sin \frac{\theta_{xy}}{2} = \frac{\sigma(x, y) - |\sigma(x, \phi) - \sigma(y, \phi)|}{2 \times \min\{\sigma(x, \phi), \sigma(y, \phi)\}} \tag{1}$$

3.4   Expressing Sign Language with 3D Character

The results after being converted to the magnetic structure of Vietnamese deaf people will be used by JASigning sign language interpretation software. JASigning software [10] is an Avatar 3D sign language speech tool developed by the University of East Anglia Computer Science. The software allows the conversion of wildcards in SiGML (a basic XML format in HamNoSys symbols) into 3D animated sign language. To allow the computer to be able to handle the syntax described by HamNoSys, East Anglian University has designed SiGML (Signing Gesture Markup Language) a modified version of XML, created by compiling HamNoSys to SiGML. Describing a SiGML symbol contains the same information as describing HamNoSys, but is in the form of a language that computers can handle. This information is synthesized, taking into consideration the description of the shapes of Avatar, the complete SiGML translation details can be ignored, as the default position of Avatar, over a period. A few seconds of each motion (SiGML simply expresses as fast, slow or normal speed), the result given in a more detailed SiGML translation, called SiGML extended, can describe a few general details that can be done with HamNoSys. Finally, the descriptions of the signs are represented by Avatar. JASigning software allows access on the website under copyright, It can be downloaded and used for evaluation purposes only. The use of software and virtual characters Anna and Marc can be contacted via the official software manufacturer's email address. Due to the difference between VSL and other sign languages, the authors have developed a dataset that includes more than 3873 words of sign language based on the database of Center for Educational Research CED [6], Vietnamese VsDic Sign Language Dictionary [12], and Institute of Educational Sciences [19].

4   Experiments

4.1   Data Collection

Our collected dataset includes 1286 samples. The word sign language representation is converted through SiGML codes designed for both hands with vertical or horizontal axis symbols, active or different dominant hands, symmetrical or asymmetric movements. The choice of non-manual expressions, such as mouth, body, shoulders, and eyes with 3873 words of sign language. Set of Vietnamese synonyms and sign language with the number of 12676 words shared by CED experts. To encourage reproducibility and further cooperation, the authors release the dataset publicly on the Github repository[1].

4.2   Evaluation Metrics

The authors use Weka tool [23] to analysis the information of dataset, conduct experiments and check the accuracy with the TER (Translation Error Rate) [20]. TER metric is the measurement published at prestigious AMTA 2006 conference. This metric performs a count of the number of edits in machine translation compared to the average size from the conversion result. The editing operations include "insert", "delete" and "replace" the words in the result string. The TER score evaluates the algorithms for the task of machine translation. The performance of translation algorithms is judged by the low TER score. While the minimum number of edits is denoted as Nb(op), Avreg Nref is the average size from references. Then, TER score is calculated as in Eq. (2).

$$TER = \frac{Nb(op)}{Avreg\ Nref} \tag{2}$$

In this paper, the authors compare the EBMT algorithm with two popular baselines, e.g. decision-tree learning machine - ID3 [17] and inductive learning algorithm - ILA [5]. Unfortunately, because of the limited allowance length of the article, we leave the explanation of these two algorithms for readers' self-interest.

4.3   Experimental Results

Phonetics in Vietnamese is controlled by signs, e.g. á à ả ã ạ â ă.. Therefore, we also investigate how our distributed framework handles this characteristic. The authors assess the performance of the model and accuracy in both cases, signed and unsigned. In the case of signed Vietnamese, the accuracy results are presented in Table 1, and the processing time of each algorithm is illustrated in Table 2. Similarly, in the case of unsigned Vietnamese, the accuracy is presented in Table 3, and the processing duration of each algorithm is illustrated in Table 4. From the results, one can conclude that there is no significant difference between the two scenarios and the EBMT works that well.

---

[1] https://github.com/raianrido/VSL.

Table 1. Experimental results on signed Vietnamese texts. The best result is in **bold**.

Algorithm	Samples	Number of rules	Accuracy (%)	Train time (s)	Test time (s)
ID3	1286	187	23.17	303.4	0.15
ILA	1286	696	74.83	14.04	1.46
EBMT	1286	721	**96.70**	0.02	2.58

Table 2. Processing time of algorithms in case of signed Vietnamese texts.

No.	Processing steps	Execution time (s)
1	Separation of words	3.23
2	Part of speech tagging	4.12
3	Assigning sequential numbers from the text	0.25

4.4 Remarks

For the original set of data, the number of samples is many and varied, consequently, the accuracy of the EBMT algorithm is higher. However, the execution time of the EBMT algorithm is longer, by matching process takes a lot of time for matching with many samples. When the dataset is large enough and diverse, the conversion Vietnamese grammar structure with the ILA and ID3 algorithms will shorten the time considerably. Due to the limited data set, the strength of the machine learning algorithm has not been fully exploited.

Table 3. Experimental results on unsigned Vietnamese texts. The best result is in **bold**.

Algorithm	Samples	Number of rules	Accuracy (%)	Train time (s)	Test time (s)
ID3	1286	187	23.26	336.24	0.01
ILA	1286	696	74.86	14.68	1.72
EBMT	1286	721	**96.84**	0.03	2.52

Table 4. Processing time of algorithms in case of unsigned Vietnamese texts.

No.	Processing steps	Execution time (s)
1	Separation of words	2.28
2	Part of speech tagging	3.89
3	Assigning sequential numbers from the text	0.28

5   Conclusion

To recapitulate, the authors have analyzed crucial steps in the workflow of Vietnamese sign language translation. The effectiveness and robustness of any VSL conversion systems primarily depend on picking the best classification algorithm. It ensures that the original messages are successfully conveyed into sign language eliminating the least meaning loss during the process. This research is the efforts of evaluating an effective classification algorithm that the authors recommend to be integrated into the workflow. The high accuracy score of 96.7% performing on a real-world dataset strongly indicates that EBMT is the appropriate choice. Considering the processing time, the EBMT algorithm is also the best solution. It takes as little as 0.02 and 0.03 s to complete the training phase in case of Vietnamese texts with signs and without signs respectively. The dataset is also publicly available for further research and cooperation.

Acknowledgment. Many thanks to the MOET's Research Project named "A study of proposing a solution to translate TV's News into 3D sign language animations for the deaf", project code no.: B2013-16-31, for funding this study. The authors would like to thank the Center for Research and Education of the Deaf and Hard of Hearing for supporting this study.

References

1. Ban, D.Q.: Ngu phap viet nam (vietnamese grammar) (2004)
2. Bao, H.T.: About handling vietnamese in information technology, vlsp-kc01/06-10 (2012)
3. Carroll, J.J.: Repetitions processing using a metric space and the angle of similarity. Centre for Computational Linguistics, Report No. 90/3 (1990)
4. Cooper, A.: Signed language sovereignties in Vietnam: deaf community responses to ASL-based tourism. In: Freidner, M., Kusters, A. (eds.) Init'sa Small World: International Deaf Spaces and Encounters, pp. 95–111 (2015)
5. Dumais, S., Platt, J., Heckerman, D., Sahami, M.: Inductive learning algorithms and representations for text categorization (1998)
6. Hanh, D.P.: Tu dien ngon ngu ky hieu viet (vietnamese sign language dictionary) (2010)
7. Hanke, T.: Hamnosys-representing sign language data in language resources and language processing contexts. LREC. **4**, 1–6 (2004)
8. Hien, D.T.: Co so cua viec day hoc cho nguoi khiem thinh bang ngon ngu ki hieu (the basis of teaching for deaf people in sign language). VNU J. Sci. Educ. Res. **29**(2) (2013)
9. Hông Phuong, L., Thi Minh Huyên, N., Roussanaly, A., Vinh, H.T.: A hybrid approach to word segmentation of Vietnamese texts. In: Martín-Vide, C., Otto, F., Fernau, H. (eds.) LATA 2008. LNCS, vol. 5196, pp. 240–249. Springer, Heidelberg (2008). https://doi.org/10.1007/978-3-540-88282-4_23
10. Jennings, V., Elliott, R., Kennaway, R., Glauert, J.: Requirements for a signing avatar. In: Proceedings of the Workshop on Corpora and Sign Language Technologies (CSLT), LREC, pp. 33–136 (2010)

11. Kennaway, R.: Experience with and requirements for a gesture description language for synthetic animation. In: Camurri, A., Volpe, G. (eds.) GW 2003. LNCS (LNAI), vol. 2915, pp. 300–311. Springer, Heidelberg (2004). https://doi.org/10.1007/978-3-540-24598-8_28

12. My, C.X.: Tu dien ngon ngu ky hieu viet vsdic (vietnamese sign language dictionary - vsdic) (2004)

13. Nagao, M.: A framework of a mechanical translation between Japanese and English by analogy principle. In: Artificial and Human Intelligence, pp. 351–354 (1984)

14. Ngon, N.C., Da, Q.L.: Application of hamnosys and avatar 3D jasigning to construction of vietnamese sign language animations. J. Sci. Technol. 1(11), 61–65 (2017)

15. Phi, L.T., Nguyen, H.D., Bui, T.Q., Vu, T.T.: A glove-based gesture recognition system for Vietnamese sign language. In: 2015 15th International Conference on Control, Automation and Systems (ICCAS), pp. 1555–1559. IEEE (2015)

16. Quach, L.D., Nguyen, C.N.: Conversion of the Vietnammese grammar into sign language structure using the example-based machine translation algorithm. In: 2018 International Conference on Advanced Technologies for Communications (ATC), pp. 27–31. IEEE (2018)

17. Quinlan, J.R.: C4.5 Programs for Machine Learning. Elsevier, Amsterdam (2014)

18. Sahoo, A.K., Mishra, G.S., Ravulakollu, K.K.: Sign language recognition: state of the art. ARPN J. Eng. Appl. Sci. 9(2), 116–134 (2014)

19. khoa hoc giao duc va Trung tam nghien cuu giao duc tre khuyet tat (Institute of educational science, V., research center for children with disabilities): Ky hieu cu chi dieu bo cua nguoi diec viet nam (gestures of vietnamese deaf people) (2002)

20. Snover, M., Dorr, B., Schwartz, R., Micciulla, L., Makhoul, J.: A study of translation edit rate with targeted human annotation. In: Proceedings of Association for Machine Translation in the Americas, vol. 200 (2006)

21. Van Hiep, N.: Vietnamese linguistics over 30 years of renovation and development. Vietnam Soc. Sci. (4), 41

22. Vietnam, U.: People with disabilities in Vietnam: key findings from the 2009 Viet Nam population and housing census (2011)

23. Witten, I.H., Frank, E., Hall, M.A., Pal, C.J.: Data Mining: Practical Machine Learning Tools and Techniques. Morgan Kaufmann, San Francisco (2016)

24. Woodward, J., Hoa, N.T.: Where sign language studies has led us in forty years: opening high school and university education for deaf people in viet nam through sign language analysis, teaching, and interpretation. Sign Lang. Stud. 13(1), 19–36 (2012)

Text Classification

Document Classification by Word Embeddings of BERT

Hirotaka Tanaka[(⊠)], Hiroyuki Shinnou[(⊠)], Rui Cao[(⊠)], Jing Bai[(⊠)],
and Wen Ma[(⊠)]

Department of Computer and Information Sciences, Ibaraki University,
4 -12 -1 Nakanarusawa, Hitachi, Ibaraki, Japan
{16t4032n,hiroyuki.shinnou.0828,18nd305g,19nd301r,
19nd302h}@vc.ibaraki.ac.jp

Abstract. Bidirectional Encoder Representations from Transformers
(BERT) is a pre-training model that uses the encoder component of
a bidirectional transformer and converts an input sentence or input sentence pair into word enbeddings. The performance of various natural
language processing systems has been greatly improved by BERT. However, for a real task, it is necessary to consider how BERT is used based
on the type of task. The standerd method for document classification by
BERT is to treat the word embedding of special token [CLS] as a feature
vector of the document, and to fine-tune the entire model of the classifier, including a pre-training model. However, after normalizing each the
feature vector consisting of the mean vector of word embeddings outputted by BERT for the document, and the feature vectors according
to the bag-of-words model, we create a vector concatenating them. Our
proposed method involves using the concatenated vector as the feature
vector of the document.

Keywords: BERT · Document classification · Word embeddings ·
BOW

1 Introduction

In recent years, many natural language processing tasks have demonstrated the
effectiveness of using pre-training models [7,8] D Various pre-training models
have been proposed; of them, Bidirectional Encoder Representations from Transformers (BERT) [2] has superior performance. BERT is a model in which 12 or
24 layers of multi-head attention used in transformer [10] are superimposed.
Parameter learning is performed under an unsupervised framework by solving
the following two tasks: the masked language model and next sentence prediction. By using the learned model, it is possible to obtain the word embeddings for
the input sentence or input sentence pair. Using BERT for actual tasks must be
considered based on the specific task. Several usage examples are provided in the
original article on BERT [2]. For example, in SST-2 (emotion analysis of a sentence), fine-tuning with the embedded representation of [CLS], a special token,

© Springer Nature Singapore Pte Ltd. 2020
L.-M. Nguyen et al. (Eds.): PACLING 2019, CCIS 1215, pp. 145–154, 2020.
https://doi.org/10.1007/978-981-15-6168-9_13

is demonstrated for input sentences as an input to the classifier. Based on this example, the above method can be used for document classification, regarding the entire document as one sentence.

However, because the input to BERT is generally a sentence, the above method may not be effective for documents. Here we classify sentences using BERT as feature-based approach, not fine-tuning. Specifically, after normalizing each feature vector, consisting of the mean vector of word embeddings outputted by BERT for the document and the feature vectors according to the bag-of-words (BOW) model, we create a vector concatenating them. In this paper, we propose a method of using this concatenated vector as the feature vector of the document.

In the experiments, the effectiveness of the proposed method is demonstrated using Japanese emotion analysis data from Webis-CLS-10[1]. In Sect. 5, we discuss feature vector weights from the BOW model and BERT. We also compare the feature vector from BERT with the feature vector obtained from the distributed representation of words.

2   Related Work

Document classification is a type of classification problem that can generally be solved using supervised learning. Therefore, many studies exist on the topic. there are many existing studies using deep learning, such as using CNN [5] and RNN [6]. However, in many natural language processing tasks including document classification, using a pre-training model is effective. There are two primary ways of using a pre-training model; the first is fine-tuning. Using this method, the information outputted from the pre-training model is used as the input of the network to solve the task, and the entire network including the pre-training model is targeted for learning. In this case, because a part of the pre-training model is already learned from a large amount of data, it is possible to learn a connected network using only a relatively small amount of data. OpenAI GPT [8] is a language model[2] that uses the decoder part of the transformer [10], which is a neural net translation whose main usage is fine-tuning. In addition, in ULMFiT [3], the pre-training model is set as a language model, and its model is fine-tuned for the target task.

Another use of the pre-training model is a feature-based. Approach, which involves using information outputted from the pre-training model as a feature for solving the target task. Distributed representations, such as word2vec, can also be regarded as a pre-training model. There are many studies that use distributed representations of words as a feature for solving tasks, including document classification. FastText constructs distributed representatios for Subword [1] and demonstrates via experiments that high-speed and high-precision document classification can be performed [4]. ELMo [7] is a model that outputs distributed representations of words with consideration of context. The entity is

---

[1] https://webis.de/data/webis-cls-10.html.

[2] The language model is also a type of pre-training model.

a two-layer bidirectional LSTM, that uses a large corpus to train language models. This becomes a pre-training model and can be used in the a feature-based approach.

BERT, which is used in this study, is an improvement of conventional pre-training models, and exceeds them in terms of performance for various tasks. Therefore, it is effective even for document classification, which is addressed in this study. However, BERT is primarily used in the form of fine-tuning, and in document classification, the input is not a sentence, but rather, a document; thus, standard fine-tuning is not effective. Therefore, in this study, we use BERT with a feature-based approach.

3 Proposed Method

3.1 BERT

The primary component of BERT is multi-head attention. Multi-head attention takes n word embeddings as input, converts each word embedding into a more appropriate embedding, and outputs it. In other words, the output consists of n transformed word embeddings.

Here, an outline of multi-head attention is provided. Because the basic is self-attention, three pairs of Q, K, V are inputs. We suppose that the word embeddings are m-dimensional. In multi-head attention, linear transformers are prepared for Q, K, V that compress m-dimensional vectors into $d_k (= m/k)$ dimensions. The entities Q, K, V are linear transformation matrices of $d_k \times d_k$. The input of multi-head attention is n m-dimensional vectors, which are converted to a matrix X of $n \times d_k$ by the above compressor, passed to Q, K, V. Then, the matrices XQ, XK, XV of $n \times d_k$ are generated. Expressing these as Q', K', V', self-attention is performed according to the following equation[3]:

$$\texttt{softmax}\left(\frac{Q'K'^T}{\sqrt{d_k}}\right)V'$$

which is a matrix of $n \times d_k$. Performing the above processing in parallel k matrices of $n \times d_k$ are created. By concatenating these, a matrix of $n \times m$ can be created. Further linear transformation of this to the same dimension produces the output of multi-head attention.

BERT is a model in which 12 or 24 layers of multi-head attention are stacked. BERT takes n@word embeddings as input and converts them into n more contextual word embeddedings.

3.2 Using Word Embedding of [CLS]

In BERT, a special token is used in an input word string; one such token [CLS]. [CLS] is placed at the first of the input word sequence, and the corresponding

---

[3] Scaled Dot-Product Attention.

word embedding is regarded as that of a sentence. In other words, the word embedding for [CLS] is used as a feature vector of a sentence for the classification of the sentence.

When using a document as the input to BERT, the document is regarded as one sentence. The word embedding of [CLS] is taken as the feature vector of the document, and the document classification can be performed by creating a classifier using this feature vector as input (see Fig. 1).

In this study, a three-layer neural network is used for the classifier (corresponding to the NN in Fig. 1). Specifically, each layer is a fully connected layer, and the input is linearly transformed into 400-, 50-, and 2-dimensional vectors, in that order. The sigmoid function is used as the activation function, and the softmax function is used for the output layer. The loss is calculated using the cross entropy error as the loss function, and the model is optimized using Adam.

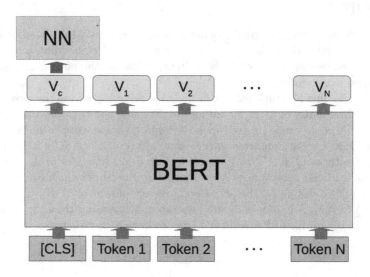

Fig. 1. Using word embedding of [CLS]

3.3 Using Word Embeddings

Input to BERT is essentially one or two sentences; even when the input is a document, it can be treated as a single sentence. However, in this case, the word embedding of [CLS] is not necessarily appropriate as the feature vector of the input document. Here, the average vector is obtained from the word embeddings outputted by BERT and is treated as the feature vector of the document (see Fig. 2). It should also be noted that in the word embeddings treated here, the word embedding for [CLS] and [SEP], special tokens of BERT, are excluded.

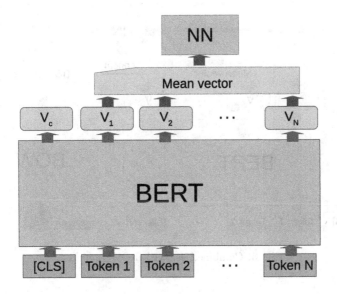

Fig. 2. Using word embeddings

3.4 Combined Use with BOW Models

The BOW is a method that uses word frequency as a feature vector of a document. The calculation method uses TF-IDF. Here, the feature vector of a document is constructed by combining the feature vector obtained by the BERT model and BOW model. For a given document d, let v_b be the feature vector obtained by the BOW model. We then divide d into morpheme units, and input the tokens to BERT to obtain a word embedding. Let v_m be the mean vector obtained from this word embeddings. These feature vectors, v_b and v_m, are normalized to unit vectors. In addition, let $[v_b; v_m]$ be the feature vector of d by concatenating v_b with v_m (see Fig. 3).

4 Experiments

4.1 Datasets

The dataset used in this experiment consists of Amazon review documents and is available on the following website: https://webis.de/data/webis-cls-10.html . We use emotion analysis data with ratings of 4 and 5 as positive, and 1 and 2 as negative.

The dataset has three domains: books, DVD, and music. There are 2,000 documents in the training data and 2,000 documents in the test data for each domain.

The feature words handled by BOW consist of 41,400 words that appear in 6,000 documents of all the training data.

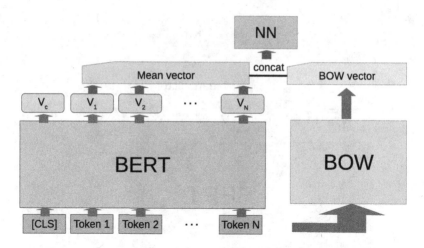

Fig. 3. Combined use with BOW models

4.2 Japanese BERT Pre-training Model

Japanese is included in the multilingual models of BERT that are available to the public. In addition, it is possible to use multilingual pre-training models for Japanese tasks. However, if these are used, the fundamental unit becomes a character; thus, the models are considered unsuitable. Therefore, in this study, we used the Japanese pre-training model available to the public on the following website by Kyoto University's Kurohashi and Kawahara laboratory as the pre-training model corresponding to Japanese:

http://nlp.ist.i.kyoto-u.ac.jp/index.php?BERT%E6%97%A5%E6%9C%AC %E8%AA%9EPretrained%E3%83%A2%E3%83%87%E3%83%AB

The text used as the input of this pre-training model has been divided into morpheme units by morphological analysis using Juman++, which is also available to the public by Kyoto University's Kurohashi and Kawahara laboratory.

4.3 Results

The experimental results are presented in Table 1 and Fig. 4. $v_{[CLS]}$ is the result when using the word embedding of [CLS] by BERT, while v_m is the result when using the mean vector obtained from the word embeddings of BERT (i.e., proposed method). The proposed method leads to superior results than the word embedding of [CLS]. v_b is the result when using the feature vector obtained by BOW, while $[v_b; v_m]$ is the result of the proposed method using the feature vector obtained by concatenating v_m and v_b. In this experiment, the proposed method led to the best results.

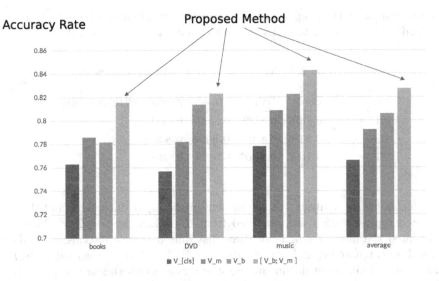

Fig. 4. Comparison of each method for each domain

Table 1. Accuracy rate of each method

	Books	DVD	Music	Average of three domains
$v_{[CLS]}$	0.7629	0.7567	0.7779	0.7658
v_m	0.7859	0.7818	0.8086	0.7921
v_b	0.7816	0.8135	0.8224	0.8058
$[v_b; v_m]$	**0.8156**	**0.8229**	**0.8427**	**0.8271**

5   Discussion

5.1   Feature Vector Weights from BOW Model and BERT

In this experiment, the proposed method $[v_b; v_m]$ produced the highest accuracy rate. In creating the vector $[v_b; v_m]$, v_b and v_m were normalized to unit vectors, and the ratio of the weight of each vector was 1:1.

Experiments were conducted in the case in which the ratio was changed to 1:2 or 2:1; the results are presented in Table 2. Adding weights to either v_b or v_m leads to superior results than not adding the weights; the accuracy rate is slightly higher if the v_m is weighted. This indicates that the information obtained from BERT is more useful than BOW.

5.2   Comparison with Distributed Representations

In this experiment, we used word embeddings by BERT. In this case, we conducted the same experiment using the distributed representation by nwjc2vec

[9] and compared the results. nwjc2vec provides distributed representation data of words obtained from the NINJAL Web Japanese Corpus (NWJC).

Table 2. Difference between v_b and v_m weights

$\|v_b\| : \|v_m\|$	Books	DVD	Music	Average of three domains
1:1	0.8156	0.8229	0.8427	0.8271
1:2	**0.8160**	0.8302	**0.8455**	**0.8306**
2:1	0.8157	**0.8307**	0.8440	0.8301

Morphological analysis was performed on the input document, and the distributed representation was obtained from the tokens by nwjc2vec. Then, we calculated the mean vector v_e of these distributed representations, normalized it, and concatenated v_b and v_e to create $[v_b; v_e]$. Then, $[v_b; v_e]$ was used as the feature vector of the input document, and the previous experiment was performed. The results are presented in Table 3.

Table 3. Comparison with proposed method using distributed representation (accuracy)

	Books	DVD	Music	Avarage of three domains
v_m	0.7859	0.7818	0.8086	0.7921
$[v_b; v_m]$	**0.8156**	0.8229	**0.8427**	**0.8271**
$[v_b; v_e]$	0.7931	**0.8288**	0.8262	0.8160

For the DVD domain, $[v_b; v_e]$ using distributed representations by nwjc2vec displays a high accuracy rate. However, in general, $[v_b; v_m]$ using word embeddings by BERT produces superior results. The distributed representations and the word embeddings by BERT express similar information; however, BERT is more useful than the distributed representations.

5.3　Combined Use of Lower Layer Word Embeddings

A neural network used in deep learning acquires a more general concept in the lower layer than in the upper layer. Here, we experiment combining use with the word embeddings of the layer that is below the top layer.

Let v_{-1} be the mean vector determined from the word embeddings of the top layer of BERT. It is equal to the vector represented by v_m. Let v_{-2} be the mean vector determined from the word embeddings of the layer below the top layer of BERT, and let v_b be the feature vector obtained by BOW. These feature vectors v_b, v_{-1}, and v_{-2} are normalized to unit vectors. In addition, let $[v_b; v_{-1}; v_{-2}]$ be the feature vector of the document by concatenating these feature vectors.

The results of the experiments using this feature vector are presented in Table 4.

Overall $[v_b; v_{-1}; v_{-2}]$ has higher accuracy than $[v_b; v_{-1}]$. Thus, combining lower-level word embeddings is effective; this point is further investigated below.

Table 4. Comparison with method using lower-level word embeddings (accuracy)

	Books	DVD	Music	Avarage of three domains
$[v_b; v_{-1}]$	**0.8156**	0.8229	0.8427	0.8271
$[v_b; v_{-1}; v_{-2}]$	0.8141	**0.8300**	**0.8474**	**0.8305**

5.4 Fine-Tuning

As mentioned in Sect. 1, fine-tuning is possible if document classification is performed using the word embedding of [CLS]. In this experiment, fine-tuning was performed on experimental data using `run_classifier.py` in the BERT source[4]. The results is presented in Table 5. v_f is the result of fine-tuning and is an improvement over $v_{[CLS]}$; however, it dose not greatly differ from v_m. The accuracy rate is much lower than $[v_b; v_m]$ (i.e., the proposed method).

Table 5. Comparison with fine-tuning (accuracy)

	Books	DVD	Music	Avarage of 3 domains
$v_{[CLS]}$	0.7629	0.7567	0.7779	0.7658
v_m	0.7859	0.7818	0.8086	0.7921
v_b	0.7816	0.8135	0.8224	0.8058
$[v_b; v_m]$	**0.8156**	**0.8229**	**0.8427**	**0.8271**
v_f	0.7894	0.7799	0.8019	0.7904

For document classification, feature-based BERT, as used in the proposed method, was demonstrated to be more effective than the fine-tuning of BERT using the word embedding of [CLS]. However, it is possible to construct a document feature vector from all word embeddings and use it to fine tune the entire model.

6 Conclusion

This study involves the classification of documents using BERT. In applying BERT to classification problems, although it is common to use the word embedding of the special token [CLS] to fine-tune the entire network including BERT,

---

[4] https://github.com/google-research/bert.

we used a feature-based approach. Specifically, after normalizing each mean vector of the word embeddings output by BERT and the feature vector by the BOW model, the connected vector was used as the feature vector of the document. Experiments using the Amazon dataset revealed the effectiveness of the proposed method. In the future, we will attempt to further improve the accuracy by testing the combination of lower level word embeddings, a fine-tuining approach, and dealing with long sequence data.

References

1. Bojanowski, P., Grave, E., Joulin, A., Mikolov, T.: Enriching word vectors with subword information. Trans. Assoc. Comput. Linguist. **5**, 135–146 (2017)
2. Devlin, J., Chang, M.W., Lee, K., Toutanova, K.: BERT: pre-training of deep bidirectional transformers for language understanding. arXiv preprint arXiv:1810.04805 (2018)
3. Howard, J., Ruder, S.: Universal language model fine-tuning for text classification. In: ACL 2018, pp. 328–339 (2018)
4. Joulin, A., Grave, E., Bojanowski, P., Mikolov, T.: Bag of tricks for efficient text classification. arXiv preprint arXiv:1607.01759 (2016)
5. Kim, Y.: Convolutional neural networks for sentence classification. In: EMNLP 2014, pp. 1746–1751 (2014)
6. Lai, S., Xu, L., Liu, K., Zhao, J.: Recurrent convolutional neural networks for text classification. In: AAAI 2015, pp. 2267–2273 (2015)
7. Peters, M., et al.: Deep contextualized word representations. In: NAACL 2018, pp. 2227–2237 (2018)
8. Radford, A., Narasimhan, K., Salimans, T., Sutskever, I.: Improving language understanding by generative pre-training. Technical report, OpenAI (2018)
9. Shinnou, H., Asahara, M., Kanako Komiya, M.S.: nwjc2vec: word embedding data constructed from NINJAL web Japanese corpus. Nat. Lang. Process. **24**(5), 705–720 (2017). (in Japanese)
10. Vaswani, A., et al.: Attention is all you need. In: Advances in Neural Information Processing Systems, pp. 5998–6008 (2017)

Deep Domain Adaptation for Low-Resource Cross-Lingual Text Classification Tasks

Guan-Yuan Chen[1,2](✉) [ID] and Von-Wun Soo[1,3] [ID]

[1] Institute of Information Systems and Applications, National Tsing Hua University, Hsinchu, Taiwan
guanyuan@gapp.nthu.edu.tw
[2] Telecommunication Laboratories, Chunghwa Telecom, Taoyuan, Taiwan
guanyuan@cht.com.tw
[3] Department of Computer Science, National Tsing Hua University, Hsinchu, Taiwan
soo@cs.nthu.edu.tw

Abstract. Recently, the data-driven machine learning approaches have shown their successes on many text classification tasks for a resource-abundant language. However, there are still many languages that lack of sufficient enough labeled data for carrying out the same specific tasks. They may be costly to obtain high-quality parallel corpus or cannot rely on automated machine translation due to unreliable or unavailable machine translation tools in those low-resource languages. In this work, we propose an effective transfer learning method in the scenarios where the large-scale cross-lingual data is not available. It combines transfer learning schemes of parameter sharing (parameter based) and domain adaptation (feature based) that are joint trained with high-resource and low-resource languages together. We conducted the cross-lingual transfer learning experiments on text classification on sentiment, subjectivity and question types from English to Chinese and from English to Vietnamese respectively. The experiments show that the proposed approach significantly outperformed the state-of-the-art models that are trained merely with monolingual data on the corresponding benchmarks.

Keywords: Cross-Lingual · Deep domain adaptation · Transfer learning

1 Introduction

Deep learning models have shown remarkable results in many text classification tasks [7,8,17,23], such as semantic analysis, subjectivity detection, etc. And most of neural text classifiers involved in these three procedures: (1) embedding words of sentences through looking up the word vector representations; (2) extracting representative features for the inputs through neural network based models, for example, CNN [9] or LSTM [6]; (3) prediction with output layers. Moreover, those methods usually require a sufficient quantity of labeled data for training.

© Springer Nature Singapore Pte Ltd. 2020
L.-M. Nguyen et al. (Eds.): PACLING 2019, CCIS 1215, pp. 155–168, 2020.
https://doi.org/10.1007/978-981-15-6168-9_14

For many languages, it can be difficult to get the plentiful labeled corpus. Therefore, most previous researches utilized machine translation (MT) to convert the high-resource language (e.g. English) corpus to the target language [3,15,20, 21,25]. However, these approaches may not succeed in the case of the MT is unreliable or even unavailable especially in such low-resource languages.

To leverage the knowledge of resource rich languages to improve the performance of target languages, in this paper, we introduce a transfer learning based framework that combine two different transfer learning method (Parameter-transfer, Feature-representation-transfer) [13] jointly train with English and target languages together, without relying on the parallel corpus or the large-scale cross-lingual data generated by MT.

First, we encode the words of the source and target language to vectors using the cross-lingual word embedding mapped from the source to the target space using Procrustes alignment with bilingual dictionaries [1,16]. We then train a couple of CNN models with one convolution layer for source and target word vector inputs. The couple of CNN models shares the same filter weights. Intuitively, since the word vectors have been aligned in a single vector space, it becomes possible to learn the common key-term for source and target language in each specific task automatically. Finally, we use adaptation layers that are embedded in a reproducing kernel Hilbert space to reduce the domain discrepancy by adding Maximum Mean Discrepancy (MMD) [4,5] losses, which can encourage the networks to learn representations that are domain invariant.

We experimentally show that the proposed cross-lingual (English-Chinese, English-Vietnamese) approach beat the state-of-the-art model [7] that train only with monolingual data. Additionally, since our architecture can utilize the main component of the neural based natural language classifiers, it may be possible to extend the monolingual classification models to cross-lingual setting without hard works.

2   Related Work

Cross-lingual text classification aim to utilize the labeled corpus from resource-rich language (e.g. English) to predict the category of samples of the low-resource language. The former researches induced the methods that utilized cross-lingual sentiment lexicons [12] or parallel bilingual data generated by machine translation [3,15,20,21,24,25]. These methods may be hard to extend to the other tasks facility or prone to fail while the outputs of machine translation is not reliable especially in the low-resource language.

Recent works have shown that the deep neural network has the abilities to learn the generalized well parameters and features across the different but related domain in computer vision field [11,13,18]. The transfer learning based method also becomes compelling in natural language processing. [26] train a NMT model with high-resource language pair then share parameters to the low-resource pair for improvement. [22] use the bilingual word embedding and a multi-task framework to improve the cross-lingual name-entity recognition (NER). [19] proposed

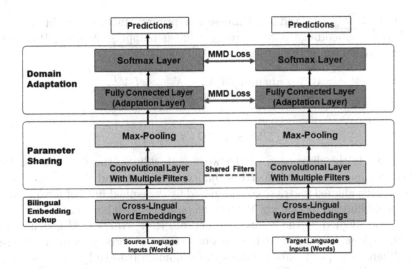

Fig. 1. An overview of our proposed architecture. Which is a couple of CNN models with cross-lingual inputs embedded through an aligned cross-lingual word embedding. The parameters of CNN filters are shared for source and target inputs. The adaptation and last output layers are added the MMD losses to minimize the discrepancy between domains.

a neural cross-lingual Spoken Language Understanding (SLU) model joint train two different languages together with the bilingual word embedding.

Although some works have demonstrated the success of transfer learning on cross-lingual NLP tasks [2], few works actually used the approach combining with sharing parameters and domain adaptation transfer schemes. In this paper, we integrated these two important transfer learning skills, with significant improvement and the scalability in many text classification tasks.

3 Methods

The architecture of the proposed model is shown in Fig. 1. It is a couple of CNN architecture for training source and target language sentences with corresponding labels $\{x_{i:n}^s, y^s\}^M$, $\{x_{i:n}^t, y^t\}^N$, where M, N are the number of source, target language examples ($M \gg N$). Either source or target language sentences is padded (if necessary) to the same length n.

3.1 Bilingual Embedding

In order to promote parameter sharing, we need to secure different word embeddings that are aligned in a single vector space, so that the words with the same meaning from two different languages could share similar vectors (for example, "*like*" and "喜歡" in Mandarin Chinese). A primary method to obtain cross-lingual word embeddings is by learning a linear mapping W using Procrustes alignment with a bilingual dictionary [1,16], such that

$$W^\star = \arg\min_{W \in O_l(\mathbb{R})} \|WX_s - X_t\|_{\mathrm{F}} = UV^T, \tag{1}$$

with

$$U\Sigma V^T = SVD\left(X_t X_s^T\right), \tag{2}$$

where, the l is the dimension of the word embeddings, the $O_l(\mathbb{R})$ is the real number $l \times l$ matrices with the orthogonality constraint, and the X_s and X_t are of the size $l \times n$ aligned matrices concatenated by the embeddings of n words from the parallel bilingual dictionary. We use the publicly available cross-lingual word embeddings[1] aligned by the mentioned method. Here, $\Lambda(x_i^d) \in \mathbb{R}^l$ is denoted as the l-dimensional aligned cross-lingual word vector of corresponding to the i-th word. Where, $d \in \{s,t\}$ indicate the specific domain (source language s, target language t). Furthermore, The $\Lambda(x_{i:n}^d) = \Lambda(x_1^d) \oplus \Lambda(x_2^d)... \oplus \Lambda(x_n^d)$ (where $\oplus$ is the symbol of concatenation) is represented as the padded sentence embedding.

3.2   Parameter Sharing

Since the word vectors of both languages have been mapped to the same vector space, intuitively, it may be possible to study methods to share the common key words (word vector weights) between these two languages. Therefore, in our approach, either source or target language sentence embeddings are sharing the same convolution filters. Thus, a convolution operation c_i with a trainable shared filter $\mathbf{w}_i \in \mathbb{R}^{hl}$ (h is the words window size of the filter $\mathbf{w}_i$) on both source and target languages can be defined as

$$c_i^d = f(\mathbf{w}_i \cdot \Lambda(x_{i:i+h-1}^d) + b_i). \tag{3}$$

where $b \in \mathbb{R}$ represents the bias value and f is an activation function. While we use the filter to calculate all possible windows of words in the sentence, we can obtain a feature map $\mathbf{c}^d = [c_1^d, c_2^d, ..., c_{n-h+1}^d] \in \mathbb{R}^{n-h+1}$ and the maximum value of this feature map $\hat{c}^d = max\{\mathbf{c}^d\}$ (calculated by the max-pooling layer). This process could be considered as to extract the significant features from the source and target language sentence embeddings by one filter. In this model, we use multiple filters to extract multiple features $\hat{\mathbf{c}}^d = [\hat{c}_1^d, \hat{c}_2^d, ..., \hat{c}_k^d] \in \mathbb{R}^k$, where k is the number of filters that are shared by these two different language inputs.

3.3   Domain Adaptation and Objective Function

One goal of domain adaptation is to minimize the cross-domain discrepancy and classification error by learning feature representations with high generalization

---

[1] https://github.com/facebookresearch/MUSE.

ability to reduce the discrepancy of marginal probability distributions between the source and target domain data points [13]. In order to achieve this goal, we consider to add the powerful two-sample test measure, the maximum mean discrepancy (MMD) [4,5] as the regularization terms to diminish the difference of the source and target domain distributions. The MMD is a distance measure between kernel embeddings of distributions in the reproducing kernel Hilbert space (RKHS) $\mathcal{H}$. The MMD of two samples of data sets (Z^s and Z^t) can be defined as

$$
\begin{aligned}
MMD^2(Z^s, Z^t) \\
= \left\| \underset{\mathbf{z}^s \in Z^s}{\mathbb{E}}[\phi(\mathbf{z}^s)] - \underset{\mathbf{z}^t \in Z^t}{\mathbb{E}}[\phi(\mathbf{z}^t)] \right\|_{\mathcal{H}}^2 \\
= \underset{\mathbf{z}^s_i, \mathbf{z}^s_j}{\mathbb{E}}[k(\mathbf{z}^s_i, \mathbf{z}^s_j)] + \underset{\mathbf{z}^t_i, \mathbf{z}^t_j}{\mathbb{E}}[k(\mathbf{z}^t_i, \mathbf{z}^t_j)] - 2\underset{\mathbf{z}^s, \mathbf{z}^t}{\mathbb{E}}[k(\mathbf{z}^s, \mathbf{z}^t)].
\end{aligned}
\tag{4}
$$

Here, $\mathbf{z}^s \in Z^s$, $\mathbf{z}^t \in Z^t$ are the source, target data point representations. $\phi(\cdot)$ is the feature map related with a reproducing kernel K:

$$
K(\mathbf{z}^s, \mathbf{z}^t) = \langle K(\mathbf{z}^s, \cdot), K(\mathbf{z}^t, \cdot) \rangle_{\mathcal{H}}.
\tag{5}
$$

The reproducing kernel K is a combination of multiple positive semi-definite kernels (details in Sect. 4.1), such that

$$
K(\mathbf{z}^s, \mathbf{z}^t) = \langle \phi(\mathbf{z}^s), \phi(\mathbf{z}^t) \rangle.
\tag{6}
$$

The computational complexity of the MMD (or MMD^2) is $O(n^2)$, where n depends on the batch size of the source and target input samples. It may not be satisfied while we calculate it on the large-scale corpus with a larger batch size setting (since larger batch size setting usually has more significant computational efficiency in comparison with the smaller ones, especially, on the neural network based models with a large-scale dataset). In order to handle this issue, we adopt the unbiased estimate of MMD [5], which can be defined as:

$$
MMD^2(Z^s, Z^t) = \frac{2}{n} \sum_{i=1}^{n/2} g_K(\mathbf{z}_i),
\tag{7}
$$

where the $\mathbf{z}_i$ defined as

$$
\mathbf{z}_i \triangleq (\mathbf{z}^s_{2i-1}, \mathbf{z}^s_{2i}, \mathbf{z}^t_{2i-1}, \mathbf{z}^t_{2i}),
\tag{8}
$$

and the

$$
\begin{aligned}
g_K(\mathbf{z}_i) \triangleq K(\mathbf{z}^s_{2i-1}, \mathbf{z}^s_{2i}) + K(\mathbf{z}^t_{2i-1}, \mathbf{z}^t_{2i}) \\
- K(\mathbf{z}^s_{2i-1}, \mathbf{z}^t_{2i}) - K(\mathbf{z}^t_{2i}, \mathbf{z}^s_{2i-1}).
\end{aligned}
\tag{9}
$$

The unbiased estimate of MMD costs $O(n)$ (it can be computed in linear time), which is desirable, while we take it on the training step with the deep learning models with the great amount of data (especially, the source domain corpus may be in a large-scale for improving the robustness).

In our case, we add the MMD regularization terms to the penultimate fully connected layer (adaptation layer) $z_{ada}^d = f(\mathbf{w}_{z_{ada}}^d \cdot \hat{\mathbf{c}}^d + b_{z_{ada}}^d)$ and the output layer $z_{out}^d = Softmax(\mathbf{w}_{z_{out}}^d \cdot z_{ada}^d + b_{z_{out}}^d)$ to form the objective function (the loss function) $\mathcal{L}$:

$$\mathcal{L} = \mathcal{L}_C(Z_{out}^s, Y^s) + \mathcal{L}_C(Z_{out}^t, Y^t) \\ + \lambda(MMD^2(Z_{ada}^s, Z_{ada}^t) + MMD^2(Z_{out}^s, Z_{out}^t)). \tag{10}$$

where, $\mathcal{L}_C(Z_{out}^s, Y^s)$ and $\mathcal{L}_C(Z_{out}^t, Y^t)$ are the average cross-entropy loss calculated with a batch of output layer results from source and target language sentences (Z_{out}^s, Z_{out}^t), and the corresponding labels (Y^s, Y^t) respectively. λ is the penalty coefficient, which determines the ratio between the classification and the MMD losses.

Table 1. The description of the dataset sizes. Where, *Classes* means the number of target classes, $Train_{src}$ and $Train_{tgt}$ mean the source language (English) and target languages (Chinese, Vietnamese) training set sizes, $Test_{src}$ and $Test_{tgt}$ mean the source and target languages testing set sizes respectively.

Dataset	Classes	$Train_{src}$	$Train_{tgt}$	$Test_{src}$	$Test_{tgt}$
SST-2	2	6920	872	1821	1821
Subj	2	8000	1000	1000	1000
TREC	6	4952	500	500	500

4   Experiments

Data Collection: For evaluation, we set up the low-resource text classification task scenarios by extending from three popular text classification corpora (benchmarks): (1) Stanford Sentiment Treebank (SST-2)[2] [17], a sentiment classification dataset with binary labels; (2) Subjectivity dataset[3] (Subj) [14], the task of the dataset is to classify sentences to be either subjective or objective; (3) TREC question dataset[4] [10], the goal of the dataset is to estimate whether a question belongs one of such 6 question types as abbreviation, entity, description, human, location and numeric value; in terms of the cross-lingual versions.

We used English corpora as the source language inputs (for increasing the comparability, we used the pre-processed dataset provided by [7][5]), and translated (by native speakers of the target language) the sentences split from English to Chinese and Vietnamese as the target low-resource language data for training, validating and testing (note that the training data for target language should be

---

[2] https://nlp.stanford.edu/sentiment/.

[3] http://www.cs.cornell.edu/people/pabo/movie-review-data/.

[4] http://cogcomp.org/Data/QA/QC/.

[5] https://github.com/harvardnlp/sent-conv-torch/tree/master/data.

significantly less than source language for setting up the low-resource text classification task scenarios). Table 1 shows the summary statistics of the datasets. We
also ensured that there are no identical data in the source and target (by translation) language training set (i.e. the situation of parallel cross-lingual inputs
will not occur).

4.1   Training Evaluation Setup

Baseline Model: First, we compare our bilingual models with a state-of-the-art
CNN based text classification model (the baseline) trained merely on monolingual data that was introduced by [7]. Then, we also conduct the ablation studies
to evaluate the effects by the two different transfer approaches: parameters sharing, domain adaptation.

Hyperparameters: For all datasets, we set up: 100 convolution filters for each
window size of 1, 2, 3, and 4 respectively; mini-batch size of 50; a total of 25 training epochs for each model. Most of these settings are based on [7] for increasing
the comparability. At the bilingual methods we set the penalty coefficient λ to 1.0,
use 5 Gaussian kernels $k(\mathbf{z}^s, \mathbf{z}^t) = e^{-||\mathbf{z}^s - \mathbf{z}^t||^2 / \gamma}$ with an equal weight and differ-
ent bandwidths (interpolated from $2^{-8}\gamma$ to $2^8\gamma$) to select the kernel k, where the
bandwidth γ is calculated from the median pairwise distances of data points [5].

Table 2. Accuracy results (rounded to 2 decimal places) of proposed models compare
to the baseline. **Sharing Filters Only**: the model that only shared the filter weights
between source and target language inputs without adding MMD losses. **MMD losses
Only**: the model added MMD losses without share parameters. **Joint**: the model used
both approaches.

Model	SST-2 (zh)	SST-2 (vi)	Subj (zh)	Subj (vi)	TREC (zh)	TREC (vi)
Baseline	0.61	0.64	0.78	0.82	0.43	0.55
Ours (Sharing Filters Only)	0.71	0.72	0.82	0.85	0.73	0.69
Ours (MMD losses Only)	0.70	0.72	0.83	0.85	0.61	0.70
Ours (Joint)	**0.76**	**0.74**	**0.86**	**0.88**	**0.78**	**0.71**

4.2   Experimental Results

Quantitative Analysis: Table 2 shows the results on three datasets in different target languages. As we can see, the model combined two proposed transfer
methods (**Joint**) that has the highest accuracy at all different settings. It is due
to the leverage of the proposed methods that may well transfer the source domain
knowledge to improve the performance on different target domain (zh, vi).
Further analysis finds that either sharing parameters or adding the domain discrepancy losses alone is conducive in increasing the accuracy on the target tasks,
and while integrating these two methods together, the performance can be further enhanced.

We also note that even though the baseline model performs fairly poor on Chinese corpus than Vietnamese, the transfer learning based model can achieve competitive or even better result in the Chinese cases. The **Joint** model also performs competitively on English accuracy: 0.83, 0.93, 0.92 on SST-2, Subj and TREC when jointly train with Chinese; and 0.83, 0.94, 0.94 on SST-2, Subj and TREC when jointly train with Vietnamese respectively (the accuracy of the baseline CNN model [7] that trained with English data only are 0.87, 0.93, 0.93 on SST-2, Subj and TREC). The accuracy drop of the proposed models on the English corpus may due to the data of Chinese and Vietnamese are significantly less than English, but the performances are still at a fair level (even better in some cases).

Qualitative Analysis: We also conducted the Qualitative Analysis by comparing the per class F1-score (%) for the baseline (the model training merely with mono-lingual data), **Sharing Filters Only** (the model that only shared the filter weights between source and target language inputs without adding MMD losses), **MMD losses Only** (the model added the terms of MMD losses but without sharing parameters), and the **Joint** (the model used both approaches) on the three text classification benchmarks on Chinese and Vietnamese.

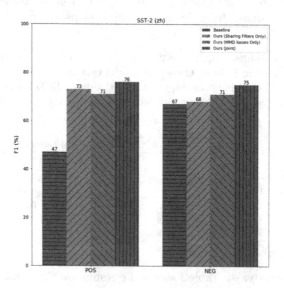

Fig. 2. The results of per class F1-score (%) for comparing models on the SST-2 (zh) testing set.

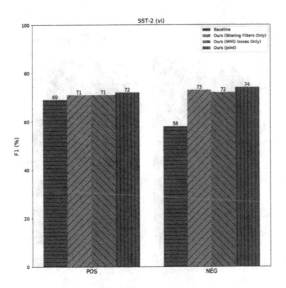

Fig. 3. The results of per class F1-score (%) for comparing models on the SST-2 (vi) testing set.

Qualitative Analysis on SST-2: Fig. 2 and Fig. 3 show the per class F1-score (%) of different models on the SST-2 (zh) and SST-2 (vi). Where the POS and NEG mean the positive and negative labels of the sentiment classification. The numbers of POS and NEG samples are 909 and 912, which is the approximately balanced case for classification. As we can see, the baseline model performed relatively poor and may produce the biased predictions on the classes. And all of the proposed methods can help to improve the performance of each category on each target languages. Especially, there are obvious improvements on the category that the baseline model performed worst (with the most +29% improvement on the POS category F1-score on the SST-2 of zh and with the most +16% improvement on the NEG category F1-score on the SST-2 of vi, while conducting the "Joint" approach). In addition, the approach using the two transfer learning schemes (**Joint**) had the greatest improvement on overall.

Qualitative Analysis on Subj: Fig. 4 and Fig. 5 show the per category F1-score (%) of different models on the Subj (zh) and Subj (vi). The numbers of Subjective and Objective samples are 485 and 515, which is the approximately balanced case for classification. The most improvements (comparing with the baseline model) of transfer learning based models are 8% on the "Objective" and "Subjective" categories on Subj (zh) and 7% on the "Objective" category on Subj (vi) with the **Joint** approach. And, either on the case of Chinese or Vietnamese, the **Joint** approach also performed best on the Subj on overall.

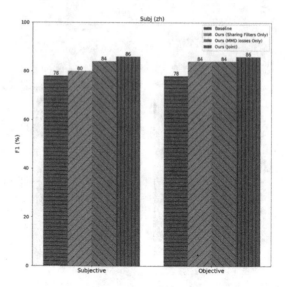

Fig. 4. The results of per class F1-score (%) for comparing models on the Subj (zh) testing set.

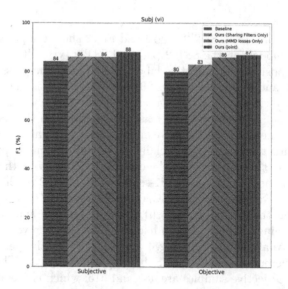

Fig. 5. The results of per class F1-score (%) for comparing models on the Subj (vi) testing set.

Qualitative Analysis on TREC: Fig. 6 and Fig. 7 show the per category F1-score (%) of different models on the TREC (zh) and TREC (vi). This task is quite different with SST-2 and Subj. First, different from the other two corpora all are the task of binary classification, there are six different question types for classification in the TREC dataset, moreover, the numbers of samples for each categories are unbalanced. Where, the DESC (description), ENTY (entity), ABBR (abbreviation), HUM (human), LOC (location), and NUM (numeric value) are the types of questions. The numbers of test samples for each question types are DESC: 138, ENTY: 94, ABBR: 9, HUM: 65, LOC: 81, and NUM: 113, which is the relatively unbalanced case for classification. Figure 6 and Fig. 7 report that besides the model training with monolingual data (baseline) perform worst on all results of all categories, it may predict extremely unsatisfactory (0% F1-score) on the category (ABBR) which has fewer samples in the dataset. We can also find that the parameter sharing based approaches (**Sharing Filters Only** and **Joint**) can overcome this problem better than other methods (**Sharing Filters Only**: +62% and +71%, **Joint**: +80% and +71% improvement of ABBR F1-score on the zh and vi TREC test set). It may due to that since the parameter sharing based approach shared the parameters between domains which could help to improve the performance on the case of rare sample categories on the target domain by leveraging these parameters that tuned with source domain data relatively well. Although the parameter sharing method can enhance the performances of the target languages significantly, while adding the MMD regularization terms (**Joint**), the results can be further improved on overall.

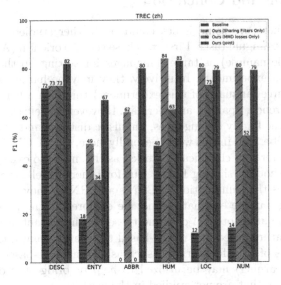

Fig. 6. The results of per class F1-score (%) for comparing models on the TREC (zh) testing set.

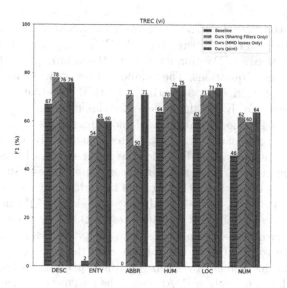

Fig. 7. The results of per class F1-score (%) for comparing models on the TREC (vi) testing set.

5   Discussion and Conclusion

Discussion: There are some issues could be further studied about transfer learning and cross-lingual tasks. First, we chose the two oriental (Asia) languages (Chinese and Vietnamese) as target languages for setting up the cross-lingual text classification experiments. Intuitively, they may be harder to bridge with English (the source language of this experiment) than the other Western Languages such as French, Spanish, and German. However, the effect of the language characteristics, the family of languages, linguistic distance and transfer learning has not been deliberated in this work. Secondly, here, we use CNN (which is simple but powerful for text classification tasks) as the model to explore the parameter sharing method by sharing filters. But for other models, such as LSTM's, GRU's, as well as the combination of CNN's and RNN's may be other alternatives for sharing parameters that are not be explored in the paper. Moreover, although we have attempted to combine these two important transfer learning methods (the feature-based and model-based transfer learning), there are some other approaches of transfer learning (such as the instance-based and relation-based transfer learning) may be utilized to help to bridge the discrepancy of different languages that are not studied in the work.

In spite of the above issues, according to the experimental results, the proposed simple (efficient) method not only works well on the important text classification problems on target languages but also achieved the competitive results on English (source language) while comparing to the strong baseline models without relying on the large-scale parallel cross-lingual corpora.

Conclusion: In this paper, we introduce a transfer learning based approach integrated the parameters sharing and domain adaptation schemes. The experimental results showed that this bilingual training method outperforms the state-of-the-art model training merely with monolingual data by leveraging the knowledge learned from the source language (English) to target languages (Chinese and Vietnamese) and sustains competitive performance in English. Furthermore, since the architecture utilizes many key components in the standard neural text classifiers, it can be easily extended to many other text classification tasks for cross-lingual settings that are left as future work to be explored.

References

1. Conneau, A., Lample, G., Ranzato, M., Denoyer, L., Jégou, H.: Word translation without parallel data. arXiv preprint arXiv:1710.04087 (2017)
2. Eriguchi, A., Johnson, M., Firat, O., Kazawa, H., Macherey, W.: Zero-shot cross-lingual classification using multilingual neural machine translation. arXiv e-prints arXiv:1809.04686, September 2018
3. Glavas, G., Franco-Salvador, M., Ponzetto, S.P., Rosso, P.: A resource-light method for cross-lingual semantic textual similarity. Knowl. Based Syst. **143**, 1–9 (2018)
4. Gretton, A., Borgwardt, K.M., Rasch, M.J., Schölkopf, B., Şmola, A.: A kernel two-sample test. J. Mach. Learn. Res. **13**(1), 723–773 (2012). http://dl.acm.org/citation.cfm?id=2503308.2188410
5. Gretton, A., et al.: Optimal kernel choice for large-scale two-sample tests. In: Pereira, F., Burges, C.J.C., Bottou, L., Weinberger, K.Q. (eds.) Advances in Neural Information Processing Systems, vol. 25, pp. 1205–1213. Curran Associates, Inc. (2012). http://papers.nips.cc/paper/4727-optimal-kernel-choice-for-large-scale-two-sample-tests.pdf
6. Hochreiter, S., Schmidhuber, J.: Long short-term memory. Neural Comput. **9**(8), 1735–1780 (1997). https://doi.org/10.1162/neco.1997.9.8.1735
7. Kim, Y.: Convolutional neural networks for sentence classification. In: Proceedings of the 2014 Conference on Empirical Methods in Natural Language Processing (EMNLP), pp. 1746–1751. Association for Computational Linguistics (2014). https://doi.org/10.3115/v1/D14-1181, http://www.aclweb.org/anthology/D14-1181
8. Lai, S., Xu, L., Liu, K., Zhao, J.: Recurrent convolutional neural networks for text classification. In: Proceedings of the Twenty-Ninth AAAI Conference on Artificial Intelligence, AAAI 2015, pp. 2267–2273. AAAI Press (2015). http://dl.acm.org/citation.cfm?id=2886521.2886636
9. Lecun, Y., Bottou, L., Bengio, Y., Haffner, P.: Gradient-based learning applied to document recognition. In: Proceedings of the IEEE, pp. 2278–2324 (1998)
10. Li, X., Roth, D.: Learning question classifiers. In: Proceedings of the 19th International Conference on Computational Linguistics - Volume 1, COLING 2002, pp. 1–7. Association for Computational Linguistics, Stroudsburg (2002). https://doi.org/10.3115/1072228.1072378
11. Long, M., Cao, Y., Wang, J., Jordan, M.I.: Learning transferable features with deep adaptation networks. In: Proceedings of the 32Nd International Conference on International Conference on Machine Learning - Volume 37, ICML 2015, pp. 97–105. JMLR.org (2015). http://dl.acm.org/citation.cfm?id=3045118.3045130

12. Mohammad, S., Salameh, M., Kiritchenko, S.: Sentiment lexicons for Arabic social media. In: Chair, N.C.C., et al. (eds.) Proceedings of the Tenth International Conference on Language Resources and Evaluation (LREC 2016). European Language Resources Association (ELRA), Paris, France, May 2016
13. Pan, S.J., Yang, Q.: A survey on transfer learning. IEEE Trans. Knowl. Data Eng. **22**(10), 1345–1359 (2010). https://doi.org/10.1109/TKDE.2009.191
14. Pang, B., Lee, L.: A sentimental education: sentiment analysis using subjectivity. In: Proceedings of ACL, pp. 271–278 (2004)
15. Shi, H., Ushio, T., Endo, M., Yamagami, K., Horii, N.: A multichannel convolutional neural network for cross-language dialog state tracking. In: 2016 IEEE Spoken Language Technology Workshop (SLT), pp. 559–564 (2016)
16. Smith, S.L., Turban, D.H.P., Hamblin, S., Hammerla, N.Y.: Offline bilingual word vectors, orthogonal transformations and the inverted softmax. CoRR abs/1702.03859 (2017). http://arxiv.org/abs/1702.03859
17. Socher, R., et al.: Recursive deep models for semantic compositionality over a sentiment treebank. In: Proceedings of the 2013 Conference on Empirical Methods in Natural Language Processing, pp. 1631–1642. Association for Computational Linguistics, Stroudsburg, October 2013
18. Tzeng, E., Hoffman, J., Zhang, N., Saenko, K., Darrell, T.: Deep domain confusion: maximizing for domain invariance. CoRR abs/1412.3474 (2014). http://arxiv.org/abs/1412.3474
19. Upadhyay, S., Faruqui, M., Tur, G., Hakkani-Tur, D., Heck, L.: (almost) zero-shot cross-lingual spoken language understanding. In: Proceedings of the IEEE ICASSP (2018)
20. Wan, X.: Using bilingual knowledge and ensemble techniques for unsupervised chinese sentiment analysis. In: Proceedings of the Conference on Empirical Methods in Natural Language Processing, EMNLP2008, pp. 553–561. Association for Computational Linguistics, Stroudsburg (2008). http://dl.acm.org/citation.cfm?id=1613715.1613783
21. Wan, X.: Co-training for cross-lingual sentiment classification. In: Proceedings of the Joint Conference of the 47th Annual Meeting of the ACL and the 4th International Joint Conference on Natural Language Processing of the AFNLP: Volume 1 - Volume 1, ACL 2009, pp. 235–243. Association for Computational Linguistics, Stroudsburg (2009). http://dl.acm.org/citation.cfm?id=1687878.1687913
22. Wang, D., Peng, N., Duh, K.: A multi-task learning approach to adapting bilingual word embeddings for cross-lingual named entity recognition. In: Proceedings of the Eighth International Joint Conference on Natural Language Processing (Volume 2: Short Papers), pp. 383–388. Asian Federation of Natural Language Processing (2017). http://aclweb.org/anthology/I17-2065
23. Zhang, X., Zhao, J., LeCun, Y.: Character-level convolutional networks for text classification. In: Proceedings of the 28th International Conference on Neural Information Processing Systems - Volume 1, NIPS 2015, pp. 649–657. MIT Press, Cambridge (2015). http://dl.acm.org/citation.cfm?id=2969239.2969312
24. Zhou, X., Wan, X., Xiao, J.: Attention-based LSTM network for cross-lingual sentiment classification. In: EMNLP (2016)
25. Zhou, X., Wan, X., Xiao, J.: Cross-lingual sentiment classification with bilingual document representation learning. In: ACL (2016)
26. Zoph, B., Yuret, D., May, J., Knight, K.: Transfer learning for low-resource neural machine translation. In: EMNLP (2016)

Multi-task Learning for Aspect and Polarity Recognition on Vietnamese Datasets

Dang Van Thin[1,3(✉)], Duc-Vu Nguyen[1,3], Kiet Van Nguyen[2,3],
Ngan Luu-Thuy Nguyen[2,3], and Anh Hoang-Tu Nguyen[2,3]

[1] Multimedia Communications Laboratory, University of Information Technology,
Ho Chi Minh City, Vietnam
{thindv,vund}@uit.edu.vn
[2] University of Information Technology, Ho Chi Minh City, Vietnam
{kietnv,ngannlt,anhnht}@uit.edu.vn
[3] Vietnam National University, Ho Chi Minh City, Vietnam

Abstract. In this paper, we present a multi-task deep neural network model based on deep learning approach to address two tasks in document-level aspect-based sentiment analysis in Vietnamese datasets: Aspect Detection and Sentiment Polarity Classification. Besides, we conduct a series of experiments with different pre-trained embeddings on two VLSP benchmark datasets for the restaurant and hotel domains. Experimental results show that our model outperforms than stable baseline approaches on both tasks for two domains and achieves the new state-of-the-art F1-scores with 64.78% for the restaurant domain and 70.90% for the hotel domain.

Keywords: Aspect-Based Sentiment Analysis · Multi-task learning · Aspect detection · Aspect Polarity · Vietnamese corpus

1 Introduction

With the rapid growth of users comments on e-commerce sites, there has been a great deal of interest in Opinion Mining with the aim of developing text mining techniques. For example, in e-commerce, customers often tend to express their positive or negative feelings are posted towards different aspects of products and services. These reviews are treated as valuable free resource and goodness measure for both organizations and consumers. In previous studies, we found that it is important to analyze customer reviews. If we deal with these reviews well enough, the extracted information will contribute to an additional base to support product improvement and customer care campaigns. Besides, market research, survey, and feedback campaigns are now widely available. Therefore, it is essential to study techniques that can analyze these comments automatically

© Springer Nature Singapore Pte Ltd. 2020
L.-M. Nguyen et al. (Eds.): PACLING 2019, CCIS 1215, pp. 169–180, 2020.
https://doi.org/10.1007/978-981-15-6168-9_15

to extract useful information. Aspect-based sentiment analysis (ABSA) is one of the tasks of exploring the sentiment on each mentioned aspect in the comment [7, 8].

The ABSA task aims to extract sentiment polarity (e.g., positive, negative, and neutral) toward a target, which is the aspect of one specific entity. The ABSA can be divided into three difference sub-tasks [7]: Aspect Detection, Opinion Target Expression, and Sentiment Polarity. However, in this paper, we focus on two sub-tasks: Aspect Detection and Aspect Polarity in Vietnamese datasets. The goal of aspect detection task is to discover one or more of the aspects in the list of pre-defined category. On the other hand, the aspect polarity task aims to identify the sentiment polarity toward the detected aspect in the review. For example, given a review *"Nhà hàng có không gian thoáng mát nhưng đồ ăn thì không được lắm "* (*"The restaurant has a cool space but the food is not very good"*), for aspect detection task, the output is *"ambience#general"* and *"food#quality"*; while the aspect polarity, the sentiment for the aspect *"ambience#general"* is *positive* and the sentiment for the aspect *"food#quality"* is *negative*. From the example, we can see that the extracted information from users comments plays an important role in the development of business organizations.

In this paper, we present a multi-task model based on deep learning framework to address two sub-tasks for aspect-based sentiment analysis problem in Vietnamese datasets for two domains: aspect detection and aspect polarity. We evaluate our model on two VLSP benchmark datasets for the restaurant and hotel domains. Besides, we also experiment with our method with various Vietnamese pre-trained word embeddings. Experimental results over two benchmark corpora illustrate that our model achieves superior performance than other approaches.

The remainder of this paper is organized as follows: the related work is introduced in Sect. 2. Section 3 presents our multi-task model, while Sect. 4 gives an overview of two datasets and detailed experiments. The result and discussion are provided in Sect. 5. Finally, Sect. 6 summaries the paper and provides suggestion directions for future research.

2   Related Work

The aspect-based sentiment analysis has attracted the research community in recent years. [3] proposed LSTM-based deep multi-task learning for aspect term extraction on sentence-level of user reviews. As our work on document-level multi-aspect sentiment classification, [17] dealt with this task as a machine comprehension problem in which pseudo question-answer pairs are constructed by a small number of aspect-related keywords and aspect ratings. Another research treated the aspect category classification and aspect term extraction on restaurant reviews by building a neural multi-task model, which is done by [16]. They treated aspect detection as multi-label classification task and aspect term extraction as a sequential labeling task. Two tasks are trained simultaneously in a BiLSTM-CNN model. Their experiment results demonstrate the effectiveness of the proposed model across three public datasets. [5] built a MATEPC model to address the aspect term extraction and aspect polarity classification task, simultaneously. They treated two tasks similarity as named entity recognition, and their model is a combination of BiLSTM and CRF. Their experimental results on two SemEval 2014 datasets proved the effectiveness of their proposed model. [9] jointly trained the aspect detection and their polarity in the end-to-end neural networks. They conducted experiments with various deep neural network and word representations on GermEval 2017 dataset. From their result, the combination of convolutional neural network with fastText embedding established the new state of the art on the GermEval dataset. [1] provided a comparative review of deep neural network model for the aspect-based sentiment analysis problem. In their paper, there is much deep learning model, and other approaches are summarized and reported for ABSA tasks.

In Vietnamese, many relevant studies laid the foundation for our research. [13] proposed an effective model for aspect-based sentiment analysis, which showed positive results. They extracted reliable word pairs (sentiment - aspect) by combining a sentiment dictionary and syntactic dependency rules and determine the sentiment of polarity based on ontology. In addition, [4] proposed a sequence-labeling scheme associated with BRNN and CRF to extract opinion target and its sentiment. They also built a Vietnamese ABSA from Youtube reviews for smartphone products. Their corpus includes 1,728 sentences for training and 370 sentences for testing. In 2018, [6] built two benchmark corpora and organized a first shared-task about aspect-based sentiment analysis in Vietnamese. Two datasets are crawled from the restaurant and hotel website. To the best of our knowledge, these are two first benchmark corpus[1] for the Vietnamese research community on aspect-based sentiment analysis problem. Based on these datasets, [11] proposed a transformation model to address two tasks on two shared-task VLSP 2018 corpora. Their model consists of two components corresponding to each task where each component is composed of multiple binary classification SVM with handcraft features. Their model achieved the best scores on two datasets for two domains in the 2018 shared-task. Then, [10] also developed a DCNN neural network to deal with aspect detection task and achieved the new scores on two VLSP datasets.

With the difference from previous works, [12] introduced an annotated corpus of sentence-level aspect category for the restaurant domain. Besides, they also translated the annotated corpus in English to Vietnamese using Google translate. After that, they conducted a series of experiments on their corpus with multiple SVM binary classifiers combined a variety of features (n-grams, word embedding, and word expansion).

---

[1] http://vlsp.org.vn/resources.

3 Our Approach

3.1 Problem Formulation

We define the task of aspect-based sentiment analysis as follow: For each domain (e.g restaurant and hotel), given a text s (a sentence or a document) and a list of tuples $\{a, p\}_{i=1}^{T}$ where p is the polarity (positive, negative or neutral) of the aspect a. The aspect a is identified by the an entity and an attribute. The text s can be assigned to one or up to T tuples. An example review for the restaurant domain is given in Table 1.

Table 1. A example of input and output of aspect-based sentiment analysis for the restaurant domain in Vietnamese.

Input	Output
Vietnamese: Đồ ăn rất ngon và nhân viên nhiệt tình. Tuy nhiên không gian nhỏ. Nhà hàng này phù hợp cho gia đình đi ăn uống. *English: The food is very delicious and the staff is enthusiastic. However, the ambience is small. This restaurant is suitable for families.*	{FOOD & QUALITY, positive} {SERVICE # GENERAL, positive} {AMBIENCE # GENERAL, negative} {RESTAURANT # GENERAL, positive}

3.2 Multi-task Model

Pre-processing. Preprocessing text is a vital step in natural language processing application. Appropriate pre-processing can help us to increase performance significantly. It is difficult to feed original data directly into deep learning model without pre-processing. Therefore, we design a chain of steps to process data in order to utilize more information from the original texts. The steps are described as follow:

– **Step 1**: HTML tags, accent marks, and icons are removed.
– **Step 2**: Urls, numbers, with special words such as price values, the hashtag is normalized.
– **Step 3**: Because a Vietnamese word can be composed more than one syllables (compound words and reduplicated words), such as "thức_ăn","nhân_viên" (staff). For that reason, it is necessary to segment the text input into meaningful words. In this paper, we use the Vietnamese Core NLP Toolkit Pyvi[2] for word segmentation.

---

[2] https://pypi.org/project/pyvi/.

Model Architecture. In this section, we describe our multi-task model to address two sub-tasks: aspect detection and aspect polarity detection in aspect-based sentiment analysis. For the aspect detection, many existing works used multiple binary classifiers [11,12] or casted it into a multi-label classification problem [10]. We propose an end-to-end architecture that automatically identifies aspect and polarity at the same time. In this paper, we model the system output as a list of A vectors, where A is the number of aspects which varies according to the application domain. In our case, A equals 12 and 34 for the restaurant and hotel domains, respectively. Each vector has four elements in which only one element is assigned 1; the other elements are 0. Four states of the vector are decoded as *None*, *Positive*, *Neutral* or *Negative* for a specific aspect. "None" indicates that the input does not contain the aspect.

The architecture of our model is illustrated in Fig. 1. It includes four main layers: Embedding layer, BiLSTM layer, CNN layer, and Fully Connected Layer. Each word w_i in the input is transformed into a low-dimensional vector $x_i \in R^d$, where d is length of the vector. The input is represented by an embedding matrix $C \in R^{d \times l}$, where l is the length of the input. Next, a Bi-directional Long Short-term Memory Network (BiLSTM) is applied on the embedding matrix C to extract contextual features of the input h = $\{h_1, h_2, ..., h_l\}$. The BiLSTM computes forward $\vec{h}_t$ and backward $\overleftarrow{h}_t$ hidden sequences from the LSTM output:

$$\vec{h}_t = \vec{f}(x_t, \vec{h}_{t-1}) \tag{1}$$

$$\overleftarrow{h}_t = \overleftarrow{f}(x_t, \overleftarrow{h}_{t+1}) \tag{2}$$

$$h_t = \vec{h}_t \oplus \overleftarrow{h}_t \tag{3}$$

where $\vec{f}$ and $\overleftarrow{f}$ are the forward and the backward functions of LSTM for one time step; $\oplus$ denotes the function of concatenation; x_t is the input at the i-th time step.

From BiLSTM layer, we put these represented features into a one-dimensional convolution layer with various kernels. Each convolutional kernel $H \in R^{d \times w}$ with a window of w word is applied on the BiLSTM output to produce a feature map c:

$$c_i = f(\sum(h[*, i : i + w - 1] \circ H) + b) \tag{4}$$

$$c = [c_1, c_2, ..., c_{l-w+1}] \tag{5}$$

where $b \in R$ is a bias term, i ranges from 1 to $(l - w + 1)$ and f is the non-linear ReLU function. To capture the most important features from the feature map, we concatenate a global max-over-time pooling over the feature map and global average pooling. Equation (6) describes the global max-pooling and Eq. (7) describes the global average pooling:

$$\hat{c}_m = \max_{1 \le j \le l-w+1} \{c_j\} \tag{6}$$

$$\hat{c}_a = \frac{1}{l - w + 1} \sum_{j=1}^{l-w+1} c_j \tag{7}$$

With k filters and different window sizes w, we have the pooling operation as follow:

$$g_{max} = [\hat{c}_{m1}, \hat{c}_{m2}, ..., \hat{c}_{mk}] \tag{8}$$

$$g_{average} = [\hat{c}_{a1}, \hat{c}_{a2}, ..., \hat{c}_{ak}] \tag{9}$$

$$g = g_{max} \oplus g_{average} \tag{10}$$

The generated feature (g) is passed to a fully connected layer to calculate the score vector $\hat{y}^{(a)}$ for each aspect $a \in T$ with softmax normalization:

$$\hat{y}^{(a)} = softmax(W^{(a)} \cdot g + b^{(a)}) \tag{11}$$

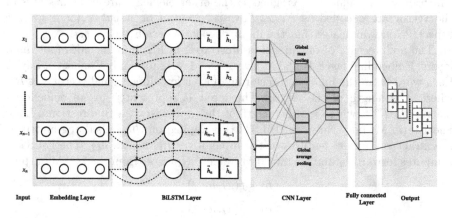

Fig. 1. The end-to-end architecture for aspect detection and corresponding polarity.

Therefore, we predict an aspect a and its polarity in one step:

$$\hat{output}^{(a)} = arg\max_i \hat{y}_i^{(a)} \ where \ i = 0, 1, 2, 3 \tag{12}$$

We take cross entropy as the loss function and it is summed over all aspects in T:

$$Loss = - \sum_{a \in T} \sum_i y_i^{(a)} \cdot \log(\hat{y}_i^{(a)}) \tag{13}$$

4   Datasets and Experiments

We evaluate the effectiveness of our model on two VLSP shared task 2018[3] benchmark datasets in the Vietnamese language for the restaurant and hotel domain [6]. The datasets were collected from e-commerce sites and manually annotated. For each domain, the whole dataset is divided into training, development, and testing dataset. There are 12 and 34 aspect categories for the restaurant and hotel domain, respectively. Table 2 shows the summary statistics of the two VLSP datasets in detail.

---

[3] http://vlsp.org.vn/resources-vlsp2018.

4.1   Experimental Setup

For our experiments, we choose our hyperparameters based on the development dataset for two domains. We set 300 as the word embedding dimension in the random case and 128 as the recurrent units of BiLSTM layer. The embedding layer will be updated in the training process. In the CNN layer, we use the filter windows of 3, 4, 5, and each filter has 128 feature maps. Two fully connected layers are set 600 and 300, respectively. The ReLu activation is used as the nonlinear function in convolution layers and fully connected layers. Dropout is used both on the embedding layer and on two fully connected layers with the probability of 0.5 and 0.2, respectively.

Table 2. Summary statistics for two VLSP Datasets after pre-processing. N denotes the Number Of Reviews, A denotes the Number Of Aspects, l denotes the Average Review Length, $|V|$ denotes the Vocabulary Size, $|I|$ denotes the Number Of Vocabulary in the Test Dataset that do not exist in the Training Dataset.

Domain	Dataset	N	A	l	$|V|$	$|I|$
Restaurant	Train	2,961	9,034	54	5,168	-
	Dev	1,290	3,408	50	3,398	1,702
	Test	500	2,419	163	3,375	1,729
Hotel	Train	3,000	13,948	47	3,908	-
	Dev	2,000	7,111	23	2,745	1,059
	Test	600	2,584	30	1,631	346

For the training, we use a mini-batch size of 50, and the fixed length of the input is the max length value of segmented review in the training dataset. We choose the number of epochs is 100. Adam optimizer is used for optimization with a starting learning rate of 0.001.

4.2   Pre-trained Word Embedding

To initialize for word embeddings in our model, we experiment the list of various pre-trained word embedding that trained on Vietnamese corpora. The list included BPEmb Subword embedding[4], News corpus word2vec embedding[5] and Wikipedia Word2vec, Elmo, Bert and Multi embedding[6] [15]. The BPEmb Subword embedding [2] is a pretrained subword unit embeddings that are trained on Wikipedia corpus. For word2vec embedding, we use two pre-trained embeddings that are trained on the different corpus: Wikipedia [15] (denoted as w2v_wiki) and News corpus [14] (denoted as w2v_news). In addition, [15] also provided the

---

[4] https://github.com/bheinzerling/bpemb.
[5] https://github.com/sonvx/word2vecVN.
[6] https://github.com/vietnlp/etnlp.

Table 3. The information of pre-trained word embeddings.

Word embedding	Corpus	Dim	Tokens	Vocab
BPEmb (BPEmb)	Wikipedia	300	-	200,000
Word2vec (w2v_news)	News	400	974,393,244	439,056
Word2vec (w2v_wiki)	Wikipedia	300	114,997,587	16,690
BERT (Bert)	Wikipedia	768	114,997,587	14,905
ELMO (Elmo)	Wikipedia	1,024	114,997,587	21,930
Multiple(Multi)	Wikipedia	2,392	114,997,587	21,590

pre-trained context embedding like as Elmo and Bert. In this work, we compare our model with various embeddings to explore the effectiveness of pre-trained embeddings for two domains.

For the pre-trained subword embedding, we calculate the word vector by mean of sub-word vectors. For out-of-vocabulary words, we initialize a vector from a uniform distribution with range $[-0.01, 0.01]$. Table 3 shows the information of pre-trained word embeddings in our experiments, and Table 4 presents the statistics of pre-trained embedding in the training dataset.

Table 4. Vocabulary statistics in the training dataset of pre-trained embeddings.

Domain	Embedding	#words in embedding	#words do not in embedding
Restaurant	BPEmb	7,873 (100%)	0
	w2v_news	6,143 (78.03%)	1,730
	w2v_wiki	3,788 (48.11%)	4,085
	Bert	3,516 (44.66%)	4,357
	Elmo	3,548 (45.06%)	4,325
	Multi	3,547 (45.05%)	4,326
Hotel	BPEmb	6,051 (100%)	0
	w2v_news	5,270 (87.09%)	781
	w2v_wiki	3,683 (60.87%)	2,368
	Bert	3,455 (57.10%)	2,596
	Elmo	3,488 (57.64%)	2,563
	Multi	3,477 (57.70%)	2,549

5   Result and Discussion

Table 5 presents the results of our experiments as well as other methods for the restaurant and hotel domain. Each model is repeated for five times. The results are reported by the mean with the standard deviation.

Table 5. Results of our model against other methods for the restaurant domain.

Restaurant				
Task	Method	Precision	Recall	F1-score
Aspect Detection	VLSP best submission [11]	79.00	76.00	77.00
	DCNN [10]	84.75	76.48	**80.40**
	Our model + Random	80.45	76.34	78.34 ± 0.37
	Our model + BPEmb	82.41	77.12	79.68 ± 0.61
	Our model + w2v_news	82.02	77.51	79.70 ± 0.54
	Our model + w2v_wiki	80.71	75.98	78.28 ± 0.52
	Our model + Bert	80.58	77.40	78.96 ± 0.3
	Our model + Elmo	79.91	77.49	78.68 ± 0.47
	Our model + Multi	80.91	77.98	79.42 ± 0.54
Aspect + Polarity	VLSP best submission [11]	62.00	60.00	61.00
	DCNN [10] •	-	-	-
	Our model + Random	64.67	61.37	62.98 ± 0.72
	Our model + BPEmb	66.87	62.59	64.66 ± 0.6
	Our model + w2v_news	66.66	63.00	**64.78** ± 0.45
	Our model + w2v_wiki	64.33	60.55	62.38 ± 0.67
	Our model + Bert	64.19	61.64	62.89 ± 0.65
	Our model + Elmo	63.08	61.17	62.11 ± 0.67
	Our model + Multi	65.21	62.84	64.00 ± 0.67

Table 6. Results of our model against other methods for the hotel domain.

Hotel				
Task	Method	Precision	Recall	F1-score
Aspect Detection	VLSP best submission [11]	75.00	64.00	69.00
	DCNN [10]	82.35	59.75	69.25
	Our model + Random	81.40	70.23	75.40 ± 0.42
	Our model + BPEmb	83.44	70.45	76.40 ± 0.12
	Our model + w2v_news	82.68	70.50	76.11 ± 0.65
	Our model + w2v_wiki	83.20	69.58	75.79 ± 0.54
	Our model + Bert	83.68	70.14	76.31 ± 0.19
	Our model + Elmo	82.37	71.84	76.75 ± 0.45
	Our model + Multi	84.03	72.52	**77.85** ± 0.66
Aspect + Polarity	VLSP best submission [11]	66.00	57.00	61.00
	DCNN [10]	-	-	-
	Our model + Random	73.22	63.17	67.82 ± 0.58
	Our model + BPEmb	75.89	64.06	69.48 ± 0.38
	Our model + w2v_news	75.14	64.07	69.16 ± 0.38
	Our model + w2v_wiki	75.47	63.12	68.75 ± 0.43
	Our model + Bert	75.94	63.65	69.25 ± 0.38
	Our model + Elmo	74.69	65.13	69.58 ± 0.54
	Our model + Multi	76.53	66.04	**70.90** ± 0.74

Main Results: Overall, our models competitive performances in these tasks for two domains. Especially in the main task (aspect with polarity), our baseline model (the model with random embedding) outperforms the state-of-the-art methods on the two datasets. In detail, our baseline model achieved the F1-score of 62.98% (increase +1.95%) and F1-score of 67.82% (increase +6.48%) for the restaurant and hotel domain, respectively. These results demonstrate that our model is effective in different domains. With using the pre-trained embedding, our best model (+ w2v_new) achieves 2.60% and 3.75% higher F1-scores than [11]. As shown in Table 5, for the hotel domain, it is very impressive that our model (+ Multi embedding) achieves at least 5.00% and 6.48% higher results than previous methods for two tasks, respectively.

Impact of Pre-trained Embedding: We also experiment our model with various pre-trained embeddings. The results are also shown in Table 5. Surprisingly, two different pre-trained embeddings achieve high F1-scores for two domains. The word2vec embedding trained on the News corpus shows the slightly higher performance than other embeddings for the restaurant domain. On the contrary, the Multi embedding that is created by concatenating a set of embeddings [15] outperforms all models for the hotel domain. The reason for the difference in results is because of the comments and vocabularies for the two domains. In the restaurant domain, the user's comment often refers to a variety of names of dishes, drinks or expresses personal opinions, while in the hotel, the user usually focuses on reviewing the overall of aspects. For that reason, the pre-trained embedding on News corpus will catch the word vectors (e.g., name of food, ...) better than other for restaurant domain. On the other hand, in the hotel domain, the vocabulary is mainly used for standard, grammatical, and less particular words; therefore, the pre-trained embeddings on Wikipedia will have better results. Based on the results of comparing the performance of embeddings in the previous work [15], our experiment also evaluated the same as [15] of Elmo, Bert, and Multi pre-trained embedding. As reported by [15], the Multi embedding can boost significantly the performance for downstream tasks. Specifically, we notice that the BPEmd subword embedding which is trained on Wikipedia corpus covert all vocabulary in the testing dataset to vector and achieved the positive results on both datasets.

6   Conclusion and Future Work

In this paper, we have proposed a multi-task model based on a deep neural network to solve two tasks at the same time of aspect-based sentiment analysis problem. Our model established the new state-of-the-art score for two benchmark datasets in Vietnamese with F1-score of 64.78% for the restaurant domain and F1-score of 70.90% for the hotel domain. Besides, we experimented our model with different pre-trained word embedding variations (such as word2vec, fastText, Bert, etc.). For the restaurant domain, our model combine with the word2vec embedding that trained on News corpus achieved the best score for two subtasks, while for the hotel domain, our model with Multi embedding[7] (is the concatenation of four embeddings: word2vec, fastText, Elmo, and Bert embedding) achieved the best F1-score on the testing dataset. For future work, we analyze the experimental results of our model on two datasets in detail. Also, we explore the effectiveness of various pre-trained embeddings with our method for two datasets.

Acknowledgment. Authors would like to thank VLSP 2018 organizers for providing two VLSP datasets for this research. We also would like to thank the anonymous reviewers for their valuable comments. This research is funded by the University of Information Technology - Vietnam National University Ho Chi Minh City under grant number B2019-26-01.

References

1. Do, H.H., Prasad, P., Maag, A., Alsadoon, A.: Deep learning for aspect-based sentiment analysis: a comparative review. Expert Syst. Appl. **118**, 272–299 (2019)
2. Heinzerling, B., Strube, M.: BPEmb: tokenization-free pre-trained subword embeddings in 275 languages. In: chair, N.C.C., et al. (eds.) Proceedings of the Eleventh International Conference on Language Resources and Evaluation (LREC 2018). European Language Resources Association (ELRA), Miyazaki, Japan (2018)
3. Li, X., Lam, W.: Deep multi-task learning for aspect term extraction with memory interaction. In: Proceedings of the 2017 Conference on Empirical Methods in Natural Language Processing, pp. 2886–2892. Association for Computational Linguistics, Copenhagen, Denmark (2017)
4. Mai, L., Le, B.: Aspect-based sentiment analysis of vietnamese texts with deep learning. In: Nguyen, N.T., Hoang, D.H., Hong, T.-P., Pham, H., Trawiński, B. (eds.) ACIIDS 2018. LNCS (LNAI), vol. 10751, pp. 149–158. Springer, Cham (2018). https://doi.org/10.1007/978-3-319-75417-8_14
5. Nguyen, H., Shirai, K.: A joint model of term extraction and polarity classification for aspect-based sentiment analysis. In: 2018 10th International Conference on Knowledge and Systems Engineering (KSE), pp. 323–328, November 2018. https://doi.org/10.1109/KSE.2018.8573340
6. Nguyen, H., et al.: VLSP shared task: sentiment analysis. J. Comput. Sci. Cybernet. **34**(4), 295–310 (2019). http://vjs.ac.vn/index.php/jcc/article/view/13160
7. Pontiki, M., et al.: Semeval-2016 task 5: aspect based sentiment analysis. In: Proceedings of the 10th international Workshop on Semantic Evaluation (SemEval 2016), pp. 19–30 (2016)
8. Pontiki, M., Galanis, D., Papageorgiou, H., Manandhar, S., Androutsopoulos, I.: Semeval-2015 task 12: Aspect based sentiment analysis. In: Proceedings of the 9th International Workshop on Semantic Evaluation (SemEval 2015). pp. 486–495 (2015)
9. Schmitt, M., Steinheber, S., Schreiber, K., Roth, B.: Joint aspect and polarity classification for aspect-based sentiment analysis with end-to-end neural networks. In: Proceedings of the 2018 Conference on Empirical Methods in Natural Language Processing, pp. 1109–1114. Association for Computational Linguistics, Brussels, October–November 2018. https://www.aclweb.org/anthology/D18-1139

---

[7] https://github.com/vietnlp/etnlp.

10. Thin, D.V., Nguyen, V.D., Nguyen, K.V., Nguyen, N.L.: Deep learning for aspect detection on Vietnamese reviews. In: 2018 5th NAFOSTED Conference on Information and Computer Science (NICS), pp. 104–109 (2018)
11. Thin, D.V., Nguyen, V., Kiet, N., Ngan, N.: A transformation method for aspect-based sentiment analysis. J. Comput. Sci. Cybernet. **34**(4), 323–333 (2019). https://doi.org/10.15625/1813-9663/34/4/13162. http://vjs.ac.vn/index.php/jcc/article/view/13162
12. Thuy, N.T.T., Bach, N.X., Phuong, T.M.: Cross-language aspect extraction for opinion mining. In: 2018 10th International Conference on Knowledge and Systems Engineering (KSE), pp. 67–72 (2018)
13. Tran, T.K., Phan, T.T.: Towards a sentiment analysis model based on semantic relation analysis. Int. J. Synthet. Emot. (IJSE) **9**(2), 54–75 (2018)
14. Vu, T., Nguyen, D.Q., Nguyen, D.Q., Dras, M., Johnson, M.: VnCoreNLP: a Vietnamese natural language processing toolkit. In: Proceedings of the 2018 Conference of the North American Chapter of the Association for Computational Linguistics: Demonstrations, pp. 56–60. Association for Computational Linguistics, New Orleans, Louisiana (2018)
15. Vu, X., Vu, T., Tran, S.N., Jiang, L.: ETNLP: a toolkit for extraction, evaluation and visualization of pre-trained word embeddings. CoRR abs/1903.04433 (2019). http://arxiv.org/abs/1903.04433
16. Xue, W., Zhou, W., Li, T., Wang, Q.: MTNA: a neural multi-task model for aspect category classification and aspect term extraction on restaurant reviews. In: Proceedings of the Eighth International Joint Conference on Natural Language Processing (Volume 2: Short Papers), pp. 151–156. Asian Federation of Natural Language Processing, Taipei (2017)
17. Yin, Y., Song, Y., Zhang, M.: Document-level multi-aspect sentiment classification as machine comprehension. In: Proceedings of the 2017 Conference on Empirical Methods in Natural Language Processing, pp. 2044–2054. Association for Computational Linguistics, Copenhagen (2017)

Evaluating Classification Algorithms for Recognizing Figurative Expressions in Japanese Literary Texts

Mateusz Babieno[1]([ID]), Rafal Rzepka[1,2]([ID]), and Kenji Araki[1]

[1] Graduate School of Information Science and Technology, Hokkaido University,
Kita-ku, Kita 14, Nishi 9, Sapporo 060-0814, Japan
{mbabieno,rzepka,araki}@ist.hokudai.ac.jp
[2] RIKEN AIP, Chuo City, Japan

Abstract. In this paper we introduce computational model for recognizing figurative expressions in Japanese language. As a part of the training data we use the set of almost 26,000 Japanese sentences comprising both similes and metaphors. These were collected manually from literary texts and hence constitute trustworthy and probably the largest existing resource of its kind. We use the data for classification task to evaluate its usability for figurativeness recognition. Precision score achieved by one of the classifiers utilized during the test shows that our model outperforms state-of-the-art methods in this aspect.

Keywords: Figurative language · Metaphor processing · Classification

1 Introduction

Enabling computers to recognize figurative use of language and making them understand its sense should be considered a necessary step in the process of further developing NLP systems and artificial intelligence in general. In many cases linguistic expressions can be understood both literally and figuratively depending on the context. To better understand the problem, try to imagine some ill-tempered person telling its humanoid robot to *get out of his face*. Robot would certainly become confused given it had not been taught the figurative sense of this expression. Example very similar to the one just presented is provided by Neuman in [10]. He rightly states that - although structurally very similar - the order *Give me the bottle* is drastically different from *Give me a break*. If the addressee was a robot, in order to make it understand what are we talking about, we would have to provide it with natural language understanding module that could differentiate between figurative and non-figurative senses.

Machine translation is another example of a field in which recognizing and understanding figurative senses of linguistic expressions becomes necessary. If one tries to translate *my sweet girl* into Japanese using probably the most popular automatic translation web service [16], he receives *watashi no amai onna no*

© Springer Nature Singapore Pte Ltd. 2020
L.-M. Nguyen et al. (Eds.): PACLING 2019, CCIS 1215, pp. 181–188, 2020.
https://doi.org/10.1007/978-981-15-6168-9_16

ko as an output. This sounds unnatural and does not convey the illocutionary meaning (cf. [1] for elaborate description of illocution).

In the introduction to our recent work [2] we have briefly discussed the problem of metaphor-related terminology vagueness. Definitions of nearly all concepts associated with figurativeness available in most popular dictionaries seem to be highly ambiguous. Becoming acquainted with a vast number of theoretical works devoted to metaphor analysis does not resolve the problem: definitions of terms related to figurative language vary from author to author. Because of it we have decided to treat as figurative every linguistic unit, be it word or multiword expression (MWE), used in non-literal sense. It follows that not only metaphors (creative and conventional ones) but also metonymies, synecdoches, similes, personifications and their opposites, idioms, proverbs and alike: all of them should be considered objects of our investigation. Only this kind of generous approach seems to allow one to avoid contradictions during identifying figurative expressions.

One of the ultimate goals of our current study is to create a model performing with precision high enough to entrust it with the task of automatic figurative expressions recognition. Getting it close to 100% would grant us with the possibility of using online texts as an exhaustless resource of such utterances. Since insufficient amount of data is still one of the biggest problems for various machine learning algorithms to encounter, creating a large corpus of figurative expressions would become a great contribution for other researchers in the field as well.

In this paper Japanese words are transcribed into alphabet using Hepburn romanization; they are indicated with italics.

2   Related Work

Recent years have seen considerable increase in works related to computational processing of figurative language. Numerous papers related to metaphor identification has been published.

Bulat et al. [4] proposed metaphor identification method using semantic representations of 541 concepts built based on their most relevant attributes (for example *is large, has four legs, has fur, has hair, has hooves, is brown* for MOOSE). This database was previously constructed and presented by McRae et al. [9]. Classification experiment conducted by the authors showed that utilizing property norm-based semantic representations of this kind can yield even better results than popular dense semantic representations (F1 score of 77% vs. 73% & 75% vs. 73%). Authors used Support Vector Machines classifier.

Tsvetkov and colleagues [13] conducted classification experiment using dataset comprising English expressions of subject-verb-object (SVO) and adjective-noun (AN) syntactic patterns (F1 score of 79% and 85% for SVO and AN types respectively). They performed pivot translation using bilingual dictionaries and show that their model yields satisfying results also for Spanish (76%, 72%), Farsi (75%, 74%) and Russian (84%, 77%). Experiment was conducted using Random Forest classifier. In [13] it is suggested that the model may serve as a tool for classification independently of the target language. Scores achieved by the algorithm in question are indeed very good but in our opinion it requires even deeper investigation in order to prove model's universality. The reason for that can be found in Dobrovol'skij and Piirainen's book [5]. Authors remind the reader of Whorfian concept of Standard Average European (SAE) according to which all languages of Europe (and that means also English, Russian and Spanish) belong to the same *sprachbund* (linguistic area) and show many common features with numerous very similar idiomatic expressions as their prominent example. Since semantically nearly identical idioms can be found also in Finnish, it can be proved that genetic affiliation or linguistic topology is of a very little importance when it comes to figurative language. On the other hand figurative expressions found in Japanese are often very different from their SAE's counterparts. It is because for a long time Japanese has been culturally isolated from Western influences that its phraseology reveals many unique cultural components. For example multiple figurative expressions containing words related to *kimono* - traditional Japanese garment - can be found. *Kare ni sode no shita o tsukau* 'offer him a bribe' includes the idiom *sode no shita* 'bribe' (lit. 'under the sleeve') evoking an image of *kimono*. What is peculiar about this piece of clothing is that it has long, wide sleeves allowing its user to hide something (also money or a gift) inside. Trying to translate this idiom into English without giving up on speaking figuratively one should say 'under the table' rather than 'under the sleeve'.

When it comes to research devoted to Japanese language processing, Hashimoto et al. [8] proposed lexical knowledge base for idioms recognition. In the introduction to this article we have already touched upon the problem, but Hashimoto et al. also point out that AI's inability to identify idioms often results in mistranslation. Japanese sentence *Kare wa mondai o kaiketsu ni hone o otta* ('He struggled to solve the problem') erroneously translated as 'He broke his bone to the resolution of a question' serve the authors as the example to illustrate the issue. In this case translation algorithm failed to recognize *hone o oru* 'to make efforts; to take pains' (lit. 'to break a bone') as being used figuratively. Another thing indicated by the authors as posing a problem in machine translation is transformation of fixed phrases, specifically phraseme's transformation occurring by e.g. adding another word in-between its constituents (eg. *kare wa yaku ni tatsu* 'he is helpful' vs. *kare wa yaku ni sugoku tatsu* 'he is very helpful'), its constituents' conjugation (e.g. *kare wa yaku ni tatta* 'he was helpful') and so forth. They point out that in many cases idioms do not allow morphological modifications (passivization, causativization, etc.) and therefore it follows that any change within their inner structure might be a hint for that they are being used literally; for example in case of *kare ga hone o orareta* 'he got his bone broken [by someone/something]' literal reading is possible while figurative is not. Unfortunately in this case such restriction does not hold for causativization, since *kare ni hone o oraseru* can be understood in both literal 'make him break a bone' and figurative 'make him struggle/make effort' senses. Authors divided idioms into 3 categories: a) non-transformable and non-ambiguous (e.g. *mizu mo shi-tataru* 'extremely handsome'), b) transformable and non-ambiguous (e.g. *yaku ni tatsu* 'to be helpful'), c) transformable and ambiguous (e.g. *hone o oru* 'to make efforts'). Rule-based classification algorithm presented by the authors achieved F1 score of 80% for idioms belonging to the class c). The model sometimes failed to distinguish true senses of ambiguous expressions which interpretations depend on larger context.

In order to resolve aforementioned problems, few years later together with Kawahara, Hashimoto has presented results for improved version of his algorithm by conducting classification experiment using large corpus of sentences collected solely from the Internet [7]. Authors gathered 146 Japanese idioms, this time only those which senses are ambiguous (e.g. above-mentioned *hone o oru* fig. 'to struggle; to make effort', lit. 'to break a bone' or *goma o suru* fig. 'to flatter', lit. 'to crash sesame'). The model was improved by using both common word sense disambiguation features (e.g. hypernyms of words in the surrounding context, etc.) and idiom-specific features (those related to morphological constraints); the features were extracted using JUMAN and KNP parsers. Improved version of the classification algorithm achieved accuracy scores of 89.25% and 88.86% depending on the features used.

3   Datasets and Annotation

3.1   Datasets for Classification Experiment

Training data prepared for the classification experiment was comprising 25,947 sentences used figuratively and corresponding number of sentences with sup-posedly literal senses. *Supposedly*, because while no genre is completely free of figurative language, it is texts belonging to encyclopedic descriptions, news arti-cles and formal discussions that are likely to be used in literal senses for the most part. First group was obtained from the digitalized dictionary of figurative Japanese expressions edited by Onai [11]. The second was collected using three different sources: Japanese Wikipedia dump [21], local assembly minutes [17] and articles from Livedoor News [19]. As for the dictionary of figurative expressions, our laboratory was provided with its abridged electronic version not available on the market, primarily comprising 25,979 items.

Our test data was initially composed of modest number of 50 sentences ran-domly taken from Japanese novels' texts freely available at Aozora Bunko digital library [15]. In the process of annotation described in greater detail in the fol-lowing subsection, 41 sentences got labeled as literal and only 9 as metaphorical. In order to make these numbers equal, we have transferred another 32 sentences with non-literal senses from train- to test-set. Eventually we have used 51,894 sentences (25,947 figurative and 25,947 literal) for training and 82 sentences (41 figurative and 41 literal) for testing.

As a preliminary step towards conducting the experiment, we have parsed sentences from both sets with JUMAN [18], removed all ASCII characters (as this format does not function as the encoding standard for Japanese text), punctuation marks, simile indicators and stopwords. Table 1 shows the full list of simile indicators as well as units we decided to consider as stopwords. The latter do not convey much meaning in isolation while being among 100 most frequent items in the dictionary.

Table 1. Simile indicators and stopwords removed during preprocessing

Simile indicators	Stopwords
yō, mitai, ppoi, gotoku, gotoshi, gotoki, kurai, marude, sazo, samo, ikanimo, atakamo, hodo	*suru, itasu, iru, oru, aru, naru, kore, sore, kono, sono, kara, koto, mono, nado, keredomo, demo, watashi, da, noda, nda, nani, gozaimasu, no, wo, ni, to, ga, wa, ka, o, go, de, mo, ka, sa, yo, he, nai, nu, ne, na, tai, tsu, rareru, reru, seru, ru*

3.2   Annotation

We have asked three Japanese native speakers to annotate 50 sentences taken randomly from Aozora Bunko digital library. They were instructed to not only decide whether a sentence have literal or non-literal sense but also - in case of the latter - to mark the expression used figuratively. Instruction was prepared in Japanese, and it can be translated into English as follows: *Read the following sets of three sentences one by one. Try to decide whether any figurative expression (metaphor, simile, etc.) is used in the middle (the second) sentence and choose the correct answer from the parenthesis. If you think that there is a figurative expression in use, indicate this word or phrase using an underline.*

As shown by Dobrzynska [6], evaluation of whether certain expression is used figuratively or literally is often context-dependent (consider for example *all men are animals* or *piece of thrash*). We have therefore provided annotators with adjacent sentences as well (the previous and the next to the target one). Odd number of annotators prevented possible problems with disagreement: sentences were labeled according to the majority's decision.

Let us briefly discuss some examples of labeling controversies. Expression *doko made mo* (lit. 'anywhere', 'wherever') from the sentence: *kenkyū wa doko made mo kenkyū de aru* 'research is a research (in every respect)' has been labeled as literal by all three annotators. To our understanding this expression should be considered as an example of underlying container metaphor.

Another example of expression in our opinion mistakenly labeled as being used in literal sense would be the following: *jikaku ga tsuyoku natta* '(one) became more aware' (lit. 'awareness became strong'). This is an example of personification and thus figurative expression, since awareness is not a living entity and therefore cannot become strong in literal sense.

Longer sentences sometimes included more than one linguistic unit used figuratively. In the following example several expressions with figurative senses can be found: *Hitode wo karizu, fūfu dake de mise o kirimawashita node, yoru no jūji kara jūniji koro made no ichiban tatekomu jikan wa me no mawaru hodo isogashiku, shōben ni tatsu hima mo nakatta* 'Because me and my wife were running the store (lit. 'cutting the store around') without anyone's help (lit. 'not borrowing a hand'), at the rush hour from 10 o'clock to around 12 o'clock in the evening it was so busy my head was spinning (lit. 'busy to the extent eyes were spinning') and I didn't even have time to take a piss'. In such cases sentences tended to get labeled as metaphorical but it is difficult to estimate which of the expressions used led the algorithm to make a correct prediction.

4 Classification Experiment

We have extracted word count vectors of 32,231 dimensions for each of 51,894 items in the training set. This was also the number of unique words in the dictionary created from all words present in the sentences used for training. Three standard state-of-the-art classifiers, namely Naïve Bayes (multinomial), Support Vector Machines (Linear SVC) and Random Forest Classifier (100 trees) were used via Scikit-learn library [20] for Python.

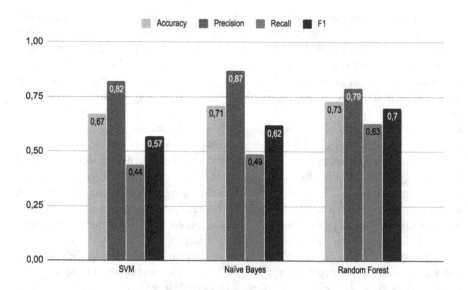

Fig. 1. Classification results

As can be noticed from Fig. 1, Random Forest has received the most balanced notes with F1 score of 70%. On the other hand, the highest precision has been achieved by Naïve Bayes which leads to the conclusion that it might be the most suitable algorithm for the task of compiling figurative language corpus in Japanese.

5   Discussion and Future Work

One of the issues which we have encountered during the process of annotation is that native speakers sometimes fail to recognize figurative nature of certain expressions, especially those which we would classify as *conventional* (also *dead* or *frozen*) metaphors. It might be true that discerning figurative expressions from the literal ones becomes even more difficult in texts produced in one's native language. One might even risk a hypothesis that in order to get more accurate results it would be better to entrust the task to a non-native speaker with sufficient level of fluency in the target language. Pretheoretical tenet according to which figurative language is somewhat deviant, unusual or that it belongs solely to the literary genres like poetry seems to be still widespread which results in belief that we rarely speak metaphorically. On the other hand, in the process of foreign language acquisition it often becomes evident for the learners that semantically similar concepts are realized in different ways - often figuratively - depending on the language. Should annotators have educational background in linguistics or semiotics is yet another problem worth further discussion.

We need to somehow resolve this problem, because - as we have seen and as Tsvetkov and colleagues [13] point out - humans may disagree whether particular expression is used metaphorically or not. It might be therefore desirable to use some more objective guidelines like those presented by Steen et al. [12] rather than depend on annotators' intuition. It is also very difficult if not impossible to find large amount of data completely free of figurative expressions. As a source of literal data we have chosen to use encyclopedic descriptions, news articles and discussions that took place during local parliamentary sessions. This way of collecting literal data is obviously not perfect, but possibly the best as it can get without supervision.

As for the results achieved in the current experiment, in comparison with our last year's approach [3], scores of SVM and Naïve Bayes have improved significantly. Precision score of Naïve Bayes got nearly 30% better. The experiment has proven that, having it trained on large amount of data and with use of high-dimensional word vector representations, Naïve Bayes might be utilized as a powerful tool for figurative speech recognition. Encouraged by promising results achieved by Tsvetkov et al.'s model [14], in this trial we have utilized Random Forest classifier as well. It yielded the most balanced results out of all three algorithms used. We plan to continue on investigating methods for increasing classification model's performance, especially in regards to its precision. If we could get it close to 100% it would enable us to enlarge existing dictionaries of figurative expressions which in turn would become an invaluable contribution to NLP-related research. Some potential applications of well performing metaphor processing models can be found in the last chapter of [14].

References

1. Austin, J.L.: How to do Things with Words, vol. 88. Oxford University Press, Oxford (1975)
2. Babieno, M., Rzepka, R., Araki, K.: Comparing Conceptual Metaphor Theory-related features using classification algorithm in searching for figurative expressions within Japanese texts. In: Proceedings of IJCAI Workshop on Language Sense on Computer, Macao, China (2019)
3. Babieno, M., Takishita, S., Rzepka, R., Araki, K.: Retrieving metaphorical sentences from Japanese literature using standard text classification methods. In: Proceedings of the 60th Language Sense Engineering SIG Conference, pp. 51–59 (2018)
4. Bulat, L., Clark, S., Shutova, E.: Modelling metaphor with attribute-based semantics. In: Proceedings of the 15th Conference of the European Chapter of the Association for Computational Linguistics: Volume 2, Short Papers, pp. 523–528. Valencia, Spain, April 2017
5. Dobrovol'skij, D., Piirainen, E.: Figurative Language: Cross-Cultural and Cross-Linguistic Perspectives. Brill, Leiden (2005)
6. Dobrzyńska, T.: Translating metaphor: problems of meaning. J. Pragmat. **24**(6), 595–604 (1995)
7. Hashimoto, C., Kawahara, D.: Compilation of an idiom example database for supervised idiom identification. Lang. Resour. Eval. **43**(4), 355 (2009)
8. Hashimoto, C., Sato, S., Utsuro, T.: Japanese idiom recognition: drawing a line between literal and idiomatic meanings. In: Proceedings of the COLING/ACL on Main Conference Poster Sessions, pp. 353–360. Association for Computational Linguistics (2006)
9. McRae, K., Cree, G.S., Seidenberg, M.S., McNorgan, C.: Semantic feature production norms for a large set of living and nonliving things. Behav. Res. Methods **37**(4), 547–559 (2005)
10. Neuman, Y., et al.: Metaphor identification in large texts corpora. PloS ONE **8**(4), e62343 (2013)
11. Onai, H.: Great Dictionary of 33800 Japanese Metaphors and Synonyms. Kodansha (2005). (in Japanese)
12. Steen, G.: A Method for Linguistic Metaphor Identification: From MIP to MIPVU, vol. 14. John Benjamins Publishing, Stoltenberg (2010)
13. Tsvetkov, Y., Boytsov, L., Gershman, A., Nyberg, E., Dyer, C.: Metaphor detection with cross-lingual model transfer. In: Proceedings of the 52nd Annual Meeting of the Association for Computational Linguistics (Volume 1: Long Papers). pp. 248–258 (2014)
14. Veale, T., Shutova, E., Klebanov, B.B.: Metaphor: a computational perspective. Synthesis Lect. Hum. Lang. Technol. **9**(1), 1–160 (2016)
15. Aozora Bunko digital library. https://www.aozora.gr.jp
16. Google Translate. https://translate.google.com/
17. Japanese Local Assembly Minutes Corpus Project. http://local-politics.jp
18. JUMAN (Morphological Analyzer for Japanese). http://nlp.ist.i.kyoto-u.ac.jp/EN/index.php?JUMAN
19. Livedoor News. http://news.livedoor.com
20. Scikit-learn. https://scikit-learn.org/stable
21. Wikipedia dumps list. https://dumps.wikimedia.org/jawiki/latest

Web Analysing

Model-Driven Web Page Segmentation for Non Visual Access

Judith Jeyafreeda Andrew[✉], Stéphane Ferrari, Fabrice Maurel, Gaël Dias, and Emmanuel Giguet

Normandie Univ, UNICAEN, ENSICAEN, CNRS, GREYC, 14000 Caen, France
{judith-jeyafreeda.andrew,stephane.ferrari,
fabrice.maurel,gael.dias,emmanuel.giguet}@unicaen.fr

Abstract. Web page segmentation aims to break a large page into smaller blocks, in which contents with coherent semantics are kept together. Within this context, a great deal of approaches have been proposed without any specific end task in mind. In this paper, we study different segmentation strategies for the task of *non visual skimming*. For that purpose, we propose to segment web pages into visually coherent zones so that each zone can be represented by a set of relevant keywords that can be further synthesized into concurrent speech. As a consequence, we consider web page segmentation as a clustering problem of visual elements, where (1) a fixed number of clusters must be discovered, (2) the elements of a cluster should be visually connected and (3) all visual elements must be clustered. Therefore, we study variations of three existing algorithms, that comply to these constraints: K-means, F-K-means, and Guided Expansion. In particular, we evaluate different reading strategies for the positioning of the initial K seeds as well as a pre-clustering methodology for the Guided Expansion algorithm, which goal is to (1) fasten the clustering process and (2) reduce unbalance between clusters. The performed evaluation shows that the Guided Expansion algorithm evidences statistically increased results over the two other algorithms with the variations of the reading strategies. Nevertheless, improvements still need to be proposed to increase separateness.

Keywords: Web page segmentation · Clustering · Reading strategies · Processing time · Non visual access

1 Introduction

For visually impaired people, accessing the web quickly and efficiently remains a challenge. While research efforts are carried out to design novel interaction models, screen reader is still the dominant technology for non visual web browsing [14].

In the TAGTHUNDER project[1], we introduce the concept of *tag thunder* as a means to produce an interactive and innovative stimulus promoting the emergence of self-adapted strategies for non visual reading [11]. The approach consists

---

[1] https://tagthunder.greyc.fr/.

© Springer Nature Singapore Pte Ltd. 2020
L.-M. Nguyen et al. (Eds.): PACLING 2019, CCIS 1215, pp. 191–205, 2020.
https://doi.org/10.1007/978-981-15-6168-9_17

in constructing oral counterparts to visual concepts (typography, layout, ...) that support the implementation of quick reading strategies such as skimming and scanning.

Skimming and scanning are two well-known reading processes, which are combined to access the document content as quickly and efficiently as possible. Scanning refers to the process of searching for a specific piece of information, and skimming is the action of passing through a document in a first glance to get an overview of its content. Skimming can easily be applied in a visual environment thanks to the visual, logical or textual document structures. Indeed, visual skimming relies on contrasted effects related to layout rendering and typographic styles. However, these effects are not available in a non visual environment. As such, reproducing the document content driven by its structure in a non visual setting is a much harder problem, but essential to be solved to improve web accessibility (e.g. visually impaired people).

In this paper, we focus on the hypothesis that successful non visual skimming strategies can take advantage of the previous identification of relevant zones with coherent semantics, that represent the coarse-grained document structure. This specific task is known as web page segmentation. Within this context, a great deal of approaches have been proposed [4,15,20], which do not focus on any specific end task, and as such are not constrained. Oppositely, we consider that non visual skimming requires three characteristics to be filled.

First, the number of zones has to be fixed in order to foster the emergence of regularities in the output and to comply with the maximum number of concurrent oral stimuli a human-being can cognitively distinguish. Indeed, we assume that each semantically coherent zone can be summarized and simultaneously synthesized into spatialized concurrent speech acts. Within this context, [7,10] have shown that the cognitive load can rise up to five different stimuli, thus limiting the number of zones resulting from the WPS process to 5. *Second*, each zone should be associated to a unique sound source spatially located in accordance with its position in the web page. Thus, each zone should be a single compact block made of contiguous web elements, and the zones should not overlap. *Third*, segmentation must be complete, which means that no web page element should remain outside a given zone, as the objective is to reveal the overall semantics of a document and not just parts of it[2].

In [2], we studied three different algorithms that comply to these constraints: the classical K-means [8], the F-K-means (a variant of K-means, which introduces the notion of force between elements instead of the euclidean distance), and the Guided Expansion algorithm (GE), which follows a propagation strategy including alignment constraints. A manual evaluation of the three algorithms had been performed by three experts measuring two clustering indicators: compactness and separateness, which was followed by a quantitative evaluation introducing different criteria for analysis. From both qualitative and quantitative evaluations, the GE proved to produce the most efficient solution over all criteria. However, as suggested in [2] the clustering process is highly sensitive to

---

[2] Oppositely to advertisement withdrawal for example.

the initial seeds positions. By following a diagonal reading strategy, we noted that most algorithms evidence an horizontal segmentation, i.e. vertical cluster are difficult to identify. Thus, in this paper we propose to use different methods to position the seeds based on reading strategies used on the web. As presented in [13] and [3], the users tend to read the web page in a "F" or "Z" strategy. As a consequence, we use this insight to position the inital $K = 5$ seeds of the tested clustering algorithms. Moreover, we study a new methodology to decrease the time complexity of the GE by introducing a simple pre-clustering technique, following the ideas of the QT algorithm [17]. As such, processing time is reduced without major performance loss, and an interesting side effect evidences the fact that more balanced clusters are obtained.

2   Related Works

Web Page Segmentation. Efforts on web page segmentation (WPS) have focused on removing noisy contents from web pages [1,18]. Later, [19] proposed a web page segmentation techniques based on the structure of the web page. Thus to be able to reflect the structure of the web page, the layout and the Document Object Model (DOM) features were used, along with some hand-crafted heuristics. This however relies on a fixed structural semantics that does not correspond to the creativity on the Web. More recently, [15] proposed Block-O-Matic. This is a pipeline strategy. This strategy combines content, geometric and logical structures of the web page. Also, [9] developed a method called HEPS (HEading-based Page Segmentation) to extract logical hierarchical structures of HTML documents. One of the main drawbacks of these approaches is that they heavily rely on the DOM. However, DOM is prone to errors due to uncontrolled page creation. Also, the number of clusters is automatically determined in these methods and thus can greatly vary from page to page. And, there could be some elements that remain unclustered. In order to overcome some of these limitations, visual-based strategies have been proposed, which mainly focus on the analysis of the visual features of the document contents as they are perceived by human readers. Notable works that follow this paradigm are VIPS [4] and the Box Clustering Segmentation (BCS) algorithm [20]. VIPS uses the DOM for the logical view of the document along with visual features of the document. But, BCS relies on a flat visual representation of the document, thus allowing great adaptability to new web contents. In particular, BCS follows a sort of hierarchical agglomerative clustering algorithm including a threshold that controls the formation of clusters. Thus the number of clusters is automatically determined and leaves some elements unclustered, with the same drawbacks as [15]. In [2], we followed the same strategy as the BCS algorithm as we exclusively rely on visual elements to segment web pages, and thus rely on a flat structure. But, we proposed three different clustering techniques (classical K-means, the F-K-means (a variant of K-means, which introduces the notion of force between elements instead of the euclidean distance), and the Guided Expansion algorithm (GE) that comply to the constraints imposed by the non visual skimming task: (1) segmentation into exactly 5 coherent zones, (2) completeness, where all visual elements belong to a given cluster and (3) connectivity of all the elements inside a cluster. In [2], we showed that the initial position of the seeds plays a crucial role in the clustering of web elements. Thus, in this paper, we propose to study variations of the algorithms used in [2] by changing the position of the initial seeds depending on different reading strategies used on the web. In order to decrease processing time and get more balanced clusters, we also introduce a modified version of the GE algorithm based on an initial pre-clustering step, which follows the ideas presented in [20] and relies on the QT algorithm [17]. Within this context, a quality area around some seeds is used to control the expansion process.

Reading Strategies. [13] propose a study on the "F" reading strategy that users use while reading the Web. The observations of [13] can be summarized as follows: (1) users first read in an horizontal movement, usually across the upper part of the content area. This initial element forms the F top bar; (2) next, users move down the page a bit and then read across in a second horizontal movement that typically covers a shorter area than the previous movement, which forms the F lower bar; (3) finally, users scan the left side of the content in a vertical movement, thus forming the F stem. In particular, the authors [13] show heat maps, which evidence the F pattern of reading on the Web. Another strategy is studied by [3]. They propose a study, which shows that users read the Web in a "Z" shape fashion when the web pages are not centered around its text content. The summary of [3] is as follows: (1) first, users scan from the top left to the top right, forming an horizontal line; (2) next, down and to the left side of the page, creating a diagonal line; (3) last, back across to the right again, forming a second horizontal line. Note that [3] and [13] also suggest other methods used by readers on the Web, but we will skip to both these ones in this study.

Evaluation. With respect to the evaluation of WPS, two strategies have been predominantly proposed - qualitative and quantitative. Qualitative evaluations is when human assessors are asked to validate the proposed segmentation against a ground truth generated by humans like in [5]. Quantitative evaluations rely on cluster correlation metrics as used in [20]. In particular, [20] uses the metrics of a general clustering problem, such as Rand Index or F-measure. However, WPS can not strictly be compared to a general clustering problem. For example, if just one visual element does not belong to its correct cluster, it may break the logical structure of the segmentation, but the quantitative metric will still remain high. Similarly, [16] create a ground truth database by segmenting web pages, and calculating specifically-tuned metrics. However, these metrics developed in [16] mostly rely on the DOM structure and thus are limited to DOM-based methodologies. In order to overcome the difficulties of quantitative evaluations based on cluster correlation metrics in non-DOM solutions, we proposed in [2] a quantitative evaluation method for the analysis of clusters based on different criteria for non-visual skimming: (1) number of cuts between zones, (2) coefficient of unbalance in terms of surface, text area and number of elements, and (3) number of nested areas. We propose the very same metrics to compare our algorithms in this paper.

3   Clustering Strategies

In this section, we briefly summarize our previous work on clustering strategies for WPS as presented in [2]. WPS for the specific task of non visual skimming can be defined as a clustering problem, where basic visual elements must be gathered into a K fixed number of clusters, where K is equal to 5. In particular, basic visual elements are first retrieved from a web page after rendering on the user's browser. DOM elements are then enriched with calculated CSS features, and each basic visual element corresponds to the last block element in each branch of the DOM tree[3]. In order to cluster the basic visual elements, we proposed three different strategies in [2]: K-means, F-K-means, and Guided Expansion. In this paper, we also propose the F-Guided Expansion algorithm, an adaptation of the Guided Expansion algorithm based on F-K-means.

K-Means. The K-means algorithm [8] is a well-established strategy when the number of clusters must be fixed a priori. Within the context of WPS, some adaptations are required. In particular, the assignment phase is based on the shortest euclidean distance between two visual elements (or between a virtual visual element in the case of the centroid), noted $dist(.,.)$. For our task, the elements to cluster are not data points in an N-dimensional space, but blocks, i.e. rectangle shapes. Thus, we use a border-to-border distance between the rectangles instead of a center-to-center distance.[4] Moreover, in order to calculate the centroid of a cluster, a virtual visual element is computed, instead of relying on the medoid, i.e. a visual element closer to a virtual center.

F-K-Means. In the previous proposal, the assignment phase is exclusively based on the geometric distance between visual objects. For this second algorithm, we propose a small variant, which takes into account the area covered by each visual basic element, the rationale being that visually bigger elements are more likely to "absorb" smaller elements than the contrary. So, if two visual elements are close to each other, their assignment function $force(b_1, b_2)$ will also depend on their differences of covered area as defined in Eq. 1, where a_{b1} (resp. a_{b2}) is the area of the visual element b_1 (resp. b_2) and $dist(.,.)$ is the shortest border-to-border euclidean distance between the basic elements. Thus, the F-K-means algorithm follows the exact same procedure as K-means, to the exception of the function used for the assignment step, which is the $force(.,.)$, i.e. the elements, which show the highest force to their centroid (a virtual visual element) are selected (See footnote 4).

$$force(b_1, b_2) = \frac{(a_{b1} * a_{b2})}{dist(b_1, b_2)} \tag{1}$$

---

[3] This is our unique use of the DOM structure.

[4] Illustration of this algorithm is presented in [2].

Guided Expansion. With the Guided Expansion (GE) algorithm, instead of assigning all visual elements to their closest centroid in a single step, only one visual element is assigned at a time to its centroid, controlled by a set of conditions that include the shortest border-to-border euclidean distance of two visual elements, the alignment between elements, and their visual similarity. The GE algorithm and its illustration is detailed in [2]. In particular, visual similarity $vsim(.,.)$ between two elements b_1 and b_2 is computed as in Eq. 2 over their respective feature vectors $\vec{b_1}$ and $\vec{b_2}$ formed by the following CSS properties of each bounding box: font-color, font-weight, font-family and background-color.

$$vsim(\vec{b_1}, \vec{b_2}) = \sum_{i=1}^{|\vec{b_1}|} \mathbb{1}_{\vec{b_1^i} = \vec{b_2^i}} \tag{2}$$

It is important to notice that a cluster is a set of visual elements, except for the first step of the algorithm. So, when the distance and the visual similarity are computed between an element and its cluster candidate, this refers to the computation of each metric between the element and all the elements in the cluster. This situation is formalized in Eqs. 3 and 4, where c_1 is the cluster candidate for b_1. However, the complexity of this algorithm is $O(n^2)$, where n is the number of visual elements in the web page. This is because until there are no unclustered elements, the element under consideration is compared with every other element to form the candidate set of elements. This will be the reason for the definition of a new algorithm detailed in Sect. 5.

$$dist(b_1, c_1) = argmin_{b_i \in c_1} dist(b_1, b_i) \tag{3}$$

$$vsim(\vec{b_1}, c_1) = argmax_{b_i \in c_1} vsim(\vec{b_1}, \vec{b_i}) \tag{4}$$

F-Guided Expansion. The F-Guided Expansion (F-GE) is a variation of the Guided Expansion algorithm presented in [2], which takes into account the area covered by each visual element. Thus, the first criterion to check between elements is the force of attraction, $force(b_1, b_2)$, between them as directed by Eq. 1, instead of the border-to-border geometric distance. Of course, this is followed by the alignment and visual similarities (Eq. 2) between elements as in the original GE algorithm.

4   Reading Strategies and Seeds Positioning

As mentioned in Sect. 2, users tend to scan/skim the Web using several reading strategies. In [2], we showed that for the algorithms mentioned in Sect. 3, the positioning of the initial seeds plays a crucial role in the clustering process. Indeed, by following a classical diagonal reading strategy, we noted that most algorithms evidence an horizontal segmentation, i.e. vertical clusters are difficult to identify. Another related issue concerns the F-K-means. If some seed is associated to a small element, this cluster will hardly expand as the $force(.,.)$

metric tends to benefit larger visual elements, thus clearly disadvantaging this algorithm compared to the other ones.

Thus, we intend to study the other reading strategies mentioned in Sect. 2. The diagonal method places the seeds on a diagonal virtually drawn on the web page from top-left to bottom-right. In particular, two seeds are positioned on each extremities, another one in the center and the two other ones between the extremities and the center of the diagonal. In this paper, we propose to place the seeds in a "F" and "Z" fashion motivated by the studies of [12,13] and [3]. The strategies are shown in Fig. 1. In Fig. 1, the blocks represent the visual blocks of the web page, the blue lines through the blocks represent the reading strategies and the red blocks indicate the chosen seeds.

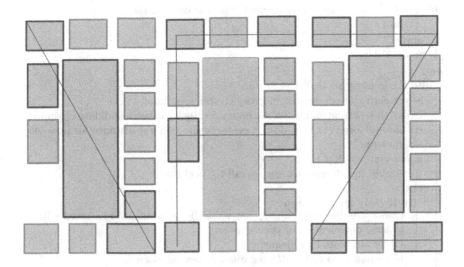

Fig. 1. Diagonal (left), F (center) and Z (right) strategies to position the seeds.

5   Pre-clustering Guided Expansion

As noticed in Sect. 3, the complexity of the Guided Expansion algorithm is $O(n^2)$. Thus, in order to decrease the time complexity and as a consequence the running time, a simple pre-clustering of the visual elements is performed. To perform this pre-clustering, we rely on the Quality Threshold algorithm [17], which clusters elements within a confidence area defined by a distance threshold. This can be viewed as a coarse-grained clustering that gathers all visual elements reliably within a small area as suggested in [20]. As such, five clusters can easily be formed using this simple pre-clustering method. The visual blocks are selected in the ordered list of visual elements[5]. For the first element in the list, the QT is applied

---

[5] Different strategies can be used to order the visual elements. In this paper, we use the order in which the visual elements appear in the DOM, using a depth-first search.

Input: The ordered list of basic visual elements; K
Output: K clusters
Threshold $\leftarrow max$(distance between two visual elements)$/10$;
$K \leftarrow 1$;
while $K \leq 5$ **do**
    Choose the first visual element, as the parent element, from the ordered list;
    Remove the first element from the ordered list;
    for *each visual element in the ordered list* **do**
        Calculate $dist(.,.)$ between the visual element and the parent element;
        if $dist(.,.) < Threshold$ **then**
            Add the visual element in the cluster of the parent element;
            Remove the visual element from the ordered list;
        end
    end
    $K \leftarrow K + 1$;
end
while *the ordered list is not empty* **do**
    Select each closest element to every cluster using $dist(.,.)$;
    Order these elements by the minimum distance to their candidate cluster;
    Remove all elements that do not evidence the smallest distance for possible assignment;
    if *there are no ties* **then**
        Assign the closest element overall to its cluster;
    end
    else if *there are ties* **then**
        Check whether the elements are vertically or horizontally aligned with at least one element of their cluster;
        Order elements by alignment;
        if *there are no ties AND one aligned element* **then**
            Assign the aligned element to its cluster;
        end
        else if *there are ties OR no aligned element* **then**
            Order elements by the maximum visual similarity to their cluster;
            Remove all elements that do not evidence the highest visual similarity for possible assignment;
            if *there are no ties* **then**
                Assign the most visually similar element to its cluster;
            end
            else if *there are ties* **then**
                Assign all elements to their cluster;
            end
        end
    end
end

Algorithm 1: Guided Expansion with QT pre-clustering.

and assigns its visual elements depending on if the border-to-border distance is within the given threshold. Then, the assigned elements are withdrawn from the list, and the same process is iterated four times based on the updated list.

This clustering obviously leaves some visual elements unclustered. So, for the visual elements that are unclustered, the Guided Expansion algorithm is used to finalize the clustering process. within this context, the five pre-clusters serve as basis for the final assignment. Note that with this simple pre-clustering step, we can reduce the complexity to $O(\alpha \times n^2)$, where $\alpha < 1$, where α depends on the size of the web page and the maximum distance between two visual elements. The Guided Expansion along with the QT pre-clustering step is given in Algorithm 1.

6 Quantitative Evaluation

In [2], we presented five criteria for a quantitative evaluation that were derived from the conclusions of a qualitative evaluation performed by 3 experts on 25 web pages. The criteria that emerged are as follows:

- Logical constraints embodied by specific HTML tag sequences such as <li> <ul> items, <title> and the following paragraph <p>, <header>, <footer> or <nav> elements should not be broken. Indeed, such breaks are likely to lead to odd clustering. As such, we propose to count one cut for each broken logical rule. This value is shown in column 1 of Table 1.
- An efficient algorithm should produce zones neither completely balanced nor too much unbalanced. To evaluate such a criterion, we test three different balance properties of the clusters: standard deviation of the surface area of the clusters, standard deviation of the number of characters within the clusters, and standard deviation of the number of visual elements within the clusters. The higher the standard deviation, the more unbalance the clusters are. The results of this property is shown in columns 2, 3 and 4 of Table 1.
- Zones should not be nested, i.e. the clustering should avoid non-rectangular clusters. To evaluate this phenomenon, the number of overlaps between the outer rectangles of all clusters is calculated, i.e. the smallest rectangle including all the elements of each cluster. So, if two clusters overlap in terms of outer rectangle, this stands for the presence of a non rectangular zone, and it is counted as a nested situation. The results of this property are shown in column 5 of Table 1.

Table 1 shows the results of the automatic evaluation for the three main criteria for a set of 150 web pages (47 tourist domain, 58 e-Commerce domain and 45 news domain[6]) segmented using all the versions of the three algorithms (K-means, F-K-means and Guided Expansion). In particular, each criterion receives the average value and the standard deviation $\pm \sigma$ for the set of 150 pages.

First, results show the superiority of the Guided Expansion algorithm over the other two algorithms in terms of number of cuts. In particular, it evidences

---

[6] All part of our project corpus.

Table 1. Automatic evaluation results for K-means, F-K-means and Guided Expansion (GE) for all reading strategies plus the pre-clustering GE. The evaluation is performed over 150 web pages. Note that D stands for Diagonal, F for F reading strategy, Z for Z reading strategy, GE P means the pre-clustering version of GE. Note also that $\pm\sigma$ stands for the standard deviation value over the 150 web pages.

Algorithm	Nb. of Cuts Avg. $\pm\sigma$	Surface area Avg. $\pm\sigma$	Text area Avg. $\pm\sigma$	Nb. of elements Avg. $\pm\sigma$	Exterior rectangle Avg. $\pm\sigma$
K-means D	2.12 ± 2.05	11.80 ± 6.46	11.40 ± 5.52	10.95 ± 8.01	5.21 ± 2.54
K-means F	2.59 ± 2.50	12.57 ± 6.54	12.52 ± 5.64	12.85 ± 9.63	4.13 ± 2.29
K-means Z	2.50 ± 2.40	13.20 ± 6.14	13.46 ± 6.02	14.85 ± 10.45	4.04 ± 2.21
F-K-means D	2.80 ± 2.76	21.14 ± 8.18	18.55 ± 7.74	22.79 ± 16.73	4.54 ± 2.20
F-K-means F	2.66 ± 2.40	20.58 ± 8.61	19.18 ± 8.63	23.87 ± 18.12	3.54 ± 1.94
F-K-means Z	2.63 ± 2.36	21.14 ± 7.82	19.40 ± 7.57	25.32 ± 18.33	3.53 ± 1.95
GE D	1.47 ± 1.85	17.34 ± 6.95	16.78 ± 6.37	19.67 ± 13.47	5.39 ± 2.22
GE F	1.43 ± 1.85	22.64 ± 7.23	22.37 ± 6.70	30.42 ± 19.93	4.91 ± 2.01
GE Z	1.34 ± 1.66	23.69 ± 7.10	22.77 ± 6.70	32.45 ± 21.82	5.26 ± 2.03
GE P	1.57 ± 1.98	12.55 ± 6.76	12.24 ± 6.35	15.04 ± 11.12	6.72 ± 2.11
F-GE D	1.75 ± 1.94	28.50 ± 8.27	27.41 ± 7.74	38.80 ± 24.62	3.46 ± 1.89
F-GE F	1.83 ± 2.08	31.12 ± 7.29	29.65 ± 7.52	43.85 ± 25.21	3.53 ± 1.89
F-GE Z	1.77 ± 1.97	31.35 ± 6.88	30.26 ± 6.75	44.90 ± 25.96	4.18 ± 2.12
F-GE P	1.80 ± 2.15	13.70 ± 7.12	12.12 ± 6.70	14.64 ± 11.07	5.92 ± 2.36

a minimum average value of 1.34 with the GE with Z reading strategy and a maximum of 1.83 with the F-GE with the F reading strategy, while K-means shows a minimum 2.12 score and F-K-means shows worst results with a minimum score of 2.63. In the case of K-means, using the F and Z reading strategies does not seem to improve the results over the diagonal strategy. But, in the case of F-Kmeans, the F and Z reading strategies give better results in terms of cuts. Thus, the three algorithms, irrespective of the reading strategies, can be sorted according to their ability to minimize the cut criterion with statistically significant values, i.e. GE is superior to K-means, which is in turn superior to F-K-means, however, the reading strategies do not seem to play a great role in this criteria. We will confirm these results in Sect. 7, where we present a complete statistical analysis of the results.

Second, balance results show similar observations whether we compare surface area, text area or number of elements between clusters. In all cases, the F-GE with diagonal, F and Z reading strategies show highest unbalance, while K-means shows the lowest unbalance. The Guided Expansion algorithm evidences some tendency to unbalanced clustering, which seems to better approximate human segmentation as explained in [2]. However, it is important to note that using a pre-clustering step with GE increases the balance between the zones in a huge way. This is because the pre-clustering step uses a threshold on distance thus restricting the number of elements in a zone and in turn producing clusters with similar sizes.

Finally, the "Exterior Rectangle" criterion, that aims to measure the number of non-rectangular shapes evidences similar results between all algorithms with around five overlaps per web page on average. Nevertheless, there is a clear statistical tendency for the F reading strategy to produce less non-rectangular zones amongst the other strategies. This is because when the seeds are placed near the border, the zones propagate only in one direction. Thus, this tends to produce rectangular zones. However, it is important to notice that the exterior rectangle criterion goes down to almost 0 for human annotators as shown in [2], who rarely proposed non-rectangular zones in their manual segmentation. As such, one might think that all algorithms are far from achieving human-like behavior. Although this is a strict reality from the figures, this difference against the manual evaluation observation may also indicate a lack of possible solutions by human annotators. Indeed, we think that acceptable segmentation can be proposed by some algorithms, although human annotators may not have thought about[7]. Thus further discussion should clearly be about the way to refine this criterion in order to distinguish between good and bad overlaps automatically.

7 Statistical Evaluation

Box Plots. In order to verify the significant difference between all tested algorithms, we first show a box plot analysis for the five criteria mentioned here before. From Fig. 2, it is witnessed that Guided Expansion a evidences a minimum of zero cuts irrespective of the strategy used to position the seeds. However, the minimum number of cuts for the other algorithms reaches the levels of the GE. From Figs. 3, 4 and 5, we can notice that the algorithms which are preprocessed with a simple clustering method tend to have more balance between the clusters. This is due to the fact that the first clustering produces initial clusters using a threshold and thus ensuring balance in the first stage of the process.

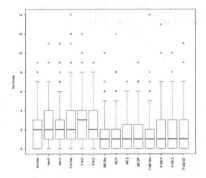 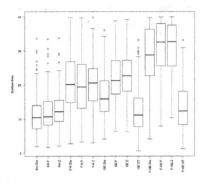

Fig. 2. Box plot for the number of cuts (column 1 in Table 1).

Fig. 3. Box plot for balance in surface area (column 2 in Table 1).

---

[7] This situation is explained in detail in [2].

From Fig. 6, it can be noticed that the more the seeds are placed near the border of the web page, the more they tend to make zones that are rectangle. Indeed, the F strategy places 5 seeds along the borders of the web page and thus witnesses less exterior rectangles i.e. more rectangular zones. Instead, the Z strategy places 4 seeds near the borders of the web page, while the diagonal strategy places 2 seeds on the borders of the web page. Thus, the order of algorithms with less nested zones can be summarized as follows: F > Z > D > PC.

Dunn Test. Once the initial ANOVA has found a significant difference in more than means, the Dunn's Test [6] can be used to pinpoint which specific means are significant from the others. Thus, the Dunn's Multiple Comparison Test is a post hoc (i.e. it's run after an ANOVA) non parametric test, which is done to determine which groups are different from others. In order to verify the differences between algorithms in terms of statistical significance, we propose to use this test. The results of the analysis with the Dunn test are shown in Table 2. Note that the algorithms within each group are not significantly different from each other. However, algorithms in different groups are significantly different

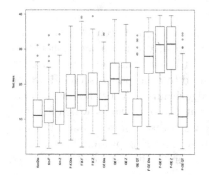

Fig. 4. Box plot for balance in text area (column 3 in Table 1).

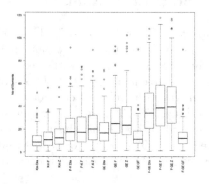

Fig. 5. Box plot for balance in visual elements (column 4 in Table 1).

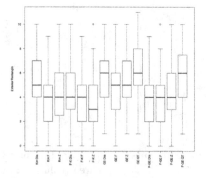

Fig. 6. Box plot for the criteria Exterior Rectangle (column 5 in Table 1)

Table 2. Dunn test analysis for the 14 algorithms over the 5 different criteria. Algorithms within a group show no statistical difference between them. Rank evidences the performance order for each criterion.

Criterion	Groups	
Cuts	1	{GE F, GE Z, GE P}
	2	{GE D}
	3	{F-GE D, F-GE P, F-GE Z}
	4	{F-GE F}
	5	{K-means D}
	6	{K-means F, F-K-means D, F-K-means Z}
	7	{F-K-means F}
	8	{K-means Z}
Surface Area	1	{K-means D}
	2	{GE P, K-means F, K-means Z}
	3	{F-GE P}
	4	{GE D}
	5	{F-K-means F}
	6	{F-K-means Z}
	7	{F-K-means D}
	8	{GE F}
	9	{GE Z}
	10	{F-GE D, F-GE F, F-GE Z}
Text Area	1	{GE P, F-GE P, K-means D, K-means F, K-means Z}
	2	{GE D, F-K-means D, F-K-means F, F-K-means Z}
	3	{GE F, GE Z}
	4	{F-GE D, F-GE F, F-GE Z}
Number of Elements	1	{K-means D}
	2	{K-means F}
	3	{GE P, F-GE P, K-means Z}
	4	{GE D, F-K-means D, F-K-means F, F-K-means Z
	5	{GE F, GE Z}
	6	{F-GE D, F-GE F, F-GE Z}
Exterior Rectangle	1	{F-GE D, K-means Z}
	2	{F-GE F}
	3	{F-K-means F, F-K-means Z}
	4	{K-means F}
	5	{F-GE Z}
	6	{F-K-means D}
	7	{GE F}
	8	{GE Z, K-means D}
	9	{GE D, F-GE P}
	10	{GE P}

from each other. Moreover, the rank of each group shows how well a given group performs for a given criterion. From this analysis, it seems that the Guided Expansion algorithm with pre-clustering (GE P) is globally the more suitable solution for WPS in the specific context of non visual information access. However, this needs to be confirmed by a qualitative analysis[8] as this algorithm shows the worst results in terms of nested clusters.

8   Conclusions

In this paper, we presented Web Page Segmentation as a clustering problem driven by the task of non visual information access. In particular, we tested three well-known algorithms, namely K-means, F-K-means and the Guided Expansion, with three reading strategies used on the web, namely diagonal, "F" and "Z". We also presented a new methodology to reduce the time complexity of the Guided Expansion algorithm by introducing a pre-clustering step based on the QT algorithm. We also tested the Guided Expansion algorithm combined with the force measure. Quantitative and statistical evaluations showed that the Guided Expansion algorithm is a good baseline, in particular in its new version including pre-clustering. Pre-clustering not only reduces the time complexity of the Guided Expansion algorithm but also improves the balance between the zones without significantly increasing the number of cuts. We also showed that the position of the initial seeds does change the results of the algorithms in a significant way. However, there are still other reading strategies used on the web that are open to exploration. As a consequence, future work needs to be endeavour to strengthen these first findings. This goes with performing a qualitative analysis and an exhaustive analysis of reading strategies. But, the automatic selection of optimal seeds seems to be the priority research direction.

References

1. Alassi, D., Alhajj, R.: Effectiveness of template detection on noise reduction and websites summarization. Inf. Sci. **219**, 41–72 (2013)
2. Andrew, J.J., Ferrari, S., Maurel, F., Dias, G., Giguet, E.: Web page segmentation for non visual skimming. In: The 33rd Pacific Asia Conference on Language, Information and Computation (2019)
3. Babich, N.: Z-shaped pattern for reading web content (2017). https://uxplanet.org/z-shaped-pattern-for-reading-web-content-ce1135f92f1c. Accessed Sept 2019
4. Cai, D., Yu, S., Wen, J.R., Ma, W.Y.: Extracting content structure for web pages based on visual representation. In: 5th Asia-Pacific Web Conference on Web Technologies and Applications (ApWeb), pp. 406–417 (2003)
5. Cai, D., Yu, S., Wen, J.R., Ma, W.Y.: Vips: a vision-based page segmentation algorithm. Technical report MSR-TR-2003-79, Microsoft, November 2003. https://www.microsoft.com/en-us/research/publication/vips-a-vision-based-page-segmentation-algorithm/

---

[8] For lack of space, we do not present this study in this paper.

6. Dunn, O.J.: Multiple comparisons among means. J. Am. Stat. Assoc. **56**(293), 52–64 (1961)
7. Guerreiro, J., Gonçalves, D.: Faster text-to-speeches: enhancing blind people's information scanning with faster concurrent speech. In: 17th International ACM SIGACCESS Conference on Computers & Accessibility (ASSETS), pp. 3–11 (2015)
8. MacQueen, J.: Some methods for classification and analysis of multivariate observations. In: 15th Berkeley Symposium on Mathematical Statistics and Probability (BSMSP), pp. 281–297 (1967)
9. Manabe, T., Tajima, K.: Extracting logical hierarchical structure of HTML documents based on headings. PVLDB **8**(12), 1606–1617 (2015)
10. Manishina, E., Lecarpentier, J.M., Maurel, F., Ferrari, S., Maxence, B.: Tag Thunder : towards non-visual web page skimming. In: 18th International ACM SIGACCESS Conference on Computers and Accessibility (ASSETS) (2016)
11. Maurel, F., Dias, G., Ferrari, S., Andrew, J.J., Giguet, E.: Concurrent speech synthesis to improve document first glance for the blind. In: 2nd International Workshop on Human-Document Interaction (HDI 2019) associated to 15th International Conference on Document Analysis and Recognition (ICDAR 2019), Sydney, Australia, 20–25 September (2019)
12. Nielsen, J.: F-Shaped pattern for reading web content. https://www.nngroup.com/articles/f-shaped-pattern-reading-web-content-discovered/
13. Pernice, K.: F-shaped pattern of reading on the web: misunderstood, but still relevant (even on mobile) (2017). https://www.nngroup.com/articles/f-shaped-pattern-reading-web-content. Accessed Sept 2019
14. Ramakrishnan, I.V., Ashok, V., Billah, S.M.: Non-visual web browsing: beyond web accessibility. In: 11th International Conference on Universal Access in Human-Computer Interaction (UAHCI), pp. 322–334 (2017)
15. Sanoja, A., Gançarski, S.: Block-o-matic: a web page segmentation framework. In: International Conference on Multimedia Computing and Systems (ICMCS), pp. 595–600 (2014)
16. Sanoja, A., Gançarski, S.: Web page segmentation evaluation. In: 30th Annual ACM Symposium on Applied Computing (SAC), pp. 753–760 (2015)
17. Xin, J., Jiawei, H.: Quality Threshold Clustering, pp. 1–2. Springer, Boston (2016). https://doi.org/10.1007/978-1-4899-7502-7_692-1
18. Yi, L., Liu, B., Li, X.: Eliminating noisy information in web pages for data mining. In: 9th ACM SIGKDD International Conference on Knowledge Discovery and Data Mining (KDD), pp. 296–305 (2003)
19. Yin, X., Lee, W.S.: Understanding the function of web elements for mobile content delivery using random walk models. In: 14th International Conference on World Wide Web (WWW), pp. 1150–1151 (2005)
20. Zeleny, J., Burget, R., Zendulka, J.: Box clustering segmentation: a new method for vision-based web page preprocessing. Inf. Process. Manag. **53**(3), 735–750 (2017)

Update Frequency and Background Corpus Selection in Dynamic TF-IDF Models for First Story Detection

Fei Wang[1,2]([✉]), Robert J. Ross[1,2], and John D. Kelleher[1,2]

[1] School of Computer Science, Technological University Dublin, Dublin, Ireland
d13122837@mydit.ie, {robert.ross,john.d.kelleher}@dit.ie
[2] ADAPT Research Centre, Dublin, Ireland

Abstract. First Story Detection (FSD) requires a system to detect the very first story that mentions an event from a stream of stories. Nearest neighbour-based models, using the traditional term vector document representations like TF-IDF, currently achieve the state of the art in FSD. Because of its online nature, a dynamic term vector model that is incrementally updated during the detection process is usually adopted for FSD instead of a static model. However, very little research has investigated the selection of hyper-parameters and the background corpora for a dynamic model. In this paper, we analyse how a dynamic term vector model works for FSD, and investigate the impact of different update frequencies and background corpora on FSD performance. Our results show that dynamic models with high update frequencies outperform static model and dynamic models with low update frequencies; and that the FSD performance of dynamic models does not always increase with higher update frequencies, but instead reaches steady state after some update frequency threshold is reached. In addition, we demonstrate that different background corpora have very limited influence on the dynamic models with high update frequencies in terms of FSD performance.

Keywords: Novelty detection · First Story Detection · Nearest neighbour · TF-IDF · Update frequency · Background corpus

1 Introduction

Novelty detection is the task of identifying data that are different in some salient respect from other predominant chunks of data in a dataset [14]. In most cases, there is not an explicit definition for novelty or sufficient novel data to form a class of novelty before detection. Instead, novelty detection is normally treated as an unsupervised learning application, i.e. no labels are available and the detection is implemented based on only the intrinsic properties of the data [16].

Online novelty detection is a special case of novelty detection, in which input data are time-ordered streams. The online characteristic brings in two additional constraints [9]: 1) the detection should be made quickly, e.g. before subsequent

© Springer Nature Singapore Pte Ltd. 2020
L.-M. Nguyen et al. (Eds.): PACLING 2019, CCIS 1215, pp. 206–217, 2020.
https://doi.org/10.1007/978-981-15-6168-9_18

data arrives; and 2) looking forward is prohibited during detection, i.e. the detection can only be made based on the data that has already arrived. One important application of online novelty detection within Natural Language Processing (NLP) is to the task of First Story Detection (FSD). In FSD, the target text documents are all stories that discuss some specific events. Given a stream of stories in chronological order, the goal of FSD is to find out the very first story for each event [2]. The stories are processed in sequence, and for each incoming candidate story, a decision is made on whether or not it discusses an event that has not been seen in previous stories; the decision making process is normally based on a novelty score, namely, if the novelty score of an incoming story is higher than a given threshold, we say the candidate is a first story.

Since it was first defined within the Topic Detection and Tracking (TDT) competition series in 1998 [1,19], hundreds of models have been proposed for the task of FSD. Nearest neighbour-based models with the traditional term vector document representations currently achieve the state of the art in FSD [11,13,16]. Because of its online characteristic, a dynamic term vector model that is incrementally updated during detection is usually adopted for FSD instead of a traditional static model [7,11,12]. However, very little previous research has investigated how a dynamic term vector model works in practice for FSD, or has investigated how to select hyper-parameters (such as the model update frequency) and background corpora for such dynamic models.

In this paper, we first theoretically analyse how a dynamic term vector model works for FSD, and then empirically evaluate the impacts of different update frequencies and background corpora on FSD performance. Our results show that dynamic models with high update frequencies outperform static models and dynamic models with low update frequencies; and, importantly, also show the FSD performance of dynamic models does not always increase along with increases in the update frequency. Moreover, we demonstrate that different background corpora have very limited influence on the dynamic models with high update frequencies in terms of FSD performance.

2   First Story Detection

As mentioned in Sect. 1, a wide variety of models have previously been investigated for FSD. These models can generally be grouped into three categories [16]: Point-to-Point (P2P) models, Point-to-Cluster (P2C) models, and Point-to-All (P2A) models. Accordingly, their novelty scores are defined as the distance of the incoming data to: an existing data point for P2P models, a cluster of existing data points for P2C models, and all the existing data points for P2A models. The P2P models are normally nearest neighbour-based [2,3,19] or approximate nearest neighbour-based [7,12] that aim at finding the most similar existing story to the incoming story. The P2C models use clusters of existing stories to represent previous events and evaluate the incoming story by comparing it with these clustered events [2,19]. The P2A models typically build a machine learning system with all the existing stories and apply this system to the incoming story to generate a novelty score [8,15,18]. Generally speaking, the nearest neighbour-based P2P models outperform the other two categories of models.

When applying a nearest neighbour-based model to FSD, the first step is to represent each text story with a document representation vector so that quantitative comparisons can be made between stories. In recent years a large number of deep learning-derived distributed document representations have been proposed that have achieved excellent performance across many NLP tasks [6]. However, for the task of FSD, the state of the art is still achieved with traditional term vector document representations [16], in which each term is represented with a single feature (dimension) in the term vector space. The most well-known term vector model is TF-IDF, short for term frequency - inverse document frequency, in which the weight of each term in a specific document is calculated as the product of the TF (term frequency) component and the IDF (inverse document frequency) component. There are many schemes of calculating these two components, but a widely-applied scheme is shown as follows [3,7]:

$$tf\text{-}idf(t,d) = tf(t,d) \times idf(t) \tag{1}$$

$$idf(t) = log\frac{N}{df(t)} \tag{2}$$

where $tf(t,d)$, representing the TF component, is the number of times the term t occurs in document d, and $idf(t)$, representing the IDF component, is the logarithmic value of the proportion of the total number of documents N divided by $df(t)$, i.e., the number of documents that contain the term t. Briefly speaking, the more a term occurs in a target document, and the less it occurs in other documents, the bigger the TF-IDF weight is for that term for that document.

From the definition of TF-IDF, we can also see that the calculation of the IDF component requires a corpus with a number of documents. However, because of the "looking back only" constraint of online detection, the target corpus for FSD detection is always unavailable for the construction of the TF-IDF model prior to detection, and thus an additional background corpus is required. Specifically, based on the background corpus, the TF-IDF model builds a term vocabulary and calculates the IDF component for each term in the vocabulary. After this step, there are two different ways to implement TF-IDF models for FSD [3,19]. The first option is to apply this fixed model, i.e., the fixed vocabulary and IDF components, to the stories in the target FSD corpus. This type of model is called a *static* TF-IDF model. In this way, any term that is unseen in the background corpus will be ignored in the detection process. The second option is to incrementally update the model, i.e., the vocabulary and IDF components, during the detection process after a number of documents arrive. This type of model is called a *dynamic* TF-IDF model. In this way, the terms that are unseen in the background corpus but have been seen up to a certain point in the target data stream are also taken into account. Because of its online nature, a dynamic model should usually be adopted for FSD [7,11,12]. Given this, in the next section, we will analyse how a dynamic TF-IDF model works for FSD and its difference from a static model in this context.

3   Dynamic Term Vector Models for First Story Detection

For a dynamic TF-IDF model we adopt an adjusted form of Eqs. 1 and 2 from earlier. Specifically we adopt the following:

$$tf\text{-}idf(t, d) = tf(t, d) \times idf(t)' \tag{3}$$

$$idf(t)' = log \frac{N'}{df(t)'} \tag{4}$$

where $tf(t, d)$ remains the same as that in Eq. 1, but the calculation of the IDF component $idf(t)'$ now makes use of an N' that captures the total number of not only the documents in the background corpus but also the stories in the target FSD corpus up to the present point, and, similarly, $df(t)'$ refers to the number of documents across both the background corpus, and the portion of target corpus to the current point, that contain the term t.

Due to the dynamic nature of the TF-IDF model, the length and features captured by a document vector now vary as we move through events, and this has potential implications to the FSD process. To illustrate, let us consider two documents (one being the candidate story and the other some story that has already been processed by our model). The comparison of these two documents is typically achieved with a distance metric; here we will assume the widely-used cosine distance:

$$cosine\_distance(\boldsymbol{d}, \boldsymbol{d}') = 1 - \frac{\boldsymbol{d} \cdot \boldsymbol{d}'}{|\boldsymbol{d}||\boldsymbol{d}'|} \tag{5}$$

where $\boldsymbol{d}$ is the TF-IDF vector for the candidate story and $\boldsymbol{d}'$ is the TF-IDF vector for a historic story that we are comparing to.

In order to better understand how a dynamic model performs for FSD, in Table 1, we unfold these two document vectors to m term features from t_1 to t_m, where m is the length of the vocabulary. In a dynamic model, the vocabulary includes both terms that were present in the background corpus and new terms that are added during the updates to the model. However, irrespective of whether a term is a new term or not, the value for a term in the TF-IDF document representation is the weight of the specific term based on the dynamic TF-IDF model.

In order to analyse how a TF-IDF representation treats both old and new terms in a document representation we group the features in our document representation into two parts: Range A which includes terms that have been present in the model for a substantial amount of time (because they were present in the background corpus or were added to the model several updates previously); and Range B which includes terms that have been added to the model recently.

In a static TF-IDF model there are only terms in Range A coming from the background corpus, and no term in Range B since the target corpus is not incorporated into the TF-IDF representation. Thus, the performance of the TF-IDF

Table 1. Two document representation vectors based on a dynamic TF-IDF model

	Range A			Range B		
	t_1	...	t_i	t_{i+1}	...	t_m
d	v_1	...	v_i	v_{i+1}	...	v_m
d'	v'_1	...	v'_i	v'_{i+1}	...	v'_m

model depends on how well the weights from the background corpus represent the terms in our target corpus. As the term weights are only calculated based on the background corpus, the selection of the background corpus has a great impact on the static TF-IDF model, and also influences the FSD performance [17]. Thus, a large-scale domain-related background corpus is normally adopted to generate realistic weights for the terms.

For a dynamic TF-IDF model, however, although it can use a large background corpus initially, new terms that are unseen in the background corpus will emerge and be incorporated into the model as detection proceeds – thus forming Range B. By definition these new terms did not occur in the background corpus, this may be because the new terms are genuinely rare in language, or else it may be because the selected background corpus was not representative of the language in the target data stream that the model is processing, or finally the new term may be a true neologism in a language. Whatever the true cause for why a particular term is a new term for a model, the weights of these new terms may be not well calibrated with respect to the weights for the terms in Range A. In Eq. 4, $df(t)'$ denotes the number of documents that contain the term t not only in the already-processed target stories, but also in the documents of the background corpus. However, by definition new terms in Range B will not have appeared in the background corpus and will only have appeared in the most recent documents in the target data stream. Therefore, the value of $df(t)'$ of a new term in Range B will be very small compared to N' in Eq. 4, and thus the TF-IDF weights for these new terms are normally very large, so we call these the rough weights with respect to the realistic weights in Range A. In the calculations of cosine distance (Eq. 5) more attention is focused on the features with larger values, and thus, the terms in Range B have a bigger effect on comparison calculations based on a dynamic TF-IDF model than they are expected to have based on the language.[1]

From the analysis above, we find that a key difference between dynamic and static TF-IDF models, when making comparisons between document vectors, is that the focus of dynamic models is more on the new terms with large rough

---

[1] It is worth noting that if looking at the whole FSD process rather than the comparison between two specific document vectors, new terms keep on being added into Range B as the updates are implemented. On the other hand, the terms already existing in Range B keep on being moved to Range A as more and more new stories arrive and the number of stories since the term's first appearance becomes large enough to generate realistic weights.

weights that emerge during detection, whereas static models focus on the existing terms whose weights are calculated only based on the background corpus.

In order to improve a static model, or indeed the static elements of a dynamic model, we can try to find a more suitable background corpus in order to generate realistic weights for the terms in the target data stream. However, for the dynamic approach it is hard to improve performance from a theoretical perspective due to the way in which weights are calculated for newly encountered terms. To overcome this limitation and try to optimise the dynamic aspects of TF-IDF modelling for FSD, in the next section we present an experimental analysis to investigate the impact of update frequency and background corpus on the model.

4 Experimental Design

In the following, we present our experimental design for evaluating the impact of the dynamic aspects of a TF-IDF model in the context of the FSD task. We focus on the impact of different update frequencies and the relevance of background corpora selection.

4.1 Target Corpus

In our experiments, we use the $TDT5$ corpus[2] as the target corpus for FSD detection. This corpus contains approximately 278 thousands newswire stories generated from April to September 2003.

4.2 Background Corpora

For the evaluation of the impact of different background corpora on the FSD results, we selected $COCA$ (The Corpus of Contemporary American English) [4] and $COHA$ (Corpus of Historical American English) [5] as the basic background corpora to be used for evaluation. The former is a comprehensive contemporary English documents collection from 1990 to present in different domains such as news, fiction, academia and so on. The latter is similar to $COCA$ in themes but covers the historical contents from 1810 to 2009. The numbers of documents in $COCA$ and $COHA$ are approximately 190,000 and 115,000 respectively. In both cases we only make use of the subsets of the two corpora that predate 2003, i.e., the year of $TDT5$'s collection. Additionally, to tease out the influence of domain relevance, we also divide the $COCA$ corpus into two distinct subsets - $COCA\_News$ and $COCA\_Except\_News$. $COCA\_News$ contains only the documents in the domain of news, which is the same as the domain of the target $TDT5$ corpus, while $COCA\_Except\_News$ contains the documents in all other domains apart from news. With these four corpora, we can investigate out how the temporality ($COCA$ vs. $COHA$) or domain specificity ($COCA\_News$ vs. $COCA\_Except\_News$) of the background corpora influence the dynamic TF-IDF models for FSD.

---

[2] https://catalog.ldc.upenn.edu/LDC2006T18.

4.3   Update Frequencies

There is no standard update frequency for a dynamic TF-IDF model. Typically, updates are implemented so as to be less frequent than every 100 documents [7], as the update process is very expensive if the update frequency is higher than every 100 documents. In our experiments, we evaluate a range of update frequencies - specifically, every 100, 500, 1000, 10000 and 100000 documents, and also implement a static TF-IDF model as a baseline. The static model can be interpreted as a dynamic model where updates are extremely infrequent. For each update frequency we build TF-IDF models for all background corpora.

4.4   FSD Evaluation

Our implementation of FSD is based on the nearest neighbour algorithm, with the cosine distance algorithm adopted as the dissimilarity measure between documents. The preprocessing of data and the evaluation of FSD results are similar to our previous research [17]. In order to reduce the effect of useless terms and different term forms, we remove terms with very high and very low document frequency, i.e., stop words and typos, for all the background and target corpora, and subsequently stem all remaining terms. Aligning with previous research [19], comparisons are only implemented with the 2000 most recent stories for each candidate story. The output of each FSD model is a list of novelty scores, one for each story in the target corpus $TDT5$. Based on these outputs, the standard evaluation process for FSD is implemented by applying multiple thresholds to sweep through all the novelty scores. For each threshold, a missing rate and a false alarm rate are calculated; then for all thresholds, the missing and false alarm rates are used to generate a DET (Detection Error Tradeoff) curve [10], which shows the trade-off between the false alarm error and the missing error in the detection results. The closer the DET curve is to the origin, the better the FSD model is said to perform. Thus, from the DET curves we calculate Area Under Curve (AUC) for each FSD model, and the model with the lowest AUC is judged to be best.

5   Results and Analysis

Below we present our experimental results in Fig. 1, and analyse the impacts of different update frequencies and background corpora on the dynamic TF-IDF models for the FSD task.

5.1   Comparisons Across Different Update Frequencies

We begin by examining the FSD performance results as influenced by update frequency. From the results shown in Fig. 1, we firstly see a trend that for each background corpus, the dynamic TF-IDF models with high update frequencies, i.e., every 100, 500 and 1000 documents, outperform the static model and dynamic

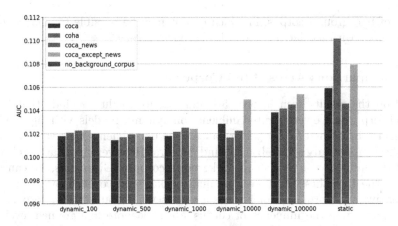

Fig. 1. Comparisons across different update frequencies and background corpora

models with low update frequencies, i.e., every 10000 and 100000 documents. As explained in Sect. 3, the dynamic models with high update frequencies focus more on the terms with large rough weights (i.e., the terms in Range B in Table 1), while the static models only focus on the terms with what we believe are realistic weights (i.e., the terms in Range A in Table 1). From this perspective, we can conclude that the terms with large rough weights play a more important role in FSD than the terms with realistic weights. Similarly, as the update frequency of a dynamic model becomes very low, the weights of most new terms are also well calibrated, and thus this dynamic model has fewer terms with rough weights, but more terms with realistic weights, which thus leads to poor FSD performance.

Secondly, we also find that for each background corpus, the FSD performance does not always improve but instead stays steady with a difference of less than 1% between models with a update frequency higher than every 1000 documents. One potential reason for this may be that as we increase the update frequency there are two counteracting processes with respect to rough weights: (a) a high update frequency means that new terms with rough weights are introduced into the model frequently, but (b) a high update frequency also means that the already-existing rough weights will themselves be updated incrementally and so may be smoothed frequently and so they don't stay rough for long. This is only our hypothesis of what might be happening.

5.2 Comparisons Across Different Background Corpora

We also make comparisons from the perspective of background corpora. From Fig. 1, it can also be seen that the differences caused by different background corpora are only noteworthy in the static model and dynamic models with low update frequencies. In the dynamic models with high update frequencies such as every 100, 500, 1000 stories, the influences are minor (less than 1%). This raises the possibility that models with high update frequencies are not affected by the choice of background corpus, in which case it may be possible to achieve good performance with a relatively small background corpus.

5.3   Comparisons Across Mini Corpora

Based on the results seen in Sect. 5.1 and 5.2, we might conclude that background corpora have very limited influence on dynamic models with high update frequencies in terms of FSD performance. The experiments thus validated our hypothesis about large-scale background corpora. However they have said little about the influence of very small background corpora. Given this, we can also propose the hypothesis that even a small background corpora can achieve as competitive a performance for FSD as a large-scale domain-related corpus.

To investigate the influence of corpus size at a more fine grained level, we extracted two small sets of documents, i.e., the first 500 stories and the last 500 stories, from each of the four background corpora to form eight very small background corpora. After that, eight dynamic TF-IDF models are built based on these corpora, and the update frequency was set to every 500 documents (the update frequency that leads to the best results in Sect. 5.1 and 5.2). The comparisons of FSD results are shown in Fig. 2 with the results of static models as the baseline.

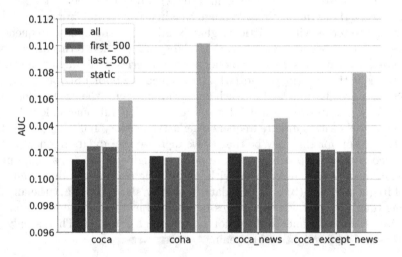

Fig. 2. Comparisons across mini background corpora with the update frequency set as every 500 stories

From the results, we can see that even based on background corpora that are quite different in scale, domain or collection time, there is no big difference (also within 1%) in the FSD results. Especially, the FSD result generated by the model based on the $First\_500\_COHA$ corpus is a little bit better than the full *COHA* corpus, even though the stories in the *First_500_COHA* corpus are collected around the year 1810 from various domains.

It is also worth mentioning that the corpus size 500 was not a crucial factor. It could have been 100, 1000 or any other number within this range. For further comparison, we also implemented pure dynamic TF-IDF models[3], i.e. the dynamic models that do not use any background corpus or with the corpus size set as 0, as shown with the tick "no_background_corpus" in Fig. 1. Unsurprisingly, the results show that the pure dynamic TF-IDF models with update frequency set as 100 or 500 do not make any big difference in FSD performance in comparison to the dynamic models based on any other background corpus with a similar update frequency, and this finding supports our conclusion that background corpora have very limited influence on dynamic models with high update frequencies in terms of FSD performance.

6   Conclusion

In this paper we empirically validated that the dynamic TF-IDF models with high update frequencies outperform the static model and the dynamic models with low update frequencies, and set out some factors that may explain this finding. However, a key element of these explanations is the observation that a high update frequency can result in new terms with relatively large weights being introduced into the TF-IDF representations. We also found that the FSD performance of dynamic models does not always improve but stays steady as the update frequency goes beyond some threshold, and that the background corpora have very limited influence on the dynamic models with high update frequencies in terms of FSD performance. Finally, we conclude that the best term vector model for FSD should be a dynamic model whose weights are initially calculated based on any small-size corpus but updated with a reasonable high frequency, e.g., for our scenario we found an update frequency of every 500 stories results in good performance.

Acknowledgement. The authors wish to acknowledge the support of the ADAPT Research Centre. The ADAPT Centre for Digital Content Technology is funded under the SFI Research Centres Programme (Grant 13/RC/2106) and is co-funded under the European Regional Development Funds.

---

[3] Actually, pure dynamic TF-IDF models should not be applied to the TDT task, because this specific task requires the detection to start from the very first story in the target data stream. However, as the first one story to be evaluated is on the 577th, if we use the stories before it as the background documents to calculate the initial TF-IDF weights, there will be no influence on the detection results. Therefore, we apply pure dynamic models with a update frequency equal to or higher than every 500 stories to the task, but just for the analysis and the proof of our hypothesis.

References

1. Allan, J., Carbonell, J. G., Doddington, G., Yamron, J., Yang, Y.: Topic detection and tracking pilot study final report (1998)
2. Allan, J., Lavrenko, V., Malin, D., Swan, R.: Detections, bounds, and timelines: UMASS and TDT-3. In: Proceedings of Topic Detection and Tracking Workshop, pp. 167–174, February 2000. sn
3. Brants, T., Chen, F., Farahat, A.: A system for new event detection. In: Proceedings of the 26th Annual International ACM SIGIR Conference on Research and Development in Information Retrieval, pp. 330–337. ACM, July 2003
4. Davies, M.: The corpus of contemporary american english as the first reliable monitor corpus of English. Liter. Linguist. Comput. **25**(4), 447–464 (2010)
5. Davies, M.: Expanding horizons in historical linguistics with the 400-million word Corpus of Historical American English. Corpora **7**(2), 121–157 (2012)
6. Goldberg, Y.: Neural network methods for natural language processing. Synthesis Lect. Hum. Lang. Technol. **10**(1), 1–309 (2017)
7. Kannan, J., Shanavas, A.M., Swaminathan, S.: Real time event detection adopting incremental TF-IDF based LSH and event summary generation. Int. J. Comput. Appl. 975, 8887
8. Leveau, V., Joly, A.: Adversarial autoencoders for novelty detection (2017)
9. Ma, J., Perkins, S.: Online novelty detection on temporal sequences. In: Proceedings of the Ninth ACM SIGKDD International Conference on Knowledge Discovery and Data Mining, pp. 613–618. ACM, August 2003
10. Martin, A., Doddington, G., Kamm, T., Ordowski, M., Przybocki, M.: The DET curve in assessment of detection task performance. National Inst of Standards and Technology Gaithersburg MD (1997)
11. Moran, S., McCreadie, R., Macdonald, C., Ounis, I.: Enhancing first story detection using word embeddings. In Proceedings of the 39th International ACM SIGIR Conference on Research and Development in Information Retrieval, pp. 821–824. ACM, July 2016
12. Petrović, S., Osborne, M., Lavrenko, V.: Streaming first story detection with application to Twitter. In: Human Language Technologies: The 2010 Annual Conference of the North American Chapter of the Association for Computational Linguistics, pp. 181–189. Association for Computational Linguistics, June 2010
13. Petrović, S., Osborne, M., Lavrenko, V. : Using paraphrases for improving first story detection in news and Twitter. In: Proceedings of the 2012 Conference of the North American Chapter of the Association for Computational Linguistics: Human Language Technologies, pp. 338–346. Association for Computational Linguistics, June 2012
14. Pimentel, M.A., Clifton, D.A., Clifton, L., Tarassenko, L.: A review of novelty detection. Sig. Process. **99**, 215–249 (2014)
15. Schölkopf, B., Platt, J.C., Shawe-Taylor, J., Smola, A.J., Williamson, R.C.: Estimating the support of a high-dimensional distribution. Neural Comput. **13**(7), 1443–1471 (2001)
16. Wang, F., Ross, R.J., Kelleher, J.D.: Bigger versus similar: selecting a background corpus for first story detection based on distributional similarity. In: International Conference on Intelligent Data Engineering and Automated Learning, pp. 107–116. Springer, Cham, November 2018
17. Wang, F., Ross, R. J., Kelleher, J.D.: Exploring online novelty detection using first story detection models. In: Proceedings of International Conference Recent Advances in Natural Language Processing 2019, pp. 1312–1320 (2019)

18. Wurzer, D., Lavrenko, V., Osborne, M.: Twitter-scale new event detection via k-term hashing. In: Proceedings of the 2015 Conference on Empirical Methods in Natural Language Processing, pp. 2584–2589, September 2015
19. Yang, Y., Pierce, T., Carbonell, J.G.: A study on retrospective and on-line event detection (1998)

A Pilot Study on Argument Simplification in Stance-Based Opinions

Pavithra Rajendran[1], Danushka Bollegala[1(✉)], and Simon Parsons[2]

[1] University of Liverpool, Liverpool, UK
{pavithra.rajendran,danushka.bollegala}@liverpool.ac.uk
[2] Kings College London, London, UK
simon.parsons@kcl.ac.uk

Abstract. Prior work has investigated the problem mining arguments from online reviews by classifying opinions based on the stance expressed explicitly or implicitly. An implicit opinion has the stance left unexpressed linguistically while an explicit opinion has the stance expressed explicitly. In this paper, we propose a bipartite graph-based approach to relate a given set of explicit opinions as simplified arguments for a given set of implicit opinions using three different features (a) sentence similarity, (b) sentiment and (c) target. Experiments are carried out on a manually annotated set of explicit-implicit opinions and show that unsupervised sentence representations can be used to accurately match arguments with their corresponding simplified versions.

Keywords: Argument mining · Argument simplification

1 Introduction

The rise of social media has allowed people to share their opinions, in the form of reviews and debates, on online portals. Argument mining [18], an emerging research area, aims to discover arguments that are present in such user-based content. This paper is a contribution to work on argument mining.

In prior work [20], opinions were extracted from a set of hotel reviews and manually annotated as explicit or implicit based on how the stance in the opinion is expressed. Stance in NLP research refers to the standpoint taken by the user, whether they are for or against the given topic. In linguistics, stance is defined as *"the expression of attitude, judgement of the user towards the standpoint taken in the content"*. According to this definition, stance can be expressed either explicitly in a sentence or must be inferred from the context.

The question we explore here is that **given a set of opinions, does classifying the opinions into explicit and implicit opinions help to identify an explicit opinion as a simplified argument for an implicit opinion?**. We consider that explicit and implicit opinions are two different ways of expressing the same argument. The difference is that an explicitly stated opinion might be easier to spot and understood by a human than an implicitly stated one,

© Springer Nature Singapore Pte Ltd. 2020
L.-M. Nguyen et al. (Eds.): PACLING 2019, CCIS 1215, pp. 218–230, 2020.
https://doi.org/10.1007/978-981-15-6168-9_19

since understanding an implicit opinion requires inference from the context. We can see examples in Table 1, which gives pairs of implicit opinions and explicit opinions that express the same argument.

Table 1. Implicit opinions with corresponding explicit opinions as their simplified arguments.

Implicit opinion	Explicit opinion
Rooms had plenty of room and nice and quiet (no noise from the hallway hardwood floors as suggested by some - all carpeted)	Room was great
We received a lukewarm welcome at check in (early evening) and a very weak offer of help with parking and our luggage	We were extremely unimpressed by the quality of service we encountered
I have been meaning to write a review on this hotel because of the fact that staying here made me dislike Barcelona (hotels really can affect your overall view of a place, unfortunately)	This hotel was just a great disappointment

Given a set of explicit and implicit opinions, we propose a bipartite graph-based approach to identify whether a given explicit opinion is a simplified argument representation of an implicit opinion. We perform experiments on a hotel review dataset with and without the implicit/explicit opinion classification using three different types of features: (a) sentence similarity using different sentence embedding representations, (b) sentiment and (c) target information. We also experiment with a dialogue-based argumentation dataset (Citizen's Dialogue), which contains speaker's arguments annotated using the rephrase relation [13]. Our results show that the semantic similarity scores obtained using the unsupervised sentence embeddings give good performance for both datasets reporting the best accuracies of 0.86 and 0.82 respectively on the hotel review and Citizen's Dialogue datasets.

2 Related Work

Existing work on argument mining can be broadly classified into monological and dialogical texts depending on the user interaction. Work on monological texts deal with persuasive essays [21], articles [1] and online reviews [23,25]. Work on dialogical texts deals with debates [11], Tweets [6], dialogues [24], and other forms of user interactions [10].

Ghosh et al. [10] annotate user comments in forums as target-callout pairs based on pragma-dialectic theory.annotate user comments in forums as target-callout pairs based on pragma-dialectic theory and also investigate on the difficulties faced in doing the annotation task. This work is useful for the research community to understand the difficulties of annotating arguments in social media texts. Boltuzic et al. [3,4] have done continuous assessment on identify premises and claim, in particular, how they are related in debates. Their definition of support and attack depends on whether the relation is explicit or not. The authors also have created a dataset consisting of 125 claim pairs containing annotated premises for filling the gap between a user claim and the main claim of a topic. An advantage of this work is the availability of the dataset that can be used for comparing whether the model proposed can be useful for other available datasets.

Habernal et al. [11] build a large corpus based on the extended Toulmin model from debate portals using a semi-supervised approach. The different components annotated to represent an argument are the following:- premise, claim, backing, rebuttal and refuta- tion. The semi-supervised approach automatically extracts features from an unlabelled corpus by clustering word embedding vectors for classifying whether a given sentence is an argument or not. An advantage of this work is the semi-supervised approach that has been evaluated in detail for in-domain and cross-domain data along with detailed error analysis. Differing from the rest, Duthie et al. [9] work on political based debate corpus to identify ethos, which is an important part of argumentation. Walker et al. [24] determine how persuasive arguments are from the audience perspective while Oraby et al. [17] classify a dialogue based on whether it is factual or emotional.

Not only does argument mining focus on annotating arguments and its components (see [14] for a detailed survey), recent work has also considered the problem of extracting relations that exist between arguments. Cabrio and Villata [7] extract abstract arguments from debates to form a bipolar argumentation framework, with the support and attack relation automatically identified using textual entailment. In this paper, the authors empirically demonstrate that, in most cases, support and attack relation satisfy entailment and contradiction relations respectively. Boltuzic et al. [3] relate arguments using implicit/explicit support and attack relations. Similarly, Bosc et al. [5] annotate the support and attack relation among arguments present in tweets. Among scientific articles, the support and attack relation were extracted by Kirschner et al. [12].

Instead of extracting relations, Carstens and Toni [8] investigate towards how relation information can help in identifying arguments. In their work, they show how in many cases, the objective statements often ignored can actually constitute an argument. Thus, they consider pairs of sentences that satisfy either the *support, attack* or *neither* relation to demonstrate the same.

Konat et al. [13] studied the rephrase relation between two arguments. According to that work, rephrased arguments must not be considered as an additional support/attack. In particular, they consider a particular dialogue corpus where the same argument is made multiple different people. We propose an unsupervised approach that uses rephrase relations for argument simplification.

3 Background

In our prior work [20], we considered a statement expressed by a sentence, which can be either positive or negative in sentiment and talks about a single target entity, to be a stance expressing an opinion. Further, opinions present in a set of hotel reviews were annotated as implicit or explicit depending on how the stance is expressed within the text. The following guidelines were given to the human-annotators:

Explicit Opinion: Direct expression of approval/disapproval towards the hotel or its aspects. Certain words or clauses have a strong intensity of expression towards a particular target. For example, *worst staff!* has a stronger intensity against the target *staff* than *the staff were not helpful.*

Implicit Opinion: Those words or clauses that do not have a strong intensity of expression towards a particular target. In the above example *staff were not helpful* is an implicit opinion. Moreover, personal facts such as *small room, carpets are dirty* etc. Some of them may also be in the form of justifications or describing an incident.

4 Bipartite Graph-Based Opinion Matching

Given a set of opinions classified as implicit/explicit, we formulate the problem of identifying simplified arguments as a maximum cost K ranked bipartite graph-matching problem. The bipartite graph is formed by mapping each implicit opinion with each of the given explicit opinions. For every implicit opinion, the top K explicit opinions with the smallest costs are considered. Three different features are explored in computing the cost function for every implicit-explicit mapping as described in the following sections.

4.1 Unsupervised Sentence Embedding

To measure sentence similarity we use both unsupervised and supervised sentence-embedding representations. First, each word is represented using pre-trained embedding vectors. Based on existing works [2,16], we perform different steps on the pre-trained word embeddings to create sentence embeddings. Mu et al. [16] perform two post-processing steps on pre-trained word embeddings. The motivation of their work is to create better word embedding representations and hence they do not focus on sentence representation. They show that word embeddings are narrowly distributed in a cone and by subtracting the mean vector and applying Principal Component Analysis (PCA), it is possible to obtain an isotropic spherical distribution, which is better at recognising similar word pairs. The two post-processing steps are described next.

Diff: Assume we are given a set V (vocabulary) of words w, which are represented by a pre-trained word embedding $w_i \in \mathbb{R}^k$ in some k dimensional vector space. The mean embedding vector, $\hat{w}$, of all embeddings for the words in V is given by:

$$\hat{w} = \frac{1}{|V|} \sum_{w \in V} w \tag{1}$$

Following [16], the mean is subtracted from each word embedding to create isotropic embeddings as follows:

$$\forall_{w \in V} \qquad \tilde{w} = w - \hat{w} \tag{2}$$

WordPCA: The mean-subtracted word embeddings given by (2) for all $w \in V$ are arranged as columns in a matrix $\mathbf{A} \in \mathbb{R}^{k \times |V|}$, and its d principle component vectors $u_1, \ldots, u_d$ are computed. Mu et al. [16] proposed an embedding which removes the l most important principle components:

$$w' = \tilde{w} - \sum_{i=1}^{l} \left(u_i^\top w \right) u_i \tag{3}$$

We use these word embeddings to create sentence embeddings:

AVG: A simple, yet surprisingly accurate, method to represent a sentence is to compute the average of the embedding vectors of the words present in that sentence. Given a sentence $\mathcal{S}$, we first represent it using the set of words $\{w | w \in S\}$. We then create its sentence embedding $s \in \mathbb{R}^k$ as follows:

$$s = \frac{1}{|S|} \sum_{w \in S} w \tag{4}$$

Depending on the pre-processing applied on the word embeddings used in (4), three different variants for sentence embeddings are possible: **AVG** (uses unprocessed word embeddings w), **Diff+AVG** (uses $\tilde{w}$) and **WordPCA+AVG** (uses w').

WEmbed: Arora et al. [2] proposed a method to create sentence embeddings as the weighted-average of the word embeddings for the words in a sentence. The weight $\psi(w)$ of a word w is computed using its occurrence probability $p(w)$ estimated from a corpus:

$$\psi(w) = \frac{a}{a + p(w)} w \tag{5}$$

$$s = \frac{1}{|S|} \sum_{w \in S} \psi(w) w \tag{6}$$

Here, a is a small constant[1]. Intuitively, frequent words such as stop words will have a smaller weight assigned to them, effectively ignoring their word embeddings when computing the sentence embeddings.

SentPCA: Given a set of sentences, Arora et al. [2] applied PCA on the matrix that contains individual sentence embeddings as columns to compute

---

[1] Set to 0.001 in our experiments.

the first principal component vector v, which is subtracted from each sentence's embedding as follows: In total we have five sentence embedding methods (**AVG**, **Diff+AVG**, **WordPCA+AVG**, **WEmbed** and **SentPCA**). In the unsupervised approach, we measure the similarity between an implicit and an explicit opinion as the cosine similarity between their corresponding sentence embeddings.

4.2 Supervised Sentence Similarity

We propose a supervised method to compute the similarity between two sentences using their sentence embeddings, created from pre-trained word embeddings as described in Sect. 4.1 using a training dataset, where each pair of sentences is manually rated for the degree of their semantic similarity. Specifically, given two sentences s_i, s_j, we first compute their sentence embeddings, respectively s_i and s_j, using one of the unsupervised sentence embedding methods described in Sect. 4.1. Next, we represent a pair of sentences using two operators: $h_\times$ (elementwise multiplication) and h_- (elementwise absolute value of the difference).[2] Intuitively, $h_\times$ captures common attributes in the two sentences, whereas h_- captures attributes unique to one of the two sentences. We then feed $h_\times$ and h_- to a neural network containing a sigmoid ($\sigma(\cdot)$) hidden layer and a softmax ($\phi(\cdot)$) output layer parametrised by a set $\theta = \{\mathbf{W}^{(\times)}, \mathbf{W}^{(-)}, \mathbf{W}^{(p)}, b^{(h)}, b^{(p)}\}$ as follows:

$$h_\times = s_i \odot s_j$$

$$h_s = \sigma\left(\mathbf{W}^\times h_\times + \mathbf{W}^{(-)}h_- + b^{(h)}\right)$$

$$h_- = |s_i - s_j|$$

$$\hat{p}_\theta = \phi\left(\mathbf{W}^{(p)}h_s + b^{(p)}\right)$$

We use the SICK [15] sentence similarity dataset that consists of pairs of sentences manually rated in an ordinal range from 1 to 5, where 1 represents the lowest and 5 represents the highest similarity. We denote this gold standard rating for s_i and s_j by $y(s_i, s_j) \in [1, K]$, where $K = 5$ for the SICK dataset. We use the class probability distribution, $\hat{p}_\theta$ to compute the expected similarity rating $\hat{y}(s_i, s_j)$ between s_i and s_j as follows:

$$\hat{y}(s_i, s_j) = r\hat{p}_\theta \qquad (7)$$

Here, the rating vector is $r = (1, 2, \ldots, K)^\top$. We would like the expected rating to be close to the gold standard rating. Following Tai et al. [22], we define a sparse target distribution p that satisfies $y = r^\top p$:

$$p_i = \begin{cases} y - \lfloor y \rfloor & \text{if } i = \lfloor y \rfloor + 1 \\ y - \lfloor y \rfloor + 1 & \text{if } i = \lfloor y \rfloor \\ 0 & \text{otherwise} \end{cases}$$

---

[2] We drop the arguments of the operators to simplify the notation.

The parameters θ of the model are found by minimising the KL-divergence between p and $\hat{p}_\theta$ subjected to ℓ_2 regularisation over the entire training dataset D of sentence pairs as follows:

$$J(\theta) = \sum_{(s_i, s_j) \in D} \mathrm{KL}\left((p^{(k)} || \hat{p}_\theta^{(k)}\right) + \frac{\lambda}{2} ||\theta||_2^2 \tag{8}$$

Here, $\lambda \in \mathbb{R}$ is the regularisation coefficient, set using validation data. The cost function of the bipartite matching problem using sentence similarity can then be defined as follows.

$$C(i, j) = \mathrm{sim}(\boldsymbol{w_i}, \boldsymbol{w_j}) \tag{9}$$

Here, sim is the cosine similarity between sentence embeddings for the unsupervised approach and the predicted similarity rating $\hat{y}$ for the supervised approach.

4.3   Sentiment and Target

Sentiment and target play an important role among these stance-bearing opinions, and we can maximise the cost function by considering these two features. For this purpose, we define the cost function as follows:

$$C(i, j) = \mathrm{sim}(\boldsymbol{s_i}, \boldsymbol{s_j}) + Q(i, j) + R(i, j) \tag{10}$$

Q and R output a threshold value if S_i and S_j have the same sentiment and target.

We focus on whether implicit/explicit classification along with linguistic structure can help in identifying an simplified argument for a given argument, without the sentiment and target information. In many cases, the target may be stated explicitly in the opinion or may mention the target implicitly. Hence, for our experiments, we make use of the dataset where the sentiment and target are manually annotated.

5   Experiments and Results

For our experiments, we use pre-trained Glove embeddings [19] with 300 dimensions[3]. For *WordPCA*, $l = 2$ is used [16]. Sentiment of an opinion and the targets present are manually annotated. Here, a domain knowledge base related to the different aspects and aspect categories is used. The threshold values for both the sentiment and target functions (given in (10)) were varied from 0 to 1 on development data and we found that 0.5 is appropriate such that the cost function is not biased towards the sentiment and target information alone.

---

[3] https://nlp.stanford.edu/projects/glove/.

5.1 Evaluation Measures

The evaluation measures used in our experiments were:

Precision@K (P@K). For every implicit opinion, the top K explicit opinions are obtained. The number of correct explicit opinions among the top K opinions are summed and divided by the total number of implicit opinions present. Thus:

$$\text{P@K} = \frac{1}{m} \sum_{i=1}^{m} \frac{n_i}{K} \tag{11}$$

where m is the total number of implicit opinions, and n_i is the number of correct explicit opinions for the corresponding i-th implicit opinion.

Averaged precision@K (Avg P@K)

$$\text{Avg P@K} = \frac{1}{K} \sum_{i=1}^{K} P@i \tag{12}$$

Here, K is the number of top explicit opinions that are considered and $P@i$ represents the precision@i score.

Mean Reciprocal Rank (MRR)

$$\text{MRR} = \frac{1}{m} = \sum_{i=1}^{i=m} \frac{1}{R_i} \tag{13}$$

where m is the total number of implicit opinions and R_i is the rank of the first correct explicit opinion for the i-th implicit opinion.

Accuracy (Acc)

$$\text{Acc} = \frac{1}{m} \sum_{i=1}^{m} l \tag{14}$$

where $l = 1$ if at least one of the correct explicit opinions is present within the top 10 explicit opinions; otherwise 0. This is because in the case of the Citizen Dialogue corpus, exactly one argument is matched as a simplified argument for another.

We randomly selected 57 implicit opinions from the implicit/explicit opinions dataset[4]. Each implicit opinion is manually tested with the three most appropriate explicit opinions that are the corresponding simplified arguments from the dataset. In total, we have 56 explicit opinions. Again, each implicit opinion was manually verified with the 56 explicit ones and any of those that represent as simplified arguments for the implicit opinion was updated. The number of explicit opinions that are simplified arguments of the implicit opinions ranged

---

[4] This dataset contains 1288 opinions manually annotated by two annotators with an inter-annotator agreement with a Cohen's kappa of 0.71.

from a minimum of 1 to a maximum of 13. On average, the number of explicit opinions as simplified arguments for an implicit opinion was 6.

A bipartite graph with the implicit and explicit opinions as nodes and edges drawn from every implicit opinion to each of the explicit opinion is considered. For each implicit opinion, the top K explicit opinions with the cost function score ranging from highest to lowest are considered as correctly predicted simplified arguments. The cost function is computed using different features as described in Sect. 4. These top K explicit opinions were then compared against the manually identified explicit opinions.

In Table 2, we report the P@K for values of $K = 10, 15$ and 20 and the Avg P@K for values of $K = 15$ and 20. We observe that using **SENTPCA** does not perform better than the simple baseline **AVG**. The results also show that **WordPCA+AVG** is the best sentence embedding representation useful for predicting the correct explicit opinions. The similarity scores obtained using this unsupervised sentence embedding representation do better than the sentiment and target functions, and we get the best performance using all three types of features.

Table 2. For a given set 57 implicit opinions and 56 explicit opinions, we compute the cosine similarity between each pair of implicit and explicit opinions using each of the methods described in Sect. 4. Moreover, sentiment and target functions are computed. Precision@K with K = 10, 15, 20 are computed and the results are present. In addition, average Precision@K with K = 15 and 20 are computed and the results are shown.

Methods	P@10	P@15	P@20	Avg P@15	Avg P@20
UNSUPERVISED					
AVG	0.15	0.22	0.30	0.13	0.16
Diff+AVG	0.15	0.21	0.27	0.12	0.15
WordPCA+AVG	**0.17**	**0.23**	**0.30**	**0.14**	**0.17**
WEmbed	0.14	0.20	0.25	0.12	0.15
SENTPCA	0.14	0.20	0.27	0.12	0.21
SUPERVISED					
AVG	0.14	0.19	0.25	0.12	0.15
Diff+AVG	0.14	0.19	0.24	0.11	0.14
WordPCA+AVG	0.14	0.21	0.25	0.12	0.15
WEmbed	0.07	0.12	0.18	0.05	0.08
SENTPCA	0.10	0.14	0.22	0.08	0.11
Sentiment	0.08	0.14	0.17	0.06	0.13
Target	0.16	0.20	0.24	0.12	0.19
Sentiment + target	0.17	0.22	0.25	0.13	0.20
WordPCA+AVG+sentiment+target	**0.28**	**0.34**	**0.39**	**0.21**	**0.26**

Table 3. We compute the sentence similarity based on the methods described in Sect. 4. Mean reciprocal rank (MRR) and accuracy (Acc) is computed. The results are reported based on the following: the information whether an opinion is implicit/explicit for the implicit/explicit dataset and the category to which an argument belongs to for the Citizen Dialogue corpus is given (With Information) or not given (Without Information).

Methods	Without Information				With Information			
	Citizen dialogue		Implicit/Explicit		Citizen dialogue		Implicit/Explicit	
	MRR	Acc	MRR	Acc	MRR	Acc	MRR	Acc
UNSUPERVISED								
AVG	0.56	0.75	0.13	0.31	0.62	0.81	0.29	0.75
Diff+AVG	0.55	0.75	0.12	0.28	0.61	0.81	0.28	0.75
WordPCA+AVG	0.59	0.80	0.07	0.24	0.64	0.86	0.25	0.82
WEmbed	0.52	0.67	0.15	0.49	0.55	0.72	0.32	0.68
SENTPCA	0.51	0.67	0.16	0.47	0.55	0.72	0.35	0.65
SUPERVISED								
AVG	0.56	0.78	0.10	0.31	0.63	0.83	0.27	0.68
Diff+AVG	0.54	0.78	0.10	0.30	0.61	0.83	0.25	0.68
WordPCA+AVG	0.57	0.76	0.06	0.24	0.63	0.80	0.26	0.74
WEmbed	0.004	0.03	0.08	0.23	0.04	0.16	0.23	0.70
SENTPCA	0.007	0.04	0.10	0.31	0.03	0.16	0.13	0.35

The Citizen's Dialogue corpus contains the rephrase relation identified among premises present in the same argument structure present within the same dialogue. As the related premises belong to the same argument structure, a premise with additional information rephrases a premise with less information but which has a similar meaning. We collected 64 argument pairs with rephrase relation from this corpus for our experiments. Firstly, we are interested to know how the implicit/explicit opinion classification helps in identifying simplified arguments for a given set of arguments. To make a fair comparison against the Citizen Dialogue corpus and to assess the adaptability of our method, we assume that there is a classification system that is able to classify a premise as being a simplified argument or not. For instance, the length of the premise could be considered as one feature. An example from the corpus is given below:

We're going to keep you informed is a simplified argument representation of *During this construction phase, we're going to be doing everything we can to keep you informed and keep you safe and keep traffic moving safely.*

We experimented on two different settings—one where the information about whether an argument is simplified or not is given and the other where the information is not given.

The results are reported in Table 3. We observe that, for both datasets, the best performances for supervised and unsupervised approach are obtained using **WordPCA+AVG**. The implicit/explicit classification significantly improves the performance for the implicit/explicit dataset. Overall, performing the two post-processing steps on pre-trained embeddings gives the best sentence embedding representation using the simple average based embedding method.

5.2 Analysis of Results

The results in the previous section provide quantitative measures of performance in identifying simplified arguments for a given set of arguments. In this section, we look in some detail at the performance of similarity measure, sentiment and target in predicting the correct answers. First, consider the results when the cost function uses all three functions—sim, Q, R (Eq. 10)—for computing the cost. These are compared with the results when the cost function uses only the sentiment and target function (Q, R in Eq. 10). We use *WordPCA+AVG* for computing the similarity measure.

We find that, in some cases, sentiment and target are not able to predict the answers correctly while in other cases, the similarity measure fails to capture the information that is explicitly provided by sentiment and target. Given the implicit opinion *"but the service is totally different with so many rooms for improvement it became not acceptable"*, the first ranked predicted explicit opinion when using all three functions (Sim + Q + R) for computing the cost was *"we were extremely unimpressed by the quality of service we encountered"*. Both the implicit and explicit opinion express the same argument about the aspect "service" and hence, the answer is correct.

For the same example, the first ranked predicted explicit opinion using the sentiment and target functions (Q + R) for computing the cost is *"the rooms are not worth the money"*. We can see that the word "rooms" in the implicit opinion has been wrongly considered to refer to hotel rooms, and this mismatch cannot be captured using the sentiment and target information alone. This mismatch means that the prediction is incorrect. The sentiment and target functions, unlike the similarity measure, do not capture any contextual information and might predict answers randomly based on the sentiment and target information.

A second example starts with the implicit opinion *"this hotel could easily be 5 star, the facilities are fantastic, the rooms are beautifully furnished and equipped with all the latest technology"*. Here the top-ranked explicit opinion using Sim + Q + R is *"the hotel rooms are nice"* is a correct match for the implicit opinion, while the top-ranked opinion using Q + R is *"the rooms are not worth the money"* which, while the aspect has been correctly determined to be hotel rooms, is completely wrong.

Both the previous examples show the similarity measure are working well. It is the elements of the cost function that makes it possible to find a good match. However there are some cases where the contextual information captured by the similarity measure is not sufficient to detect a good match. This is where the domain knowledge information that identifies different aspects as the same target is not captured by the similarity measure. For example the implicit opinion *"the laundry came back promptly"* is correctly matched with the explicit opinion *"the service was great"* by the sentiment and target functions, but the similarity measure does not recognise these opinions as being similar. This might be

because both sentences are quite short, and many of the words they contain—
"came", "was", "back" and so on—are common words that are not good features
for opinion matching. It is also possible that the embeddings of the words "laundry" and "service" were not available or were not present as close word pairs.
Understanding the performance of the similarity measure is something we will
investigate more.

6 Conclusion

We proposed an unsupervised bipartite graph-based approach to automatically
predict among opinions, where one opinion can be represented as a simplfiied
argument of the another without changing the context. Three different features:
sentence similarity, sentiment and target are used for computing the cost function. Our experimental results on two different datasets show that unsupervised
sentence representations help in matching arguments with their corresponding
simplified arguments. Moreover, we observe that the weighted-averaged sentence
embeddings, useful for similarity tasks, do not give the best performance. The
best performance is achieved when sentences are represented using averaged word
vectors, where the word vectors are post-processed using **WordPCA**. This, in
combination with sentiment and target gives a precision@10 of 0.28.

References

1. Al-Khatib, K., Wachsmuth, H., Kiesel, J., Hagen, M., Stein, B.: A news editorial
   corpus for mining argumentation strategies. In: CicLing, pp. 3433–3443 (2016)
2. Arora, S., Liang, Y., Ma, T.: A simple but tough-to-beat baseline for sentence
   embeddings. In: ICLR (2017)
3. Boltužić, F., Šnajder, J.: Back up your stance: recognizing arguments in online
   discussions. In: ACL, pp. 49–58 (2014)
4. Boltuzic, F., Šnajder, J.: Fill the gap! analyzing implicit premises between claims
   from online debates. In: ArgMining@ACL, pp. 124–133 (2016)
5. Bosc, T., Cabrio, E., Villata, S.: Dart: a dataset of arguments and their relations
   on Twitter. In: LREC, pp. 1258–1263 (2016)
6. Bosc, T., Cabrio, E., Villata, S.: Tweeties squabbling: positive and negative results
   in applying argument mining on social media. In: COMMA, pp. 21–32 (2016)
7. Cabrio, E., Villata, S.: Combining textual entailment and argumentation theory
   for supporting online debates interactions. In: ACL, pp. 208–212 (2012)
8. Carstens, L., Toni, F.: Towards relation based argumentation mining. In: ArgMin-
   ing@ACL, pp. 29–34 (2015)
9. Duthie, R., Budzynska, K., Reed, C.: Mining ethos in political debate. In:
   COMMA, pp. 299–310 (2016)
10. Ghosh, D., Muresan, S., Wacholder, N., Aakhus, M., Mitsui, M.: Analyzing argu-
    mentative discourse units in online interactions. In: ACL, pp. 39–48 (2014)
11. Habernal, I., Gurevych, I.: Exploiting debate portals for semi-supervised argumen-
    tation mining in user-generated web discourse. In: EMNLP, pp. 2127–2137 (2015)
12. Kirschner, C., Eckle-Kohler, J., Gurevych, I.: Linking the thoughts: analysis of
    argumentation structures in scientific publications. In: ArgMining@EMNLP, pp.
    1–11 (2015)

13. Konat, B., Budzynska, K., Saint-Dizier, P.: Rephrase in argument structure. In: FLA Workshop@COMMA 2016, pp. 32–39 (2016)
14. Lippi, M., Torroni, P.: Argument mining: a machine learning perspective. In: TAFA, pp. 163–176 (2015)
15. Marelli, M., Bentivogli, L., Baroni, M., Bernardi, R., Menini, S., Zamparelli, R.: Semeval-2014 task 1: evaluation of compositional distributional semantic models on full sentences through semantic relatedness and textual entailment. In: SemEval@ COLING, pp. 1–8 (2014)
16. Mu, J., Bhat, S., Viswanath, P.: All-but-the-top: simple and effective postprocessing for word representations, CoRR abs/1702.01417 (2017)
17. Oraby, S., Reed, L., Compton, R., Riloff, E., Walker, M.A., Whittaker, S.: And that's a fact: distinguishing factual and emotional argumentation in online dialogue. In: ArgMining@HLT-NAACL, pp. 116–126 (2015)
18. Palau, R.M., Moens, M.F.: Argumentation mining: the detection, classification and structure of arguments in text. In: ICAIL, pp. 98–107 (2009)
19. Pennington, J., Socher, R., Manning, C.: Glove: global vectors for word representation. In: Proceedings of the 2014 Conference on Empirical Methods in Natural Language Processing (EMNLP), pp. 1532–1543 (2014)
20. Rajendran, P., Bollegala, D., Parsons, S.: Contextual stance classification of opinions: a step towards enthymeme reconstruction in online reviews. In: ArgMining@ACL 2016 (2016)
21. Stab, C., Gurevych, I.: Parsing argumentation structures in persuasive essays. Comput. Linguist. **43**(3), 619–659 (2017)
22. Tai, K.S., Socher, R., Manning, C.D.: Improved semantic representations from tree-structured long short-term memory networks. In: ACL, pp. 1556–1566 (2015)
23. Villalba, M.P.G., Saint-Dizier, P.: A framework to extract arguments in opinion texts. IJCINI **6**(3), 62–87 (2012)
24. Walker, M.A., Anand, P., Lukin, S.M., Whittaker, S.: Argument strength is in the eye of the beholder: audience effects in persuasion. In: EACL, pp. 742–753 (2017)
25. Wyner, A., Schneider, J., Atkinson, K., Bench-Capon, T.J.M.: Semi-automated argumentative analysis of online product reviews. In: COMMA, pp. 43–50 (2012)

Automatic Approval of Online Comments with Multiple-Encoder Networks

Vu Dang[(⊠)]

FPT Technology Research Institute, FPT University, Hanoi, Vietnam
vudh5@fpt.com.vn

Abstract. In modern online publishing, user comments are an integral part of any media platform. Between the high volume of generated comments and the need for moderation of inappropriate content, human approval becomes a serious bottleneck with negative consequences for both operating cost and user experience. To alleviate this problem we present a text classification model for automatic approval of user comments on text articles. With multiple textual input from both the comment in question and the host article, the model uses a neural network with multiple encoders. Different choices for encoder networks and combination methods for encoder outputs are investigated. The system is evaluated on news articles from a leading Vietnamese online media provider, and is currently on a test run with said newspaper.

Keywords: Neural network · Text classification · Online media

1 Introduction

For the vast majority of modern-day media platforms, the social component has become more important than ever. User discourse is now central to drawing and maintaining user interest, sometime even more so than the media content itself. On the other hand the proliferation of fake news and social predation has placed online media under intense scrutiny, and content providers are expected to moderate their own users. As the amount of user-generated content for the typical provider far outstrips the capacity of its editorial staff, human moderation puts great stress on editorial resources while unable to maintain timely approval of user content.

In this work we describe the work done on automatic approval of comments on news articles for VnExpress, a leading online media provider in Vietnam. The task is formulated as a supervised text classification problem using examples of accepted and rejected comments during actual operation of the editorial staff. Since the acceptability of a comment is dependent on the topic and content of the host article, the model input includes such information as the title, summary and category of the article in addition to the comment itself. The model is trained to predict a binary label for each comment and article, corresponding to acceptance or rejection.

© Springer Nature Singapore Pte Ltd. 2020
L.-M. Nguyen et al. (Eds.): PACLING 2019, CCIS 1215, pp. 231–242, 2020.
https://doi.org/10.1007/978-981-15-6168-9_20

The problem of text classification has seen much progress thanks to recent developments in deep learning. Instead of applying generic machine learning models to traditional text features such as n-grams or POS, more and more specialized neural networks have been designed to work directly on words and characters from the input text. Recurrent neural networks, in particular the Long Short-Term Memory (LSTM) [5], have long proved their suitability to sequential data such as text. On the other hand convolutional neural networks (CNNs), initially popularized in computer vision applications by such seminal works as [9,11], were later adapted to handle text input with considerable success. [3,8] introduced relatively simple CNNs for a variety of natural language processing tasks including text classification. Further works went into specialized nuances of CNNs for text classification; for example [7] explores CNNs with direct convolutions over one-hot word encodings, while [10] proposes a new network structure combining recurrent and convolutional connections. [15] stacked LSTM and convolutional layers for speech recognition, and [21] uses essentially the same architecture for text classification. More recently attention mechanisms have also found success with both image [14,18] and text [1]. Self-attention networks [13,19] offer relatively simple yet effective models to represent sequential input, while the Transformer model [16] set new standards for neural machine translation using purely attentional elements.

In this work we propose a multi-encoder network for classifying the acceptability of each comment in its context, with an encoder for each text input channel. We consider various design choices for the overall model as well as different encoder architectures, namely the CNN, LSTM-CNN with attention, LSTM with self-attention, and Transformer encoder. Our model is undergoing a trial run at VnExpress with a view to replacing most of the current manual moderation effort. The remainder of this paper is structured as follows: In Sect. 2 we further elaborate the problem setting and the input data. In Sect. 3 we describe the proposed model. Section 4 presents a number of variants to the model and Sect. 5 reports the result of comparative evaluation between them. Finally Sect. 5 concludes the paper.

2   Problem Description

This work aims to automate the moderation process for comments on online news articles. In the current process, user comments can be submitted to the article page at any time but will only be displayed once a human editor has manually approved the content. The moderation follows a set of guidelines where a rejection can be issued for a variety of reasons such as:

- Political incorrectness.
- Vulgarity, pornography or other offensive content.
- Baseless accusation or defamation.
- Advertisement or PR.
- Antisocial content.
- Personal attacks.

As such approval depends on not just the comment itself but also the context. For example a comment accusing a suspect can be approved if the host article has already established that said suspect was found guilty in court, but has to be rejected if the crime is still under investigation. Similarly a comment praising a particular product can be displayed in an article about said product, but will be rejected as advertisement in the context of an unrelated article. Thus we frame the automation task as a classification problem using both the comment and its context as input, where the output label indicates approval or rejection. The classifier is trained on past comments and articles using actual editors' approvals and rejections as ground truth. Its output will be used to decide the approval of a comment if the confidence meets a certain threshold, to be established by the editors after a trial run. On the other hand if the confidence is low, the comment and provisional decision will be passed to an editor to review.

In particular the following input features are available:

- A text string representing the content of the comment.
- A text string representing the title of the host article.
- A text string representing the summary of the host article. This summary is produced by an editor for preview purposes as part of the standard operating procedure.
- A discrete variable representing the top-level category of the host article, for example *politics*, *sport* or *travel*.

Dataset statistics are presented in Table 1. The text inputs are segmented into words using a word tokenizerl. This step is necessary since modern Vietnamese language is written as a sequence of syllables with no specific notation for word boundaries.

Table 1. Dataset statistics

Accepted comments	274154
Rejected comments	186474
Unique articles	16087
Unique categories	16
Average comment length (words)	30.6
Average title length (words)	12.6
Average summary length (words)	22.1

3   Proposed Model

3.1   Model Structure

The general structure of our proposed model is illustrated in Fig. 1, with one encoder network for each text input channel and a dense layer with softmax on top. The encoder outputs are concatenated, then element-wise multiplication is performed with a category-specific weight vector before going to the dense layer.

Let $\mathbf{x}$ be the word sequence representing the comment itself, and $\mathbf{t}$ and $\mathbf{s}$ represent the title and summary of the host article respectively. Let F_x, F_t and F_s be the corresponding encoder functions with the same output dimension d, and denote by $\oplus$ the concatenation operator. Then the combined encoder output is a $3d$-dimensional vector

$$E(\mathbf{x}, \mathbf{t}, \mathbf{s}) = F_x(\mathbf{x}) \oplus F_t(\mathbf{t}) \oplus F_s(\mathbf{s})$$

Let c be the category of the host article and e_c be an $3d$-dimensional vector corresponding to c (which can be regarded as an embedding from the set of categories to $\mathbb{R}^{3d}$), and denote by $\circ$ the Hadamard or element-wise product. Furthermore let W and b be the weight and bias of the final dense layer, then the final model output is

$$\text{softmax}(W(E(\mathbf{x}, \mathbf{t}, \mathbf{s}) \circ e_c) + b)$$

The most natural approach to incorporate category information would be to append an embedding vector or a one-hot vector to the combined encoder output. However this approach only allows the category information to influence the next layer additively, while we have observed that the nature of the moderation process is highly dependent on the category of the host article. For example comments on politics are generally rejected for completely different reasons than comments on lifestyle articles. Thus we opted to use a Hadamard product of embeddings, which allows the input category to have direct influence over all features coming into the last classifier layer.

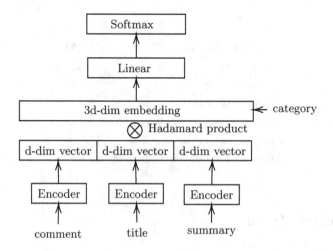

Fig. 1. Proposed classification model for comment approval

3.2 Self-attention Encoder

For the encoders we use the self-attention network in [13]. The network consists of a bi-directional LSTM layer followed by a self-attention mechanism. Let $H = (h_1 \ldots h_n)$ be the hidden states of the LSTM layer, then the embedding M of the whole input sequence is a linear combination of these states, namely $M = AH$. The combination weights are computed as in Fig. 2 (taken from [13]). In short the weight matrix A is the output of a two-layer perceptron without bias (using ReLU activation instead of tanh as in [13]), with a softmax applied along the second dimension (i.e. along the sequence). The rows of A act like different filters which allow the mechanism to attend to multiple parts of the input sequence, and the embedding M is a $r \times 2u$ matrix where r is the number of filters and $2u$ is the output dimension of the bi-directional LSTM. Finally we use a fully connected layer to reduce M to a d-dimensional embedding vector. For regularization we apply dropout to the input and output layers.

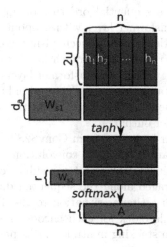

Fig. 2. Structure of the self-attention weights

4   Model Variants

In this section we present various options to the model and evaluate them on the collected dataset.

4.1 Baselines

As baseline we evaluate a simple application of logistic regression, SVM and gradient boosted trees. Each input text channel is represented by a bag-of-words from unigrams, bigrams and trigrams, with one-hot encoding for article category. These representations are concatenated into a long vector to feed into the classifier. The results for these classifiers are compared to the results of our proposed model with different encoders in Sect. 5.

4.2 Encoders

Here we list the different networks which were used as encoders in our model:

- **CNN:** We implement a simple convolutional network in the style of [8]. The network consists of multiple convolutional layers applied in parallel to the model input. The convolutional outputs go through adaptive pooling before concatenation, with a dense layer on top.
- **LSTM-CNN with Attention:** The first stage of this network is a bi-directional LSTM layer with a soft attention mechanism. The second stage has the same structure as the aforementioned CNN encoder and acts on the output of the first stage.
- **Self-attention:** This is the proposed encoder as detailed in Sect. 3.
- **Transformer:** The Transformer network is an influential architecture using purely attentional elements, first proposed in [16]. The core of this network is the Multi-Head Attention mechanism (Fig. 3), with added positional embeddings to account for the sequential ordering of input tokens. Our encoder consists of a single Transformer layer with dropout, where residual input is added to the output of the attention mechanism before layer norm is applied. In our preliminary experiment this is much faster to train and less prone to overfitting than stacking multiple Transformer layers as in [16], which makes sense as our model contains three encoders and only has to predict a binary label. The output sequence is averaged along the sequential dimension to produce the final encoder output.
- **ConvS2S:** This encoder is derived from ConvS2S [4], a purely convolutional network for machine translation. Each convolutional block consists of a convolution with Gated Linear Units and a residual connection (Fig. 4). The output sequence is averaged along the sequential dimension to produce the final encoder output. In a similar situation to the Transformer encoder, we found that using a single block for each encoder resulted in faster training and less overfitting than stacking multiple blocks per encoder.

4.3 Model Configurations

We experimented with various ways to put the encoder outputs together with the category information as below:

- **Input concatenation:** As a baseline we simple concatenate all input channels (title, summary and comment) with the one-hot representation of the category to form a single sequence. The model is reduced to a simple classifier consisting of a single encoder which feeds into a dense layer with softmax.
- **Feature concatenation:** We concatenate the encoder outputs and the one-hot representation of the category into a single sequence before the dense layer with softmax.
- **Shared encoder:** This is the same configuration as the proposed model, except that all the encoders share the same weights.

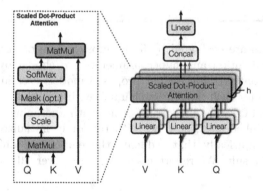

Fig. 3. Multi-head attention from the Transformer architecture

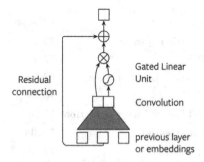

Fig. 4. Convolutional block from the ConvS2S architecture

- **Feature multiplication:** In stead of concatenating the encoder outputs, we take the Hadamard product of all the encoder outputs and an embedding of the category (all being of the same dimensionality). This product is the input to the final dense layer with softmax.

4.4   Pretrained Embeddings

We evaluated our model with both randomly initialized word embeddings and pretrained embeddings. The pretrained embeddings were obtained from [17], in particular we used the 300-dimensional fastText [2] word vectors. When using the pretrained embeddings we initialize all pretrained words to their corresponding embedding vectors and the remaining words to random embeddings. Then we experimented with two options:

- Fixing the embedding vectors of pretrained words to their initial pretrained values and only fine-tuning the randomly initialized embeddings.
- Fine-tuning all embedding vectors during training.

5   Evaluation

The model variants are evaluated by 5-fold cross validation, using 10% of the training set for validation and hyperparameters are chosen by a simple grid search. The best configuration for our proposed models uses independent self-attention encoders for each input text channel, with a hidden dimension of 200 and dropout rate of 0.2. The best result comes with finetuning 300-dimensional pretrained word embeddings; however with randomly initialized word embeddings we found empirically that reducing the embedding dimension to 100 achieves the best result. We report macro-F1 score over all five folds in percentage points.

Table 2. Evaluation of different encoders and baselines

	F1 score
Logistic regression	73.07
SVM	72.42
Gradient boosted trees	68.59
CNN	76.58
LSTM-CNN with Attention	78.28
Self-attention	**78.34**
Transformer	77.93
ConvS2S	77.12

Table 2 shows the F1 score for different encoders. The best results by a considerable margin belong to the Self-attention and LSTM-CNN with Attention, and thus we perform further evaluations with these two encoders. The results are reported in Fig. 5. All model variants outperform the baseline with input concatenation by a large margin, with the proposed version clearly achieving the best results for both encoder options.

Table 3 reports our experiment with pretrained embeddings for the two best encoders. Fixed embeddings fall behind by a large margin with both encoders, while for the LSTM-CNN with Attention the random embeddings actually outperformed fine-tuned pretrained embeddings. This is because the optimal embedding dimension found in our experiments for this encoder is 200 while the embedding vectors are 300-dimensional, and raising the dropout rate could not compensate for the increased model capacity. On the other hand the Self-attention encoder achieves the best result by fine-tuning pretrained embeddings.

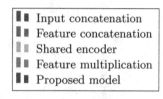

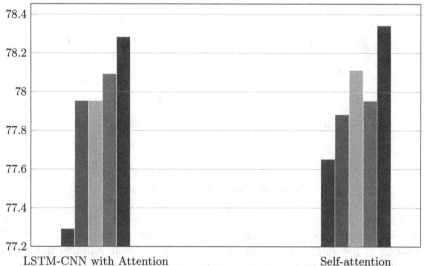

Fig. 5. Evaluation of model configurations

Table 3. Evaluation of pretrained embeddings

Configuration	Embedding	F1 score
Self-attention	Random	78.34
Self-attention	Fixed	75.00
Self-attention	Fine-tuned	**78.42**
LSTM-CNN with Attention	Random	78.28
LSTM-CNN with Attention	Fixed	75.16
LSTM-CNN with Attention	Fine-tuned	77.85

5.1 Performance Considerations

In practical use we are mainly interested in the inference time since training can be scheduled while the system has no control over when new comments arrive. Figure 6 show the average inference time over 100 batches for different batch sizes from 1 to 1000, run on a single Tesla V100-SXM2 GPU. While the average batch time for the Self-attention encoder stays essentially constant, the LSTM-CNN with Attention encoder slows down considerably with larger batches as it uses up the GPU memory.

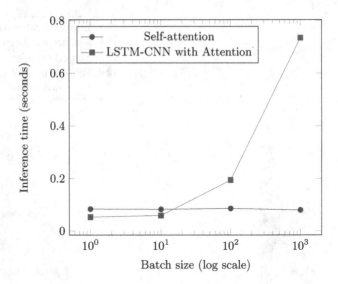

Fig. 6. Inference time by batch size

In principle this means the Self-attention encoder scales much better with the volume of incoming comments. However in practice it requires thoughtful implementation to make use of this large-batch advantage: the system cannot just wait for a full batch before inferencing since the wait time might be too long, but running each single comment through the model as soon as they come in would be highly inefficient. Currently we implement a fixed waiting time where comments which arrive within this time will be processed as a batch. The waiting time is chosen empirically; in fact setting it close to 0.1 s makes sure that the system has time to finish a batch before the next one arrives, while the impact on user experience is negligible. In the rare case that there are too many incoming comments during the waiting time, we simply create extra model instances on other worker machines to distribute the load.

6   Conclusion

In this paper we formulated the task of automatic moderation of online comments to news articles as a classification problem. We proposed a neural network model for the task and explored various design choices for our model including different encoders, configurations and the use of pretrained word embeddings. Evaluation on operational data shows the consistent superiority of the LSTM with self-attention as the encoder for our model and the importance of article category. The best-performing model variant is currently being trialed by the online newspaper VnExpress, to be optimized for deployment. We also discussed some practical consideration regarding batching, which arose during the trial run.

Since the proposed model uses one encoder per input text channel, it can be easily extended to richer formulations of the task, for example by adding related articles or other comments in the same conversation. Given the multimodal nature of digital media, the natural next step in our work is to incorporate images and videos where applicable and we are investigating specific choices of encoders for each mode of input. A more challenging problem is the need to justify the decisions taken by the automatic system, which will speed up the process of manual review considerably in case the model's prediction does not meet the confidence threshold. Attention mechanisms have shown some potential in explaining neural networks even though their practical usefulness in this regard is being debated (for example see [6]). In addition certain neural network architectures have been specifically designed to provide evidence for their predictions such as [12,20] as part of the growing interest in explainable machine learning. These advances provide promising directions for a deeper approach to the problem of automatic content moderation.

References

1. Bahdanau, D., Cho, K., Bengio, Y.: Neural machine translation by jointly learning to align and translate. arXiv e-prints abs/1409.0473, September 2014. https://arxiv.org/abs/1409.0473
2. Bojanowski, P., Grave, E., Joulin, A., Mikolov, T.: Enriching word vectors with subword information. Trans. Assoc. Comput. Linguist. 5, 135–146 (2017)
3. Collobert, R., Weston, J., Bottou, L., Karlen, M., Kavukcuoglu, K., Kuksa, P.: Natural language processing (almost) from scratch. J. Mach. Learn. Res. 12, 2493–2537 (2011). http://dl.acm.org/citation.cfm?id=1953048.2078186
4. Gehring, J., Auli, M., Grangier, D., Yarats, D., Dauphin, Y.N.: Convolutional sequence to sequence learning. In: Proceedings of the 34th International Conference on Machine Learning - Volume 70, ICML 2017, pp. 1243–1252. JMLR.org (2017). http://dl.acm.org/citation.cfm?id=3305381.3305510
5. Hochreiter, S., Schmidhuber, J.: Long short-term memory. Neural Comput. 9(8), 1735–1780 (1997). https://doi.org/10.1162/neco.1997.9.8.1735
6. Jain, S., Wallace, B.C.: Attention is not explanation. CoRR abs/1902.10186 (2019)
7. Johnson, R., Zhang, T.: Effective use of word order for text categorization with convolutional neural networks. In: NAACL HLT 2015, The 2015 Conference of the North American Chapter of the Association for Computational Linguistics: Human Language Technologies, Denver, Colorado, USA, 31 May–5 June 2015, pp. 103–112 (2015), http://aclweb.org/anthology/N/N15/N15-1011.pdf
8. Kim, Y.: Convolutional neural networks for sentence classification. In: Proceedings of the 2014 Conference on Empirical Methods in Natural Language Processing, a Meeting of SIGDAT, a Special Interest Group of the ACLEMNLP 2014, Doha, Qatar, 25–29 October 2014, pp. 1746–1751 (2014), http://aclweb.org/anthology/D/D14/D14-1181.pdf
9. Krizhevsky, A., Sutskever, I., Hinton, G.E.: ImageNet classification with deep convolutional neural networks. In: Proceedings of the 25th International Conference on Neural Information Processing Systems - Volume 1, NIPS 2012, pp. 1097–1105. Curran Associates Inc., Red Hook (2012). http://dl.acm.org/citation.cfm?id=2999134.2999257

10. Lai, S., Xu, L., Liu, K., Zhao, J.: Recurrent convolutional neural networks for text classification. In: Proceedings of the Twenty-Ninth AAAI Conference on Artificial Intelligence, AAAI 2015, pp. 2267–2273. AAAI Press (2015). http://dl.acm.org/citation.cfm?id=2886521.2886636

11. Lecun, Y., Bottou, L., Bengio, Y., Haffner, P.: Gradient-based learning applied to document recognition. In: Proceedings of the IEEE, pp. 2278–2324 (1998)

12. Lei, T., Barzilay, R., Jaakkola, T.: Rationalizing neural predictions. In: Proceedings of the 2016 Conference on Empirical Methods in Natural Language Processing, Austin, Texas, pp. 107–117. Association for Computational Linguistics, November 2016. https://doi.org/10.18653/v1/D16-1011. https://www.aclweb.org/anthology/D16-1011

13. Lin, Z., et al.: A structured self-attentive sentence embedding (2017)

14. Mnih, V., Heess, N., Graves, A., Kavukcuoglu, K.: Recurrent models of visual attention. In: Ghahramani, Z., Welling, M., Cortes, C., Lawrence, N.D., Weinberger, K.Q. (eds.) Advances in Neural Information Processing Systems 27, pp. 2204–2212. Curran Associates, Inc. (2014). http://papers.nips.cc/paper/5542-recurrent-models-of-visual-attention.pdf

15. Sainath, T.N., Vinyals, O., Senior, A., Sak, H.: Convolutional, long short-term memory, fully connected deep neural networks. In: 2015 IEEE International Conference on Acoustics, Speech and Signal Processing (ICASSP). pp. 4580–4584, April 2015. https://doi.org/10.1109/ICASSP.2015.7178838

16. Vaswani, A., et al.: Attention is all you need. In: Guyon, I., et al. (eds.) Advances in Neural Information Processing Systems 30, pp. 5998–6008. Curran Associates, Inc. (2017). http://papers.nips.cc/paper/7181-attention-is-all-you-need.pdf

17. Vu, X., Vu, T., Tran, S.N., Jiang, L.: ETNLP: a toolkit for extraction, evaluation and visualization of pre-trained word embeddings. CoRR abs/1903.04433 (2019)

18. Xu, K., et al.: Show, attend and tell: neural image caption generation with visual attention. In: Proceedings of the 32nd International Conference on International Conference on Machine Learning - Volume 37, ICML2015, pp. 2048–2057. JMLR.org (2015). http://dl.acm.org/citation.cfm?id=3045118.3045336

19. Yang, B., Wang, L., Wong, D.F., Chao, L.S., Tu, Z.: Convolutional self-attention networks. CoRR abs/1904.03107 (2019). http://arxiv.org/abs/1904.03107

20. Zhang, Y., Marshall, I., Wallace, B.C.: Rationale-augmented convolutional neural networks for text classification. In: Proceedings of the 2016 Conference on Empirical Methods in Natural Language Processing, Austin, Texas, pp. 795–804. Association for Computational Linguistics, November 2016. https://doi.org/10.18653/v1/D16-1076. https://www.aclweb.org/anthology/D16-1076

21. Zhou, C., Sun, C., Liu, Z., Lau, F.C.M.: A C-LSTM neural network for text classification. CoRR abs/1511.08630 (2015)

Question and Answering, Dialog Analyzing

Is the Simplest Chatbot Effective in English Writing Learning Assistance?

Ryo Nagata[1,2(✉)], Tomoya Hashiguchi[3], and Driss Sadoun[4,5]

[1] Konan University, 8-9-1 Okamoto, Higashinada, Kobe, Hyogo 658-8501, Japan
nagata-pacling2019@ml.hyogo-u.ac.jp
[2] Japan Science and Technology Agency, PRESTO, 4-1-8 Honcho, Kawaguchi, Saitama 332-0012, Japan
[3] University of Hyogo, 7-1-28 Minatojima-minamimachi, Chuo-ku, Kobe, Hyogo 650-0047, Japan
aa19j508@ai.u-hyogo.ac.jp
[4] PostLab, 113 rue de la République, 13002 Marseille, France
driss.sadoun@postlab.fr
[5] ERTIM, INALCO, 7 rue de Lille, 75007 Paris, France

Abstract. While writing plays a central role in writing learning, non-native learners often find difficulty in writing English, which hinders them from engaging in writing exercises. This paper examines the hypothesis that even the simplest chatbot (such as ELIZA) has a positive effect on assisting learners in writing more. We empirically show such a tendency by comparing words that learners produce by using a standard editor and a chatbot-based writing system. We further look into the writing results, showing that the chatbot-based system has good effects on word usage and self-revision. Finally, we propose a new writing exercise combining the chatbot-based system with the conventional method.

Keywords: Writing learning · Learning assistance · Chatbot · Feedback · EFL/ESL

1 Introduction

While writing plays a central role in writing learning, non-native learners of English often find difficulty in writing. This is especially true for beginner to intermediate English as Foreign Language (EFL) learners; they tend to be able to write only a small amount. They have to think about what to write in English, considering other English-related requirements including lexical choice, sentence construction, and writing organization. Because of this high demand, it often happens that they cannot arrive at what they want to write. Even if they somehow manage to, they might not know how to realize it in English. These circumstances make it hard for them to keep motivated in writing learning.

It is favorable to work on as many writing exercises and to write as many amounts as possible in order to improve one's writing skill. Recent studies on

L.-M. Nguyen et al. (Eds.): PACLING 2019, CCIS 1215, pp. 245–256, 2020.
https://doi.org/10.1007/978-981-15-6168-9_21

corrective feedback such as Nagata et al. [11] and Sheen et al. [17] show that learners can improve their writing accuracy even without corrective feedback but with self-revision of their writings. Considering these, one of the keys to success in writing learning assistance is to let learners write more.

The NLP community has mainly focused on English writing learning assistance by means of f grammatical error correction/correction as Sect. 2 will discuss. Although their performance has dramatically improved in recent years, they do not directly aim to help learners write more, but rather remedy problems in what has been written. Accordingly, learning assistance to help learners write more will complement grammatical error detection and correction and together they will be beneficial in writing learning. Besides grammatical error correction becomes effective only when learners write a certain amount.

In view of this background, in this paper, we explore how we can facilitate writing learning by using a chatbot-based writing learning system. Specifically, we introduce and examine the following hypothesis:

Hypothesis:
Even the simplest chatbot (such as ELIZA [18]) has positive effects on facilitating writing learning.

To achieve it, we develop a chatbot based on ELIZA for English writing assistance. Questions and responses from a chatbot will likely guide learners in what and how to write; they can be hints for what to write and also be examples that learners may adopt in their own writing. This will likely help learners write more. This also might have other positive effects on writing learning such as acquiring new words and expressions and/or new grammatical items. Besides, positive reactions such as praise and surprise from a chatbot might keep learners motivated in writing exercises.

The reason why we use ELIZA is that it is a very simple dialog system, which can be a good baseline system (or a starting point) to develop more sophisticated writing assistance systems based on dialog systems.

In this paper, we explore the above hypothesis by having learners write essays using a standard editor and a writing assistance system based on ELIZA. We then compare the amount of writing that they produce under the two conditions. We further investigate the resulting essays to show:

1. They tend to be able to write more by using the chatbot-based system.
2. The chatbot-based system has advantages and disadvantages in learning assistance.
3. How we can make the chatbot-based system more sophisticated.

2   Related Work

The work most related to the present work in NLP is probably grammatical error detection/correction. The former is a task of detecting where errors exist in a given sentence whereas the latter is of retrieving its corrected form. In the beginning, researchers focused on specific error types – the most frequent ones such as errors in article [5,6], preposition [4], and number [10]. The state-of-the-art methods typically solve the problems as sequence labeling (e.g., [8,13,14]) or Machine Translation (MT) with DNNs (e.g., [7,12]). As a result, performance has dramatically improved.

In second language acquisition, there has be a great amount of work on corrective feedback (CF). Most written CF research has compared different types of written feedback to investigate whether certain types of CF produce a more positive effect than others (e.g., the work [1,15,17]). While the discussion whether a certain type of CF is better or not is still ongoing, researchers [11,17] show that learners can improve their writing skill by self-revision even without explicit CF (although direct CF is more effective).

There has been also work on the investigation of learner characteristics and feedback in tutorial dialogue. Marineau et al. [9] investigate speech acts in tutorial dialogue. Boyer et al. [3] show that tutorial strategies intended to maximize learner motivational outcomes may not be the same strategies that maximize cognitive outcomes. Others such as Core et al. [2] investigate the correlation between tutorial dialogues and learning.

3 Method

The chatbot-based system used in this work is an implementation of ELIZA as shown in Fig. 1. The utterances[1] on the left-hand side are from the chatbot. As in this figure, writing begins with a chatbot's utterance to the learner. The learner writes by responding to the chatbot's utterances. The resulting exchange is saved in a log file.

A teacher of English and two NLP researchers collaboratively created two scripts for the chatbot; the topics were *self-introduction* and *hobby*. They designed the scripts so that learners can write easily. They set 30 and 28 keywords for *self-introduction* and *hobby*, respectively. These keywords correspond to 83 and 79 response templates, respectively.

The other condition uses a standard editor, whose interface is shown in Fig. 2. It allows the user to input and revise text. Accordingly, learners perform writing just as in normal English writing exercises.

We had 19 learners of English write essays using the two systems (ten participated in the experiment in 2016 and nine in 2017). They were all college students. Each learner wrote an essay on the two topics (*self-introduction* and *hobby*). To be precise, they wrote an essay on one of the two topics using the chatbot-based system and another on the other topic using the standard editor. They were given 50 minutes to write an essay. They did not use other sources of information such as a dictionary for writing.

---

[1] The input and output of the chatbot-based system are not actually utterances but rather typed and written sentences. Nevertheless, the term *utterance* is used in this paper for illustration purposes.

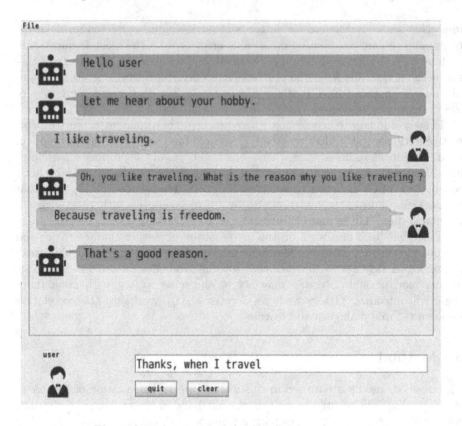

Fig. 1. Sample picture of the Chatbot-based system.

Here, note that there are four combinations of topics and systems (e.g., first, *self-introduction* with the chatbot-based system, and then *hobby* with the standard editor and so on). To reduce the order effect, each learner was randomly assigned to one of the four combinations, keeping the total number of learners equal under each combination. To be precise, five out of 19 were assigned to the combination first, *self-introduction* with the chatbot-based system, and another five were assigned to the combination first, *self-introduction* with the standard editor. The rest were assigned equally to the remaining combinations. All learners wrote two essays on different days. In the end, one of nine in 2017 did not finish the two essays. Consequently, we obtained 36 essays from 18 learners.

In 2017, we had the learners answer a questionnaire at the end of each session. Each learner freely wrote about his or her impression/opinion on the system (either chatbot-based system or the standard editor) that he or she used to write.

To measure the writing amount, we preprocessed the obtained data. We merged the same or similar sentences into one if they were responses to the same question from the chatbot; note that we implemented the chatbot in such a way that it may ask the exact same question several times. It is not appropriate to count them all for the purpose of measuring how much learners write. When

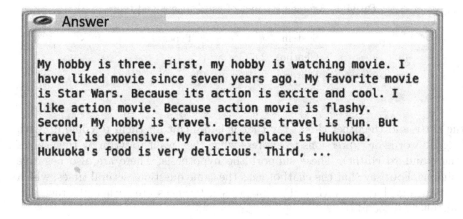

My hobby is three. First, my hobby is watching movie. I have liked movie since seven years ago. My favorite movie is Star Wars. Because its action is excite and cool. I like action movie. Because action movie is flashy. Second, My hobby is travel. Because travel is fun. But travel is expensive. My favorite place is Hukuoka. Hukuoka's food is very delicious. Third, ...

Fig. 2. The standard editor used for writing.

we found two or more similar sentences with the above condition, we kept the longest one and deleted the rest. We performed this preprocessing manually. In addition, we removed the following symbols: .,!?/()-:''. We used the Stanford-CoreNLP tool[2] to tokenize the obtained data.

4 Results

Table 1 summarizes the results. Here, # *tokens* denotes the average number of word tokens produced by using the standard editor and the chatbot-based system. Similarly, # *types* denotes the number of word types. Note that repeated sentences are excluded as already explained in Sect. 3.

Table 1 shows that, on average, the number of word tokens that the learners produced using the chatbot-based system is approximately 1.2 times more than that using the standard editor. Figure 3 depicts the insight where the x and y axes correspond to the numbers of word tokens using the standard editor and the chatbot-based system, respectively; each point corresponds to each learner. Figure 3 shows that when using the chatbot-based system, most of the learners wrote equal to or more than when using the standard editor. However, the difference is not statistically significant at the 0.05 level (two tailed t-test, $p = 0.064$).

Similarly, we performed the statistical test for word types, obtaining a larger value of $p = 0.270$. This is not so surprising considering that the chatbot-based system is designed to help learners write more but not to help them use a variety of words. However, some of the learners in their writings adopted words or expressions from the chatbot's utterances as Sect. 5 will shortly discuss.

The obtained questionnaires are summarized as follows. Three out of the eight say that it is easier to write using the chatbot-based system. Five (two of

---

[2] https://stanfordnlp.github.io/CoreNLP/.

Table 1. Average number of word tokens and types.

System	# Tokens	# Types
Chatbot	247	115
Editor	204	105

the three and the other three) say that it is difficult for them to come up with English words or expressions to express what they want to write by themselves (the standard editor). These support the hypothesis. There are also negative opinions. Four say that the chatbot asks the same questions several times, which is annoying. Three say that the chatbot's utterances are rather abstract and difficult to respond to.

5   Discussion

Section 4 has shown that learners tend to be able to write more using the chatbot-based system although the difference is not statistically significant. This section explores its advantages and disadvantages in writing learning[3].

Looking into the obtained essays reveals that there exist words unique to the writings with the chatbot-based system. Table 2 shows the unique words together with their occurrences[4]. Some of the words such as *lyrics* and *elder* have relatively high word levels.

Because these words are unique to the writings with the chatbot-based system, the learners likely borrowed or adopted them from utterances of the chatbot-based system. Indeed, in many cases, the chatbot first introduced these words in its utterances and the learner adopted them in his or her next response. Table 3 shows the ratios of the learners that satisfy the condition. For example, the first line in Table 3 shows that learners produced the word *born* only in the writings with the chatbot-based system and that they always adopted it in their response right after seeing the utterance of the chatbot containing *born* as in *I was born in California.* or *Where were you born?*. Table 3 shows that all learners often adopted the unique words from the chatbot's utterances. This suggests that the chatbot is effective in letting learners use words they do not know or tend not to use by themselves. Considering these findings, it will likely let learners use (and consequently practice) new words in an actual context to include the words in the script of the chatbot, which should be a good practice for them.

More generally, utterances or responses from the chatbot might have good effects on writing learning. As a clue suggesting this, we found that the use of the definite and indefinite articles was very different under the two conditions as shown in Table 4. Table 4 reveals that (i) the ratio of the definite article to the

---

[3] Because of the copyright issue, the actual content cannot be included in the paper. Instead, their bi-grams are shown in the appendix.

[4] Table 2 includes only words whose occurrences are more than three.

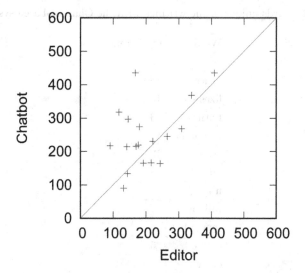

Fig. 3. Scatter plot of the number of word tokens produced by using the standard editor (x-axis) and the Chatbot-based system (y-axis).

indefinite article is much higher in the standard editor than in the chatbot-based system and (ii) the error rates for the definite and indefinite articles are higher in the standard editor. One of the possible reasons might be the difference in the writing modes (the standard writing and the dialog-based writing). At the same time, it seems that the chatbot had a positive influence on the learners' article choices because its utterances often contained articles (especially indefinite articles). These positive instances might have facilitated the correct use of the indefinite articles when the learners used the chatbot-based system. Besides, some of them corrected article errors by self-revision in the course of the dialog as the next paragraph will describe. In contrast, when using the standard editor, seemingly they more often suffered from the fossilization [16] to the definite article as in *I go to the school by the train*. These observations suggest that the chatbot's utterances can be good examples when learners perform writing exercises[5].

In contrast, as indicated in the questionnaires, one of the disadvantages of the chatbot-based system is that it asks the same question several times, which might lower learner's motivation. It is only natural for the system to do so because of the simple implementation of ELIZA where the same rule is reused when all rules are exhausted for a keyword in the script. One can overcome the problem by adding a function memorizing what has been questioned and avoiding asking the same question.

At the same time, asking the same question twice actually has a good aspect. When the learners were asked the same question, they tended to review, evaluate, and revise what they had written previously. We looked into all data and

---

[5] Bi-grams concerning articles are shown in the appendix.

Table 2. Words unique in the writings with the Chatbot-based system.

Word	Occurrences
yes	29
lyrics	18
likes	15
born	13
nickname	12
yesterday	12
an	8
n't	7
talking	7
remember	6
active	5
elder	4
sports	4

Table 3. Ratios of the learners adopting the unique words from the proceeding Chatbot's utterance.

Word	Ratio (%)
born	100
nickname	89
active	56
an	56
lyrics	56
elder	44
sports	44

found that ten out of the 18 learners improved their original sentences in second responses when they had the same question. The revisions range over spelling errors, subject-verb agreement, missing article errors, and addition of details (to what they had written). Considering these, basically asking the same questions should be avoided, but it will likely be better to ask the same question occasionally.

Another drawback in the chatbot-based system is that learners have almost no opportunity to consider the organization of their writing. This is simply because the chatbot leads the writing activity and it asks questions randomly when it finds no keyword in the learner's writing. It would be beneficial to add a function to control the organization. However, it would be difficult to implement such a function.

Table 4. Statistics on definite and indefinite articles.

System	Definite		Indefinite	
	Ratio	Error rate	Ratio	Error rate
Chatbot	0.40	0.12	0.60	0.25
Editor	0.72	0.26	0.28	0.31

Even if it were possible, it would not necessarily mean that learners consider the organization of their writing[6].

To get around the problem, we propose the following writing learning exercise based on the chatbot: (1) write using the chatbot-based system just as in this work; (2) review the result; (3) examine organization by grouping and expanding the present writing, and removing unnecessary information; (4) write once more in the standard way based on (3). Through (1) to (4), learners can perform a more thorough writing activity, examining the organization of their writing.

6   Conclusions

In this paper, we have examined the hypothesis that even the simplest chatbot (such as ELIZA) has a positive effect on helping learners write more. We have shown such a tendency in the experiments although the difference in the writing amount is not statistically significant. We have looked into the writing results, showing that the chatbot-based system has good effects on word use and self-revision. We have also discussed advantages and disadvantages of the chatbot-based system. Finally, we have proposed a new writing exercise method combining the chatbot-based system with the conventional method.

Acknowledgements. We would like to thank the three anonymous reviewers for their useful comments on this paper. This work was partly supported by Konan Premier Project and Japan Science and Technology Agency (JST), PRESTO Grant Number JPMJPR1758, Japan

A   Bi-grams Obtained from the Experimental Results

The contents of the writing results cannot be provided because of the copyright issue. Instead, this appendix shows bi-grams and their statistics obtained from the writing results. Table 5 shows top ten bi-grams obtained from the experimental results. Table 6 shows top ten bi-grams starting with the indefinite article that were obtained from the experimental results. Table 7 shows top ten bi-grams starting with the definite article that were obtained from the experimental results.

---

[6] Rather, they might write without considering organization because the system guides the writing activity although the resulting writing is well organized. What is important here is to let learners consider organization for a writing learning exercise themselves.

Table 5. Bi-grams obtained from the experimental results.

Chatbot		Standard Editor	
Bi-gram	# Occurrences	Bi-gram	# Occurrences
i like	77	i like	38
i do	62	, i	35
do n't	47	i 'm	30
i have	47	i have	28
, i	46	want to	23
my friend	33	is very	22
i was	31	it is	21
do you	30	i want	20
my friends	28	my hobby	17
it is	28	years old	17

Table 6. Bi-grams starting with indefinite article.

Chatbot		Standard Editor	
Bi-gram	# Occurrences	Bi-gram	# Occurrences
an active	5	A sound	3
a very	2	A nikujyaga	3
a dog	2	A student	2
a game	2	A little	1
a test	2	A cake	1
a good	2	A dog	1
a movie	2	A game	1
a salt-bread	2	A harf	1
a mother	2	A good	1
a player	1	A movie	1

Table 7. Bi-grams starting with definite article.

Chatbot		Standard Editor	
Bi-gram	# Occurrences	Bi-gram	# Occurrences
the way	6	The plant	6
the reason	2	The first	5
the question	2	The most	4
the same	2	The best	3
the point	2	The sound	3
the world	2	The story	2
the soccer	2	The world	2
the second	1	The live	2
the flowers	1	The leaf	2
the game	1	The road	2

References

1. Bitchener, J., Young, S., Cameron, D.: The effect of different types of corrective feedback on ESL student writing. J. Second Lang. Writ. **14**(3), 191–205 (2005)
2. Core, M.G., Moore, J.D., Zinn, C.: The role of initiative in tutorial dialogue. In: Proceedings of 10th Conference on European Chapter of the Association for Computational Linguistics, vol. 1, pp. 67–74 (2003). DOI: https://doi.org/10.3115/1067807.1067818
3. Elizabeth Boyer, K., Phillips, R.D. Wallis, M., Vouk, M.C. Lester, J.: Learner characteristics and feedback in tutorial dialogue. In: Proceedings of 12th Workshop on Innovative Use of NLP for Building Educational Applications, pp. 53–61 (2008)
4. Felice, R.D., Pulman, S.G.: A classifier-based approach to preposition and determiner error correction in L2 English. In: Proceedings of 22nd International Conference on Computational Linguistics, pp. 169–176 (2008)
5. Han, N.R., Chodorow, M., Leacock, C.: Detecting errors in English article usage with a maximum entropy classifier trained on a large, diverse corpus. In: Proceedings of 4th International Conference on Language Resources and Evaluation, pp. 1625–1628 (2004)
6. Han, N.R., Chodorow, M., Leacock, C.: Detecting errors in English article usage by non-native speakers. Nat. Lang. Eng. **12**(2), 115–129 (2006)
7. Junczys-Dowmunt, M., Grundkiewicz, R., Guha, S., Heafield, K.: Approaching neural grammatical error correction as a low-resource machine translation task. In: Proceedings of 2018 Conference of the North American Chapter of the Association for Computational Linguistics: Human Language Technologies, vol. 1 (Long Papers), pp. 595–606 (2018). https://doi.org/10.18653/v1/N18-1055
8. Kaneko, M., Sakaizawa, Y., Komachi, M.: Grammatical error detection using error- and grammaticality-specific word embeddings. In: Proceedings of 8th International Joint Conference on Natural Language Processing, vol. 1: Long Papers, pp. 40–48 (2017). http://www.aclweb.org/anthology/I17-1005

9. Marineau, J., et al.: Classification of speech acts in tutorial dialog. In: Proceedings of Workshop on Modeling Human Teaching Tactics and Strategies of ITS, pp. 65–71 (2000)

10. Nagata, R., Kawai, A., Morihiro, K., Isu, N.: A feedback-augmented method for detecting errors in the writing of learners of English. In: Proceedings of 44th Annual Meeting of the Association for Computational Linguistics, pp. 241–248 (2006)

11. Nagata, R., Nakatani, K.: Evaluating performance of grammatical error detection to maximize learning effect. In: Proceedings of 23rd International Conference on Computational Linguistics, poster volume, pp. 894–900 (2010)

12. Napoles, C., Callison-Burch, C.: Systematically adapting machine translation for grammatical error correction. In: Proceedings of 12th Workshop on Innovative Use of NLP for Building Educational Applications, pp. 345–356 (2017). https://doi.org/10.18653/v1/W17-5039, http://aclweb.org/anthology/W17-5039

13. Rei, M.: Semi-supervised multitask learning for sequence labeling, In: Proceedings of 55th Annual Meeting of the Association for Computational Linguistics, vol. 1: Long Papers, pp. 2121–2130 (2017). https://doi.org/10.18653/v1/P17-1194

14. Rei, M., Yannakoudakis, H.: Compositional sequence labeling models for error detection in learner writing. In: Proceedings of 54th Annual Meeting of the Association for Computational Linguistics, vol. 1: Long Papers, pp. 1181–1191 (2016). https://doi.org/10.18653/v1/P16-1112

15. Robb, T., Ross, S., Shortreed, I.: Salience of feedback on error and its effect on EFL writing quality. TESOL Q. **20**(1), 83–93 (1986)

16. Selinker, L.: Interlanguage. Int. Rev. Appl. Linguist. Lang. Teach. **10**(3), 209–231 (1972)

17. Sheen, Y.: The effect of focused written corrective feedback and language aptitude on ESL learners' acquisition of articles. TESOL Q. **41**, 255–283 (2007). https://doi.org/10.1002/j.1545-7249.2007.tb00059.x

18. Weizenbaum, J.: ELIZA – a computer program for the study of natural language communication between man and machine. Commun. ACM **9**(1), 36–45 (1966)

Towards Task-Oriented Dialogue in Mixed Domains

Tho Chi Luong[2(✉)] and Phuong Le-Hong[1,2]

[1] College of Science, Vietnam National University, Hanoi, Vietnam
phuonglh@vnu.edu.vn
[2] FPT Technology Research Institute, FPT University, Hanoi, Vietnam
tholc2@fpt.com.vn

Abstract. This work investigates the task-oriented dialogue problem in mixed-domain settings. We study the effect of alternating between different domains in sequences of dialogue turns using two related state-of-the-art dialogue systems. We first show that a specialized state tracking component in multiple domains plays an important role and gives better results than an end-to-end task-oriented dialogue system. We then propose a hybrid system which is able to improve the belief tracking accuracy of about 28% of average absolute point on a standard multi-domain dialogue dataset. These experimental results give some useful insights for improving our commercial chatbot platform FPT.AI, which is currently deployed for many practical chatbot applications.

Keywords: Task-oriented dialogue · Multi-domain belief tracking · Mixed-domain belief tracking · Natural language processing

1 Introduction

In this work, we investigate the problem of task-oriented dialogue in mixed-domain settings. Our work is related to two lines of research in Spoken Dialogue System (SDS), namely *task-oriented dialogue system* and *multi-domain dialogue system*. We briefly review the recent literature related to these topics as follows.

Task-oriented dialogue systems are computer programs which can assist users to complete tasks in specific domains by understanding user requests and generating appropriate responses within several dialogue turns. Such systems are useful in domain-specific chatbot applications which help users find a restaurant or book a hotel. Conventional approach for building a task-oriented dialogue system is concerned with building a quite complex pipeline of many connected components. These components are usually independently developed which include at least four crucial modules: a natural language understanding module, a dialogue state tracking module, a dialogue policy learning module, and a answer generation module. Since these systems components are usually trained independently, their optimization targets may not fully align with the overall system evaluation

© Springer Nature Singapore Pte Ltd. 2020
L.-M. Nguyen et al. (Eds.): PACLING 2019, CCIS 1215, pp. 257–266, 2020.
https://doi.org/10.1007/978-981-15-6168-9_22

criteria [1]. In addition, such a pipeline system often suffers from error propagation where error made by upstream modules are accumuated and got amplified to the downstream ones.

To overcome the above limitations of pipeline task-oriented dialogue systems, much research has focused recently in designing end-to-end learning systems with neural network-based models. One key property of task-oriented dialogue model is that it is required to reason and plan over multiple dialogue turns by aggregating useful information during the conversation. Therefore, sequence-to-sequence models such as the encoder-decoder based neural network models are proven to be suitable for both task-oriented and non-task-oriented systems. Serban et al. proposed to build end-to-end dialogue systems using generative hierarchical recurrent encoder-decoder neural network [2]. Li et al. presented persona-based models which incorporate background information and speaking style of interlocutors into LSTM-based seq2seq network so as to improve the modeling of human-like behavior [3]. Wen et al. designed an end-to-end trainable neural dialogue model with modularly connected components [4]. Bordes et al. [5] proposed a task-oriented dialogue model using end-to-end memory networks. At the same time, many works explored different kinds of networks to model the dialogue state, such as copy-augmented networks [6], gated memory networks [7], query-regression networks [8]. These systems do not perform slot-filling or user goal tracking; they rank and select a response from a set of response candidates which are conditioned on the dialogue history.

One of the significant effort in developing end-to-end task-oriented systems is the recent Sequicity framework [9]. This framework also relies on the sequence-to-sequence model and can be optimized with supervised or reinforcement learning. The Sequicity framework introduces the concept of *belief span* (bspan), which is a text span that tracks the dialogue states at each turn. In this framework, the task-oriented dialogue problem is decomposed into two stages: bspan generation and response generation. This framework has been shown to significantly outperform state-of-the-art pipeline-based methods.

The second line of work in SDS that is related to this work is concerned with multi-domain dialogue systems. As presented above, one of the key components of a dialogue system is dialogue state tracking, or belief tracking, which maintains the states of conversation. A state is usually composed of user's goals, evidences and information which is accumulated along the sequence of dialogue turns. While the user's goal and evidences are extracted from user's utterances, the useful information is usually aggregated from external resources such as knowledge bases or dialogue ontologies. Such knowledge bases contain slot type and slot value entries in one or several predefined domains. Most approaches have difficulty scaling up with multiple domains due to the dependency of their model parameters on the underlying knowledge bases. Recently, Ramadan et al. [10] has introduced a novel approach which utilizes semantic similarity between dialogue utterances and knowledge base terms, allowing the information to be shared across domains. This method has been shown not only to scale well to multi-domain dialogues, but also outperform existing state-of-the-art models in single-domain tracking tasks.

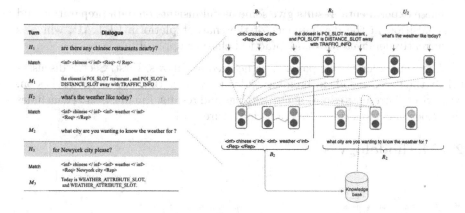

Fig. 1. Sequicity architecture.

The problem that we are interested in this work is task-oriented dialogue in mixed-domain settings. This is different from the multi-domain dialogue problem above in several aspects, as follows:

- First, we investigate the phenomenon of alternating between different dialogue domains in subsequent dialogue turns, where each turn is defined as a pair of user question and machine answer. That is, the domains are mixed between turns. For example, in the first turn, the user requests some information of a restaurant; then in the second turn, he switches to the a different domain, for example, he asks about the weather at a specific location. In a next turn, he would either switch to a new domain or come back to ask about some other property of the suggested restaurant. This is a realistic scenario which usually happens in practical chatbot applications in our observations. We prefer calling this problem mixed-domain dialogue rather than multiple-domain dialogue.
- Second, we study the effect of the mixed-domain setting in the context of multi-domain dialogue approaches to see how they perform in different experimental scenarios.

The main findings of this work include:

- A specialized state tracking component in multiple domains still plays an important role and gives better results than a state-of-the-art end-to-end task-oriented dialogue system.
- A combination of specialized state tracking system and an end-to-end task-oriented dialogue system is beneficial in mix-domain dialogue systems. Our hybrid system is able to improve the belief tracking accuracy of about 28% of average absolute point on a standard multi-domain dialogue dataset.

– These experimental results give some useful insights on data preparation and acquisition in the development of the chatbot platform FPT.AI[1], which is currently deployed for many practical chatbot applications.

The remainder of this paper is structured as follows. First, Sect. 2 discusses briefly the two methods in building dialogue systems that our method relies on. Next, Sect. 3 presents experimental settings and results. Finally, Sect. 4 concludes the paper and gives some directions for future work.

2   Methodology

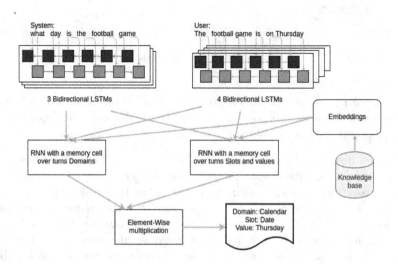

Fig. 2. Multi-domain belief tracking with knowledge sharing.

In this section, we present briefly two methods that we use in our experiments which have been mentioned in the previous section. The first method is the Sequicity framework and the second one is the state-of-the-art multi-domain dialogue state tracking approach.

2.1   Sequicity

Figure 1 shows the architecture of the Sequicity framework as described in [9]. In essence, in each turn, the Sequicity model first takes a bspan (B_1) and a response (R_1) which are determined in the previous step, and the current human question (U_2) to generate the current bspan. This bspan is then used together with a knowledge base to generate the corresponding machine answer (R_2), as shown in the right part of Fig. 1.

The left part of that figure shows an example dialogue in a mixed-domain setting (which will be explained in Sect. 3).

---

[1] http://fpt.ai/.

2.2 Multi-domain Dialogue State Tracking

Figure 2 shows the architecture of the multi-domain belief tracking with knowledge sharing as described in [10]. This is the state-of-the-art belief tracker for multi-domain dialogue.

This system encodes system responses with 3 bidirectional LSTM network and encodes user utterances with $3 + 1$ bidirectional LSTM network. There are in total 7 independent LSTMs. For tracking domain, slot and value, it uses 3 corresponding LSTMs, either for system response or user utterance. There is one special LSTM to track the user affirmation. The semantic similarity between the utterances and ontology terms are learned and shared between domains through their embeddings in the same semantic space.

3 Experiments

In this section, we present experimental settings, different scenarios and results. We first present the datasets, then implementation settings, and finally obtained results.

3.1 Datasets

We use the publicly available dataset KVRET [6] in our experiments. This dataset is created by the Wizard-of-Oz method [11] on Amazon Mechanical Turk platform. This dataset includes dialogues in 3 domains: calendar, weather, navigation (POI) which is suitable for our mix-domain dialogue experiments. There are 2,425 dialogues for training, 302 for validation and 302 for testing, as shown in the upper half of Table 1.

In this original dataset, each dialogue is of a single domain where all of its turns are on that domain. Each turn is composed of a sentence pair, one sentence is a user utterance, the other sentence is the corresponding machine response. A dialogue is a sequence of turns. To create mix-domain dialogues for our experiments, we make some changes in this dataset as follows:

- We keep the dialogues in the calendar domain as they are.
- We take a half of dialogues in the weather domain and a half of dialogues in the POI domain and mix their turns together, resulting in a dataset of mixed weather-POI dialogues. In this mixed-domain dialogue, there is a turn in the weather domain, followed by a turn in POI domain or vice versa.

We call this dataset *the sequential turn dataset*. Since the start turn of a dialogue has a special role in triggering the learning systems, we decide to create another and different mixed-domain dataset with the following mixing method:

- The first turn and the last turn of each dialogue are kept as in their original.
- The internal turns are mixed randomly.

We call this dataset *the random turn dataset*. Some statistics of these mixed-domain datasets are shown in the lower half of the Table 1.

Table 1. Some statistics of the datasets used in our experiments. The original KVRET dataset is shown in the upper half of the table. The mixed dataset is shown in the lower half of the table.

Dataset	KVRET		
Dialogues	Train: 2,425 ; Test: 302 ; Dev.: 302		
Domains	calendar	weather	POI

Dataset	MIXED DOMAINS			
Domains	calendar	weather	POI	Mixed weather – POI
Train	828	398	400	400
Test	102	50	50	50
Dev.	102	50	50	50

3.2 Experimental Settings

For the task-oriented Sequicity model, we keep the best parameter settings as reported in the original framework, on the same KVRET dataset [9]. In particular, the hidden size of GRU unit is set to 50; the learning rate of Adam optimizer is 0.003. In addition to the original GRU unit, we also re-run this framework with simple RNN unit to compare the performance of different recurrent network types. The Sequicity tool is freely available for download.[2]

For the multi-domain belief tracker model, we set the hidden size of LSTM units to 50 as in the original model; word embedding size is 300 and number of training epochs is 100. The corresponding tool is also freely available for download.[3]

3.3 Results

Our experimental results are shown in Table 2. The first half of the table contains results for task-oriented dialogue with the Sequicity framework with two scenarios for training data preparation. For each experiment, we run our models for 3 times and their scores are averaged as the final score. The *mixed training* scenario performs the mixing of both the training data, development data and the test data as described in the previous subsection. The *non-mixed training* scenario performs the mixing only on the development and test data, keeps the training data unmixed as in the original KVRET dataset. As in the Sequicity framework, we report entity match rate, BLEU score and Success F1 score. **Entity match rate** evaluates task completion, it determines if a system can generate all correct constraints to search the indicated entities of the user. **BLEU** score evaluates the language quality of generated responses. **Success F1** balances the recall and precision rates of slot answers. For further details on these metrics, please refer to [9].

---

[2] https://github.com/WING-NUS/sequicity.
[3] https://github.com/osmanio2/multi-domain-belief-tracking.

Table 2. Our experimental results. **Match.** and **Succ. F1** are Entity match rate and Success F1. The upper half of the table shows results of task-oriented dialogue with the Sequicity framework. The lower half of the table shows results of multi-domain belief tracker.

Sequicity		Case 1 - sequential turn			Case 2 - random turn		
		Match.	BLEU	Succ. F1	Match.	BLEU	Succ. F1
mixed training	GRU	0.6367	**0.1930**	**0.7358**	0.6860	0.1862	**0.7562**
	RNN	0.7354	0.1847	0.7129	0.6591	0.1729	0.7105
non-mixed training	GRU	0.7399	0.1709	0.7055	**0.7488**	0.1820	0.7173
	RNN	0.7706	0.1453	0.6156	0.6995	0.1580	0.6633
		Domain - accuracy	Slot - accuracy	Value - accuracy			
Belief tracker	Multi-dom.	0.8253	**0.9329**	**0.9081**			
	Sequicity		0.7171	0.5644			

In the first series of experiments, we evaluate the Sequicity framework on different mixing scenarios and different recurrent units (GRU or RNN), on two mixing methods (sequential turn or random turn), as described previously. We see that when the training data is kept unmixed, the match rates are better than those of the mixed training data. It is interesting to note that the GRU unit is much more sensitive with mixed data than the simple RNN unit with the corresponding absolute point drop of about 10%, compared to about 3.5%. However, the entity match rate is less important than the Success F1 score, where the GRU unit outperforms RNN in both sequential turn and random turn by a large margin. It is logical that if the test data are mixed but the training data are unmixed, we get lower scores than when both the training data and test data are mixed. The GRU unit is also better than the RNN unit on response generation in terms of BLEU scores.

We also see that the task-oriented dialogue system has difficulty running on mixed-domain dataset; it achieves only about 75.62% of Success F1 in comparison to about 81.1% (as reported in the Sequicity paper, not shown in our table). Appendix A shows some example dialogues generated automatically by our implemented system.

In the second series of experiments, we evaluate the belief tracking components of two systems, the specialized multi-domain belief tracker and the Sequicity bspan component. As shown in the lower half of the Table 2, Sequicity capability of belief tracking is much worse than that of the multi-domain belief tracker. The slot accuracy gap between the tools is about 21.6%, the value accuracy gap is about 34.4%; that is a large average gap of 28% of accuracy. This result suggests a future work on combining a specialized belief tracking module with an end-to-end task-oriented dialogue system to improve further the performance of the overall dialogue system.

3.4   Error Analysis

In this subsection, we present an example of erroneous mixed dialogue with multple turns. Table 3 shows a dialogue in the test set where wrong generated responses of the Sequicity system are marked in bold font.

Table 3. A mixed dialogue example in the test set with erroneous generated responses. The last two columns show respectively the system's generated bspan and the gold bspan or belief tracker.

Turn	Dialogue	Domain	Generated bspan	Gold bspan
1. Human	Car please send me the fastest route to go back home	POI	EOS_Z1	home fastest EOS_Z1
1. Machine	There is **heavy traffic** on the way to **Pizza Hut**	POI		
2. Human	What is the weather like this week	Weather info	EOS_Z1	this_week EOS_Z1
2. Machine	What city are you interested in?	Weather info		
3. Human	Cleveland	Weather info	cleveland EOS_Z1	cleveland EOS_Z1
3. Machine	**It will be clear sky on monday**	Weather info		
4. Human	What is the address?	POI	cleveland EOS_Z1	home fastest EOS_Z1
4. Machine	**Cleveland** is located at **5672 barringer street**	POI		
5. Human	Please send me the fastest route on screen and let 's go there, thank you!	POI	cleveland EOS_Z1	home fastest EOS_Z1
5. Machine	There is **moderate traffic** on our way but i sent the route to your screen, drive carefully!	POI		
6. Human	Thank		cleveland EOS_Z1	EOS_Z1
6. Machine	You are welcome			

In the first turn, the system predicts incorrectly the bspan, thus generates wrong slot values (heavy traffic and Pizza Hut). The word Pizza Hut is an arbitrary value selected by the system when it cannot capture the correct value home in the bspan. In the second turn, the machine is not able to capture the value this_week. This failure does not manifest immediately at this turn but it is accumulated to make a wrong answer at the third turn (monday instead of this_week).

The third turn is of domain weather and the fourth turn is switched to domain POI. The bspan value cleveland is retained through cross domain, resulting in an error in the fourth turn, where cleveland is shown instead of home. This example demonstrates a weakness of the system when being trained on a mixed-domain dataset. In the fifth turn, since the system does not recognize the value fastest in the bspan, it generates a random and wrong value moderate traffic. Note that the generated answer of the sixth turn is correct despite of the wrong predicted bspan; however, it is likely that if the dialogue continues, this wrong bspan may result in more answer mistakes. In such situations, multi-domain belief tracker usually performs better at bspan prediction.

4   Conclusion

We have presented the problem of mixed-domain task-oriented dialogue and its empirical results on two datasets. We employ two state-of-the-art, publicly available tools, one is the Sequicity framework for task-oriented dialogue, and another is the multi-domain belief tracking system. The belief tracking capability of the specialized system is much better than that of the end-to-end system. We also show the difficulty of task-oriented dialogue systems on mixed-domain datasets through two series of experiments. These results give some useful insights in combining the approaches to improve the performance of a commercial chatbot platform which is under active development in our company. We plan to extend this current research and integrate its fruitful results into a future version of the platform.

A   Example Dialogues

The following is three example dialogues generated by our system. The first dialogue is in single-domain. The next two dialogues are in mixed-domains.

Turn	Dialogue 1	Domain
Human	What time is my next dinner event and who will be attending?	Calendar
Machine	Your dinner is on MONDAY at 5 PM	Calendar
Human	That will do, thanks	
Machine	You are welcome	

Turn	Dialogue 2	Domain
Human	Please give me the address and directions to home within 5 miles of here, avoiding all heavy traffic	POI
Machine	HOME is located at 5671 BARRINGER STREET	POI
Human	What is the forecast for the next 7 days?	Weather info
Machine	What city are you interested in?	Weather info
Human	Cleveland, will it be windy?	Weather info
Machine	It will not be windy in Cleveland this week	Weather info
Human	Thank you	
Machine	You're welcome	

Turn	Dialogue 3	Domain
Human	Are there any chinese restaurants nearby?	POI
Machine	The closest is TAI PAN restaurant, and TAI PAN is 5 miles away with NO TRAFFIC	POI
Human	What's the weather like today?	Weather info
Machine	What city are you wanting to know the weather for?	Weather info
Human	for Newyork city please? weather info	
Machine	Today is CLEAR SKIES, and WARM	Weather info
Human	Ok, good, set the direction to there	POI
Machine	The address is 830 Almanor Ln, i sent it on your screen	POI
Human	Ok thanks	
Machine	You're welcome!	

References

1. Liu, B., Tur, G., Hakkani-Tur, D., Shah, P., Heck, L.: Dialogue learning with human teaching and feedback in end-to-end trainable task-oriented dialogue systems. In: Proceedings of NAACL (2018)
2. Serban, I., Sordoni, A., Bengio, Y., Courville, A.C., Pineau, J.: Building end-to-end dialogue systems using generative hierarchical neural network models. In: Proceedings of AAAI (2016)
3. Li, J., Galley, M., Brockett, C., Spithourakis, G.P., Gao, J., Dolan, B.: A persona-based neural conversation model. In: Proceedings of ACL (2016)
4. Wen, T.H., et al.: A network-based end-to-end trainable task-oriented dialogue system. In: Proceedings of EACL (2017)
5. Bordes, A., Boureau, Y.L., Weston, J.: Learning end-to-end goal-oriented dialogue. In: Proceedings of ICLR (2017)
6. Eric, M., Manning, C.D.: A copy-augmented sequence-to-sequence architecture gives good performance on task-oriented dialogue. In: Proceedings of EACL (2017)
7. Liu, F., Perez, J.: Gated end-to-end memory networks. In: Proceedings of EACL (2017)
8. Seo, M.J., Hajishirzi, H., Farhadi, A.: Query-regression networks for machine comprehension. In: Preprint ArXiv (2016)
9. Lei, W., Jin, X., Ren, Z., He, X., Kan, M.Y., Yin, D.: Sequicity: simplifying task-oriented dialogue systems with single sequence-to-sequence architectures. In: Proceedings of ACL (2018)
10. Ramadan, O., Budzianowski, P., Gasic, M.: Large-scale multi-domain belief tracking with knowledge sharing. In: Proceedings of ACL (2018)
11. Kelley, J.F.: Iterative design methodology for user-friendly natural language office information applications. ACM Trans. Inf. Syst. 2(1), 26–41 (1984)

Timing Prediction of Facilitating Utterance in Multi-party Conversation

Tomonobu Sembokuya and Kazutaka Shimada[(✉)]

Department of Artificial Intelligence, Kyushu Institute of Technology,
680-4 Iizuka, Fukuoka 820-8502, Japan
shimada@pluto.ai.kyutech.ac.jp

Abstract. Supporting consensus-building in multi-party conversations is a very important task in intelligent systems. To conduct smooth, active, and productive discussions, we need a facilitator who controls a discussion appropriately. However, it is impractical to assign a good facilitator to each group in the discussion environment. The goal of our study is to develop a digital facilitator system that supports high-quality discussions. One role of the digital facilitator is to generate facilitating utterances in the discussions. To realize the system, we need to predict the timing of facilitating utterances. To apply a machine learning technique to our model, we construct a data set from the AMI corpus, first. For the construction, we use some rules based on the annotation of the corpus. Then, we generate a prediction model with verbal and non-verbal features extracted from discussions. We obtained 0.75 on the F-measure. We compared our model with a baseline method. Our model outperformed the baseline (0.7 vs. 0.5 on the AUC value). In addition, we introduce additional features about the role of participants in the AMI corpus. By using the additional features, the F measure increased by 2 points. The experimental results show the effectiveness of our model.

Keywords: Multi-party conversation · Timing prediction · Facilitation

1 Introduction

In collaborative work, people need to discuss several topics for decision-making on a meeting, namely multi-party conversation. It is a very important task in intelligent systems to support consensus-building in conversations with multiple participants. Participants in a discussion often struggle to identify the most suitable solution for a decision on a meeting agenda because there are generally many alternatives and criteria related to making a decision. As a result, they often fail to make a satisfying decision. It leads to the failure of the discussion. To conduct smooth, active and productive discussions, they need an effective facilitator who controls the discussion appropriately. However, it is impractical to assign an effective facilitator to each group in the discussion environment due to a lack of human resources. Although a project manager needs to appropriately

© Springer Nature Singapore Pte Ltd. 2020
L.-M. Nguyen et al. (Eds.): PACLING 2019, CCIS 1215, pp. 267–279, 2020.
https://doi.org/10.1007/978-981-15-6168-9_23

handle a discussion in business meetings, he/she might not have remarkable facilitation skills, such as asking questions to gain additional information and asking follow-up questions to further expand participants' understanding. Ordinary people in a group discussion might subconsciously need help from others to generate a good decision. Therefore, a system that supports consensus-building plays a very important role in discussion.

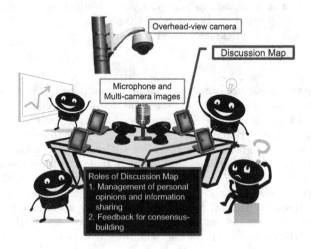

Fig. 1. Overview of our digital facilitator system with discussion maps. The goal of our study is to develop a system that behaves like a facilitator by using several modalities, such as speech and image inputs.

The goal of our study is to construct a system that cooperatively supports consensus-building and management of conversation for high-quality discussion. The system is referred to as a digital facilitator: a collaborative agent for participants of discussions. Figure 1 shows an overview of our system. We are developing a prototype system to support real discussions [6]. The system estimates the current state of a discussion and then generates sentences and charts that describe it. This is a part of our digital facilitator system. However, the generation timing in the current system depends on participants' clicks on the system: a passive control of the system. Therefore we need facilitator's knowledge, behavior, and patterns to realize a good digital facilitator: an active control from the system.

In this paper, we focus on timing about intervention or facilitation by the digital facilitator. We define an utterance by a participant that behaves like a chairperson on the discussion as "Facilitating Utterance." We propose a prediction model of the timing of such utterances by using a machine learning technique. The contributions of this paper are as follows:

- We design a guideline for constructing training data from the AMI corpus for a timing prediction task of facilitating utterances. It is based on dialogue act tags and social role tags in the corpus.

- We propose a timing prediction model using verbal and non-verbal features for facilitating utterances. We compare the effectiveness of the features experimentally.

2 Related Work

Shiota et al. [14] have reported an analysis of characteristics of facilitators in two multi-party conversations corpora: the AMI corpus [3] and the Kyutech corpus [19]. In the analysis, they generated decision tree models to classify each participant into a facilitator and a non-facilitator in the corpora. Omoto et al. [11] have reported the analysis of facilitating behavior of the exemplary facilitator from measured non-verbal and para-linguistic data. They defined four actions for facilitation: convergence, divergence, conflict, and concretization, and then analyzed the conversations on the basis of these factors. However, these studies analyzed conversations from a macro perspective. We need to determine the timing of facilitation as a function of our digital facilitator system, namely a micro perspective.

Lala et al. [7] have proposed an approach to attentive listening, which integrates continuous backchannels with responsive dialogue to user statements to maintain the flow of conversation in spoken dialogue tasks. They constructed a prediction model based on a logistic regression approach. The task is that a backchannel would occur in 500 ms or not. In addition, they improved their system by incorporating a statement response model on the four different response types and a flexible turn-taking model. They evaluated the system with the autonomous android, Erica, as a pilot study. Skantze [16] has proposed a turn-taking model using LSTM for spoken dialogue systems. The model predicted the speech activity for an upcoming fixed time window. They also evaluated how the hidden layer in the network can be used as a feature vector for turn-taking decisions in human-robot interaction data. The target of these studies is a dialogue with two persons. On the other hand, our task is to predict facilitating utterances in multi-party conversations and discussions.

3 Data Construction

We need a data set for a prediction model based on machine learning. For the purpose, we utilize the AMI corpus and the tag sets.

3.1 AMI Meeting Corpus

The AMI corpus [3] is one of the most famous meeting corpora. It consists of scenario and non-scenario meetings. In this paper, we handle scenario meetings.

In the scenario task, participants pretended members in a virtual company, which designs remote controls. Each participant played each role: project manager (PM), industrial designer (ID), user-interface designer (UI), and marketing expert (ME).

The AMI corpus contains numerous annotations, such as topic tags and dialogue acts. In this paper, we focus on the dialogue act tags. The dialogue acts denote speakers' intentions, such as "inform" and "backchannel." The number of dialogue act tags is 15.

Some researchers annotated social role tags for 59 meetings on the scenario portion of the AMI corpus [12,17]. Each meeting was segmented into short clips by long pauses: pauses longer than 1 s. One social role was assigned to each speaker in each segment by annotators. Each annotator for the tagging was asked to watch the entire video segment and assign a speaker to a role on the basis of a list of specified guidelines. The number of social role tags is five, and the roles are summarized as follows:

- Protagonist: a speaker that takes the floor, drives the conversation, asserts its authority, and assumes a personal perspective.
- Supporter: a speaker that shows a cooperative attitude demonstrating attention and acceptance as well as providing technical and relational support.
- Neutral: a speaker that passively accepts ideas from the others without expressing his/her ideas.
- Gatekeeper: a speaker that acts as a group moderator. He/she mediates and encourages the communication.
- Attacker: a speaker who deflates the status of others, expresses disapproval, and attacks other speakers.

In this paper, we handle the 59 meetings with the social roles.

3.2   Facilitating Utterance

Our task is to predict the timing of facilitating utterances in conversations. Therefore, we need to determine which utterances correspond to facilitating utterances in the conversations to apply a machine learning method for the prediction. For the purpose, we configure tree-based rules to determine the facilitating utterances. Figure 2 shows the flowchart of the determination process.

First, we check whether each utterance is spoken by a participant with the Gatekeeper tag. If so, move to the next step. If not so, we regard the utterance as a Non-facilitating utterance. From the definition in Sect. 3.1, the Gatekeeper tag is the most important factor for the judgment of facilitating utterances. Although project managers (PM) often have a similar role with Gatekeeper in discussions, we focus on the Gatekeeper tags only. The reason is that participants with other roles (ID, UI, and ME) often behave like a facilitator.

Next, we focus on specific dialogue act tags. Table 1 shows examples of the dialogue act tags in the AMI corpus. We use "Inform", "Suggest", "Offer", and "Elicit-*" for the determination. We can correctly remove utterances with these tags as backchannel utterances of Gatekeepers, e.g., "Uh, I see."

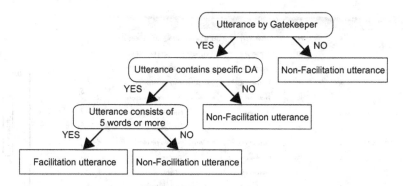

Fig. 2. Flowchart for the determination process. The data set is constructed by using social role tags and dialogue act tags in the AMI corpus.

Finally, we select utterances with five words or more as facilitating utterances. By using this rule, we can remove utterances for giving information from the facilitating utterance list, e.g., the utterance "No." with the Inform tag. The threshold, five or more, was determined experimentally[1].

4   Method

We explain our method and the task in this section. Figure 3 shows an overview of our method. The 1st and 14th utterances with the orange color in the figure are instances of facilitating utterances by the rules in Sect. 3.2.

We regard utterances within S_p sec. from the current utterance as one cluster. Then, we assign the "+1" label to the cluster that contains a facilitating utterance, otherwise -1. Our model predicts the label by using features extracted from utterances in the previous S_a range. We concatenate features of five S_a.

We use Support Vector Machines (SVMs) as the classifier. We utilize LIB-SVM [4] for the implementation. The parameters are default settings, and the kernel is RBF. For SVMs, we extract the following features:

f1) Average of word embedding
    The word embedding is a vector representation of each individual word that is pre-trained by some of the syntactic and semantic relationships in the language [9]. We use a model trained from Wikipedia and Web news. We utilize fastText [2] for the implementation. We calculate the average value of the embedding vectors of words that appear in S_a.

$$Emb_{ave}(t, S_a) = \frac{\sum_{x_i \in Words(t,S_a)} Emb(x_i)}{S_a} \tag{1}$$

where $Emb(x_i)$ is the embedding of a word x_i based on fastText. t denotes time in the discussion and the unit of S_a is seconds.

---

[1] Five is the mode value of utterances with the Gatekeeper and specific DA tags.

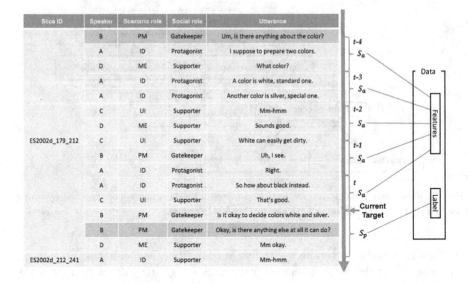

Fig. 3. Overview of our timing prediction model. Our model extracts features from the previous utterances in each S_a and concatenates the features for the prediction model. The label denotes $+1$ if the S_p contains a facilitating utterance and -1 if the S_p does not contain a facilitating utterance.

f2) Average of words in each S_a

Participants tend to frequently utter his/her thoughts and opinions in heated discussion. They also tend to not frequently utter his/her thoughts and opinions in non-hated discussion. It indicates that the number of utterances is one important feature for the prediction. Here, we utilize the average number of words in S_a as a feature for the prediction model.

$$NumWords_{rate}(t, S_a) = \frac{NumWords(t, S_a)}{S_a} \qquad (2)$$

where $NumWords$ is the number of words in S_a in t.

f3) Ratios of overlap and silence

In a similar situation to f2, overlaps and silences occur in discussion. To capture the tendencies of each participant about the two characteristics, we introduce ratios of overlaps and silences as the features. The feature values are also the average values of overlap length and silence length in S_a.

$$SilentTime_{ratio}(t, S_a) = \frac{\sum_{sl_i \in Silents(t, S_a)} sl_i}{S_a} \qquad (3)$$

$$OverlapTime_{ratio}(t, S_a) = \frac{\sum_{ol_i \in (t, S_a)} ol_i}{S_a} \qquad (4)$$

where sl_i and ol_i are silence length and overlap length in S_a, respectively.

Table 1. Dialogue acts on AMI meeting corpus.

DA tag	Meaning
Backchannel	Response such as "yeah"
Stall	Filled pauses
Fragment	Utterance that does not convey a speaker intention
Inform	Giving information
Suggest	Expressing an intention relating to the actions of another individual, the group as a whole, or a group in the wider environment
Offer	Expressing an intention relating to his or her own actions
Assess	Comment that expresses an evaluation
Elicit-*	Requests about the DA; e.g., if * is "Inform", it denotes a request that someone else give some information

f4) Number of long silences

Long silences often indicate non-heated discussion as compared with short silences. Therefore, we detect silences that are longer than a threshold and then utilize the frequency of the long silences as the feature[2].

$$NumSilents_T(t, S_a) = \sum_{sl_i \in Silents(t, S_a)} \begin{cases} 1 & (sl_i \geq T) \\ 0 & (sl_i < T) \end{cases} \tag{5}$$

where t is a threshold for a long time silence.

f5) Number of speaker changes

In heated discussion, speaker changes occur frequently. Therefore, we utilize the number of speaker changes in S_a as the feature.

5   Experiment

We evaluated our timing prediction model with the dataset described in Sect. 3. We set $S_p = 30$ s because the average time of each segment in social role annotation in the previous studies was approximately 30 s. We evaluated the dataset with 10-fold cross-validation. We analyzed our method and the data in terms of types of features, length of S_a, and types of dialogue acts. In addition, we compared our model with a baseline.

---

[2] For overlaps, we do not handle this feature because the overlap length is usually shorter as compared with the silence length in discussion.

Table 2. Comparison of features.

Features	P	R	F
Verbal	0.74	0.76	0.75
Non-Verbal	0.73	0.55	0.63
All	0.74	0.76	0.75

5.1 Discussion About Features

To discuss the effectiveness of features, we categorize the features into two types: verbal and non-verbal features. Here, the verbal features are f1 and f2, and the non-verbal features are f3, f4 and f5 described in Sect. 4. We generated three prediction models: a model with verbal features, a model with non-verbal features, and a model with all features. Then, we evaluated the models with precision (P), recall (R), and F-values. We set $S_a = 30$ in this experiment.

Table 2 shows the experimental result. From the result, the non-verbal features were not essentially effective for the prediction. The values of the model with verbal features and the model with all features were the same in all criteria. However, instances that the models predicted incorrectly were not completely the same. Some instances were predicted correctly by using the ratio of overlaps in f3. Therefore, non-verbal features are not always counterproductive to the prediction. We need to discuss more effective nonverbal features through detailed error analysis.

To achieve better accuracy, we need to apply other features that are obtained from speech information for the prediction model. Prosodic features, such as pitch and volume, were often used in studies about participant's role recognition [13,18]. As a verbal feature, we utilized the average vector based on word embedding. However, the average vector lost some information, such as fillers in utterances. We need to discuss the effectiveness of some specific words and phrases for the improvement of the model.

5.2 Discussion About S_a

In Sect. 5.1, we set $S_a = 30$. However, it is not clear which S_a is appropriate for the prediction model. Therefore, we compared different settings about S_a ($S_a = 5, 10, 20$, and 30.) We used all features in the comparison because the setting was the best in Sect. 5.1.

Table 3. Comparison of S_a.

S_a [sec]	P	R	F
5	0.71	0.71	0.71
10	0.74	0.72	0.73
20	0.75	0.75	0.75
30	0.74	0.76	0.75

Table 3 shows the experimental result. We obtained better results for $S_a = 20$ and 30, as compared with the smaller values of S_a.

We analyzed the difference in the results between $S_a = 5$ and $S_a = 30$ in detail. First, we discuss the case that the setting $S_a = 30$ was better than that of $S_a = 5$. For the case that $S_a = 30$ was better, a long silence often appeared in the S_a. It indicates the decrease of the number of utterances in the S_a. It led to the decrease of the number of words. Since the verbal features were effective in our model, long S_a was important to capture the features. On the other hand, the setting $S_a = 5$ was unfitted and unsuitable in the case that S_a contained a long silence because the model cannot adequately capture the verbal features due to a small number of words.

Next, we discuss another situation, that is $S_a = 5$ was better. In this situation, the problem was also opposed to the setting that $S_a = 30$ was better. In other words, $S_a = 5$ performed well in the case that S_a contained many words and no long silence in S_a. Information about important words in S_a vanished by the average vector generated from too many words. Thus, the optimal length of S_a depends on the tendency in each S_a.

The goal of this model is to detect the timing of facilitating utterances for our digital facilitator system. In other words, we need to ensure real-time prediction in discussion. Thus, the small S_a is essentially suitable although the current result is the opposite. We need to investigate the optimal S_a through the addition of effective features described in Sect. 5.1.

5.3   Discussion About Dialogue Act

As we used Dialogue Act (DA) tags in the data construction, DA tags are closely related to social roles because DA tags denote the intention of each utterance. Hence, we compared the recall rate of each DA tag in the model with all features[3] and $S_a = 30$.

Table 4 shows the experimental result about some specific DA tags that are closely related to facilitating utterances, such as Inform and Elicit-*. "Num" in

Table 4. Recall about each DA.

DA tag	Num	R
Inform	1305	0.75
Suggest	311	0.76
Offer	122	0.78
Elicit-Inform	178	0.79
Elicit-Offer-or-Suggestion	31	0.77
Elicit-Assessment	104	0.83
Elicit-Comment-about-Understanding	2	0.5

[3] Note that our model did not use any DA tags as features.

the table denotes the number of instances with each DA tag in the experimental data set. We almost obtained balanced results, namely 0.75 or more. Although the recall rate of "Elicit-Comment-about-Understanding" was not enough, the reason was the number of instances in the data set.

The DA tag with the best recall rate was "Elicit-Assessment." In other words, the previous utterances of "Elicit-Assessment" contained much information for the prediction. The guideline of the AMI corpus [1] said,

> In an ELICIT-ASSESSMENT, the speaker attempts to elicit an assessment (or assessments) about what has been said or done so far. Sometimes a speaker seems to be making a suggestion and eliciting an assessment about it at the same time.

This definition entails a part of the roles of facilitating utterances. However, utterances with the DA tags listed in Table 4 are not always facilitating utterances. Therefore, it is insufficient to predict the timing of facilitating utterances by using only the DA tags although they are important features for the prediction of facilitating utterances. Moreover, DA tags cannot be applied to a real-time prediction model easily due to the need for annotation by human annotators. On the other hand, our model was able to predict facilitating utterances in a relatively high accuracy without DA tag information.

5.4 Comparison with a Baseline

We compared our model with a baseline. The baseline was based on a simple and naive assumption; each S_p always contains one or more facilitating utterances. In other words, the baseline always produced +1 for each S_p, namely the majority of the label in the data set. For the evaluation, in addition to P, R, and F-values, we introduce the Area Under the Curve (AUC) of the Receiver Operating Characteristic (ROC). The ROC curve is created by plotting the true positive rate (TPR) against the false positive rate (FPR) at various threshold settings. The AUC is the area under the ROC curve. Therefore the AUC of the baseline becomes 0.5 from the definition.

Table 5 shows the experimental result. The F-value of the baseline was more than that of our model due to the perfect Recall rate. However, our method outperformed the baseline in terms of the AUC value. It shows the effectiveness and appropriateness of our method as compared with the baseline.

Table 5. Comparison with a baseline.

Model	P	R	F	AUC
Baseline	0.65	1.00	0.79	0.50
Ours	0.74	0.76	0.75	**0.70**

Table 6. The effectiveness of the additional features.

Features	P	R	F	AUC
f1 to f5	**0.74**	0.76	0.75	0.70
f1 to f9	0.73	**0.83**	**0.77**	**0.71**

5.5   Additional Features About PM

In this section, we introduce additional features into the method in Sect. 4. Our target in this paper is utterances of Gatekeeper. The ratio of Gatekeeper's utterances about Project Manager (PM) in the scenario task was approximately 50%. Therefore, utterances of PM are strongly correlated with facilitating utterances.new Thus, we add four new features focusing on PM. The features are as follows:

f6) Average of PM's words in each S_a
    This is a feature focusing on the specific role, PM, about f2 in Sect. 4. We restrict the numerator of Eq. (2) to the number of utterances of PM.
f7) Ratio of silence of PM
    This is a restricted feature of f3 in Sect. 4. In a similar way to f6, we also restrict the numerator of Eq. (3) to the number of utterances of PM.
f8) Number of silences of PM
    This is also a restricted feature of f4 in Sect. 4. We count only the number of long silences of PM, unlike in the case of Eq. (5).
f9) Number of PM's words within a certain time
    This is the number of speaker changes to PM only in S_a, namely a restricted feature of f5 in Sect. 4.

We evaluated the effectiveness of the additional features. We compared the method with the best result, namely all features in Sect. 4 and $S_a = 30$ (the last rows in Table 3, 4, and 5). Table 6 shows the result. Although the precision rate slightly decreased as compared with the method without the additional features, the recall rate substantially increased (0.76 vs. 0.83) and the F measure and AUC also increased by 2 points and 1 point, respectively. This result shows the effectiveness of features focusing on the role, PM. On the other hand, it's not always true that participants in conversations have a role like the AMI corpus. In other words, the generality of features focusing on PM is not high. Therefore, utilization of roles in meetings to the prediction is a highly important issue.

6   Conclusion

In this paper, we proposed a model for predicting the timing of facilitating utterances for the digital facilitator. We defined an utterance by a participant that behaves like a chairperson on the discussion as facilitating utterances. We designed a guideline for constructing training data from the AMI corpus for a timing prediction task of facilitating utterances. It was based on dialogue act tags and social role tags in the corpus.

We applied verbal and non-verbal features for the timing prediction model of facilitating utterances. We evaluated our model in terms of types of features, length of S_a, types of dialogue acts, and comparison with a baseline. The verbal features were effective for the prediction. The non-verbal features also performed a certain function for the prediction. Our model also outperformed a simple baseline in terms of the AUC value. As a whole, the experimental results show the effectiveness of our prediction model.

To achieve better accuracy, we need to apply other features that are obtained from speech information, such as pitch and volume. We also need to discuss the effectiveness of some specific words and phrases for the improvement of the model. In the experiment, the S_a of the best F-value was 30. However, the optimal length of S_a depends on the tendency in each S_a. The dynamic setting of the S_a is interesting future work.

In this paper, we handled the AMI corpus as the data set. We have also developed a multi-party conversation corpus, the Kyutech corpus [19]. Shiota et al. [14] reported the difference between the AMI corpus and the Kyutech corpus, for the facilitators' behavior. Therefore, analysis and evaluation of our corpus for this timing prediction are important future work.

One big issue of our work is the validity of the rules for the determination of facilitating utterances in Section 3.2. We automatically constructed the training data by using the rules. The rules were based on annotated tags in the AMI corpus and our heuristics and were intuitively plausible. However, we need to discuss the validity more deeply through manual data analysis.

There are many approaches and studies to build mutually agreeable solutions and a consensus, such as multi-agent systems for negotiation [15] and a large-scale online discussion [5]. Studies about facilitation robots [8] and human communication skills [10] are also related to our work. The knowledge from these studies would lead to the improvement of our work.

Acknowledgment. This work was supported by JSPS KAKENHI Grant Number 17H01840.

References

1. Guidelines for Dialogue Act and Addressee Annotation Version 1.0 (2005)
2. Bojanowski, P., Grave, E., Joulin, A., Mikolov, T.: Enriching word vectors with subword information. Trans. Assoc. Computat. Linguistics **5**, 135–146 (2017)
3. Carletta, J.: Unleashing the killer corpus: experiences in creating the multi-everything ami meeting corpus. Lang. Res. Eval. **41**(2), 181–190 (2007)
4. Chang, C.-C., Lin, C.-J.: Libsvm: a library for support vector machines. ACM Trans. Intell. Syst. Technol. (TIST) **2**(3), 27 (2011)
5. Ito, T., Imi, Y., Ito, T., Hideshima, E.: COLLAGREE: a facilitator-mediated large-scale consensus support system. In: Proceedings of the 2nd Collective Intelligence Conference (2014)

6. Kirikihira, R., Shimada, K.: Discussion map with an assistant function for decision-making: a tool for supporting consensus-building. In: Egi, H., Yuizono, T., Baloian, N., Yoshino, T., Ichimura, S., Rodrigues, A. (eds.) International Conference on Collaboration Technologies, pp. 3–18. Springer, Cham (2018). https://doi.org/10.1007/978-3-319-98743-9_1

7. Lala, D., Milhorat, P., Inoue, K., Ishida, M., Takanashi, K., Kawahara, T.: Attentive listening system with backchanneling, response generation and flexible turn-taking. In: Proceedings of the 18th Annual SIGdial Meeting on Discourse and Dialogue, pp. 127–136 (2017)

8. Matsuyama, Y., Akiba, I., Fujie, S., Kobayashi, T.: Four-participant group conversation: a facilitation robot controlling engagement density as the fourth participant. Comput. Speech Lang. **33**(1), 1–24 (2015)

9. Mikolov, T., Chen, K., Corrado, G., Dean, J.: Efficient estimation of word representations in vector space. arXiv preprint arXiv:1301.3781 (2013)

10. Okada, S., et al.: Estimating communication skills using dialogue acts and nonverbal features in multiple discussion datasets. In: Proceedings of the 18th ACM International Conference on Multimodal Interaction, pp. 169–176 (2016)

11. Omoto, Y., Toda, Y., Ueda, K., Nishida, T.: Analyses of the facilitating behavior by using participant's agreement and nonverbal behavior. J. Inf. Process. Soc. Japan **52**(12), 3659–3670 (2011). (in Japanese)

12. Sapru, A., Bourlard, H.: Automatic social role recognition in professional meetings using conditional random fields. In: Proceedings of Interspeech (2013)

13. Sapru, A., Bourlard, H.: Automatic recognition of emergent social roles in small group interactions. IEEE Trans. Multimedia **17**(5), 746–760 (2015)

14. Shiota, T., Yamamura, T., Shimada, K.: Analysis of facilitators' behaviors in multi-party conversations for constructing a digital facilitator system. In: International Conference on Collaboration Technologies, pp. 145–158 (2018)

15. Sierra, C., Jennings, N.R., Noriega, P., Parsons, S.: A framework for argumentation-based negotiation. In: Singh, M.P., Rao, A., Wooldridge, M.J. (eds.) ATAL 1997. LNCS, vol. 1365, pp. 177–192. Springer, Heidelberg (1998). https://doi.org/10.1007/BFb0026758

16. Skantze, G.: Towards a general, continuous model of turn-taking in spoken dialogue using LSTM recurrent neural networks. In: SIGdial Conference (2017)

17. Vinciarelli, A., Valente, F., Yella, S.H., Sapru, A.: Understanding social signals in multi-party conversations: automatic recognition of socio-emotional roles in the AMI meeting corpus. In: 2011 IEEE International Conference on Systems, Man, and Cybernetics, pp. 374–379 (2011)

18. Weninger, F., Krajewski, J., Batliner, A., Schuller, B.: The voice of leadership: models and performances of automatic analysis in online speeches. IEEE Trans. Affect. Comput. **3**(4), 496–508 (2012)

19. Yamamura, T., Shimada, K., Kawahara, S.: The Kyutech corpus and topic segmentation using a combined method. In: Proceedings of the 12th Workshop on Asian Language Resources, pp. 95–104 (2016)

Evaluating Co-reference Chains Based Conversation History in Conversational Question Answering

Angrosh Mandya[✉], Danushka Bollegala, and Frans Coenen

Department of Computer Science, University of Liverpool, Liverpool, UK
{angrosh,danushka,coenen}@liverpool.ac.uk

Abstract. This paper examines the effect of using co-reference chains based conversational history against the use of entire conversation history for conversational question answering (CoQA) task. The QANet model is modified to include conversational history and NeuralCoref is used to obtain co-reference chains based conversation history. The results of the study indicates that in spite of the availability of a large proportion of co-reference links in CoQA, the abstract nature of questions in CoQA renders it difficult to obtain correct mapping of co-reference related conversation history, and thus results in lower performance compared to systems that use entire conversation history. The effect of co-reference resolution examined on various domains and different conversation length, shows that co-reference resolution across questions is helpful for certain domains and medium-length conversations.

Keywords: Co-reference based conversation history · Conversational question answering · QANet

1 Introduction

In recent times, the focus of Machine Comprehension (MC) has shifted from answering questions that most likely have an answer in the contextual passage [5,7] to answering more difficult questions that are conversational in nature, with answers often absent in the contextual passage [3,6,9]. The **Co**nversational **Q**uestion **A**nswering (CoQA) dataset is developed for measuring the ability of systems to answer such conversation-style questions. An important aspect of this dataset is the presence of large amounts of co-reference links between questions. Almost half of the CoQA questions (49.7%) contain explicit co-reference markers (e.g. *he, she, it*) that refer back to previous questions [6]. For example, for the sample conversation in Table 1, the pronoun *'she'* in q_2 an q_3 refers back to the name of the cat (*'cotton'*) in q_1.

A key characteristic of CoQA systems such as DrQA, PGNet, DrQA+PGNet [6], Bidaf++ [9], FlowQA [3] is to use previous conversational history to provide contextual information essential for answering the current question.

L.-M. Nguyen et al. (Eds.): PACLING 2019, CCIS 1215, pp. 280–292, 2020.
https://doi.org/10.1007/978-981-15-6168-9_24

Table 1. Example conversation from CoQA Dataset

Once upon a time, in a barn near a farm house, there lived a little white kitten named Cotton. Cotton lived high up in a nice warm place above the barn where all of the farmer's horses slept. But Cotton wasn't alone in her little home, but shared her hay bed with her mommy and 5 other sisters.

q_1 : What color was Cotton?
a_1 : white

q_2 : Where did she live?
a_2 : in a barn

q_3 : Did she live alone?
a_3 : No

q_4 : Who did she live with?
a_4 : with her mommy and 5 sisters

Q_5 : What color were her sisters?
A_5 : orange and white

For example, to answer q_3 in Table 1, the CoQA model [3,6,9] uses previous set of questions and answers $\{q_2, a_2\}$ and $\{q_2, a_2, q_1, a_1\}$ to input one and two conversation histories, respectively as contextual information. A major drawback of this method, is that the CoQA model can easily miss out on key information vital for answering conversational questions. For instance, to answer q_4 in Table 1, using $\{q_3, a_3, q_2, a_2\}$, does not provide key input *'cotton'* as contextual information useful for answering q_4. However, identifying the link between pronoun *"she"* in q_4 and *"cotton"* in Q_1 through resolving co-reference chains in $\{q_3, q_2, q_1\}$, can allow us to use $\{q_1, a_1\}$ as inputs to the CoQA system rather than $\{q_3, a_3, q_2, a_2\}$. Thus, resolving co-reference chains in conversation history and providing more relevant contextual information can be useful for improving the performance of CoQA systems.

Based on this motivation, we focus on examining the usefulness of resolving co-reference chains in conversation history for the CoQA task. The main contribution of this paper is not to propose a state-of-the-art (SOTA) model for CoQA but to provide an empirical analysis of the effect of using co-reference chains in CoQA. To this end, we conduct several experiments using co-reference based conversation history to examine its influence against using the entire conversation history. To identify co-reference chains, we use NeuralCoref[1], a neural network

---

[1] https://github.com/huggingface/neuralcoref.

based co-reference resolution tool. For our experiments, we modify QANet [10], a SOTA model for MC to include the conversational history as an input to the model.

The empirical results presented in this paper shows that even though co-reference links are present in large number across conversational questions in CoQA, the abstract nature of questions in CoQA renders it difficult to map a given question to co-reference related conversation history, resulting in lower performance compared to systems that use entire conversation history.

2   Related Work

The CoQA dataset was proposed by [6] for evaluating convesational question-answering systems. The dataset provides human style conversational questions and preserves the naturalness of the answers evident in typical conversations. Besides developing the CoQA dataset, [6] also evaluated several standard MC models such as sequence-to-sequence (seq2seq), pointer-generator network (PGNet), Document Reader Question Answering (DrQA) system and a combined DrQA+ PGNet model for CoQA as baseline models. Following the availability of CoQA dataset, several models have been proposed for CoQA. The BiDAF++ model [9] based on the Bidirectional Attention Flow (BiDAF) model [7] augmented with self-attention [1] was proposed to compute similarities between the context and conversation history. A Flow mechanism was used to add intermediate representations obtained during the process of answering previous questions [3]. SDNet, a contextual attention-based deep neural network [11] was proposed to leverage inter-attention and self-attention for CoQA. *Google SQuad 2.0+MMFT (ensemble)*[2], the latest model listed on CoQA Leaderboard currently outperforms human performance on CoQA.

QANet [10], the SOTA model for MC was proposed to combine CNNs and self-attention networks to model local interactions and global interactions, respectively. QANet is shown to outperform SOTA MC models such as BiDAF [7], R-Net [8], Reinforced Mnemonic Reader [2] on SQuAD 1.0 dataset [5], both in terms of speed and accuracy.

As stated previously, the focus of this paper is not to propose SOTA for CoQA but to investigate the influence of co-reference links in answering conversational questions. Since QANet provides an efficient and faster means for MC, we propose to modify the QANet model in the context of CoQA. The modification of QANet to use similarity between context and conversation history is similar to the method proposed in BiDAF++[9]. Although various models [3,6,9,11] have been proposed for CoQA, none of the studies have specifically focused on examining the influence of co-reference links in CoQA. To the best knowledge of the authors, this is the first study that provides an extensive empirical analysis of co-reference chains in CoQA. To this end, we use the modified QANet model to examine the performance of using co-reference chains based conversation history against using the available previous conversation history.

---

[2] Listed on March 29, 2019.

3   Problem Formulation

Given a context passage c, conversational history $(q_1, a_1, ...q_{i-1}, a_{i-1})$ and question q_i, the task is to predict the answer $\hat{a}_i$.

$$p(\hat{a}_i|q_i) = f(c_i, q_1, a_1, ..., q_{i-1}, a_{i-1}) \tag{1}$$

However, instead of using the available $(q_1, a_1, ...q_{i-1}, a_{i-1})$, we propose to use the set of co-reference chains based conversation history $(q_k, a_k, ...q_{k-1}, a_{k-1})$, defined as the set of previous question-answer pairs that have co-reference links to the current question q_i.

$$p(\hat{a}_i|q_i) = f(c_i, q_k, a_k, ..., q_{k-1}, a_{k-1}) \tag{2}$$

Given two questions q_i and q_j, we say that there exists a co-reference link between q_i and q_j, if a word $u \in q_i$ refer to the same *person* or *thing* $v \in q_j$. Thus, the question-answer pair $\{q_j, a_j\}$ forms the co-reference chains based conversation history for q_i. For example, in Table 1, given q_4 and q_1, we consider a co-reference link between words 'she' $\in q_4 \rightarrow$ 'Cotton' $\in q_1$, thus providing $\{q_1, a_1\}$ as the co-reference chains based conversation history for q_4. To evaluate the use of such conversation history, the QANet model is modified for CoQA as explained in the following section.

4   QANet Model for CoQA

The architecture of the modified QANet model for CoQA is described in Fig. 1. We briefly describe the main components of the model. For a detailed explanation of QANet model, please refer [10].

4.1   Input Embedding Layer

The embedding for each word w is obtained by concatenating its word embedding with the character embeddings. The hyper-parameters of QANet [10] are retained, with word embedding initialized using $p_1 = 300$ dimensional pre-trained GloVe embeddings [4] and character embedding as a trainable vector of dimensionality $p_2 = 200$.

4.2   Embedding Encoding Layer

The embedded input comprising c, q_i, and $\{q_1, a_1, ...q_i, a_i\}$ is provided as input to the encoding layer that consists of a stack of convolution, self-attention and feed-forward layers. The default network settings of the residual block are retained in the encoding layer. The encoding layer receives as the input a vector of dimensionality $p_1 + p_2 = 500$ for each individual word and maps to one-dimensional convolution of dimensionality $d = 128$.

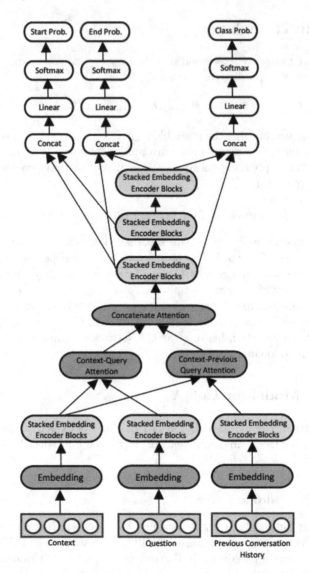

Fig. 1. Modified QANet model for CoQA. For residual network details, please refer [10]

4.3 Attention Layer

The main modification of QANet model for CoQA is in the attention layer, which besides computing similarity between c and q_i (context-query attention), also computes similarity between c and conversation history ($q_1, a_1, ...q_{i-1}, a_{i-1}$) (context-conversation history attention).

Let $\mathbb{C}, \mathbb{Q}$ and $\mathbb{R}$ be the encoded context, current question and conversation history, respectively. The similarities between each pair of words between c and

q_i, and c and $(q_1, a_1, .. q_{i-1}, a_{i-1})$ is computed using similarity matrices $S_1 \in \mathbb{R}^{n \times m_1}$ and $S_2 \in \mathbb{R}^{n \times m_2}$, where n is the length of c, and m_1, m_2 are the lengths of q_i and $(q_1, a_1, .. q_{i-1}, a_{i-1})$, respectively. Each row of S_1 and S_2 is normalised using the softmax function to obtain matrices $\bar{S}_1$ and $\bar{S}_2$. The context-query and context-conversation history attention are computed as $A_1 = \bar{S}_1 \cdot Q^T \in \mathbb{R}^{n \times d}$ and $A_2 = \bar{S}_2 \cdot R^T \in \mathbb{R}^{n \times d}$. The tri-linear function [7] is used as the similarity function: $f(q_i, c) = \mathbf{W}_0[q_i, c, q_i \odot c]$, $f(q_j, c) = \mathbf{W}_1[q_j, c, q_j \odot c]$, where $\odot$ is the element-wise multiplication and $\mathbf{W}_0, \mathbf{W}_1$ are trainable vectors. To compute query-context attention (B_1) and conversation history-context attention (B_2), column normalized matrices $\bar{\bar{S}}_1$ and $\bar{\bar{S}}_2$ of $\bar{S}_1$ and $\bar{S}_2$ are computed using softmax function and B_1 and B_2 are obtained by $B_1 = \bar{S}_1 \bar{\bar{S}}_1^T C^T$ and $B_2 = \bar{S}_2 \bar{\bar{S}}_2^T C^T$.

4.4 Model Encoding Layer

The input to the model encoding layer is $[c, a_1, c \odot a_1, c \odot b_1, a_2, c \odot a_2, c \odot b_2]$, where a_1, a_2, b_1, b_2 is a row of attention matrix A_1, A_2, B_1, B_2, respectively. The default settings of QANet are retained to share weights each of the 3 repetitions (M_0, M_1, M_2) of the model encoder.

4.5 Output Layer

The span selection method [7,8] is used to predict the probability of each position in the context as being the start or end of an answer span. Specifically, the start and the end position probabilities are modelled as: $p^1 = \text{softmax}(\mathbf{W}_1[\mathbf{M}_0; \mathbf{M}_1])$ and $p^2 = \text{softmax}(\mathbf{W}_2[\mathbf{M}_0; \mathbf{M}_2])$. Simultaneously, we also output the probability p^c of belonging to one of the four classes {yes, no, unknown, span}: $p^c = \text{softmax}(\mathbf{W}_3[\mathbf{M}_0; \mathbf{M}_2])$, where $\mathbf{W}_1, \mathbf{W}_2, \mathbf{W}_3$ are trainable variables and $\mathbf{M}_0, \mathbf{M}_1, \mathbf{M}_2$ are the output of the model encoder from bottom to top, respectively.

The loss function to learn the start and end probabilities is defined as the negative sum of the log probabilities of the predicted distributions of true start and end indices, averaged over all training examples:

$$L_0(\theta) = -\frac{1}{N} \sum_i^N \log(p_{y_i^1}^1) + \log(p_{y_i^2}^2) \tag{3}$$

The loss function to learn class probabilities is defined as the negative sum of the question belonging to a particular class, averaged over all training examples:

$$L_1(\theta) = -\frac{1}{N} \sum_i^N \log(p_{y_i^c}^c) \tag{4}$$

Here y_i^1, y_i^2, y_i^c are respectively the groundtruth start and end positions, and the class of example i and θ contains all trainable parameters. The total loss is:

$$L = L_0(\theta) + L_1(\theta) \tag{5}$$

4.6 Inference

In the inference stage, for each question q_i, we first use p^c to predict whether q_i is answerable. If it is answerable, we predict the span (s, e) with the maximum p^1, p^2, otherwise we predict the class as the answer for q_i.

5 Experiments

5.1 Evaluation Metric

We conduct experiments on the CoQA dataset [6]. However, because the test set in CoQA is not publicly available and the main objective of this paper is to primarily investigate the effect of co-reference resolution in CoQA and not compete with systems listed on the CoQA leaderboard, we report our results only on the development set and not on the test set. To this end, we randomly choose 80% of CoQA training data as our train set and the remaining 20% as the development set, to develop the model. The learnt model is tested on the CoQA development set. Further, following [6], we report macro-average F1 score as the evaluation metric.

5.2 Implementation

The original settings of the QANet model [10] is retained while the modifying the QANet model for CoQA. The co-reference chains were derived employing NeuralCoref[3], a pipeline extension for spaCy 2.0 that annotates and resolves co-reference clusters using a neural network.

5.3 Results

The following explains the key results of this study

Using Co-reference Chains Based History Vs. Using Available Previous History. In order to examine the influence of co-reference chains in answering conversation questions, the following models were evaluated:

- **qanet-1-ccq** and **qanet-2-ccq**, model that uses previous one and two co-reference chain linked questions, respectively;
- **qanet-1-ccqa** and **qanet-2-ccqa**, model that uses previous one and two co-reference chain linked questions and answers, respectively;
- **qanet-1-pqa** and **qanet-2-pqa**, that uses previously available one and two questions and answers, respectively;

---

[3] https://github.com/huggingface/neuralcoref.

The overall performance of different models on the development set of the CoQA, in Table 2 shows that models using the entire previous conversation history (QANET-1-PQA and QANET-2-PQA) performs slightly better than models that use co-reference chains based conversation history (QANET-1-CCQ, QANET-2-CCQ, QANET-1-CCQA, QANET-2-CCQA). Interestingly for two domains "Children Stories" and "Literature", the co-reference chains based model (QANET-2-CCQA) achieves the best performance, indicating that the set of question-answer pairs identified based on co-reference resolution is helpful in answering conversational questions, particularly for these two domains. However, for other three domains the model using the available previous conversation history (QANET-2-PQA) achieves the highest performance. Though not conclusive, these results indicate that co-reference chains based conversation history can be helpful for CoQA in some cases.

Table 2. F1 scores of QANet based models for different domains in CoQA Development Set.

	Child.	Liter.	Mid-High.	News	Wiki.	Overall
QANET-1-CCQ	62.4	56.7	63.1	66.9	67.4	63.4
QANET-2-CCQ	61.3	57.4	63.5	68.5	69.2	63.9
QANET-1-CCQA	65.7	59.3	64.6	70.2	68.2	65.3
QANET-2-CCQA	**66.8**	**60.1**	62.8	71.5	70.2	66.2
QANET-1-PQA	64.9	57.8	65.8	74.1	73.7	67.2
QANET-2-PQA	65.2	58.9	**66.2**	**75.5**	**73.9**	**67.9**

Absence of Contextual Information. The main reason for the poor performance of co-reference chain based models can be attributed to the absence of contextual information necessary for answering conversational questions, for co-reference based models. As seen in Table 3, NeuralCoref facilitates identification of co-reference chain linked questions for about 80% of questions in the CoQA development set. This means that for the rest 20% of the questions, the contextual information in terms of previous questions and answers is not available for co-reference based models. Thus, these models have to entirely rely on the information available in the current question to answer it, resulting in a lower performance compared to QANET-1-PQA and QANET-2-PQA, which have conversation history for all questions, except the first. To address the problem of questions without co-reference chains based previous questions, we conducted experiments using the available previous conversation history for those questions where co-reference related previous conversation history was not available. However, the results (not reported here) showed that the inclusion did not help in improving the performance.

The co-reference chains based questions obtained for a sample paragraph in CoQA development set provided in Table 6 shows that there are no co-reference chains based previous questions for q_2 to q_5. The problem of not identifying co-reference linked previous questions for q_2 to q_5 is not because of the poor performance of NeuralCoref, but rather due to missing clues in q_2 to q_5 that does not help NeuralCoref in identifying co-reference links in previous questions. Further, as may be seen in Table 6, questions q_2 to q_5 are quite abstract and change the topic of discussion, without providing any information about the change in the topic. This further makes it difficult to identify co-reference links in previous questions. These aspects further establish the complex nature of questions in CoQA dataset. The above results indicates that even though there are a high number of questions with co-reference links, connecting a given question to more relevant previous questions is quite challenging (Table 4).

Incorrect Contextual Information. The poor performance of co-reference chains-based models can also be attributed to combining incorrect contextual information with the current question. For example for questions q_8 to q_{10} in Table 1, the same question (q_5) is used as the co-reference chains-based previous question. However, information provided by q_5 is not very helpful in answering questions q_8 to q_{10}, and thus results in lower performance of the model. Although, experiments were conducted to include questions within a certain window in the question sequence, the performance (not reported here) almost remained the same.

Table 3. Number of co-reference chain linked questions for various domains in CoQA Development Set

	Child.	Liter.	Mid-High.	News	Wiki.	Total
TQ	1425	1630	1653	1649	1626	7983
TQ_coref_links	1181	1274	1385	1313	1223	6376
(%)	82.87	78.15	83.78	79.62	79.33	80.70

Table 4. Co-reference chains based questions obtained using NeuralCoref for a sample paragraph in domain "Children Stories" in CoQa development set.

	Questions in sequence	Co-reference chains based questions
1	What was the name of the fish?	-
2	What looked like a birds belly?	-
3	Who said that?	-
4	Was Sharkie a friend?	-
5	Did they get the bottle?	-
6	What was in it?	Did they get the bottle?
7	Did a little boy write	Did they get the bottle? the note?
8	Who could read the note?	Did they get the bottle?
9	What did they do with	Did they get the bottle? the note?
10	Did they write back?	Did a little boy write the note? Did they get the bottle?
11	Were they excited ?	Did a little boy write the note? Did they get the bottle?

Paragraphs with Higher Proportion of Co-reference based Conversation History. The performance of QANet model on conversations that have higher proportion of co-reference chains-based conversation history (80% and 60% questions have conversation history) (shown in Table 5), achieves a slightly better F1-score of 66.5 and 65.9, respectively, against a lower F1-score of 65.3 achieved with considering all conversations with co-reference linked questions. Although, there is a slight improvement the difference is not significant, indicating that even a lower percentage of questions that do not have any previous history can affect the model's performance. Further, it is also important to note that the errors induced by the co-reference resolution system can be compounding in nature and thus, can significantly lower the performance.

Table 5. F1 scores of co-reference based QANet models for conversations with different percentage of co-reference chains in CoQA development set.

	Overall
QANET-80%-CON-CCQA	66.5
QANET-60%-CON-CCQA	65.9
QANET-ALL-CON-WITH-CCQA	65.3

Table 6. Replacing co-referenced pronouns in questions with referenced words from previous questions.

	Questions in sequence	Questions with replaced pronouns
1	What color was Cotton?	What color was Cotton?
2	Where did she live?	Where did Cotton live?
3	Did she live alone?	Did Cotton live alone?
4	Who did she live with?	Ho did Cotton live with ?

Using Answers with Questions. The results provided in Table 2 also indicates that co-reference chain based conversation history alone is not sufficient for answering conversational questions. As seen in Table 2, the QANET-1-CCQ QANET-2-CCQ models which uses coreference-chain based questions alone perform poorly in comparison to the models QANET-1-CCQA, QANET-2-CCQA, QANET-1-PQA, QANET-2-PQA which employs both questions and corresponding answers together. Thus, it is useful to use previous answers along with questions, to augment the contextual information necessary to answer conversational questions.

Replacing Co-referenced Pronouns in Questions. Experiments were also conducted to evaluate the performance co-reference based models by replacing co-referenced pronouns in the current question with referenced words in previous questions. For example, using NeuralCoref facilitates identification of co-reference link between the pronoun "she" $\in q_2, q_3, q_4$ and the noun "Cotton" $\in q_1$ (Table 1). Using this co-reference link, the pronoun "she" is replaced with noun "Cotton" as shown in Table 6.

The performance of the QANet model using current question alone and using questions with replacing co-reference pronouns is provided in Table 7. The results in Table 7 shows that it is difficult to obtain a comparable score using current question alone and thus, contextual information in terms of conversation history plays an important role in achieving optimum performance for CoQA. However, interestingly a small improvement (F1-score of 58.88 vs. 57.30) is achieved when co-referenced pronouns in questions are replaced with either *person* or *thing* that it refers to in the previous questions. The replacement of co-referenced pronouns

Table 7. F1-scores of model using current question with replacing co-reference pronouns for the domain of "Children Stories" in CoQA dataset.

	QANET_REG_QUEST	QANET_COREF_REP_QUEST
Yes	80.62	54.84
No	34.57	66.60
Unknown	37.50	48.48
Span	56.69	58.64
Overall	57.30	58.88

Table 8. F1-scores of model using current question with replacing co-reference pronouns on different conversation length for domain "Children Stories" in CoQA dataset.

	QANET_REG_QUEST	QANET_COREF_REP_QUEST
Conversation length ≤ 14		
Yes	82.81	54.68
No	35.71	66.32
Unknown	41.66	63.63
Span	54.42	58.75
Overall	55.76	58.95
Conversation length > 14		
Yes	79.16	54.94
No	33.86	66.77
Unknown	33.33	33.33
Span	58.68	58.54
Overall	58.59	58.83

particularly seem to help in answering "No", "Unknown", and "Span prediction" type questions.

The minimum, maximum and the average number of questions for paragraphs in the domain of "Children Stories" in CoQA development set are 10, 25 and 14, respectively. Therefore the performance of QANET_REG_QUEST and QANET_COREF_REP_QUEST was examined on two groups: (a) paragraphs with ≤14 questions; and (b) paragraphs with >14 questions as shown in Table 8. As Table 8 shows, replacing co-referenced pronouns in paragraphs with ≤14 questions significantly helps in answering question types such as "no" (66.32 vs. 35.71), "unknown" (63.3 vs. 41.66) and "span prediction" (58.75 vs. 54.42). These results indicates that more accurate co-reference links are obtained in conversations with lower to medium (around 14) number of questions. However, as the length of conversations increase, there seems to be little effect of using co-reference links, which is most likely due to poor co-reference links between questions.

Comparison with CoQA Baseline Models. As mentioned previously, the objective of this paper is not to compete against SOTA for CoQA task. However, it needs to be noted that the QANet models using one and two conversation history (QANET-1-PQA and QANET-2-PQA) achieves an F1-Score of 67.2 and 67.9, respectively on the CoQA development set. These results are slightly better than the performance of baseline models: Seq2Seq (27.5); PGNet (45.4); DrQA (54.7); DrQA+PGNet (66.2), obtained on the CoQA development set [6]. The results of QANET-1-PQA and QANET-2-PQA are also comparable with scores of BiDAF++w/0-ctx (63.4); BiDAF++w/1-ctx (68.6); BiDAF++w/2-ctx (68.7) [9] on CoQA development set. The modified QANet model described in this paper follows a similar approach of BiDAF++ to combine context with conversation history, indicating the usefulness of QANet in the context of CoQA.

6   Conclusion

We presented in this paper an empirical analysis of using co-reference chains based conversation history for CoQA. The results presented in this paper shows that although there exists a large proportion of co-reference links across questions in CoQA, the abstract nature of questions renders it difficult to map together co-reference related questions for large number of questions, resulting in lower performance in comparison to models that use previously available conversation history. The results also show that using co-reference related questions can help in conversations which have fewer questions.

References

1. Clark, C., Gardner, M.: Simple and effective multi-paragraph reading comprehension. arXiv preprint arXiv:1710.10723 (2017)

2. Hu, M., Peng, Y., Huang, Z., Qiu, X., Wei, F., Zhou, M.: Reinforced mnemonic reader for machine reading comprehension. arXiv preprint arXiv:1705.02798 (2017)
3. Huang, H.Y., Choi, E., Yih, W.T.: Flowqa: grasping flow in history for conversational machine comprehension. arXiv preprint arXiv:1810.06683 (2018)
4. Pennington, J., Socher, R., Manning, C.: Glove: global vectors for word representation. In: Proceedings of the 2014 Conference on Empirical Methods in Natural Language Processing (EMNLP), pp. 1532–1543 (2014)
5. Rajpurkar, P., Zhang, J., Lopyrev, K., Liang, P.: Squad: 100,000+ questions for machine comprehension of text. arXiv preprint arXiv:1606.05250 (2016)
6. Reddy, S., Chen, D., Manning, C.D.: Coqa: a conversational question answering challenge. arXiv preprint arXiv:1808.07042 (2018)
7. Seo, M., Kembhavi, A., Farhadi, A., Hajishirzi, H.: Bidirectional attention flow for machine comprehension. arXiv preprint arXiv:1611.01603 (2016)
8. Wang, W., Yang, N., Wei, F., Chang, B., Zhou, M.: Gated self-matching networks for reading comprehension and question answering. In: Proceedings of the 55th Annual Meeting of the Association for Computational Linguistics, vol. 1: Long Papers, pp. 189–198 (2017)
9. Yatskar, M.: A qualitative comparison of COQA, squad 2.0 and QUAC. arXiv preprint arXiv:1809.10735 (2018)
10. Yu, A.W., et al.: Qanet: combining local convolution with global self-attention for reading comprehension. arXiv preprint arXiv:1804.09541 (2018)
11. Zhu, C., Zeng, M., Huang, X.: Sdnet: contextualized attention-based deep network for conversational question answering. arXiv preprint arXiv:1812.03593 (2018)

Speech and Emotion Analyzing

Multiple Linear Regression of Combined Pronunciation Ease and Accuracy Index

Katsunori Kotani[1]([✉]) and Takehiko Yoshimi[2]

[1] Kansai Gaidai University, Hirakata, Osaka, Japan
kkotani@kansaigaidai.ac.jp
[2] Ryukoku University, Otsu, Shiga, Japan

Abstract. This study proposes an index for measuring the ease and accuracy of pronunciation that is derived by combining the ease and the accuracy of learners' pronunciation, assesses the reliability and validity of the combined index, and develops a measurement method. The ease is a scale score for the ease of pronunciation that learners subjectively judged, and the accuracy demonstrates learners' pronunciation performance. A previous study proposed an index regarding ease of pronunciation, and other previous studies independently proposed indices related to accuracy. These two types of indices should be combined because they have different aspects and compensate for each other. To develop the proposed measurement method, a learner corpus of pronunciation is compiled, and index reliability and validity are assessed using the classical test theory. The assessments demonstrate that the proposed index is moderately reliable and not valid. The proposed measurement method is developed using multiple linear regression analysis. The dependent variable is an index consisting of both pronunciation ease, which is subjectively judged by learners of foreign languages, and pronunciation accuracy, which is defined as the similarity between a reference sentence and learner pronunciation. The independent variables are defined as sentence linguistic features and learner features, such as foreign language proficiency test scores. The experimental results demonstrate that the measured ease and accuracy of pronunciation has moderate correlation with the observed ease and accuracy ($r = 0.69$), and a significant contribution is observed in all the linguistic and learner features ($p < 0.05$).

Keywords: Pronunciation practice material · Multiple linear regression model · Analysis for pronunciation difficulty and accuracy · English as a foreign language

1 Introduction

Effective teaching of English as a foreign language is required to ensure that learners are highly motivated [7,12]. Learners' motivation will sustain if they are provided with materials that are interesting and with a difficulty level appropriate for the learners' proficiency. A promising language resource is the Internet.

© Springer Nature Singapore Pte Ltd. 2020
L.-M. Nguyen et al. (Eds.): PACLING 2019, CCIS 1215, pp. 295–306, 2020.
https://doi.org/10.1007/978-981-15-6168-9_25

One advantage of this resource is that learners can select materials depending on their interest; however, the disadvantage of this approach is that the difficulty level is often unclear to learners, while the choice of proper materials by teachers is time and effort consuming. However, if the difficulty of the materials is measured automatically, learners can independently access materials appropriate for their proficiency levels without teachers' assistance.

For materials for practicing learners' pronunciation skills, a previous study [11] developed a method of measuring the ease of pronunciation using multiple linear regression analysis. A dependent variable was learners' subjective evaluation of how easily they were able to pronounce a sentence (hereafter, pronounceability in terms of ease: EASE). EASE was subjectively determined by learners on a five-point Likert scale. Although this index has some advantages, EASE had the disadvantage of the possibility that it might include learners' biases due to learners' over/underestimation. Another previous study [1] developed a method of measuring EASE using support vector machines and multiple linear regression analysis. The dependent variable was automatic evaluation results for pronunciation errors. Although this index is free from learners' and evaluators' biases, the index fails to reflect learners' subjective judgment of the pronounceability.

These problems were solved using objective evaluation of how accurately learners were able to pronounce a sentence (hereafter, pronounceability in terms of accuracy: ACC) [10,13,15]. ACC was defined as the similarity, or the normalized edit distance, between a reference sentence and a transcription of learners' pronunciation. The primary advantage of ACC was that it avoided learner bias from subjective evaluations. Another advantage was that ACC was able to identify pronunciation problems that learners failed to notice.

This study proposes a third pronounceability index by combining EASE and ACC (hereafter, pronounceability in terms of EASE and ACC: EASE&ACC). The goal of this combined index is to make the most of the advantages of EASE and ACC because EASE and ACC compensate for each other. The former demonstrates the pronounceability explicit to learners as judged by learners. The latter explicates the pronounceability that learners fail to recognize. In this study, the combined index was derived by multiplying EASE by ACC.

This study assessed EASE&ACC by answering the following research questions:

1. How stable is EASE&ACC as an evaluation index?
2. To what extent does EASE&ACC help classify learners based on English proficiency?
3. How accurately can EASE&ACC be measured based on linguistic and learners' features?

The first two questions were answered within the framework of the classical test theory [2]. The third question was answered using multiple linear regression analysis considering EASE&ACC as a dependent variable and linguistic and learner features as independent variables.

2   Compilation of Phonetic Learner Corpus

A data instance for multiple linear regression comprised speech sounds from learners reading aloud, transcriptions of the speech sounds, a reference sentence, EASE&ACC, the linguistic features of the reference sentence, and learners' features (i.e., learners' scores on an English proficiency test).

2.1   Collection of Pronunciation Data

The speech sounds were collected by recording learners' reading-aloud English texts. The reading-aloud task proceeded as follows. First, they listened to the reference texts read aloud by a native speaker of American English who was a voice actor (female, 35 years old). Second, they read a sentence aloud. Then, they determined its EASE on a five-point Likert scale (1: easy; 2: somewhat easy; 3: average; 4: somewhat diffiult; and 5: diffiult). In this task, learners could read a sentence twice. They were asked to complete each task as quickly as possible, and to stop working when the task was completed. They were prohibited from using dictionaries or any other reference books, and from returning to revise a sentence after moving on to another sentence.

The texts used for reading aloud were selected from those distributed by the International Phonetic Association encompassing basic English sounds [5,8]. This enabled us to analyze which types of English sounds influence learners' pronunciation. The texts were originally appropriated from Aesop's Fables. The title of Text I was "The North Wind and the Sun," and Text II was "The Boy Who Cried Wolf." Text I contained five sentences, and Text II contained 10 sentences. The sentences in Texts I and II are shown in Figs. 1 and 2, respectively. Deterding [5] reported that Text I failed to encompass certain sounds, such as the initial and medial /z/ and syllable initial /θ/, and added another text, Text II, to include the English pronunciation of these sounds. In addition, the semantic properties of the texts were supposed to be neglected, because the contents were familiar in the area where this study was carried out.

The pronunciation data were collected from 50 learners at universities – 28 males and 22 females with a mean age of 20.8 years and with a standard deviation (SD) of 1.3 – who were compensated for their participation. All learners were asked to submit valid scores from a Test of English for International Communication (TOEIC) taken in the current or previous year. The 50 learners were classified into three levels based on the TOEIC scores: below 490 (beginner level) $(N = 16)$, 490 to below 730 (intermediate level) $(N = 16)$, and 730 or above (advanced level) $(N = 18)$. In the study sample, the mean TOEIC score was 607.7, and the SD was 186.2; the minimum score was 295 and the maximum was 900.

2.2   Annotation of Pronunciation Data

The combined index EASE&ACC was derived by a simple multiplication of EASE and ACC. It ranged from 0 to 5, and as it approached 0, pronounceability increased.

The North Wind and the Sun were disputing which was the stronger, when a traveller came along wrapped in a warm cloak.

They agreed that the one who first succeeded in making the traveller take his cloak off should be considered stronger than the other.

Then the North Wind blew as hard as he could, but the more he blew the more closely did the traveller fold his cloak around him; and at last the North Wind gave up the attempt.

Then the Sun shone out warmly, and immediately the traveller took off his cloak.

And so the North Wind was obliged to confess that the Sun was the stronger of the two.

Fig. 1. Sentences in Text I.

There was once a poor shepherd boy who used to watch his flocks in the fields next to a dark forest near the foot of a mountain.

One hot afternoon, he thought up a good plan to get some company for himself and also have a little fun.

Raising his fist in the air, he ran down to the village shouting 'Wolf. Wolf.'

As soon as they heard him, the villagers all rushed from their homes, full of concern for his safety, and two of his cousins even stayed with him for a short while.

This gave the boy so much pleasure that a few days later he tried exactly the same trick again, and once more he was successful.

However, not long after, a wolf that had just escaped from the zoo was looking for a change from its usual diet of chicken and duck.

So, overcoming its fear of being shot, it actually did come out from the forest and began to threaten the sheep.

Racing down to the village, the boy of course cried out even louder than before.

Unfortunately, as all the villagers were convinced that he was trying to fool them a third time, they told him, 'Go away and don't bother us again.'

And so the wolf had a feast.

Fig. 2. Sentences in Text II.

The pronunciation was manually transcribed by two native speakers of American English who had one-year transcription experience. They satisfied hiring conditions that one was trained to replicate interviews and/or meetings, and that one was unaccustomed to English spoken by learners. The latter condition was assigned because the familiarity with learners' pronunciation might lead to correct errors unconsciously. They shared the transcription task as follows: one (25-year-old male) listened to the learners' pronunciation and transcribed it, and the other (26-year-old female) reviewed the transcription. They read the reference texts before carrying out their tasks. They were instructed to replicate learners' pronunciation without adding, deleting, or substituting any expressions to improve grammaticality and/or acceptability (except the addition of punctuation such as commas and periods). When their judgments were different, the reviewer's judgement was prioritized to the transcriber's judgment.

ACC was calculated by dividing the Levenshtein edit distance between learners' transcribed sentences and the reference sentences by the number of characters in a longer sentence. It reflected the differences between the reference sentences and the transcriptions of the learners' pronunciation due to the substitution, deletion, or insertion of characters. It ranged from 0 to 1, with 1 indicating greater differences. Before measuring the edit distance, symbols such as commas and periods were deleted, and capital letters in expressions were changed to lowercase in the transcription and reference data.

The linguistic features comprised sentence length, mean word length, multiple-syllable words, and word difficulty. They were automatically derived from a sentence that learners read aloud. Sentence length [3] was derived as the number of words in a sentence. Mean word length [3] was derived by dividing the number of syllables in a sentence by the number of words in the sentence. The number of multiple-syllable words in a sentence [6] was derived by calculating $\sum_{i=1}^{N}(S_i - 1)$, where N denoted the number of words in a sentence and S_i denoted the number of syllables in the i-th word. This subtraction ignored single-syllable words. Word difficulty was derived as the rate of words not listed in a basic vocabulary list [9] relative to the total number of words in a sentence.

Learners' features were determined using the TOEIC scores. Although TOEIC comprises listening and reading tests, Chauncey Group International [16] reported a strong correlation between TOEIC scores and Language Proficiency Interview results, an established direct assessment of oral language proficiency developed by the Foreign Service Institute of the U.S. Department of State.

2.3 Properties of Phonetic Learner Corpus

The phonetic learner corpus included 750 instances: 15 sentences read aloud by 50 learners. The frequency distribution of EASE&ACC is shown in Table 1. Most instances (77.2%) appeared below 0.5, and the distribution did not follow normal distribution according to the Kolmogorov-Smirnov test ($K = 8.14$, $p < 0.05$). The median EASE&ACC was 0.13, and the range was 3.14.

The frequency distribution of EASE and that of ACC are shown in Table 2 and Table 3, respectively. EASE occurred most frequently in 3, and thus learners

Table 1. Frequency distribution of EASE&ACC.

Class of EASE&ACC	Frequency	Relative frequency (%)
$0 \leq$ EASE&ACC < 0.5	579	77.2
$0.5 \leq$ EASE&ACC < 1.0	57	7.6
$1.0 \leq$ EASE&ACC < 1.5	25	3.3
$1.5 \leq$ EASE&ACC < 2.0	31	4.1
$2.0 \leq$ EASE&ACC < 2.5	47	6.3
$2.5 \leq$ EASE&ACC < 3.0	6	0.8
$3.0 \leq$ EASE&ACC < 3.5	5	0.7
$3.5 \leq$ EASE&ACC < 4.0	0	0.0
$4.0 \leq$ EASE&ACC < 4.5	0	0.0
$4.5 \leq$ EASE&ACC ≤ 5.0	0	0.0

subjectively judged the pronounceability neither as difficult nor as easy. On the other hand, ACC occurred most frequently in $0 \leq$ ACC < 0.1, which meant that learners' pronunciation contained few errors. These distribution provided a piece of evidence for combining EASE and ACC as a pronounceability index.

Table 2. Frequency distribution of EASE.

Class of EASE	Frequency	Relative frequency (%)
EASE = 1	48	6.4
EASE = 2	127	16.9
EASE = 3	357	47.6
EASE = 4	191	25.5
EASE = 5	27	3.6

The means and SDs of the linguistic features derived from Texts I and II demonstrated similar properties at the sentence level as shown in Table 4.

3 Verification of Reliability and Validity

The reliability and validity of EASE&ACC were verified within the framework of the classical test theory. The reliability of EASE&ACC was examined through internal consistency, referring to whether EASE&ACC demonstrates similar results for sentences with similar pronounceability. The internal consistency was tested in terms of Cronbach's α [4]. Cronbach's α is a reliability coefficient ranging from 0 (absence of reliability) to 1 (absolute reliability), with values over 0.90 indicating strong reliability, values over 0.80 indicating good reliability, and values over 0.70 indicating acceptable reliability [14].

Table 3. Frequency distribution of ACC.

Class of ACC	Frequency	Relative frequency (%)
$0 \leq$ ACC <0.1	514	68.5
$0.1 \leq$ ACC < 0.2	76	10.1
$0.2 \leq$ ACC < 0.3	8	1.1
$0.3 \leq$ ACC < 0.4	50	6.7
$0.4 \leq$ ACC < 0.5	1	0.1
$0.5 \leq$ ACC < 0.6	48	6.4
$0.6 \leq$ ACC < 0.7	3	0.4
$0.7 \leq$ ACC < 0.8	50	6.7
$0.8 \leq$ ACC < 0.9	0	0.0
$0.9 \leq$ ACC ≤ 1.0	0	0.0

Table 4. Means and SDs of linguistic features of Texts I and II.

Feature	Text I		Text II	
	Mean	SD	Mean	SD
Sentence length (words)	22.6	8.3	21.6	7.6
Mean word length (syllables)	1.3	0.1	1.2	0.1
Multiple-syllable words (words)	6.4	2.8	5.7	3.0
Word difficulty	0.3	0.1	0.2	0.1

The reliability coefficient of EASE&ACC indicated acceptable reliability as shown in Table 5. The reliability exceeded the baseline. Hence, this suggested the reliability of an evaluation criterion. However, the reliability was weaker than that of EASE and that of ACC. Future study will examine whether the combination method, i.e., multiplication, decreased reliability by comparing other methods, such as weighted summation.

Table 5. Cronbach's α reliability coefficients of pronounceability.

Pronounceability	Cronbach's α
EASE&ACC	0.71
EASE	0.92
ACC	0.85

Construct validity was examined from the viewpoint of distinctiveness. If EASE&ACC appropriately reflects learners' proficiency, EASE&ACC should demonstrate a statistically significant difference among learners at different proficiency levels. The 750 instances were classified into three levels based on the

TOEIC scores: below 490 (beginner level) ($N = 240$), 490 to below 730 (intermediate level) ($N = 240$), and 730 or above (advanced level) ($N = 270$).

Table 6 shows the means and SDs of EASE&ACC for the three proficiency levels. The distinctiveness of EASE&ACC was investigated using analysis of variance (ANOVA). ANOVA did not show statistically significant differences between the three levels of learners for EASE&ACC as depicted in Table 7 ($F(1, 748) = 2.83$, $p > 0.05$).

Table 6. Means and SDs of EASE&ACC at the three proficiency levels.

Level	N	Mean	SD
Beginner level	240	0.49	0.68
Intermediate level	240	0.45	0.69
Advanced level	270	0.39	0.63

EASE&ACC failed to demonstrate construct validity depending on TOEIC-based proficiency. Hence, this suggested that EASE&ACC was not useful for developing a measurement method for pronounceability depending on learners' proficiency levels. One remaining possibility is the development of a pronounceability measurement method that does not account for learners' proficiency, that is, a method that outputs a single pronounceability value for learners at any proficiency level.

Table 7. ANOVA table.

	Df	Sum sq.	Mean sq.	F	Sig.
Between groups	1	1.2	1.25	2.83	0.09
Within groups	748	329.7	0.44		
Total	749	330.9			

Note, too, that construct validity was observed in an ANOVA of ACC ($p < 0.05$) but not in that of EASE ($p > 0.05$). Future study will examine the influence of the combination method of EASE and ACC.

4 Pronounceability Measurement

EASE&ACC was measured through a multiple linear regression equation. Even though higher classification performance was observed in other methods such as support vector machines, multiple linear regression was employed in order to analyze the strengths of independent variables by standardized regression coefficients. The independent variables were the linguistic and learner features described in Sect. 2.2. The entire phonetic learner corpus was used as the training data for multiple linear regression.

A significant regression equation was found ($F(5, 744) = 87.3$, $p < 0.05$) with an adjusted squared correlation coefficient R^2 of 0.37. This low squared correlation is likely due to the biased class distribution, where 77.2% of the instances were concentrated in the lowest EASE&ACC class ($0 \leq$ EASE&ACC < 0.5) as shown in Table 1.

Then, the EASE&ACC was measured again using training data that excluded biased class distribution. The training data consisted of 371 instances, that is, 200 instances randomly selected from the 579 instances of the lowest EASE&ACC class (EASE&ACC < 0.5) and 171 instances from the other classes ($0.5 \leq$ EASE&ACC). A significant regression equation was found ($F(5, 365) = 60.7$, $p < 0.05$) with an adjusted squared correlation coefficient R^2 of 0.45.

The contribution of linguistic and learner features can be observed using standardized partial regression coefficients; this contribution increases with the absolute value of the coefficients. The standardized partial regression coefficients are summarized in Table 8. A significant contribution was observed in all the features ($p < 0.05$). The degree of contribution ranged in the following order: Multiple-syllable words, Mean word length, Word difficulty, Sentence length, and TOEIC. The positive correlation, i.e., a coefficient with a plus sign, was observed in Mean word length, Word difficulty, and Sentence length. The negative correlation, i.e., a coefficient with a negative sign, was observed in TOEIC, and Multiple-syllable words. Since EASE&ACC should have positive correlation with any linguistic features, the negative relation is acceptable with TOEIC but not with Multiple-syllable words, which should be explored in the future study.

Table 8. Standardized partial regression coefficients of the independent variables.

Feature	Standardized partial regression coefficient
Sentence length	0.41*
Mean word length	0.66*
Multiple-syllable words	−0.83*
Word difficulty	0.46*
TOEIC	−0.09*

*significant at $p < 0.05$.

The pronounceability measurement method was evaluated using leave-one-out cross validation, considering one instance as test data and $N - 1$ instances as training data ($N = 371$). A scatter plot of the correlation between the observed and measured EASE&ACC is depicted in Fig. 3. The correlation analysis was carried out between the observed and measured EASE&ACCs. The Spearman's rank correlationn coefficient between the observed and measured EASE&ACC was 0.69, which exhibited moderate correlation.

The pronounceability measurement method was also examined based on measurement errors in the cross validation. Errors were calculated as the absolute

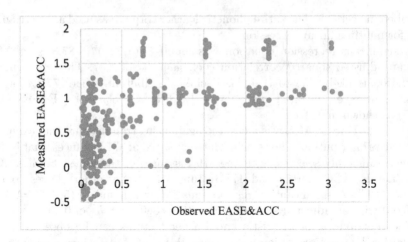

Fig. 3. Scatter plot of measured and observed EASE&ACCs.

Table 9. Frequency distribution of errors between observed and measured EASE&ACCs.

Class of Errof	Frequency	Relative frequency (%)
$0 \leq$ ERR < 0.1	48	12.9
$0.1 \leq$ ERR < 0.2	66	17.8
$0.2 \leq$ ERR < 0.3	54	14.6
$0.3 \leq$ ERR < 0.4	28	7.6
$0.4 \leq$ ERR < 0.5	29	7.8
$0.5 \leq$ ERR < 0.6	29	7.8
$0.6 \leq$ ERR < 0.7	26	7.0
$0.7 \leq$ ERR < 0.8	8	2.2
$0.8 \leq$ ERR < 0.9	16	4.3
$0.9 \leq$ ERR < 1.0	17	4.6
$1.0 \leq$ ERR < 1.1	21	5.7
$1.1 \leq$ ERR < 1.2	10	2.7
$1.2 \leq$ ERR < 1.3	7	1.9
$1.3 \leq$ ERR < 1.4	4	1.1
$1.4 \leq$ ERR < 1.5	0	0.0
$1.5 \leq$ ERR < 1.6	3	0.8
$1.6 \leq$ ERR < 1.7	2	0.5
$1.7 \leq$ ERR < 1.8	0	0.0
$1.8 \leq$ ERR < 1.9	1	0.3
$1.9 \leq$ ERR < 2.0	1	0.3
$2.0 \leq$ ERR < 2.1	1	0.3

values of the differences between the observed and measured values. The distribution of errors is shown in Table 9. Most of the errors were observed in the lower error range $(0.1 \leq \text{ERR} < 0.2)$.

5   Conclusion

This study proposed a pronounceability index that combined the ease and accuracy of learners' pronunciation, assessed the reliability and validity of the combined index using the classical test theory, and constructed a pronounceability measurement method using multiple linear regression.

The results showed that the proposed index was moderately reliable and not valid. The results also demonstrated that the measured pronounceability had moderate correlation with the observed pronounceability.

Future studies should work to examine combination methods, extend the phonetic learner corpus, introduce new features, evaluate the measurement performance between different learning algorithms, and assess the learner appropriateness of materials selected using the pronounceability measurement method. In addition, the proposed method should be examined empirically, whether materials chosen by the method enhance the learning effect. Results from this assessment are to determine whether the method should use the combined index as in this study, or non-combined indices, i.e., EASE and ACC.

References

1. Bang, J., Lee, G.G.: Determining sentence pronunciation difficulty for non-native speakers. In: Proceedings of Speech and Language Technology in Education, pp. 132–136. International Speech Communication Association, Grenoble (2013)
2. Brown, J.D.: Testing in Language Programs. Prentice-Hall, New Jersey (1996)
3. Chall, J.S., Dial, H.E.: Predicting listener understanding and interest in newscasts. Educ. Res. Bull. **27**(6), 141–153+168 (1948)
4. Cronbach, L.J.: Essentials of Psychological Testing, 3rd edn. Harper & Row, New York (1970)
5. Deterding, D.: The North Wind versus a Wolf: Short texts for the description and measurement of English pronunciation. J. Int. Phonetic Assoc. **36**(2), 187–196 (2006)
6. Fang, I.E.: The easy listening formula. J. Broadcast. **11**(1), 63–68 (1966)
7. Hwang, M.H.: How strategies are used to solve listening difficulties: Listening proficiency and text level effect. English Teach. **60**(1), 207–226 (2005)
8. International Phonetic Association: Handbook of the International Phonetic Association: A Guide to the Use of the International Phonetic Alphabet. Cambridge University Press, Cambridge (1999)
9. Kiyokawa, H.: A formula for predicting listenability: The listenability of English language materials 2. Wayo Women's Univ. Lang. Literature **24**, 57–74 (1990)
10. Kotani, K., Yoshimi, T.: Assessment of an index for measuring pronunciation difficulty. In: Proceedings of the 5th Workshop on Natural Language Processing Techniques for Educational Applications, pp. 119–124. Association for Computational Linguistics, Melbourne (2018)

11. Kotani, K., Yoshimi, T.: Machine learning classification of pronunciation difficulty for learners of English as a foreign language. Res. Corpus Linguist. **6**, 1–8 (2018)
12. Lai, D.: A study on the influencing factors of online learners' learning motivation. Higher Educ. Soc. Sci. **9**(4), 26–30 (2015)
13. Li, W., Siniscalchi, S.M., Chen, N.F., Lee, C.H.: Improving non-native mispronunciation detection and enriching diagnostic feedback with DNN-based speech attribute modeling. In: Proceedings of 2016 IEEE International Conference on Acoustics, Speech and Signal Processing, pp. 6135–6139. The Institute of Electrical and Electronic Engineers, Shanghai (2016)
14. Sheridan, B., Puhl, L.: Evaluating an indirect measure of student literacy competencies in higher education using Rasch measurement. In: In Engelhard, G., Wilson, M. (eds.) Objective Measurement: Theory into Practice, vol. 3, pp. 19–44. Norwood, New Jersey (1996)
15. Wieling, M., Bloem, J., Mignella, K., Timmerneister, M., Nerbonne, J.: Measuring foreign accent strength in English: Validating levenshtein distance as a measure. Lang. Dyn. Change **4**(2), 253–269 (2014)
16. Wilson, K.M.: Relating TOEIC Scores to Oral Proficiency Interview Ratings. Princeton, Educational Testing Services (1993)

Rap Lyrics Generation Using Vowel GAN

Tomoya Miyano[✉] and Hiroaki Saito

Graduate school of Science and Technology, Keio University, Tokyo, Japan
miyano_tomoya380@keio.jp

Abstract. Despite the success of recent rap and poetry generations using neural models, many of them do not consider vowels of the entire lyrics. Also, in many cases it is virtually impossible to generate completely new lyrics, because only existing rap lyrics are used as data sets. This paper proposes a new method of rap lyrics generation using a large amount of text such as novels in addition to rap lyrics. We divided the generation of rap lyrics into two steps; first, Generative Adversalial Net (GAN) generates rhymes and flows. Second, sequence-to-sequence converts them into rap lyrics. In addition, this method refers to the generation style of rap songs. In other words, they determine the music and rhythm first and apply the words second. We evaluated our method based on BLEU that can be measured mechanically.

Keywords: Rap song · GAN · Sequence-to-sequence

1 Introduction

Poetry has often been the subject of research on computer-based creative text generation. For example, in Sonnet, rhyme positions and stress patterns are predetermined, and a complex generation mechanism that makes use of its characteristics has been considered [1,2].

Unlike the Sonnet, the target of this paper is not the fixed rhyme position or stress pattern, thus a more flexible generation method is required. We assume that the vowels of the rap lyrics represent rhyme and flow. Therefore, we convert existing rap lyrics into vowels and use them for generation to automate rhyme and flow determination. Next, we learn to convert rhymes and flows into rap lyrics using novels. By using the proposed method, flexible rap lyrics generation is realized only with neural models. Figure 1 shows part of a novel from data sets. Figure 2 shows phoneticized sentences of Fig. 1, and Fig. 3 shows vowels extracted from Fig. 2.

© Springer Nature Singapore Pte Ltd. 2020
L.-M. Nguyen et al. (Eds.): PACLING 2019, CCIS 1215, pp. 307–318, 2020.
https://doi.org/10.1007/978-981-15-6168-9_26

call me ishmael.
some years ago never mind how long precisely
having little or no money in my purse.
and nothing particular to interest me on shore.
i thought i would sail about a little
and see the watery part of the world.
it is a way i have of driving off the spleen
and regulating the circulation.

Fig. 1. A novel data

k'O:l m,i:; 'ISmeIl
s,Vm j'i@3z a#g'oU n'Ev3 m'aInd h,aU l'0N prI2s'aIsli
h,avIN l'It@+++L_:_: O@ n'oU m'VnI; In maI p'3:s
and n'VTIN p3t'IkjUl3 tU 'Intr@st m,i:; ,0n S'o@
'aI T'O:t 'aI wUd s'eIl a#b,aUt a# l'It@L_:_:
and s'i: D@ w'O:t@ri p'A@t 0vD@ w'3:ld
It Iz a# w'eI 'aI hav 0v dr'aIvIN '0f D@ spl'i:n_:_:
and r'Egju:l,eItIN D@ s,3:kjUl'eIS@n

Fig. 2. Phoneticized representation from Fig. 1 data

O: i: I eI
V i@ 3 a oU E 3 aI aU 0 I2 aI i
a I I @L O@ oU V I I aI 3:
a V I 3 I U 3 U I @ i: 0 o@
aI O: aI U eI a aU a I @L
a i: @ O: @ i A@ 0 @ 3:
I I a eI aI a 0 aI I 0 @ i:
a E u: eI I @ 3: U eI @

Fig. 3. Extraction from Fig. 2 data

2   Related Works

The early study of poetry generation began with rule based models. Gervás
made the generation system guided by a set of construction heuristics obtained
from formal literature on Spanish poetry [3]. After that, a number of systems
were proposed using various methods for poetry generation.

Oliveira built a platform using grammatical and semantic templates [4,5].
Agirrezabal *et al.* extracted the POS tag sequences from some verse corpora,
calculated the probability of each sequence, and replaced each word with other

words according to its POS tag [6]. Greene *et al.* applied unsupervised learning to reveal word stress patterns in corpus of raw poetry, and used those word stress patterns, in addition to rhyme and discourse models, to generate English love poetry [7].

Yan *et al.* proposed a system that retrieves candidate terms out of a large poem corpus, and ordered these terms to fit into poetry formats, satisfying tonal and rhythm requirements [8]. Poetry generation using translation techniques has also been proposed. Zhou *et al.* proposed a system that takes the first sentence as input and generates an N-best list of second sentences as output using a phrase based SMT model [10].

In recent research, neural models are often used. The system using RNN or LSTM has achieved great results. Yi *et al.* took the generation of Chinese classical poem lines as a sequence-to-sequence learning problem, and built a system based on the RNN encoder-decoder structure to generate quatrains, with a topic word as input [11]. Ghazvininejad *et al.* combines LSTM and FSA for generating poetry considering of stress patterns and rhyming [1], which Lau *et al.* later made entirely a neural model with attention [2].

As research on the generation of rap lyrics, learning of rap battle and estimation of the next line from the previous line were also proposed [12,13]. The use of GAN in text generation is considered to be difficult to update the generative model, but a mechanism using reinforcement learning has been devised [9]. In early research, evaluation was not performed well, but in recent years evaluations using Crowdsourcing and evaluations by experts were conducted [2].

3   Architecture

We propose an architecture with two-step components using neural models. GAN generates the vowel sequences and sequence-to-sequence converts them into rap lyrics. This section details the tool, the data set, the neural models, and the generation procedure. Figure 4 shows the overall system configuration.

3.1   Tool

eSpeak[1] is a compact open source speech synthesizer for English and other languages, for Linux and Windows. We use this tool to convert rap lyrics and novel sentences into phonemes. The vowels for data sets were extracted from those phonemes.

These are the English vowels used by the English spelling-to-phoneme translations. In some varieties of English, different vowels may have the same sound, but they are kept separate because they may differ in another variety.

---

[1] http://espeak.sourceforge.net/.

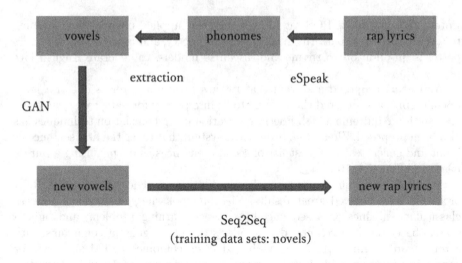

Fig. 4. Our generation process

English Vowels[2]:
[@, 3, 3:, @L, @2, @5, a, aa, A:, a#, A@, E, e@, I, I2, i, i:, i@, 0, V, u:, U, U@, O:, O@, o@, aI, eI, OI, aU, oU, aI@, aU@]

3.2 Data Set

Rap Lyrics. From the lyrics of the famous rap songs, 100 songs were selected for vowels generation. We used only 16 lines of lyrics from each song, thus each rap lyrics datasets were about the same length. These rap lyrics were converted to phonemes by eSpeak, and were extracted vowels for training data sets of GAN.

Novels. Project Gutenberg[3] is a site for collecting out-of-copyright texts. In this paper, 30 Mbyte novels were collected from Project Gutenberg to create data pairs of texts and vowels. In order to make new rap lyrics to be generated into appropriate length, novel's sentences were separated by commas, periods, question marks, colons or semicolons and only 4 to 20 words were adopted.

We remove characters that interfere with learning, such as () or " from collected sentences and converted all letters to the lower case. In order to make target data from the adopted words for learning, they are converted to phonemes using eSpeak and only vowels were extracted. As a result, the data set of the texts is 26 Mbytes, and the number of data pairs is 607,643.

---

[2] http://espeak.sourceforge.net/phonemes.html.
[3] https://www.gutenberg.org.

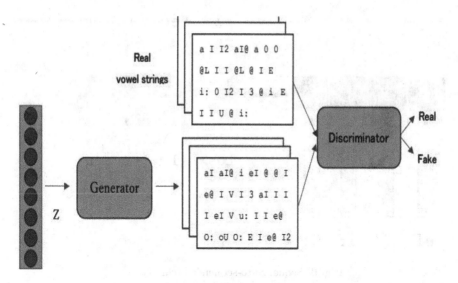

Fig. 5. GAN architecture

3.3 GAN

GAN determines vowels as the first step of rap lyrics generation. Since vowels correspond to flow and rhyme, the proposed model can express rap lyrics in a more flexible form than models for Sonnet generation. The Generator model outputs vowels, and the Discriminator model determines whether it is real or not. The weight of Generator model is updated by the correctness of the decision of Discriminator model. Figure 5 shows the GAN architecture. Generator generates vowels by inputting random noise. The other is Discriminator, which is given a real data or a fake data generated by Generator, and determines its authenticity.

3.4 Sequence-to-Sequence

Sequence-to-sequence converts vowels into new rap lyrics as a second step in generation. Using novels extracted from Project Gutenberg as a training data set, we create a text and vowels data sets in the same way as with rap lyrics. In the training, sequence-to-sequence model learns to restore the vowels obtained by the above process to the original novel sentences.

3.5 Generation Procedure

Sequence-to-sequence outputs new rap lyrics character by character because a word may contain several rhymes. And in sonet generation research, generation is often performed in reverse to put rhyme words at the end of a line. The same technique was adopted in this paper (NR). Furthermore, the pattern of reverse input (RR) was also tested, and two of them were compared (Fig. 6).

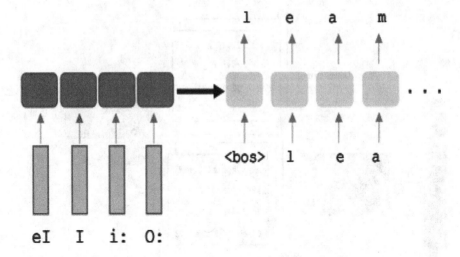

Fig. 6. Sequence-to-sequence architecture

4 Evaluation

Generally speaking, it is difficult to evaluate rap lyrics generated by machine learning. In previous papers, subjective assessment by people was used in many cases. The problems with manual evaluation are high cost and ambiguity and high variability. In this paper, we do not use vague evaluation methods, but use mechanical processing evaluation methods for generated rap lyrics. First of all, we consider the evaluation of GAN and sequence-to-sequence separately.

4.1 GAN

The evaluation of the generated vowels is difficult even for people. Therefore we make a graph of generator/discriminator loss during learning, and measure the degree of achievement.

Figure 7 shows the loss of GAN's generator and discriminator. Figure 8 shows generation examples of GAN at epoch 10000. Although the loss of the generator has not dropped much, the Generator seems to be able to learn to rhyme at the end of the line.

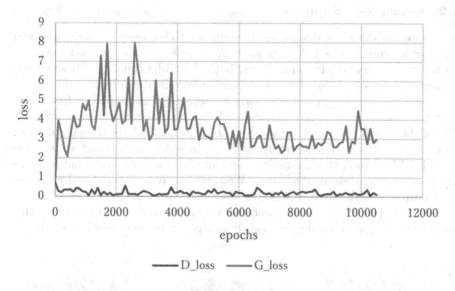

Fig. 7. Generation and discrimination loss

```
i@O:OeIU@o@
U@i@aIO:o@
O:u:aU@i@U@i@aIU@O@
eIo@aU@O:eIo@oUO@O:i@eIO:O@
i@O@i@O@o@o@eIo@i@aIaU@aIo@aU@
eIoUo@oUeIi@eIaU@aIO@aU@
aIO@i@eIo@O:OaIaIi@i@U@i@O@oUoUeIaU@aUO@aU@
U@eIaU@u:O@O@eIOO:OIO:aIO@aU@
u:aU@u:oUaIeIo@O@oU o@u:aIaIaIeIo@O@oU
o@u:O:eIeIO@i@eIO:oU
oUOIi@aI@aU@eIeIo@oU
oUO:aIaU@u:i@aIaI
aIaU@u:O@i@O:i@aIaI
O@oUo@aIi@i@O@o@oUu:O@i@O:i@aIeI
OIaIaIO:oUeIi@u:O:o@i@O:i@o@aI
```

Fig. 8. Generation examples of GAN at epoch 10000

4.2   Sequence-to-Sequence

The vowels of the rap lyrics output by sequence-to-sequence are compared with the vowels generated by GAN. As you can see in Fig. 9, if sequence-to-sequence correctly converts vowels to rap lyrics, both should be equal. We measure how correctly sequence-to-sequence converts input vowels to rap lyrics.

There are three types of comparison methods proposed. The first is the matched vowels count, which calculates the number of correct vowels from the end of the line. The second is the error rate, which calculates the rate at which each vowel of the generated rap lyrics is not included in the input vowels. The third is BLEU score, which is mainly used for machine translation, but it is used for vowels comparison in this paper. We used NLTK to calculate BLEU score. This output is normalized to a real number between 0 and 1.

We collected 100 lines of generated vowels by GAN, and based on this, sequence-to-sequence model for each epoch generates new rap lyrics. We use

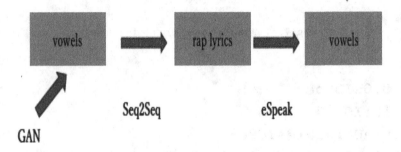

Fig. 9. Vowel restoration

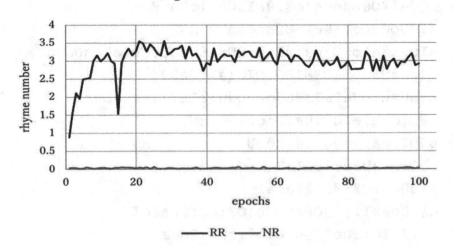

Fig. 10. Matched vowels count

the three evaluation methods listed above, and each average value was taken and graphed. Just in case, we chose a good vowel series.

This is to avoid that the rap lyrics can not be generated successfully by sequence-to-sequence because the vowel sequences obtained by GAN are not appropriate. Figure 10, Fig. 11 and Fig. 12 show that the generation which uses both input and output in reverse order (RR) scored higher in all evaluation methods. They also show that the score increased as learning progressed.

Figure 13 and Fig. 14 show a generation example of vowels by GAN and a generation example of rap lyrics converted from those vowels.

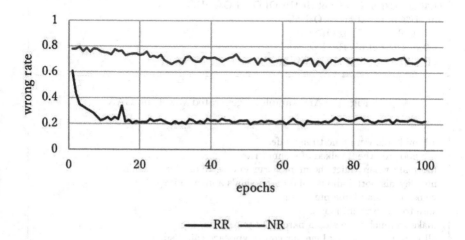

Fig. 11. Error rate

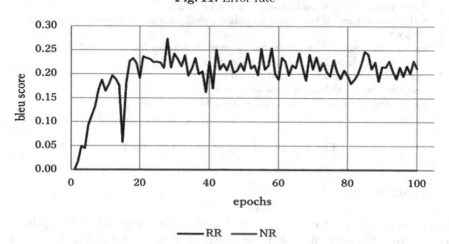

Fig. 12. BLEU score

```
oU O: aI aU@ u: i@ aI aI
aI aU@ u: O: 0 O: i@ aI aI
O: oU o@ eI i@ i@ O@ O@ oU U O@ i: O@ i@ aI eI
aU eI aI O@ oU aI i@ V O@ aI i@ O: i@ aI aI
i@ O: i@ eI U@ o@
U@ i@ aI U@ o@
O: u: aU@ i@ U@ i@ aI U@ O@
eI o@ aU@ O: eI o@ oU O@ U@ i@ aI O: O:
i@ O: i@ o@ o@ O@ eI o@ i@ eI aU@ aI O@ aU@
eI oU o@ oU eI i@ eI aU@ aI O@ aU@
aI O: i@ eI o@ O@ 0 aI aI i@ i@ U@ i@ o@ oU aI@ eI aU@ aU O@ aU@
U@ eI aU@ u: O@ O: eI i@ U@ OI O: aI O@ aU@
u: aU@ u: oU aI eI o@ O@ oU
o@ u: eI aI aI eI o@ O@ oU
o@ u: O: eI eI o@ i@ eI O: oU
oU OI i@ oU aU@ eI eI o@ oU
```

Fig. 13. An example of generated vowels by GAN

```
all my hoofs were true to my wife.
my poor fooltbeuf talked of steum i rite.
too malk where quiet the more warm so cosette or glorian might say.
now lay his portelodious child on horshia's audience i ely.
whose penasian lame pure sour.
sure to write youre more.
make so much pure gloom parean by pure form.
all were no more they lone our cordiaryppeace call'd small.
for to be able to see yours for five airs to state our miles or hours.
they wrote four whole nations save our life for hours.
with a woversband or face which by polian ardoriant who break pierre stones out
for hours.
my farefore wound or rustian makes cornury all time for hours.
grouped hour do so like great board nor spoke.
more soon jane night finds maist forlorn clothes.
four fuge alway halebore to make all grown.
whole ideal movn our flames claims youre told.
```

Fig. 14. An example of generation result

5   Discussion

In our proposed architecture, we aimed to generate completely new rap lyrics using novels. Since we did not use existing rap lyrics as a training data set in the second step, it is impossible to output those in a cut-and-paste fashion. And the output line is based on novels, thus the meaning is also maintained to some extent. The connection between lines is limited to rhymes and flows, a new mechanism will be needed to pursue meaning. For the entire rap lyrics, an architecture that considers not only vowels but also meaning is required. For example, there could be a method that uses the previous line as input in addition to the vowels. Research on the meaning of sentences is still developing and will have to go through some breakthroughs. Also, rather than generating vowels and rap lyrics sequentially, more sophisticated methods to generate simultaneously are expected.

It is difficult to evaluate imaginative research represented by rap lyrics generation. Although manual subjective evaluation is often performed, the variation is large and the cost is also high. In this study, we worked on the skill of rap generation by mechanical scoring. However, this is a score only for rhymes, thus we can't say that it is a perfect evaluation method for rap lyrics. For not only generation, but also evaluation, we need a way to compare the quality of structures and poetry.

6 Conclusion

This paper proposed a new method of rap generation using a large amount of text such as novels in addition to rap lyrics. Rap lyrics generation is performed in two steps. GAN generates rhymes and flows, and sequence-to-sequence converts them into rap lyrics. Although it is in the early stages of research, it showed the possibility of generating rap lyrics that is not a quote of existing rap songs.

References

1. Ghazvininejad, M., Shi, X., Choi, Y., Knight, K.: Generating topical poetry. In: Proceedings of the 2016 Conference on Empirical Methods in Natural Language Processing, pp. 1183–1191 (2016)
2. Lau, H.J., Cohn, T., Baldwin, T., Brooke, J., Hammond, A.: Deep-speare: a joint neural model of poetic language, meter and rhyme. In: Proceedings of the 56th Annual Meeting of the Association for Computational Linguistics (ACL 2018) (2018)
3. Gervás, P.: WASP: evaluation of different strategies for the automatic generation of Spanish verse. In: Proceedings of the AISB 2000 Symposium on Creative & Cultural Aspects of AI, pp. 93–100 (2000)
4. Oliveira, H.: Automatic generation of poetry: an overview. In: Proceedings of 1st Seminar of Art, Music, Creativity and Artificial Intelligence (2009)
5. Oliveira, H.: PoeTryMe: a versatile platform for poetry generation. In: Computational Creativity, Concept Invention, and General Intelligence, vol. 1 (2012)
6. Agirrezabal, M., Arrieta, B., Astigarraga, A., Hulden, M.: POS-tag based poetry generation with WordNet. In: Proceedings of the 14th European Workshop on Natural Language Generation, pp. 162–166 (2013)
7. Greene, E., Bodrumlu, T., Knight, K.: Automatic analysis of rhythmic poetry with applications to generation and translation. In: Proceedings of the 2010 Conference on Empirical Methods in Natural Language Processing, pp. 524–533 (2010)
8. Yan, R.: i, Poet: automatic poetry composition through recurrent neural networks with iterative polishing schema. In: IJCAI, pp. 2238–2244 (2016)

9. Yu, L., Zhang, W., Wang, J., Yu, Y.: SeqGAN: sequence generative adversarial nets with policy gradient. In: 31st AAAI Conference on Artificial Intelligence (2017)

10. Zhou, M., Jiang, L., He, J.: Generating Chinese couplets and quatrain using a statistical approach. In: Proceedings of the 23rd Pacific Asia Conference on Language, Information and Computation, vol. 1 (2009)

11. Yi, X., Li, R., Sun, M.: Generating Chinese classical poems with RNN encoder-decoder. In: Sun, M., Wang, X., Chang, B., Xiong, D. (eds.) CCL/NLP-NABD -2017. LNCS (LNAI), vol. 10565, pp. 211–223. Springer, Cham (2017). https://doi.org/10.1007/978-3-319-69005-6_18

12. Malmi, E., Takala, P., Toivonen, H., Raiko, T., Gionis, A.: DopeLearning: a computational approach to rap lyrics generation. In: Proceedings of the 22nd ACM SIGKDD International Conference on Knowledge Discovery and Data Mining, pp. 195–204. ACM (2016)

13. Wu. D., Addanki, K.: Learning to rap battle with bilingual recursive neural networks. In: 24th International Joint Conference on Artificial Intelligence (2015)

Emotion Recognition for Vietnamese Social Media Text

Vong Anh Ho[1], Duong Huynh-Cong Nguyen[1], Danh Hoang Nguyen[1],
Linh Thi-Van Pham[2], Duc-Vu Nguyen[3], Kiet Van Nguyen[1],
and Ngan Luu-Thuy Nguyen[1(✉)]

[1] University of Information Technology, VNU-HCM, Ho Chi Minh City, Vietnam
{15521025,15520148,15520090}@gm.uit.edu.vn, {kietnv,ngannlt}@uit.edu.vn
[2] University of Social Sciences and Humanities, VNU-HCM,
Ho Chi Minh City, Vietnam
vanlinhpham888@gmail.com
[3] Multimedia Communications Laboratory, University of Information Technology,
VNU-HCM, Ho Chi Minh City, Vietnam
vund@uit.edu.vn

Abstract. Emotion recognition or emotion prediction is a higher approach or a special case of sentiment analysis. In this task, the result is not produced in terms of either polarity: positive or negative or in the form of rating (from 1 to 5) but of a more detailed level of analysis in which the results are depicted in more expressions like sadness, enjoyment, anger, disgust, fear and surprise. Emotion recognition plays a critical role in measuring brand value of a product by recognizing specific emotions of customers' comments. In this study, we have achieved two targets. First and foremost, we built a standard **V**ietnamese **S**ocial **M**edia **E**motion **C**orpus (UIT-VSMEC) with exactly 6,927 emotion-annotated sentences, contributing to emotion recognition research in Vietnamese which is a low-resource language in natural language processing (NLP). Secondly, we assessed and measured machine learning and deep neural network models on our UIT-VSMEC corpus. As a result, the CNN model achieved the highest performance with the weighted F1-score of 59.74%. Our corpus is available at our research website (https://sites.google.com/uit.edu.vn/uit-nlp/corpora-projects).

Keywords: Emotion recognition · Emotion prediction · Vietnamese · Machine learning · Deep learning · CNN · LSTM · SVM

1 Introduction

Expressing emotion is a fundamental need of human and that we use language not just to convey facts, but also our emotions [7]. Emotions determine the quality of our lives, and we organize our lives to maximize the experience of positive emotions and minimize the experience of negative emotions [4]. Thus, Paul Ekman [3] proposed six basic emotions of human including enjoyment,

L.-M. Nguyen et al. (Eds.): PACLING 2019, CCIS 1215, pp. 319–333, 2020.
https://doi.org/10.1007/978-981-15-6168-9_27

sadness, anger, surprise, fear, and disgust through facial expression. Nonetheless, apart from facial expression, many different sources of information can be used to analyze emotions since emotion recognition has emerged as an important research area. And in recent years, emotion recognition in text has become more popular due to its vast potential applications in marketing, security, psychology, human-computer interaction, artificial intelligence, and so on [11].

In this study, we focus on the problem of recognizing emotions for Vietnamese comments on social network. To be more specific, the input of the problem is a Vietnamese comment from social network, and the output is a predicted emotion of that comment labeled with one of these: enjoyment, sadness, anger, surprise, fear, disgust and other. Several examples are shown in Table 1.

Table 1. Examples of emotion-labeled sentences.

No.	Vietnamese sentences	English translation	Emotion
1	Ảnh đẹp quá!	The picture is so beautiful!	Enjoyment
2	Tao khóc..huhu.. Tao rớt rồi	I'm crying..huhu.. I failed the exam.	Sadness
3	Khuôn mặt của tên đó vẫn còn ám ảnh tao.	The face of that man still haunts me.	Fear
4	Cái gì cơ? Bắt bỏ tù lũ khốn đó hết!	What the fuck? Arrest all those goddamn bastards!	Anger
5	Thật không thể tin nổi, tại sao lại nhanh đến thế??	It's unbelievable, why can be that fast??	Surprise
6	Những điều nó nói làm tao buồn nôn.	What he said makes me puke.	Disgust

In this paper, our two key contributions are summarized as follows.

- One of the most primary contributions is to obtain the UIT-VSMEC corpus, which is the first corpus for emotion recognition for Vietnamese social media text. As a result, we achieved 6,927 emotion annotated-sentences. To ensure that only the best results with high consistency and accuracy are reached, we built a very coherent and thorough annotation guideline for the dataset. The corpus is publicly available for research purpose.
- The second one, we tried using four learning algorithms on our UIT-VSMEC corpus, two machine learning models consisting of Support Vector Machine (SVM) and Random Forest versus two deep learning models including Convolutional Neural Network (CNN) and Long Short-Term Memory (LSTM).

The structure of the paper is organized thusly. Related documents and studies are presented in Sect. 2. The process of building corpus, annotation guidelines, and dataset evaluation are described in Sect. 3. In Sect. 4, we show how to apply SVM, Random Forest, CNN, and LSTM for this task. The experimental results are analyzed in Sect. 5. Conclusion and future work are deduced in Sect. 6.

2   Related Work

There are some related work in English and Chinese. In 2007, the SemEval-2007 Task 14 [20] developed a dataset for emotion recognition with six emotion classes (enjoyment, anger, disgust, sadness, fear and surprise) including 1,250 newspaper headline human-annotated sentences. In 2012, Mohammad [11] published an emotion corpus with 21,052 comments from Tweets annotated also by six labels of emotion (enjoyment, anger, disgust, sadness, fear and surprise). In 2017, Mohammad [10] again published a corpus annotated with only four emotion labels (anger, fear, enjoyment and sadness) for 7,079 comments from Tweets. In 2018, Wang [22] put out a bilingual corpus in Chinese and English for emotion recognition including 6,382 sentences tagged by five different emotions (enjoyment, sadness, fear, anger and surprise). In general, corpus for emotion recognition task use some out of six basic emotions of human (enjoyment, sadness, anger, disgust, fear and surprise) based on Ekman's emotion theory [3].

In terms of algorithms, Kratzwald [9] tested the efficiency of machine learning algorithms (Random Forest and SVM) and deep learning algorithms (Long Short-Term Memory (LSTM) and Bidirectional Long Short-Term Memory (BiLSTM)) combined with pre-trained word embeddings on multiple emotion corpora. In addition, the BiLSTM combined with pre-trained word embeddings reached the highest result of 58.2% of F1-score compared to Random Forest and SVM with 52.6% and 54.2% respectively on the General Tweets corpus [9]. Likewise, Wang [21] proposed the Bidirectional Long Short-Term Memory Multiple Classifiers (BLSTM-MC) model on a bilingual corpus in Chinese and English that achieved the F1-score of 46.7%, ranked third in the shared task NLPCC2018 - Task 1 [22].

In Vietnamese, there are several related work in sentiment analysis such as aspect-based sentiment analysis in the VLSP shared task [15] and sentiment analysis on students' feedback [16–18]. However, after a comprehensive research, we learned that neither a corpus nor a work for the emotion recognition for Vietnamese text-based is currently on deck. Hence, we present the UIT-VSMEC corpus for this task then we test and compare the results based on F1-score measurement between machine learning (SVM and Random Forest) and deep learning models (CNN and LSTM) on our corpus as the first results.

3   Corpus Construction

In this section, we present the process of developing the UIT-VSMEC corpus in Sect. 3.1, annotation guidelines in Sect. 3.2, corpus evaluation in Sect. 3.3 and corpus analysis in Sect. 3.4.

3.1   Process of Building the Corpus

The overview of corpus-building process which includes three phases is shown in Figure 1 and the detailed description of each phase is presented shortly thereafter.

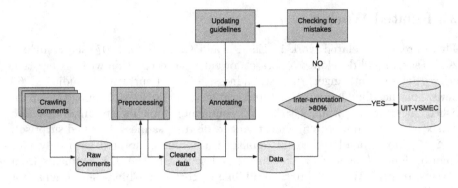

Fig. 1. Overview of corpus-building process.

Phase 1 - Collecting Data: We collect data from Facebook which is the most popular social network in Vietnam. In particular, according to a survey of 810 Vietnamese people conducted by [2] in 2018, Facebook is used the most in Vietnam. Moreover, as reported by the statistic of the number of Facebook users by [14], there are 58 million Facebook users in Vietnam, ranking as the 7th country with the most users in the world. As the larger the number of the users and the more interaction between them, the richer the data. And to collect the data, we use Facebook API to get Vietnamese comments from public posts.

Phase 2 - Pre-processing Data: To ensure users' privacy, we replace users' names in the comments with PER tag. The rest of the comments is kept as it is to retain the properties of a comment on social network.

Phase 3 - Annotating Data: This step is divided into two stages: **Stage 1**, building annotation guidelines and training for four annotators. Data tagging is repeated for 806 sentences along with editing guidelines until the consensus reaches more than 80%. **Stage 2**, 6,121 sentences are shared equally for three annotators while the another checking the whole 6,121 sentences and the consensus between the two is above 80%.

3.2   Annotation Guidelines

There are many suggestions on the number of basic emotions of human. According to studies in the field of psychology, there are two outstanding viewpoints in basic human emotions: "Basic Emotions" by Paul Ekman and "Wheel of Emotions" by Robert Plutchik [7]. Paul Ekman's studies point out that there are six basic emotions which are expressed through the face [3]: enjoyment, sadness, anger, fear, disgust and surprise. Eight years later, Robert Plutchik gave an another definition of emotion. In this concept, emotion is divided into eight basic ones which are polarized in pairs: enjoyment - sadness, anger - fear, trust - disgust and surprise - anticipation [7]. Despite the agreement on basic emotions, psychologists dissent from the number of which are the most basic that may have in humans, some ideas are 6, 8 and 20 or even more [10].

To that end, we chose six labels of basic emotions (enjoyment, sadness, anger, fear, disgust and surprise) for our UIT-VSMEC corpus based on six basic human emotions proposed by Ekman [3] together with Other label to mark a sentence with an emotion out of the six above or a sentence without emotion, considering most of the automated emotion recognition work in English [7,13] are all established from Ekman's emotion theory (1993) and with a large amount of comments in the corpus, a small number of emotions makes the manual tagging process more convenient.

Based on Ekman's instruction in basic human emotions [5], we build annotation guidelines for Vietnamese text with seven emotion labels described as follows.

- **Enjoyment**: For comments with the states that are triggered by feeling connection or sensory pleasure. It contains both peace and ecstasy. The intensity of these states varies from the enjoyment of helping others, a warm uplifting feeling that people experience when they see kindness and compassion, an experience of ease and contentment or even the enjoyment of the misfortunes of another person to the joyful pride in the accomplishments or the experience of something that is very beautiful and amazing. For example, the emotion of the sentence "Nháy mắt thôi cũng đáng yêu, kkk" (English translation: "Just the act of winking is so lovely!") is Enjoyment.
- **Sadness**: For comments that contain both disappointment and despair. The intensity of its states varies from discouragement, distraughtness, helplessness, hopelessness to strong suffering, a feeling of distress and sadness often caused by a loss or sorrow and anguish. The Vietnamese sentence "Lúc đấy khổ lắm... kỉ niệm: (" (English translation: "It was hard that time..memory :)") has an emotion of Sadness, for instance.
- **Fear**: For comments that show anxiety and terror. The intensity of these states varies from trepidation - anticipation of the possibility of danger, nervousness, dread to desperation, a response to the inability to reduce danger, panic and horror - a mixture of fear, disgust and shock. A given sentence "Chuyện này làm tao nổi hết da gà" (English translation: "This story causes me goosebumps") is a Fear labeled-sentence.
- **Anger**: For comments with states that are triggered by a feeling of being blocked in our progress. It contains both annoyance and fury and varies from frustration which is a response to repeated failures to overcome an obstacle, exasperation - anger caused by strong nuisance, argumentativeness to bitterness - anger after unfair treatment and vengefulness. For example, "Biến mẹ mày đi!" (English translation: "You fucking get lost!") is labeled with Angry.
- **Disgust**: For comments which show both dislike and loathing. Their intensity varies from an impulse to avoid something disgusting or aversion, the reaction to a bad taste, smell, thing or idea, repugnance to revulsion which

is a mixture of disgust and loathing or abhorrence - a mixture of intense disgust and hatred. As "Làm bạn với mấy thể loại này nhục cả người" (English translation: "Making friends with such types humiliates you") has an emotion of Disgust.

- **Surprise**: For comments that express the feeling caused by unexpected events, something hard to believe and may shock you. This is the shortest emotion of all emotions, only takes a few seconds. And it passes when we understand what is happening, and it may become fear, anger, relief or nothing ... depends on the event that makes us surprise. "Trên đời còn tồn tại thứ này sao??" (English translation: "How the hell in this world this still exists??") is annotated with Surprise.

- **Other**: For comments that show none of those emotions above or comments that do not contain any emotions. For instance, Mình đã xem rất nhiều video như này rồi nên thấy cũng bình thường (English translation: "I have seen a lot of videos like this so it's kinda normal") is neutral, so its label is Other.

3.3   Corpus Evaluation

We use the A_m agreement measure [1] to evaluate the consensus between annotators of the corpus. This agreement measure was also utilized in the UIT-VSFC corpus [16]. A_m is calculated by the following formula.

$$A_m = \frac{P_o - P_e}{1 - P_e}$$

where, P_o is the observed agreement which is the proportion of sentences with both of the annotators agreed on the classes pairs and P_e is the expected agreement that is the proportion of items for which agreement is expected by chance when the sentences are seen randomly.

Table 2 presents the consensus of the entire UIT-VSMEC corpus with two separate parts. The consensus of 806 sentences which can be seen as the first stage of annotating data mentioned in Sects. 3 and 3.1. A_m agreement level is high with 82.94% by four annotators. And the annotation agreement in the second stage (Sects. 3 and 3.1) where X_1, X_2, X_3 are 3 annotators independently tagging data and Y is the checker of one another. The A_m and P_o of annotation pair X_3-Y are highest of 89.12% and 92.81%.

Table 2. Annotation agreement of the UIT-VSMEC corpus (%)

Stage	Annotators	P_0	P_e	A_m
1	X_1-X_2-X_3-Y (806 sentences)	92.66	56.98	82.94
2	X_1-Y (2,032 sentences)	88.00	18.61	85.25
	X_2-Y (2,112 sentences)	86.27	31.23	80.03
	X_3-Y (1,977 sentences)	92.81	33.95	89.12

3.4   Corpus Analysis

After building the UIT-VSMEC corpus, we obtained 6,927 human-annotated sentences with one of the seven emotion labels. Statistics of emotion labels of the corpus is presented in Table 3.

Table 3. Statistics of emotion labels of the UIT-VSMEC corpus

Emotion	Number of sentences	Percentage (%)
Enjoyment	1,965	28.36
Disgust	1,338	19.31
Sadness	1,149	16.59
Anger	480	6.92
Fear	395	5.70
Surprise	309	4.46
Other	1,291	18.66
Total	6,927	100

Through Table 3, we concluded that the comments got from social network are uneven in number among different labels in which the enjoyment label reaches the highest number of 1,965 sentences (28.36%) while the surprise label arrives at the lowest number of 309 sentences (4.46%).

Besides, we listed the number of the sentences of each label up to their lengths. Table 4 shows the distribution of emotion-annotated sentences according to their lengths. It is easy to see that most of the comments are from 1 to 20 words accounting for 81.76%.

Table 4. Distribution of emotion-annotated sentences according to the length of the sentence (%)

Length	Enjoyment	Disgust	Sadness	Anger	Fear	Surprise	Other	Overall
1–5	5.16	2.84	1.70	0.69	0.94	0.85	1.87	**14.05**
6–10	8.98	4.22	4.98	1.41	1.42	2.25	7.20	**30.38**
10–15	5.87	3.99	4.11	1.40	1.27	0.94	5.00	**22.58**
16–20	4.17	3.05	2.51	1.14	0.85	0.24	2.79	**14.75**
21–25	1.96	1.93	1.50	0.66	0.40	0.15	1.11	7.71
26–30	1.08	1.31	0.95	0.45	0.27	0.01	0.53	4.6
>30	1.23	1.97	0.84	1.17	0.55	0.02	0.15	5.93
Total	28.36	19.31	16.59	6.92	5.70	4.46	18.66	100

4 Methodology

In this paper, we use two kinds of methodologies to evaluate the UIT-VSMEC corpus including two machine learning models (Random Forest and SVM) and two deep learning models (CNN and LSTM) as the first models described as follows.

4.1 Machine Learning Models

The authors in [9] proposed SVM and Random Forest algorithms for emotion recognition. Under which, we also tested three more machine learning algorithms including Decision Tree, kNN and Naive Bayes on 1,000 emotion-annotated sentences extracted from the UIT-VSMEC corpus by Orange3. Consequently, Random Forest achieved the second best result after SVM which is displayed in Table 5. It is the main reason why we chose SVM and Random Forest for experiments on the UIT-VSMEC corpus.

Table 5. Experimental results by Orange3 of machine learning models on 1,000 emotion-annotated sentences from the UIT-VSMEC corpus (%)

Method	Accuracy	Weighted F1
Random Forest	35.8	**32.8**
SVM	37.6	**37.0**
Decision Tree	30.5	29.6
kNN	28.9	27.1
Naïve Bayes	20.8	19.2

Random Forest. Random Forest is a versatile machine learning algorithm when used for classification problems, predicting linear regression values and multi-output tasks. The idea of Random Forest is to use a set of Decision Tree classification, each of which is trained on different parts of the dataset. After that, Random Forest will get back all the classification results of the seedlings from which it chooses the most voted one to give the final result. Despite of its simplicity, Random Forest is one of the most effective machine learning algorithms today [19].

Support Vector Machine (SVM). We use the SVM machine learning algorithm as a baseline result for this emotion recognition problem. According to the authors in [13], SVM is an effective algorithm for classification problems with high features. Here, we use SVM model supported by scikit-learn library.

4.2 Deep Learning Models

Long Short-Term Memory (LSTM). LSTM is also applied for the UIT-VSMEC corpus for various reasons. To begin with, LSTM is considered as the state-of-the-art method of almost sequence prediction problems. Moreover, through the two competitions WASSA-2018 [8] and SemEval-2018 Task 1 [12] for emotion recognition task, we acknowledged that LSTM was effectively used the most. Furthermore, LSTM has advantages over conventional neural networks and Recurrent Neural Network (RNN) in many different ways due to its selective memory characteristic in a long period. This is also the reason why the authors in [9] chose to use it in his paper. Therefore, we decided to use LSTM on the same problem on our corpus.

LSTM consists of four main parts: Word embeddings input, LSTM cell network, fully connected and softmax. With the input, each cell in the LSTM network receives a word vector represented by word embeddings with the form $[1 \times n]$ where n is the fixed length of the sentence. Then cells calculate the values and gets the results as vectors in LSTM cell network. These vectors will go through fully connected and the output values will then pass through softmax function to give an appropriate classification for each label.

Convolutional Neural Network (CNN). We use Convolutional Neural Network (CNN) algorithm which is proposed in [6] to recognize emotions in a sentence. CNN is the algorithm that achieves the best results in four out of the seven major problems of Natural Language Processing which includes both emotion recognition and question classification tasks (text classification, language model, speech recognition, title generator, machine translation, text summarization and Q&A systems) [6,24].

A CNN model consists of three main parts: Convolution layer, pooling layer and fully connected layer. In convolution layer - the Kernel, we used 3 types of filters of different sizes with total 512 filters to extract the high-level features and obtain convolved feature maps. These then go through the pooling layer which is responsible for reducing the spatial size of the convolved feature and decreasing the computational power required to process the data through dimensionality reduction. The convolutional layer and the pooling layer together form the i-th layer of a Convolutional Neural Network. Moving on, the final output will be flattened and fed to a regular neural network in the fully connected layer for classification purposes using the softmax classification technique.

5 Experiments and Error Analysis

5.1 Corpus Preparation

We at first built a normalized corpus for comparison and evaluation where spelling errors have been corrected and acronyms in various forms have been converted back to their original words, seeing it is impossible to avoid such

Table 6. Vietnamese abbreviations in the dataset.

No.	Abbreviation	Vietnamese meaning	English meaning
1	"dc" or "dk" or "duoc"	"được"	"ok"
2	"ng" or "ngừi"	"người"	"people"
3	"trc" or "trk"	"trước"	"before"
4	"cg" or "cug" or "cũg"	"cũng"	"also"
5	"mk" or "mik" or "mh"	"mình"	"I"

problems of text on social networks when it does not distinguish any type of users. Table 6 shows some examples being encountered the most in the dataset.

We then divided the UIT-VSMEC corpus into the ratio of 80:10:10, in which 80% of the corpus is the training set, 10% is the validation one and the rest is the test set. The UIT-VSMEC corpus is an imbalanced-labels corpus, therefore, to ensure that sentences in low-volume labels are distributed fully in each set, we use stratified sampling method utilizing train_test_split() function supported by scikit learn library to distribute them into training, validation and test sets. The result is presented in Table 7.

Table 7. Statistics of emotion-labeled sentences in training, validation and test sets.

Emotion	Train	Dev	Test	Total
Enjoyment	1,573	205	187	1,965
Disgust	1,064	141	133	1,338
Sadness	938	92	119	1,149
Anger	395	38	47	480
Fear	317	38	47	395
Surprise	242	36	31	309
Other	1,019	132	140	1,291
All	5,548	686	693	6,927

5.2   Experimental Settings

In this paper, to represent words in vector form, we use two different methods word embeddings and bag of words. For the two machine learning models SVM and Random Forest, we use bag of words in conjunction with TF-IDF. For the two other deep learning models LSTM and CNN, we utilize pre-trained word embeddings including word2vec [1] and fastText [2] used as its main techniques.

---

[1] https://github.com/vncorenlp/VnCoreNLP.
[2] https://fasttext.cc/docs/en/crawl-vectors.html.

With machine learning models SVM and Random Forest, grid-search method is utilized to get the most appropriate parameters for the task. In particular, with SVM we use word of tag (1, 3) combined with bag of char (1,7) features and loss function hinge, and to reduce overfitting we apply the l2-regularization technique with lambd. = 1e−4. About Random Forest model, the number of decision trees is 256 and the depth of the trees is 64.

For LSTM model, we use the many-to-one architecture due to the classification requirement of the problem. To select proper parameters for emotion recognition in Vietnamese, we add two drop-out classes of 0.75 and 0.5 respectively to increase processing time as well as to avoid overfitting.

Regarding deep learning CNN model, we apply three main kernels: 3, 4 and 5 with a number of each is 128. Besides, drop-out of 0.95 and l2 of 0.01 are adopted to avoid overfitting. Properties and models are developed from Yoon Kim's work [6].

5.3 Experimental Results

In this section, we present the results of two experiments. Firstly, we test and compare the results of each model on the UIT-VSMEC corpus. Secondly, we evaluate the influence of Other label on this corpus after implementing these machine learning and deep learning models on the corpus yet without Other label. All models are evaluated by accuracy and weighted F1-score metrics.

Table 8. Experimental results of the UIT-VSMEC corpus.

Corpus	Algorithm	Accuracy(%)	Weighted F1-Score(%)
Original	RandomForest+BoW	50.64	40.11
	SVM+BoW	58.00	56.87
	LSTM+word2Vec	53.39	53.30
	LSTM+fastText	54.25	53.77
	CNN+word2Vec	**59.74**	**59.74**
	CNN+fastText	56.85	56.79
Without Other label	RandomForest+BoW	50.64	49.14
	SVM+BoW	63.12	62.45
	LSTM+word2Vec	61.70	61.09
	LSTM+fastText	62.06	61.83
	CNN+word2Vec	**66.48**	**66.34**
	CNN+fastText	63.47	62.68

Through this, we concluded that, when removing Other label, the weighted F1-score reaches higher results with the same methods. Firstly, it is because of the decrease in number of emotion labels in the UIT-VSMEC corpus from 7 to 6 (anger, enjoyment, surprise, sadness, fear and disgust). Secondly, sentences not affected by noise data from Other label gives better results. To conclude, Other label does affect the performance of these algorithms. This will be our focus in building data in the future. Apart from that, we evaluate the learning curves of the four models proposed with the original dataset (the seven labels dataset) in which Random Forest and SVM utilize BoW feature, CNN and LSTM utilize word2vec embeddings. To conduct this experiment, we keep the test and the validation sets while putting training set in stages from 2,000 with 500 sentence-jump until the end of the set (5,548 sentences).

5.4   Error Analysis

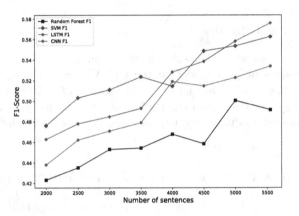

Fig. 2. Learning curves of the classification models on the UIT-VSMEC corpus.

As can be seen in Fig. 2, when the size of the training set increases, so does the weighted F1-scores of the four models despite the slightly drop of 0.092% in Random Forest when the number of the set grows from 5,000 to 5,500 sentences. In the meanwhile, compared to LSTM, the two deep learning models reach significant higher results, principally CNN combined with word2vec. Thus, we take this point to continue expanding the corpus as well as improving the performance of these models.

To demonstrate the performance of classification models, we use confusion matrix to visualize the ambiguity between actual labels and predicted labels. Fig. 3 is the confusion matrix of the best classification model (CNN + word2vec) on the UIT-VSMEC corpus. As can be seen, the model performs well on classifying enjoyment, fear and sadness labels while it confuses between anger and disgust labels as their ambiguities are at the highest percentage of 39.1%. There are two reasons causing this confusion. Primarily, it is the inherent vagueness of the definitions of anger and disgust construed through [23]. Secondly, the data limitation of these labels is an interference for the model to execute at its best. We noted this in our next step to continue building a thorough corpus for the task.

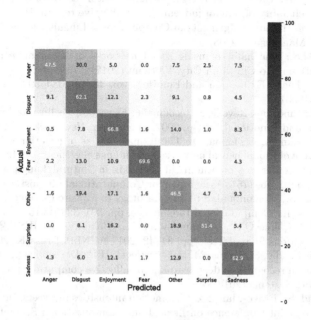

Fig. 3. Confusion matrix of the best classification model on the UIT-VSMEC corpus.

6   Conclusion and Future Work

In this study, we built a human-annotated corpus for emotion recognition for Vietnamese social media text for research purpose and achieved 6,927 sentences annotated with one of the seven emotion labels namely enjoyment, sadness, anger, surprise, fear, disgust and other with the annotation agreement of over 82%. We also presented machine learning and deep neural network models used for classifying emotions of Vietnamese social media text. In addition, we reached the best overall weighted F1-score of 59.74% on the original UIT-VSMEC corpus with CNN using the word2vec word embeddings. This paper is the very first effort of emotion recognition for Vietnamese social media text.

In the future, we want to improve the quantity as well as the quality of the corpus due to its limitation of comments expressing emotions of anger, fear and surprise. Besides, we aim to conduct experiments using other machine learning models with distinctive features as well as deep learning models with various word representations or combine both methods on this corpus.

Acknowledgment. We would like to give our thanks to the NLP@UIT research group and the Citynow-UIT Laboratory of the University of Information Technology - Vietnam National University Ho Chi Minh City for their supports with pragmatic and inspiring advice.

References

1. Bhowmick, P.K., Basu, A., Mitra, P.: An agreement measure for determining inter-annotator reliability of human judgements on affective tex. In: Proceedings of the Workshop on Human Judgements in Computational Linguistics, pp. 58–65. COLING 2008, Manchester (2008)
2. company, J.S.: The habit of using social networks of Vietnamese people 2018. brands vietnam, Ho Chi Minh City, Vietnam (2018)
3. Ekman, P.: Facial Expression and Emotion. vol. 48, pp. 384–392. American Psychologist (1993)
4. Ekman, P.: Emotions Revealed: Recognizing Faces and Feelings to Improve Communication and Emotional Life, p. 2007. Macmillan, New York (2012)
5. Ekman, P., Ekman, E., Lama, D.: The Ekmans' Atlas of Emotion (2018)
6. Kim, Y.: Convolutional neural networks for sentence classifications. In: Proceedings of the 2014 Conference on Empirical Methods in Natural Language Processing (EMNLP), pp. 1746–1751. Association for Computational Linguistics, Doha (2014)
7. Kiritchenko, S., Mohammad, S.: Using hashtags to capture fine emotion categories from Tweets. In: Computational Intelligence, pp. 301–326 (2015)
8. Klinger, R., Clerc, O.D., Mohammad, S.M., Balahur, A.: IEST:WASSA-2018 Implicit Emotions Shared Task. pp. 31–42. 2017 AFNLP, Brussels (2018)
9. Kratzwald, B., Ilic, S., Kraus, M.S., Feuerriegel, H.P.: Decision support with text-based emotion recognition: deep learning for affective computing, pp. 24–35. Decision Support Systems (2018)
10. Mohammad, S., Bravo-Marquez, F.: Emotion intensities in tweets. In: Proceedings of the Sixth Joint Conference on Lexical and Computational Semantics (*SEM), pp. 65–77. Association for Computational Linguistics, Vancouver (2017)
11. Mohammad, S.M.: #Emotional Tweets. In: First Joint Conference on Lexical and Computational Semantics (*SEM), pp. 246–255. Association for Computational Linguistics, Montreal (2012)
12. Mohammad, S.M., Bravo-Marquez, F., Salameh, M., Kiritchenko, S.: SemEval-2018 task 1: affect in tweets. In: Proceedings of International Workshop on Semantic Evaluation, pp. 1–17. New Orleans, Louisiana (2018)
13. Mohammad, S.M., Xiaodan, Z., Kiritchenko, S., Martin, J.: Sentiment, emotion, purpose, and style in electoral tweets, pp. 480–499. Information Processing and Management: an International Journal (2015)
14. Nguyen: Vietnam has the 7th largest number of Facebook users in the world. Dan Tri newspaper (2018)
15. Nguyen, H.T.M., et al.: VLSP shared task: sentiment analysis. J. Comput. Sci. Cybern. **34**, 295–310 (2018)
16. Nguyen, K.V., Nguyen, V.D., Nguyen, P., Truong, T., Nguyen, N.L.T.: UIT-VSFC: vietnamese students' feedback corpus for sentiment analysis. In: 2018 10th International Conference on Knowledge and Systems Engineering (KSE), pp. 19–24. IEEE, Ho Chi Minh City (2018)
17. Nguyen, P.X.V., Truong, T.T.H., Nguyen, K.V., Nguyen, N.L.T.: Deep learning versus traditional classifiers on vietnamese students' feedback corpus. In: 2018 5th NAFOSTED Conference on Information and Computer Science (NICS), pp. 75–80. Ho Chi Minh City (2018)
18. Nguyen, V.D., Nguyen, K.V., Nguyen, N.L.T.: Variants of long short-term memory for sentiment analysis on vietnamese students' feedback corpus. In: 2018 10th International Conference on Knowledge and Systems Engineering (KSE), pp. 306–311. IEEE, Ho Chi Minh City (2018)

19. Pedregosa, F., et al.: Scikit-learn: machine learning in python. J. Mach. Learn. Res. **12**, 2825–2830 (2011)
20. Strapparava, C., Mihalcea, R.: SemEval-2007 Task 14: affective text. In: Proceedings of the 4th International Workshop on Semantic Evaluations (SemEval-2007), pp. 70–74. Association for Computational Linguistics, Prague (2007)
21. Wang, T., Yang, X., Ouyang, C., Guo, A., Liu, Y., Li, Z.: A multi-emotion classification method based on BLSTM-MC in code-switching text. In: Zhang, M., Ng, V., Zhao, D., Li, S., Zan, H. (eds.) NLPCC 2018. LNCS (LNAI), vol. 11109, pp. 190–199. Springer, Cham (2018). https://doi.org/10.1007/978-3-319-99501-4_16
22. Wang, Z., Li, S., Wu, F., Sun, Q., Zhou, G.: Overview of NLPCC 2018 Shared Task 1: Emotion Detection in Code-Switching Text. In: Zhang, M., Ng, V., Zhao, D., Li, S., Zan, H. (eds.) NLPCC 2018. LNCS (LNAI), vol. 11109, pp. 429–433. Springer, Cham (2018). https://doi.org/10.1007/978-3-319-99501-4_39
23. Zhang, S., Wu, Z., Meng, H.M., Cai, L.: Facial expression synthesis using PAD emotional parameters for a chinese expressive avatar. In: Paiva, A.C.R., Prada, R., Picard, R.W. (eds.) ACII 2007. LNCS, vol. 4738, pp. 24–35. Springer, Heidelberg (2007). https://doi.org/10.1007/978-3-540-74889-2_3
24. Zhang, Y., Wallace, B.C.: A Sensitivity Analysis of (and Practitioners' Guide to) Convolutional, pp. 253–263. 2017 AFNLP, Taipei, Taiwan (2017)

Effects of Soft-Masking Function on Spectrogram-Based Instrument - Vocal Separation

Duc Chung Tran[1](✉)[iD] and M. K. A. Ahamed Khan[2][iD]

[1] Computing Fundamental Department and FPT Technology Research Institute, FPT University, Hoa Lac Hi-Tech Park, Hanoi 155300, Vietnam
chungtd6@fe.edu.vn
[2] Faculty of Engineering Technology and Built Environment, UCSI University, Kuala Lumpur, Malaysia
mohamedkhan@ucsiuniversity.edu.my

Abstract. This paper presents an analysis of effects of soft-masking function on spectrogram-based instrument - vocal separation for audio signals. The function taken into consideration is of 1st-order with two masking magnitude parameters: one for background and one foreground separation. It is found that as the masking magnitude increases, the signal estimations are improved. The background signal's spectrogram becomes closer to that of the original signal while the foreground signal's spectrogram represents better the vocal wiggle lines compared to the original signal spectrogram. With the same increase in the masking magnitude (up to ten-fold), the effect on background signal spectrogram is more significant compared to that of foreground signal. This is evident through the significant ($\approx$three times) reduction of background signal's root-mean-square (RMS) values and the less significant reduction (approximately one-third) of foreground signal's RMS values.

Keywords: Soft-mask · Spectrogram · Noise · Reduction · Cancellation · Audio · Signal · Instrument · Vocal · Separation

1 Introduction

Signal - noise separation is one of the most important tasks in signal processing, in particular, image and audio processing as it helps to remove unwanted signals from the source [27]. While image processing is of interests of the majority researchers [18,19,28], many scientists are working toward audio processing [21]. This is because audio is the main mean of communication in daily life.

In reality, it is difficult to completely remove noise from an audio source. Multiple attempts have been made to reduce noise level to improve speech quality [4,22]. Typically, noises are reduced by applying filters or combination of filters such as Weiner, Bayesian, exponentially weighted moving average (EWMA) [7,8], even by applying deep neural networks [25], or by separating background

© Springer Nature Singapore Pte Ltd. 2020
L.-M. Nguyen et al. (Eds.): PACLING 2019, CCIS 1215, pp. 334–344, 2020.
https://doi.org/10.1007/978-981-15-6168-9_28

(noise / instrument) and foreground (vocal) elements [24]. However, each approach has its own limitations such as requiring extensive calculation power from hosting computer in neural network-based approaches. Although [24] is effective in source separation, recent reported works have not discussed the effect of setting parameters for the first-order soft-masking function used in the proposed algorithm.

In this work, audio noise canceling technique based on spectrogram analysis with first-order soft-masking function is studied and analyzed. The main contribution of this work is the analysis of the effects of the soft-masking function on spectrogram-based instrument and vocal separation which is not discussed else where in the literature.

The remaining of the paper is organized as follows: Sect. 2 details related researches in noise canceling based on spectrogram in instrument, vocal separation in recent years. Section 3 presents the proposed research methodology. Section 4 discusses results and presents the analysis on the different settings of masking parameters in soft-masking-based foreground and background spectrogram generations. Finally, Sect. 5 presents conclusions and some suggestions for future works.

2   Related Works

With the development of the audio signal processing techniques, source separation is becoming an important aspect of studies in recent years [10,15,30]. By separating the audio source signals, one can analyze and assess its background noise and provide estimation of vocal source [21]. Usually the vocal source contains useful information of the audio signal as it carries the communication messages between the corresponding sender and receiver. In a noisy environment, often there are multiple audio source signals. If the ambient environment is noisy, especially when interference exists with high energy [9], it will be difficult to hear the audio signal communicated between humans, therefore the analysis of the communication messages become extremely difficult. In such cases, source separation techniques can be used to support audio noise removal application [23]. It is expected that the extracted information will contain only the vocals which analysts want to research after the vocals being separated from the noise.

However, in any environment, audio noises can come from the surrounding sounds generated by talking persons, running instruments or even the recording instrument itself. Since complete noise removal is impossible to achieve, in recent studies, there are several approaches for noise filtering and reduction including: using Weiner filters [1,32], Bayesian filter [17], infinite impulse response (IIR) filter [26], nonnegative matrix factorization (NMF) informed beam forming [25], frost beam former [11], auto regression gains [14], sub-band adaptive filtering [6], acoustic map [9], etc. Each method has its own advantages and disadvantages.

Deep learning approaches like [27] for noise removal requires extensive computer computational effort which is not suitable for low processing devices or embedded systems. While the unsupervised learning approach [31] shall lead to false positive or unwanted noise reduction effects since dropout and background noise training is not monitored. Even approaches using combination of traditional filters like Bayesian, Wiener with Kalman [17] require more processing power from the computer running the algorithms. This in-turn shall affect real-time processing feature of the proposed approaches. One possible approach for reducing noise in audio signal is to use spectrogram instrument - vocal separation [5, 24]. However the discussion for the first-order soft-masking function used in the separation was not discussed. This research gap will be addressed in this study.

3   Methodology

In this work, the instrument - vocal separation algorithm presented in Fig. 1 is studied.

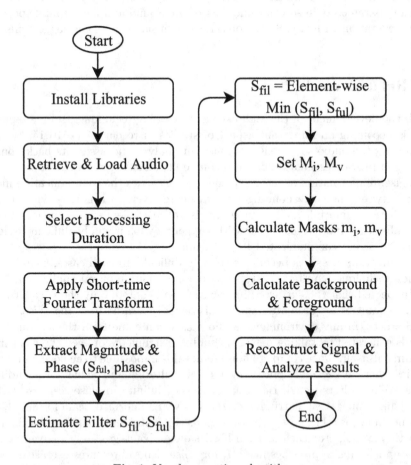

Fig. 1. Vocal separation algorithm

At first, the required development libraries are imported into the program. These libraries are Python-based: matplotlib, numpy, LibROSA, IPython, requests, scipy, and thinkx [29]. Second, a sample audio file is retrieved automatically from a GitHub repository. The source file is then unzipped for getting the data file. Then the file is loaded to cloud system's memory for further processing. Next, an audio signal duration is selected for being processed at later steps.

The short-time Fourier transform (STFT) [20] is then applied to the signal under consideration. Its magnitude (S_{ful}) and phase (*phase*) are extracted from the transformed signal. Following, an estimation of the signal is performed to create a signal S_{fil} which is approximately similar to the original signal S_{ful}. This will be used for background calculation later.

For representing signal's loudness or strength over time, often spectrogram is used in audio analysis [2, 3, 16]. At different frequencies, the signal are represented in different forms. The composition of all forms in all frequencies constructs the audio signal. Using the spectrogram, one can see the energy level of the signals over time, thus it gives a great visual sensation for the analyst when assessing a particular audio. In particular, spectrograms are two-dimensional graphs with an additional dimension that is represented by colors at the plotted pixels. The vertical axis represents frequencies range of the audio signal while the horizontal line represents the timing. The color on the pixel plotted on the spectrogram indicates the signal amplitude level. The darker the color, the less amplitude value of the signal at that particular time and frequency and vice-versa. In this work, spectrograms of the estimated foreground and background of audio signals will be analyzed.

The magnitude and phase estimation is based on nearest-neighbors filtering technique. By using this filter, each data point (e.g., spectrogram column) is replaced by its aggregated nearest neighbors in feature space. This is useful in de-noising a spectrogram or feature matrix. It is possible to recover the non-local means method [5] by providing a weighted recurrence input matrix and specifying the aggregation function as an average calculation function. Similarly, setting the aggregation function to median calculation produces sparse de-noising as in REPET-SIM [23].

Therefore, the selected aggregation function is "median" (taking per-frequency median value of the aggregated similar frames) while the matrix calculation is "cosine". For avoiding the local continuity biasing, the similar frames are constrained to be separated by at least 0.5 s. This helps to suppress sparse or non-repetitive deviations from the average spectrum and to discard well vocal elements in the signal.

In order to ensure the estimated signal's amplitude at any frame not exceeding the original signal magnitude, minimum element-wise operator is applied to the estimated signal and the original one.

$$S_{fil} = min(S_{ful}, S_{fil}) \tag{1}$$

Here, it is possible to use margin values for reducing the differentiation between the vocal and instrumentation masks. It should be noted that two separate margins (M_i, M_v) are used for background and foreground separation respectively.

For calculating the instrument and vocal masks, soft-masking function f_m is used. It is the time-frequency masking function which is calculated based on repeating spectrograms. The function takes three parameters as input: X - input array corresponding to positive mask elements, X_{ref} - reference or background array, and p - power coefficient [23]. Its equation is described as follows.

$$f_{m(X, X_{ref}, p)} = \frac{X^p}{X^p + X_{ref}^p}. \tag{2}$$

Since, the input parameters for the soft-masking function for instrument and vocal masks are: $(S_{fil}, M_i(S_{ful} - S_{fil}), p)$ and $(S_{ful} - S_{fil}, M_v S_{fil}, p)$, one has

$$m_i = f_m(S_{fil}, M_i(S_{ful} - S_{fil}), p) \tag{3}$$

$$m_v = f_m(S_{ful} - S_{fil}, M_v S_{fil}, p) \tag{4}$$

Applying Eq. 2 for Eqs. 3 and 4, one has

$$m_i = \frac{S_{fil}^p}{S_{fil}^p + (M_i(S_{ful} - S_{fil}))^p} \tag{5}$$

$$m_v = \frac{[S_{ful} - S_{fil}]^p}{(S_{ful} - S_{fil})^p + (M_v S_{fil})^p} \tag{6}$$

Simplifying the above equations, one has

$$m_i = \frac{S_{fil}^p}{S_{fil}^p + M_i^p(S_{ful} - S_{fil})^p} \tag{7}$$

$$m_v = \frac{(S_{ful} - S_{fil})^p}{M_v^p S_{fil}^p + (S_{ful} - S_{fil})^p} \tag{8}$$

In order to separate instrument and vocal components (S_{bg} and S_{fg} respectively) from the original signal, the masks are multiplied with the original spectrum.

$$S_{bg} = m_i S_{ful} \tag{9}$$

$$S_{fg} = m_v S_{ful} \tag{10}$$

From Eqs. 7 and 9 one has

$$S_{bg} = \frac{S_{fil}^p}{S_{fil}^p + M_i^p(S_{ful} - S_{fil})^p} S_{ful} \tag{11}$$

From Eqs. 8 and 10 one has

$$S_{fg} = \frac{(S_{ful} - S_{fil})^p}{M_v^p S_{fil}^p + (S_{ful} - S_{fil})^p} S_{ful} \tag{12}$$

This work aims to analyze characteristics of the first-order mask as it is effective and support fast processing of the audio signal. Therefore, $p = 1$.

From Eqs. 7 and 9 one has

$$S_{bg} = \frac{S_{fil}S_{ful}}{S_{fil}(1 - M_i) + M_iS_{ful}} \tag{13}$$

From Eqs. 8 and 10 one has

$$S_{fg} = \frac{(S_{ful} - S_{fil})S_{ful}}{(M_v - 1)S_{fil} + S_{ful}} \tag{14}$$

After calculating the background and foreground signals, the audio signal are reconstructed (by multiplying the signal with the original phase data obtained at the beginning of the experiment) and analyzed. For analyzing the findings, the audio feature, RMS values, are calculated.

The entire code in this work are developed online on Google Colab platform [13]. The programming language used is Python (version 3) [12].

4   Results and Analysis

For ease of analysis, a trimmed audio signal (noisy telephone call) with duration of 3 s was used. The obtained full spectrum of the audio signal is presented in Fig. 2. From the figure, it is seen that the first 0.25 s of the audio, there is almost no signal. While the audio starts from about 0.25 s to 3 s. The peak frequency in the spectrogram is at slightly above 4,000 Hz. The wiggle lines on the spectrogram represents the vocals which composed of the signals at multiple frequencies. It should be noted that the lines at different frequencies follow similar pattern.

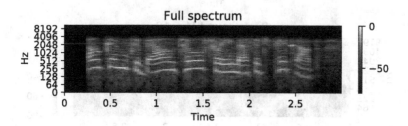

Fig. 2. 3-second signal - full spectrum

From Eqs. 13 and 14, it is seen that when $M_v = M_i = 1$, $S_{bg} = S_{fil}$ and $S_{fg} = S_{ful} - S_{fil} = S_{ful} - S_{bg}$. In this case, the background signal equals the filter mask while the foreground signal is the difference between the original signal and the mask.

When $0 < M_i < 0.5$, more weightage is placed on S_{ful} while less weightage is placed on S_{fil}. On the other hand, when $0.5 < M_i < 1$, more weightage is placed on S_{fil} and less is placed on S_{ful}.

When $M_i > 1$, denominator of S_{bg} is still positive, thus S_{bg} is non-negative. Since $S_{fil} < S_{ful}$, $S_{fil}(1 - M_i) > S_{ful}(1 - M_i)$, thus $S_{fil}(1 - M_i) + M_i S_{ful} > S_{ful}(1 - M_i) + M_i S_{ful} = S_{ful}$. Therefore, $S_{bg} <= S_{fil}$. Because S_{fil} and S_{ful} are constants, the more the value of M_i, the more the value of S_{bg}'s denominator, thus, the less the value of S_{bg}.

Based on Eq. 14, S_{fg} numerator is always positive, thus for its denominator to be positive, $M_v >= 1$.

For supporting the above analysis, the spectrograms of background and foreground signals at different values of M_i and M_v are presented in Figs. 3 and 4 respectively.

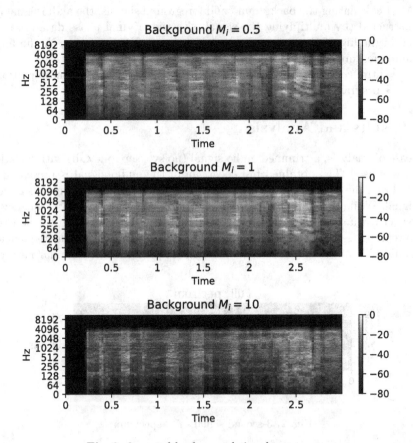

Fig. 3. 3-second background signal spectrum

In Fig. 3, there is little difference between the estimated background signals when $M_i = 0.5$ and when $M_i = 1$. While when M_i is significantly large, i.e., equals to 10, the estimated signal has significantly lower energy compared to the first two cases.

In Fig. 4, when $M_v = 0.5$ and when $M_v = 1$ the differences in spectrogram signals can be observed at low frequencies (below 128 Hz), at about 2.5 s. This can be seen clearer when $M_v = 10$ which is much more higher than the previous two values. Therefore, it is observed that there is little difference in the estimated foreground signals when M_v is increasing.

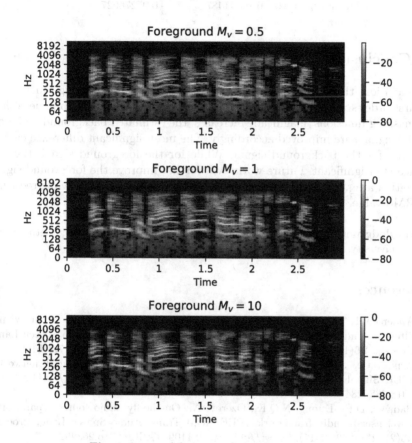

Fig. 4. 3-second foreground signal spectrum

In order to analyze the noise-ness of the estimated background and foreground signals, the RMS feature is analyzed. The results are presented in Table 1. Based on the obtained results, when M_i increases, the RMS values of the background signals are decreasing, this means the decrease in signal peak-to-peak amplitude. In this case, the estimated background signals are closer to that of the original audio signal (full spectrogram). Therefore, the estimation of the background signal is improved when M_i increases. Similarly, when M_v increases, the estimation of foreground signal is also improved because its RMS value decreases. As a side effect, this reduces the vocal loudness and softens voice which can affect clearness when hearing.

Table 1. RMS features of background and foreground signals

	Background signal	Foreground signal
M_i or $M_v = 0.5$	0.01895677	0.07939214
M_i or $M_v = 1$	0.01368008	0.07613512
M_i or $M_v = 10$	0.00641487	0.05270627

5   Conclusions and Future Works

In conclusion, this research has analyzed the effect of soft-masking function on the first-order spectrogram-based instrument - vocal separation technique. When the masking functions' magnitude increases, the estimated background and foreground signals are improved accordingly. The most significant difference can be observed for the background signal. While for the foreground signal, the difference is insignificant. Future works will further improve the foreground signal estimation after the separation phase for retrieving clearer vocal while reducing the RMS value.

Acknowledgment. The authors would thank FPT University, Hanoi, Vietnam and UCSI University, Kuala Lumpur, Malaysia for supporting this research.

References

1. Andersen, K.T., Moonen, M.: Robust speech-distortion weighted interframe wiener filters for single-channel noise reduction. IEEE/ACM Trans. Audio Speech Lang. Process. **26**(1), 97–107 (2018). https://doi.org/10.1109/TASLP.2017.2761699
2. Arık, S.O., Jun, H., Diamos, G.: Multi-head convolutional neural networks. IEEE Signal Process. Lett. **26**(1), 94–98 (2019). https://doi.org/10.1109/LSP.2018.2880284
3. Badawy, D.E., Duong, N.Q.K., Ozerov, A.: On-the-fly audio source separation-a novel user-friendly framework. IEEE/ACM Trans. Audio Speech Lang. Process. **25**(2), 261–272 (2017). https://doi.org/10.1109/TASLP.2016.2632528
4. Braun, S., Habets, E.A.P.: Linear prediction-based online dereverberation and noise reduction using alternating kalman filters. IEEE/ACM Trans. Audio Speech Lang. Process. **26**(6), 1119–1129 (2018). https://doi.org/10.1109/TASLP.2018.2811247
5. Buades, A., Coll, B., Morel, J.: A non-local algorithm for image denoising. In: 2005 IEEE Computer Society Conference on Computer Vision and Pattern Recognition (CVPR 2005), vol. 2, pp. 60–65, June 2005. https://doi.org/10.1109/CVPR.2005.38
6. Cheer, J., Daley, S.: An investigation of delayless subband adaptive filtering for multi-input multi-output active noise control applications. IEEE/ACM Trans. Audio Speech Lang. Process. **25**(2), 359–373 (2017). https://doi.org/10.1109/TASLP.2016.2637298

7. Chung, T.D., Ibrahim, R.B., Asirvadam, V.S., Saad, N.B., Hassan, S.M.: Adopting ewma filter on a fast sampling wired link contention in wirelesshart control system. IEEE Trans. Instrum. Meas. **65**(4), 836–845 (2016). https://doi.org/10.1109/TIM. 2016.2516321

8. Chung, T.D., Ibrahim, R., Asirvadam, V.S., Saad, N., Hassan, S.M.: Wireless HART: Advanced EWMA Filter Design for Industrial Wireless Networked Control Systems, 1st edn. Taylor & Francis Group, LLC, Abingdon (2017)

9. Crocco, M., Martelli, S., Trucco, A., Zunino, A., Murino, V.: Audio tracking in noisy environments by acoustic map and spectral signature. IEEE Trans. Cybernet. **48**(5), 1619–1632 (2018). https://doi.org/10.1109/TCYB.2017.2711497

10. Duong, T.T.H., Duong, N.Q.K., Nguyen, P.C., Nguyen, C.Q.: Gaussian modeling-based multichannel audio source separation exploiting generic source spectral model. IEEE/ACM Trans. Audio Speech Lang. Process. **27**(1), 32–43 (2019). https://doi.org/10.1109/TASLP.2018.2869692

11. Ekpo, S.C., Adebisi, B., Wells, A.: Regulated-element frost beamformer for vehicular multimedia sound enhancement and noise reduction applications. IEEE Access **5**, 27254–27262 (2017). https://doi.org/10.1109/ACCESS.2017.2775707

12. Foundation, P.S.: Python software foundation (2019). https://www.python.org/

13. Google: Welcome to colaboratory (2019). https://colab.research.google.com

14. He, Q., Bao, F., Bao, C.: Multiplicative update of auto-regressive gains for codebook-based speech enhancement. IEEE/ACM Trans. Audio Speech Lang. Process. **25**(3), 457–468 (2017). https://doi.org/10.1109/TASLP.2016.2636445

15. Itakura, K., Bando, Y., Nakamura, E., Itoyama, K., Yoshii, K., Kawahara, T.: Bayesian multichannel audio source separation based on integrated source and spatial models. IEEE/ACM Trans. Audio Speech Lang. Process. **26**(4), 831–846 (2018). https://doi.org/10.1109/TASLP.2017.2789320

16. Koluguri, N.R., Meenakshi, G.N., Ghosh, P.K.: Spectrogram enhancement using multiple window savitzky-golay (MWSG) filter for robust bird sound detection. IEEE/ACM Trans. Audio Speech Lang. Process. **25**(6), 1183–1192 (2017). https:// doi.org/10.1109/TASLP.2017.2690562

17. Laufer, Y., Gannot, S.: A bayesian hierarchical model for speech enhancement with time-varying audio channel. IEEE/ACM Trans. Audio Speech Lang. Process. **27**(1), 225–239 (2019). https://doi.org/10.1109/TASLP.2018.2876177

18. Xia, L., Chung, T.D., Kassim, K.A.A.: An automobile detection algorithm development for automated emergency braking system. In: 2014 51st ACM/EDAC/IEEE Design Automation Conference (DAC), pp. 1–6, June 2014. https://doi.org/10. 1145/2593069.2593083

19. Liu, Y., Jaw, D., Huang, S., Hwang, J.: Desnownet: context-aware deep network for snow removal. IEEE Trans. Image Process. **27**(6), 3064–3073 (2018). https:// doi.org/10.1109/TIP.2018.2806202

20. Luis-Valero, M., Habets, E.A.P.: Low-complexity multi-microphone acoustic echo control in the short-time fourier transform domain. IEEE/ACM Trans. Audio Speech Lang. Proces. **27**(3), 595–609 (2019). https://doi.org/10.1109/TASLP. 2018.2885786

21. Mahé, G., Jaïdane, M.: Perceptually controlled reshaping of sound histograms. IEEE/ACM Trans. Audio Speech Lang. Proces. **26**(9), 1671–1683 (2018). https:// doi.org/10.1109/TASLP.2018.2836143

22. Marquardt, D., Doclo, S.: Interaural coherence preservation for binaural noise reduction using partial noise estimation and spectral postfiltering. IEEE/ACM Trans. Audio Speech Lang. Proces. **26**(7), 1261–1274 (2018). https://doi.org/10. 1109/TASLP.2018.2823081

23. Rafii, Z., Pardo, B.: Online repet-sim for real-time speech enhancement. In: 2013 IEEE International Conference on Acoustics, Speech and Signal Processing, pp. 848–852, May 2013. https://doi.org/10.1109/ICASSP.2013.6637768

24. Raguraman, P.R.M., Vijayan, M.: Librosa based assessment tool for music information retrieval systems. In: 2019 IEEE Conference on Multimedia Information Processing and Retrieval (MIPR), pp. 109–114, March 2019. https://doi.org/10.1109/MIPR.2019.00027

25. Shimada, K., Bando, Y., Mimura, M., Itoyama, K., Yoshii, K., Kawahara, T.: Unsupervised speech enhancement based on multichannel nmf-informed beamforming for noise-robust automatic speech recognition. IEEE/ACM Trans. Audio Speech Lang. Proces. 27(5), 960–971 (2019). https://doi.org/10.1109/TASLP.2019.2907015

26. Sienko, M.: Loop-filter design and analysis for delta-sigma modulators and oversampled IIR filters. IEEE Trans. Circuits Syst. I Regul. Pap. 65(12), 4121–4132 (2018). https://doi.org/10.1109/TCSI.2018.2838021

27. Stallmann, C.F., Engelbrecht, A.P.: Gramophone noise detection and reconstruction using time delay artificial neural networks. IEEE Trans. Syst. Man Cybernet. Syst. 47(6), 893–905 (2017). https://doi.org/10.1109/TSMC.2016.2523927

28. Tan, W.R., Chan, C.S., Aguirre, H.E., Tanaka, K.: ArtGAN: artwork synthesis with conditional categorical GANs. In: 2017 IEEE International Conference on Image Processing (ICIP), pp. 3760–3764, September 2017. https://doi.org/10.1109/ICIP.2017.8296985

29. L.D. Team: Librosa (2019). https://librosa.github.io/librosa/

30. Torcoli, M., Herre, J., Fuchs, H., Paulus, J., Uhle, C.: The adjustment/satisfaction test (a/st) for the evaluation of personalization in broadcast services and its application to dialogue enhancement. IEEE Trans. Broadcast. 64(2), 524–538 (2018). https://doi.org/10.1109/TBC.2018.2832458

31. Xu, Y., Huang, Q., Wang, W., Foster, P., Sigtia, S., Jackson, P.J.B., Plumbley, M.D.: Unsupervised feature learning based on deep models for environmental audio tagging. IEEE/ACM Trans. Audio Speech Lang. Process. 25(6), 1230–1241 (2017). https://doi.org/10.1109/TASLP.2017.2690563

32. Zheng, C., Deleforge, A., Li, X., Kellermann, W.: Statistical analysis of the multichannel wiener filter using a bivariate normal distribution for sample covariance matrices. IEEE/ACM Trans. Audio Speech Lang. Process. 26(5), 951–966 (2018). https://doi.org/10.1109/TASLP.2018.2800283

Parsing and Segmentation

Japanese Predicate Argument Structure Analysis with Pointer Networks

Keigo Takahashi[(✉)], Hikaru Omori, and Mamoru Komachi

Tokyo Metropolitan University, 6-6 Asahigaoka, Hino, Tokyo 191-0065, Japan
{takahashi-keigo,omori-hikaru}@ed.tmu.ac.jp
komachi@tmu.ac.jp

Abstract. Recently, neural network models with sequence labeling were adopted for Japanese predicate argument structure analysis (PASA). However, the sequence labeling approach can assign the same argument to multiple arguments. Thus, we propose a novel neural PASA method using pointer networks to alleviate the problem of multiple assignments. Experimental results show that our single model can achieve state-of-the-art performance on the NAIST Text Corpus without using syntactic features.

Keywords: Natural language processing · Japanese predicate argument analysis · Neural network · Pointer network

1 Introduction

Predicate argument structure analysis (PASA) is a task that recognizes the semantic structure of predicates and arguments. Intra-sentential PASA is a subtask in PASA that identifies the predicate and intra-sentential arguments. In Japanese intra-sentential PASA, there are three core arguments: nominative (NOM), accusative (ACC) and dative (DAT). Each core label has at most one noun phrase for each predicate in this task, although there are cases of multiple predicates in sentences. Our research focuses on identifying core arguments in Japanese intra-sentential PASA[1] as shown in Table 1. This is an example with intra-sentential zero anaphora, which omits the core ACC argument.

Most Japanese intra-sentential PASA models are based on a sequence labeling approach. However, because sequence labeling dose not have any specific restrictions on the assignment of core arguments, it can label the same core argument as multiple arguments. We refer to these as multi-assignments. Multi-assignments in Japanese PASA are divided into two categories: multiple assignments of the same core argument (called **sentence-wise multi-assignments**), an example of which is shown in Table 2a; and multiple assignments of the same argument to different labels (called **label-wise multi-assignments**, where the label refers to the core argument), an example of which is shown in Table 2b.

---

[1] Japanese intra-sentential PASA includes intra-sentential zero anaphora; however, this and exophora are excluded here.

© Springer Nature Singapore Pte Ltd. 2020
L.-M. Nguyen et al. (Eds.): PACLING 2019, CCIS 1215, pp. 347–359, 2020.
https://doi.org/10.1007/978-981-15-6168-9_29

Table 1. Example of Japanese predicate argument structure analysis with intrasen-
tential zero anaphora.

私が**友人**に**お餅**を**あげる**と、**友人**はその場で**食べた**。
When **I gave** a piece of **rice cake** to my **friend**, my **friend ate** there.

Predicate	NOM	DAT	ACC
gave	I	rice cake	friend
ate	friend	rice cake [zero anaphora]	-

Table 2. Example of multi-assignment failures of Japanese predicate argument structure analysis.

(a) Example of sentence-wise multi-assignment failure of Japanese predicate argument structure
analysis. NOM is assigned to multiple arguments.

私は**友人**に**お餅**を**あげた**。
I gave a piece of **rice cake** to my **friend**.

Predicate	NOM	DAT	ACC
gave	I, friend	rice cake	-

(b) Example of label-wise multi-assignment failure of Japanese predicate argument structure
analysis. The subject "I" is assigned to multiple labels.

私は**友人**に**お餅**を**あげた**。
I gave a piece of **rice cake** to my **friend**.

Predicate	NOM	DAT	ACC
gave	I	rice cake	I

We propose two novel methods to alleviate these problems: the use of pointer
networks [27] to solve sentence-wise multi-assignments and restricted decoding
to solve label-wise multi-assignments.

A pointer network is a neural network model that outputs a pointer to indicate a specific position of the input sequence. Using pointer networks is a natural solution to alleviate the problem of sentence-wise multi-assignments because
there is at most one core argument for each label in Japanese PASA.

Restricted decoding is a method that prevents label-wise multi-assignments.
It is useful for label-wise multi-assignments because using pointer networks exclusively does not consider the interaction between core arguments.

Our main contributions are summarized as follows:

- To the best of our knowledge, this is the first attempt to formulate Japanese
  PASA as a sequence pointing task, via pointer networks, rather than as a
  sequence labeling task.
- Our single model achieved state-of-the-art performance in terms of the overall
  F-score using the NAIST Text Corpus but without using syntactic features.

2   Related Work

There are two categories of research on Japanese PASA. Most proposals employ supervised methods that learn Japanese PASA models from annotated corpora.

Conventional Models. Iida et al. [12] proposed using syntactical features with a pairwise comparison model for Japanese zero anaphora resolution. Imamura et al. [13] presented a discriminative model for argument identification in Japanese PASA. Hayashibe et al. [10] exploited positional and type features for Japanese PASA. Yoshikawa et al. [28] proposed a Markov logic network-based approach to Japanese PASA that jointly considers the dependency between multiple predicate relations. Ouchi et al. [18] proposed a graph-based model to account for multiple predicate interactions.

Our research differs from these studies as they predict each core argument independently. Moreover, our research considers the relationships between arguments by using pointer networks to identify the most plausible candidate without relying on hand-crafted syntactic features.

Another task similar to PASA is semantic role labeling (SRL). In English SRL research, Hacioglu [9] exploited dependency tree for SRL with a support vector machine (SVM). Pradhan et al. [22] proposed the use of syntactic features generated by different parsers with SVM. Punyakanok et al. [23] revealed the importance of syntactic parsing for an SRL system. Akbik and Li [1] proposed an easy-first decoding algorithm with k-nearest neighbors classification for English SRL.

These studies show that syntactic features are important for the SRL task. However, our proposed model did not use syntactic features; instead, it achieved state-of-the-art performance in Japanese PASA with pointer networks.

Neural Models. Neural network-based methods have been adopted widely in Japanese PASA. Matsubayashi and Inui [15] adopted dependency path embedding for Japanese PASA to consider dependency path information. Ouchi et al. [19] proposed a grid recurrent neural network (RNN) to improve Japanese PASA by considering the global assignments of arguments. Matsubayashi and Inui [16] built upon this research by applying self-attention and pooling to the grid RNN such that the interaction of predicates can take the same argument as a different core argument. Furthermore, Omori and Komachi [17] proposed a multi-task learning method to leverage event-noun argument structure analysis (ENASA) and improve Japanese PASA.

Our research differs from these studies by predicting sentence-wise likelihoods instead of using the label-wise likelihood of each candidate. Additionally, previous studies [16,19] do not consider core argument relationships, but focus instead on multiple predicate relationships. We adopt the opposite approach by considering multiple predicate relationships rather than core argument relationships. Furthermore, our study did not use any syntactical features such as dependency features, which are used in the study by Omori and Komachi [17].

However, neural network models have also been studied in English SRL. Collobert et al. [5] attempted to use a neural network model for SRL. FitzGerald et al. [8] proposed the use of a graphical model that represented the global assignment of arguments to their semantic roles via a neural network. He et al. [11] presented a deep bi-directional long short-term memory (LSTM) model for SRL tasks with beam decoding. Strubell et al. [24] applied multitask-learning with a Linguistically-Informed Self-Attention (LISA) neural network model for English SRL.

Japanese PASA tasks are similar to these, and we attempt to address Japanese PASA using a neural network-based pointer network model with easy-first restricted decoding.

Pointer Network Models. There are many studies that used pointer networks for similar tasks. Cheng et al. [3] proposed a pointer network-based model for an English implicit argument prediction task, which is to identify the correct filler for an implicit argument role of a predicate, with document and query encoders. Zhai et al. [29] applied pointer networks to an English sequence chunking task using a bi-directional LSTM model. Li et al. [14] presented a pointer network-based approach to an English text segmentation task using a bi-directional gated recurrent unit (GRU) model.

Our research is the first attempt to apply pointer networks to Japanese PASA. In previous studies [16,19], a sequence labeling approach was adopted for this; we use pointer networks instead to alleviate the problem of sentence-wise multi-assignments in Japanese PASA.

3   Japanese Predicate Argument Structure Analysis (PASA) with Sequence Labeling

Our research focuses on intra-sentential Japanese PASA. We follow the task settings of Ouchi et al. [19] in that the predicate word is known when the core arguments of a predicate are predicated.

In this section, we describe the baseline method of using sequence labeling. Figure 1 shows the architecture for the baseline. Each model comprises an input, RNN, and output layer.

3.1   Input Layer

We use the following input features as inputs for the model, where an input feature is the concatenation of the surface, distance, and predicate features converted to a vector by an embedded projection matrix, and vectors are concatenated as x_t, where t is the time step of the sentence sequence. The details of the embedded matrix are in Experiments section.

Surface Features. We convert the surface of the predicate and argument candidate using an embedded projection matrix.

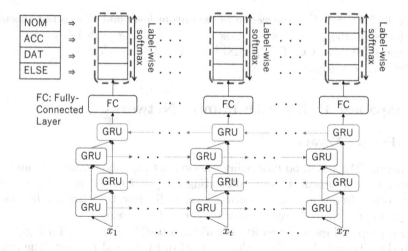

Fig. 1. Predicate argument structure analysis as a sequence labeling task.

Distance Features. We use two scalar features: (1) the word distance between the predicate and the argument candidate, and (2) the distance between the bunsetsu[2] with the predicate and that with the argument candidate.

Predicate Feature. We use a binary feature indicating whether the argument is a predicate.

3.2 RNN Layer

We use the GRU [4] to form two layers of a stacked bi-directional GRU. The index of each layer is $l \in [1, \cdots, L]$. We apply residual connections between layers following Ouchi et al. [19] and Matsubayashi and Inui [16]. The hidden state $h_t^l \in \mathbb{R}^{d^h}$ is calculated as follows:

$$h_t^l = \begin{cases} \mathrm{gru}^l(h_t^{l-1}, h_{t-1}^l) & (l = \text{odd}) \\ \mathrm{gru}^l(h_t^{l-1}, h_{t+1}^l) & (l = \text{even}) \end{cases} \qquad (1)$$

where $\mathrm{gru}^l(\cdot)$ denotes the l-th layer GRU function. In addition, $h_t^0 = x_t$.

3.3 Output Layer

The input of the output layer is the tensor h_t^L, which is the output from the last layer of the RNN. Then, we obtain the output vector o_t^{base} using the softmax function:

$$o_t^{\text{base}} = \mathrm{softmax}(W^{\text{base}} h_t^L + b^{\text{base}}) \qquad (2)$$

---

[2] Functional chunk.

where $W^{\text{base}} \in \mathbb{R}^{4 \times d^{\text{h}}}$ is the weight of the output layer and b^{base} is the bias term

Moreover, the output of the baseline model is the probability vector that represents [NOM, ACC, DAT, ELSE], where ELSE is a category that is "not applicable" to the core arguments.

4 Japanese PASA with Pointer Networks

4.1 Pointer Networks

Our neural PASA method uses pointer networks to assign at most one candidate for each core argument, thereby alleviating the problem of sentence-wise multi-assignments. In our model, the output layer is different from the baseline model but the input and RNN layers are the same as the baseline.

Our proposed method predicts the likelihood o_t^{ptr} of each word using pointer networks. Figure 2 shows the architecture of our proposed method. The output is calculated as follows:

$$s_t^{\text{ptr},i} = W^{\text{ptr},i} h_t^{\text{ptr},i} + b^{\text{ptr},i} \tag{3}$$

$$o^{\text{ptr},i} = \text{softmax}(\phi \oplus s_1^{\text{ptr},i} \oplus \ldots \oplus s_T^{\text{ptr},i}) \tag{4}$$

where i indicates the core argument. The output layer is built for each core argument. Here, $W^{\text{ptr},i} \in \mathbb{R}^{d^{\text{ptr}} \times d^{\text{hidden}}}$ denotes the weight, d^{ptr} denotes the dimensions, and b^{ptr} denotes the bias term. Furthermore, $s_t^{\text{ptr},i}$ is a scalar value. It is concatenated with another scalar value of ϕ, which represents the non-existence of the core argument in the sentence where ϕ is, in our experiment, a constant that equals 0.0. Originally, pointer networks models were proposed to solve a sequence to sequence problem [3], however, we apply pointer networks straight to the task. Thus, we pool all the output of the sequence generated by RNN layer and decode them as a sequence.

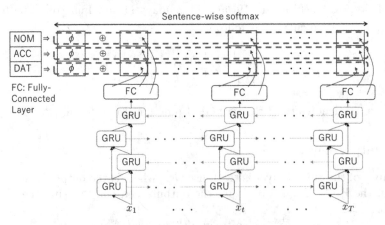

Fig. 2. Predicate argument structure analysis as a sequence pointing task using pointer networks.

Algorithm 1. Global Argmax Algorithm

1: Input:
2: s: a concatenated matrix of probability vectors of NOM, ACC, and DAT.
3: s_{ij}: an element of a concatenated matrix, where i denotes a label and j denotes an index in a sentence.
4: Output:
5: *results*: indexes of the arguments in a sentence for each case.
6:

7: Do (global argmax):
8: *results* $\Leftarrow [-1, -1, -1]$
9: **while** $-1 \in results$ **do**
10:     $i, j \Leftarrow \arg\max_{s_{ij} \in s}(s)$
11:     $s_{ij} \Leftarrow -\infty$
12:     **if** $results[i] = -1$ **then** $results[i] \Leftarrow j$

4.2 Decoding

We attempt the following decoding algorithms.

No Restriction. "No restriction" applies no restriction on decoding. This means the decision to pick up the arguments for each case is independent. Label-wise multi-assignments may occur in this case.

Global-argmax. To prevent label-wise multi-assignments, we propose restricted decoding denoted "global-argmax." The pseudocode is shown in Algorithm 1, from Line 7 to Line 12. The global-argmax assigns core arguments in descending order of the highest likelihood. This algorithm is based on the easy-first decoding proposed in Akbik and Li [1] for English SRL. Global-argmax decoding predicts only unassigned core arguments to prevent label-wise multi-assignments.

Local-argmax. We also experimented with another restricted decoding denoted "local-argmax." It determines arguments in the order ACC, NOM, and DAT, which is based on yet another heuristic of the easy-first strategy. It exploits prior knowledge about Japanese PASA.

5 Experiments

5.1 Setting

We used NAIST Text Corpus (NTC) 1.5 for our experiments. We divided the dataset into training, development, and test sets, following Taira et al. [25]. Table 3 shows the number of instances in each split. We used morphological and dependency information automatically annotated with NTC[3]. We also took the

---

[3] Dependency information was used only for evaluation.

Table 3. Statistics of NAIST Text Corpus 1.5.

	Training	Development	Test
Sentences	23,218	4,628	8,816
Predicates	62,489	12,724	23,981
NOM (dep)	37,615	7,520	14,230
NOM (zero)	11,546	2,552	4,764
ACC (dep)	24,997	5,105	9,532
ACC (zero)	1,802	394	783
DAT (dep)	5,855	1,637	2,547
DAT (zero)	359	112	211

Table 4. Hyperparameters for our models and search space.

Parameters	Ours	Search space
Optimizer	SGD	Adam, Adagrad, Adadelta, SGD, and RMSProp
Learning ratio	0.2	0.4, 0.3, 0.2, 0.1, 0.05, 0.01, 0.005, 0.001, 0.0001, and 0.00001
Batch size	2	2 to 200 (The step size is two.)
Gradient clip	2	0 to 20 (The step size is one.)
Dropout ratio	0.4	0.0 to 0.7 (The step size is 0.1.)

coreference chain into account such that if an argument co-refers to other noun phrases, we only label a noun phrase that directly modifies the predicate as a gold standard; otherwise, we do not label it.

The surface features were converted by pre-trained word embedding[4] based on GloVe [20] with retrofitting [7] word embedding.

We use the F1 score to evaluate our model with the same method used by Ouchi et al. [19], Matsubayashi and Inui [16], and Omori and Komachi [17].

Our models were implemented using PyTorch version 1.0.1.post2. The hyperparameters are shown in Table 4. The hyperparameters were tuned with Hyperopt [2] version 0.1.2. All the hyperparameters in the baseline and our proposed model are the same.

5.2   Results

We show the results of our experiment in Table 5. Our model achieved a higher score than models in previous studies [16,17,19] in terms of overall performance without using syntactic features.

---

[4] http://cl.asahi.com/api_data/wordembedding.html.

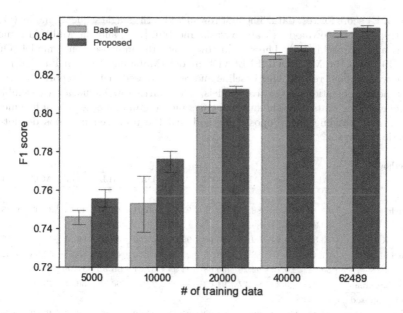

Fig. 3. Learning curves of F1 scores. These F1 scores are the averages of the results of five models. The error bars indicate 95% confidence intervals.

6 Discussion

Decoding Algorithms. In our research, the results of decoding algorithms did not significantly affect the scores shown in Table 5 of our proposed models. The "Local-argmax" score is slightly higher than others in a single model; however, this decoding algorithm is heuristic and requires domain knowledge to implement. The result of "Global-argmax" achieved the higher score than "No restriction," but both restricted decoding algorithms did not improve over the baseline by a large margin in a single model. Furthermore, the scores are same in an ensemble model. This means that the decoding algorithms did not significantly affect the model performance.

The percentage of label-wise multi-assignment is 0.23% (54 cases) for "No restriction" decoding; with the "Global-argmax" decoding method it did not occur with the single model.

Learning Curve. A comparison between our baseline and proposed methods demonstrate that pointer networks can improve the performance of argument identification in both Dep and Zero relationships. Figure 3 shows the overall F1 score obtained by varying the amount of training data. Our proposed model consistently outperformed the baseline. The results show the efficiency of the sequence pointing method, in terms of the amount of training data required, compared to sequence labeling methods.

Table 5. F1 score for test data. The first row includes the single sequence model (Ouchi et al. [19]), the BASE model (Matsubayashi and Inui [16]), and the Multi-ALL model (Omori and Komachi [17]). The second row shows the multi-sequence model (Ouchi et al. [19]) and the MP_POOL_SELFATT model (Matsubayashi and Inui [16]). The third row shows our results. Our baseline and our proposed method report the average and standard deviation of the five models. The fourth row includes an ensemble of ten models using Matsubayashi and Inui [16]'s method and an ensemble of five models using on our baseline and proposed method. All the proposed methods do not use syntactic features.

Method	All		Dep				Zero			
	ALL	SD	ALL	NOM	ACC	DAT	ALL	NOM	ACC	DAT
Single model										
Ouchi+ 17	81.15	-	88.10	88.32	93.89	65.91	46.10	49.51	35.07	9.83
M & I 18	83.39	±0.13	89.90	-	-	-	54.37	-	-	-
O & K 19	83.82	±0.10	90.15	90.68	95.06	67.56	53.50	56.37	45.36	8.70
Ouchi+ 17	81.42	-	88.17	88.75	93.68	64.38	47.12	50.65	32.35	7.52
M & I 18	83.94	±0.12	90.26	90.88	94.99	67.57	**55.55**	**57.99**	**48.9**	**23**
Our baseline	84.22	±0.20	90.77	91.07	94.93	73.17	53.70	56.25	45.32	11.86
Our proposed										
- no restriction	84.43	±0.22	90.87	91.00	95.15	73.69	53.95	56.55	45.60	11.36
- global-argmax	84.46	±0.22	90.91	91.03	95.19	**73.74**	53.97	56.57	45.60	11.26
- local-argmax	**84.48**	±0.22	**90.96**	**91.13**	**95.24**	73.34	53.70	56.31	45.29	12.18
Ensemble model										
M & I 18	85.34	-	91.26	91.84	95.57	70.8	**58.07**	**60.21**	**52.5**	**26**
Our baseline	85.60	-	91.60	**92.03**	95.41	74.49	56.46	59.06	47.97	7.91
Our proposed										
- no restriction	85.77	-	91.70	91.98	95.59	75.91	57.11	59.39	50.25	13.90
- global-argmax	**85.83**	-	91.77	91.93	95.65	**76.02**	57.14	59.42	50.32	13.90
- local-argmax	85.81	-	**91.80**	92.01	**95.68**	75.63	56.67	59.10	49.30	12.31

Label-wise Multi-Assignments. Table 6 shows an unsuccessful example of our baseline model when it did not predict the core argument. Although it is difficult to understand this sentence alone even someone is a native Japanese speaker, the contextual meaning is "Takahashi and Saito, who act in a drama, are both 33 years old." The baseline model assigned multi-core arguments as NOM. By contrast, our proposed model correctly predicted NOM because of the use of pointer networks, which assigns only one candidate for each case. In this example, "act (演じる)" is a predicate that requires one NOM. The baseline model predicts two animate words of "Takahashi (高橋)" and "Saito (斎藤)" as NOM; however, the gold label for the predicate is only one animate word.

Incorrect Case Assignments. Table 7 shows another example of an incorrectly assigned argument. Our proposed method predicted the wrong argument, i.e., "choice (選択)" as NOM. This category accounts for 12.16% of the total assignments. The correct answer is "parents (両親)," which does not directly modify the predicate; thus, it was difficult to predict.

Table 6. Error of predicting core argument.

演じる高橋、斎藤が三十三歳。
Takahashi and Saito who **act** are 33 years old.

Type	NOM	DAT	ACC
Gold	Saito	-	-
Our baseline	Takahashi, Saito	-	-
Our proposed	Saito	-	-

Table 7. Error of assigning a wrong word.

両親の選択は**誤**っていなかったようだ。
It seems that the choice of parents was not **wrong**.

Type	NOM	DAT	ACC
Gold	parents	choice	-
Our baseline	choice	-	-
Our proposed	choice	-	-

7 Conclusion

We presented a novel Japanese PASA model using pointer networks. We showed that our model, which does not use syntactic features, outperforms methods based on sequence labeling. In future research, we plan to incorporate a self-attention-based [26] large-scale pre-trained deep language representation model such as ELMo [21] or BERT [6] to leverage the structure of a sentence. We will also apply multitask learning with a multi-head self-attention LISA model [24] to Japanese PASA.

References

1. Akbik, A., Li, Y.: K-SRL: instance-based learning for semantic role labeling. In: Proceedings of COLING pp. 599–608 (2016)
2. Bergstra, J., Yamins, D., Cox, D.: Making a science of model search: hyperparameter optimization in hundreds of dimensions for vision architectures. In: Proceedings of ICML, pp. 115–123 (2013)
3. Cheng, P., Erk, K.: Implicit argument prediction as reading comprehension. arXiv preprint arXiv:1811.03554 (2018)
4. Cho, K., et al.: Learning phrase representations using RNN encoder-decoder for statistical machine translation. In: Proceedings of EMNLP, pp. 1724–1734 (2014)
5. Collobert, R., Weston, J., Bottou, L., Karlen, M., Kavukcuoglu, K., Kuksa, P.: Natural language processing (almost) from scratch. J. Mach. Learn. Res. **12**, 2493–2537 (2011)
6. Devlin, J., Chang, M.W., Lee, K., Toutanova, K.: BERT: pre-training of deep bidirectional transformers for language understanding. In: Proceedings of NAACL, pp. 4171–4186 (2019)

7. Faruqui, M., Dodge, J., Jauhar, S.K., Dyer, C., Hovy, E., Smith, N.A.: Retrofitting word vectors to semantic lexicons. In: Proceedings of NAACL. pp. 1606–1615 (2015)
8. FitzGerald, N., Täckström, O., Ganchev, K., Das, D.: Semantic role labeling with neural network factors. In: Proceedings of EMNLP, pp. 960–970 (2015)
9. Hacioglu, K.: Semantic role labeling using dependency trees. In: Proceedings of COLING, pp. 1273–1276 (2004)
10. Hayashibe, Y., Komachi, M., Matsumoto, Y.: Japanese predicate argument structure analysis exploiting argument position and type. In: Proceedings of IJCNLP, pp. 201–209 (2011)
11. He, L., Lee, K., Levy, O., Zettlemoyer, L.: Jointly predicting predicates and arguments in neural semantic role labeling. In: Proceedings of ACL, pp. 364–369 (2018)
12. Iida, R., Inui, K., Matsumoto, Y.: Exploiting syntactic patterns as clues in zero-anaphora resolution. In: Proceedings of ACL-COLING, pp. 625–632 (2006)
13. Imamura, K., Saito, K., Izumi, T.: Discriminative approach to predicate-argument structure analysis with zero-anaphora resolution. In: Proceedings of ACL-IJCNLP, pp. 85–88 (2009)
14. Li, J., Sun, A., Joty, S.: Segbot: a generic neural text segmentation model with pointer network. In: Proceedings of IJCAI, pp. 4166–4172 (2018)
15. Matsubayashi, Y., Inui, K.: Revisiting the design issues of local models for Japanese predicate-argument structure analysis. In: Proceedings of IJCNLP, pp. 128–133 (2017)
16. Matsubayashi, Y., Inui, K.: Distance-free modeling of multi-predicate interactions in end-to-end Japanese predicate argument structure analysis. In: Proceedings of COLING. pp. 94–106 (2018)
17. Omori, H., Komachi, M.: Multi-task learning for Japanese predicate argument structure analysis. In: Proceedings of NAACL, pp. 3404–3414 (2019)
18. Ouchi, H., Shindo, H., Duh, K., Matsumoto, Y.: Joint case argument identification for Japanese predicate argument structure analysis. In: Proceedings of ACL-IJCNLP, pp. 961–970 (2015)
19. Ouchi, H., Shindo, H., Matsumoto, Y.: Neural modeling of multi-predicate interactions for Japanese predicate argument structure analysis. In: Proceedings of ACL, pp. 1591–1600 (2017)
20. Pennington, J., Socher, R., Manning, C.: GloVe: Global vectors for word representation. In: Proceedings of HLT-EMNLP, pp. 1532–1543 (2014)
21. Peters, M.E., Neumann, M., Iyyer, M., Gardner, M., Clark, C., Lee, K., Zettlemoyer, L.: Deep contextualized word representations. In: Proceedings of NAACL, pp. 2227–2237 (2018)
22. Pradhan, S., Hacioglu, K., Ward, W., Martin, J.H., Jurafsky, D.: Semantic role chunking combining complementary syntactic views. In: Proceedings of CoNLL, pp. 217–220 (2005)
23. Punyakanok, V., Roth, D., Yih, W.T.: The importance of syntactic parsing and inference in semantic role labeling. Computat. Linguistics **34**(2), 257–287 (2008)
24. Strubell, E., Verga, P., Andor, D., Weiss, D., McCallum, A.: Linguistically-informed self-attention for semantic role labeling. In: Proceedings of EMNLP, pp. 5027–5038 (2018)
25. Taira, H., Fujita, S., Nagata, M.: A Japanese predicate argument structure analysis using decision lists. In: Proceedings of EMNLP, pp. 523–532 (2008)
26. Vaswani, A., et al.: Attention is all you need. In: Proceedings of NIPS, pp. 5998–6008 (2017)

27. Vinyals, O., Fortunato, M., Jaitly, N.: Pointer networks. In: Proceedings of NIPS, pp. 2692–2700 (2015)
28. Yoshikawa, K., Asahara, M., Matsumoto, Y.: Jointly extracting Japanese predicate-argument relation with Markov logic. In: Proceedings of IJCNLP, pp. 1125–1133 (2011)
29. Zhai, F., Potdar, S., Xiang, B., Zhou, B.: Neural models for sequence chunking. In: Proceedings of AAAI, pp. 3365–3371 (2017)

An Experimental Study on Constituency Parsing for Vietnamese

Luong Nguyen-Thi[1(✉)] and Phuong Le-Hong[2]

[1] Dalat University, Lam Dong, Vietnam
luongnt@dlu.edu.vn
[2] College of Science, Vietnam National University, Hanoi, Vietnam
phuonglh@vnu.edu.vn

Abstract. This paper presents an experimental study in Vietnamese constituency syntactic parsing. We first compare results of two recent constituency parsers on a Vietnamese corpus, a shift-reduce constituency parser and a neural parser. We then integrate distributed word representations into the shift-reduce parser to improve its F_1 score. We also report a new state-of-the-art parsing score for Vietnamese with F_1 score of 80% in the neural parser. Finally, we perform error analysis on a sample of 100 sentences for the neural parser, which helps categorize ambiguous and difficult constructions inherent to the problem of Vietnamese syntactic parsing.

Keywords: Constituency parsing · Shift-reduce · Self-attention · Distributed word representation

1 Introduction

Constituency parsing has been an attractive method to syntactic parsing over the years. Various methods have been proposed to improve parsing accuracy. In 2013, Zhu et al. [5] proposed a simple yet effective extension to the shift-reduce process. Their parser gives a high accuracy in constituency parsing, with F_1 score of 90.3%. In the past years, neural network approaches improved constituency parsing [1–4]. In particular, Kitaev and Klein [6] use a self-attentive architecture which can enhanced a state-of-the-art constituency parser. Their parser achieved a new state-of-the-art results, with F_1 score of 95.13% for English.

For Vietnamese, there are some studies on constituency parsing [7–11,13]. In 2007, Nguyen Quoc The and Le Thanh Huong [7] use the Lexicalized Probabilistic Context-free Grammar for Vietnamese syntactic parsing. The Vietnamese treebank [9] contains over 10,000 sentences in Penn treebank format in 2009. A year later, Phuong Le-Hong et al. [10] presented a method that extracts Lexicalized Tree-Adjoining Grammars (LTAG) from treebanks automatically. At the same time, Le Thanh Huong et al. [11] proposed a parser based on a probabilistic Head-Driven Phrase Structure Grammar that can handle constraints about syntactic of words and semantic of words that one may produce. In 2015, Phuong

© Springer Nature Singapore Pte Ltd. 2020
L.-M. Nguyen et al. (Eds.): PACLING 2019, CCIS 1215, pp. 360–373, 2020.
https://doi.org/10.1007/978-981-15-6168-9_30

Le-Hong et al. [12] presented the development of a grammar and a syntactic parser for Vietnamese language. They also presented the construction and evaluation of a deep syntactic parser based on LTAG for the Vietnamese language. In 2016, Nguyen Thi Quy et al. [13] built a treebank for Vietnamese syntactic parsing consisting of 20,000 sentences.

In this paper, we introduce an empirical study of Vietnamese constituency parsing. The remainder of the paper is organized as follows. The next Sect. 2 introduces some preliminary concepts of constituency parsing and distributed word representation. Then, we present experiments on the shift-reduce method for constituency parsing and propose the use of distributed word representations in Sect. 3. We also present experiments on the self-attentive method. Last Sect. 4 discusses related work and concludes the paper.

2 Background

2.1 Constituency Parsing

Constituency parsing extracts a constituency-based parse tree from a sentence that represents its syntactic structure according to a phrase structure grammar. For example, consider the constituency analysis of a sentence in the VietTreeBank *"Mảnh đất của đạn bom không còn người nghèo."* (The land of bomb bullets has not had poor people anymore) in Fig. 1:

(S (NP-SUB (Nc-H Mảnh) (N đất) (PP (E-H của) (NP (N-H đạn) (N-H bom)))) (VP (R không) (V-H còn) (NPDOB (N-H người) (A nghèo))) (. .))

S for sentence, the top-level structure in this example. NP for the noun phrase *"Mảnh đất của đạn bom"*. VP for the verb phrase *"không còn người nghèo"*.

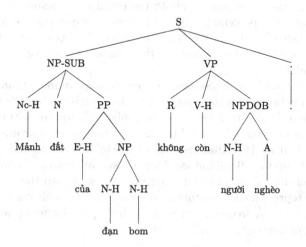

Fig. 1. An example of constituency tree

2.2   Distributed Word Representations

Last few years, many tasks in natural language processing used distributed word representations to improve their accurate. A word is represented by the distribution of many real value components of a vector, such a representation is called a word distribution representation. The word unit is not considered to be the smallest unit in the problem of natural language processing, but then the real value of each vector becomes the smallest processing unit. The real value can hold information about not only syntactic but also semantic [14]. Previously, the words using in NLP problems are often encoded by one-hot vectors. Directions of one-hot vectors and the size of the vocabulary are the same. This vector representation is easy to understand and simple however it contains two major problems. The first problem is data sparseness and the second problem is that it is not able to catch the semantic between words. This limitation has prompted some machine learning methods to create better representations, one of these methods is distributed word representations. Those representations are shown to be more effective in supporting many tasks in natural language processing, for example machine translation and speech recognition.

3   Experiments

3.1   Dataset

Distributed Word Representations: We use the similar database in our research [15]. To create distributed word representations, we use the dataset consisting of 7.3 GB of text from 2 million articles collected from a Vietnamese news portal. The text is first normalized to lower case. All special characters are removed except these common symbols: the comma, the semi-colon, the colon, the full stop and the percentage sign. All numeral sequences are replaced with the special token <number>, so that correlations between a certain word and a number are correctly recognized by the neural network or the log-bilinear regression model.

Each word in the Vietnamese language may consist of more than one syllable with spaces in between, which could be regarded as multiple words by the unsupervised models. Hence it is necessary to replace the spaces within each word with underscores to create full word tokens. The tokenization process follows the method described in [16]. After removal of special characters and tokenization, the articles add up to 969 million word tokens, spanning a vocabulary of 1.5 million unique tokens. We train the unsupervised models with the full vocabulary to obtain the representation vectors, and then prune the collection of word vectors to the 5,000 most frequent words, excluding special symbols and the token <number> representing numeral sequences.

Constituency Treebank in Vietnamese: We use Vietnamese constituency treebank from [9]. This treebank is one result of a national project which aims to develop basic resources and tools for Vietnamese language and speech processing[1]. The raw texts of the treebank are collected from the social and political sections of the Youth online daily newspaper. The corpus is divided into three sets corresponding to three annotation levels: word-segmented, part-of-speech-tagged and syntax-annotated set. The syntax-annotated corpus, a subset of the part-of-speech-tagged set, is currently composed of 10,471 sentences (225,085 tokens). Sentences range from 2 to 105 words, with an average length of 21.75 words. We use a training set of about 8,000 sentences and a test set of about 2,000 sentences. Table 1 shows some common constituency tags in Vietnamese. Table 2 describes the occurrence of constituency labels in Vietnamese constituency treebank. Example NP (noun phrase) appears 58,119 times in this Vietnamese constituency treebank.

Table 1. Some constituency tags in Vietnamese

No.	Category	Description
1	S	Simple declarative clause
2	NP	Noun phrase
3	VP	Verb phrase
4	PP	Preposition phrase
5	CC	Coordinating conjunction
6	QP	Quantitative phrase
7	AP	Adjective phrase
8	XP	Unknown phrase
9	SQ	Question
10	SBAR	Subordinate clause

Table 2. The occurrence of constituency tags

No.	Constituency tag	Count
1	NP	58,119
2	VP	38,906
3	S	20,502
4	PP	13,427
5	AP	7,050
6	SBAR	5,256
7	NP-SUB	2,142
8	QP	428
9	SQ	305
10	XP	281
11	Other	341
Sum		**146,757**

---

[1] VLSP Project, https://vlsp.hpda.vn/demo/.

3.2 Constituency Parsers

We experiment with 2 parsers: a Shift-Reduce Constituency Parser which is built by the NLP group of Stanford university [17] and the Berkeley neural parser based on Constituency Parsing with a Self-Attentive Encoder from ACL 2018 [6]. In addition to the baseline results of these parsers, we propose integration of distributed word representation features into the Shift-Reduce Consituency Parser that we call Shift-ReduceD. We used the default settings in training stage of these parsers.

The base feature set[2] in Shift-Reduce Constituency parser ϕ_0 is as in [17]. Feature sets ϕ_c, ϕ_s, ϕ_g consist of the features defined in ϕ_0, plus the six indicator word features that get the distributed word representation by respectively CBOW, Skip-Gram, GloVe method: the first four words on the stack s, the first two words on the queue q as shown in Table 3. The symbol s_i represents a stack item, q_i represents a queue item and w denotes a word. As such, ϕ_c has the distributed word representation by CBOW method of the first four words on stack:$s_0 w, s_1 w, s_2 w, s_3 w$ and the first two words on the queue: $q_0 w, q_1 w$.

Table 3. Feature sets for use in the Shift-ReduceD

System	Feature set	Feature templates
Shift-Reduce	ϕ_0	Baseline
Shift-ReduceD	ϕ_s	$\phi_0 \cup \{v_s(s_0 w), v_s(s_1 w),$ $v_s(s_2 w), v_s(s_3 w),$ $v_s(q_0 w), v_c(q_1 w)\}$
	ϕ_c	$\phi_0 \cup \{v_c(s_0 w), v_c(s_1 w),$ $v_c(s_2 w), v_c(s_3 w),$ $v_c(q_0 w), v_c(q_1 w)\}$
	ϕ_g	$\phi_0 \cup \{v_g(s_0 w), v_g(s_1 w),$ $v_g(s_2 w), v_g(s_3 w),$ $v_g(q_0 w), v_g(q_1 w)\}$

3.3 Evaluation Metrics

The evaluation metrics used for parsing accuracy are the recall, precision and F_1-score over parse trees as follows:

$$Recall = \frac{\#(correct constituents)}{\#(constituents in the gold parse tree)} \qquad (1)$$

$$Precision = \frac{\#(correct constituents)}{\#(constituents in the output parse tree)} \qquad (2)$$

$$F_1 = \frac{2 * Recall * Precision}{Recall + Precision} \qquad (3)$$

---

[2] https://nlp.stanford.edu/software/srparser.html.

3.4   Errors Analysis

Overall Accuracy. Table 4 shows the parsing accuracy for four features sets $\phi_0, \phi_s, \phi_c, \phi_g$ in the Shift-Reduce constituency parser.

Table 4. Parsing accuracy in Shift-Reduce constituency parser

Feature set	LR	LP	F_1
ϕ_0	70.40%	71.84%	71.11%
ϕ_s	69.97%	72.50%	71.21%
ϕ_c	70.27%	72.38%	71.31%
ϕ_g	70.01%	72.55%	71.26%

The integration of distributed word representation into the shift-reduce constituency parser increases the F_1-score accuracy about 0.2%.

Next, we experiment Berkeley neural parser based on Constituency Parsing with a Self-Attentive Encoder [6]. This parser uses CharLSTM feature that a bidirectional LSTM is operated independently over the characters in each word and uses the LSTM result instead of part-of-speech tag embeddings. Embeddings from Language Models (ELMo) also use in this parser. ELMo is proposed by Peters et al. [18] which models captures both the compicated properties of syntactical and semantics word usage as well as the way these uses linguistic context. The authors used a deep bidirectional language model in learning process to create these vectors. A variety of natural language processing tasks that use EMLo in performance have the state of the art results. Table 5 presents the parsing accuracy in Berkeley neural parser for Vietnamese. It is interesting that the use ELMo feature results in about 1.44% point higher than the use CharLSTM feature with F_1-score of 80.96% compared to 79.51%.

Table 5. Parsing accuracy in Berkeley neural parser

Feature set	LR	LP	F_1
With CharLSTM	78.47%	80.57%	79.51%
With EMLo	80.53%	81.39%	**80.96%**

Figure 2 shows the impact of sentence length in Berkeley neural parser. F_1-score reduces when sentence length increases. With sentences shorter than 10 words, F_1-score is around 88%. However, the score decreased about 5% when sentence length is longer than 40 words.

Comparing the parsing accuracy between the Shift-Reduce constituency parser and the Berkeley neural parser, we see that F_1 score in the Shift-Reduce constituency parser is lower about 9.5% than the F_1-score in the Berkeley neural

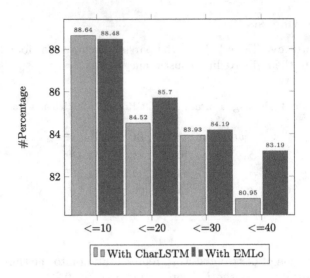

Fig. 2. F_1-score: impact of sentence length in Berkeley neural parser

parser in Table 6. Especially, with Vietnamese constituency treebank from [9], F_1-score in the Berkeley neural parser increases about 11.6% compared to the result of [12] in 2015.

Table 6. Parsing results

System	F_1
LTAG (Phuong Le-Hong et al. [12])	69.33%
Shift-reduce constituency parser	71.11%
Shift-reduce constituency parser with distributed word representations	71.31%
Berkeley neural parser	**80.96%**

To analyze error and ambiguous constructions in constituency parser, we use the output from Berkeley neural parser. The results are achieved by a manual analysis that uses 100 sentences from the Berkeley neural parser output.

Impact of Tagging Error. Table 7 indicates the number of occurrences label errors. Some errors are due to several node errors such as VP attachment, PP attachment, NP attachment, where a single word phrase typically causes a single node error. Ratio is the average number of node errors divided by an error, i.e Nodes Involved/Occurrences. For example, PP attachment is any occurrence in that the transformation related shifting a prepositional phrase or the invalid bracket over a prepositional phrase. In particular, single word phrase is the highest contributor to errors in Berkeley neural parser.

Table 7. Phrase errors in Vietnamese constituency parsing

Error type	Occurrences	Nodes involved	Ratio
Single word phrase	36	36	1.0
VP attachment	31	94	3.0
PP attachment	30	60	2.0
NP internal structure	29	43	1.5
Clause attachment	26	44	1.7
NP attachment	20	51	2.6
Coordination	17	39	2.3
Other	17	34	2.0

Ambiguous Constructions in Vietnamese. In this step, we use an analysis tool which are built by Kummerfeld et al. [19] to analyze the error types. This tool was built to classify English parsing errors. We also use directly this tool to the Vietnamese parsing output.

Figure 3 shows that VP attachment error caused by ambiguous construction "V-H NP R V-H VP". The phrase *"Làm nghề này dễ bị sập hầm"* in this figure is predicted for VP attachment errors. However, this phrase is a clause in gold tree.

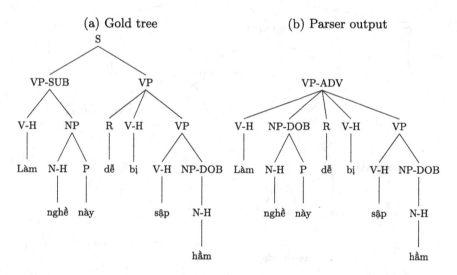

Fig. 3. Example illustrating an ambiguity between clause attachment and VP attachment in Vietnamese

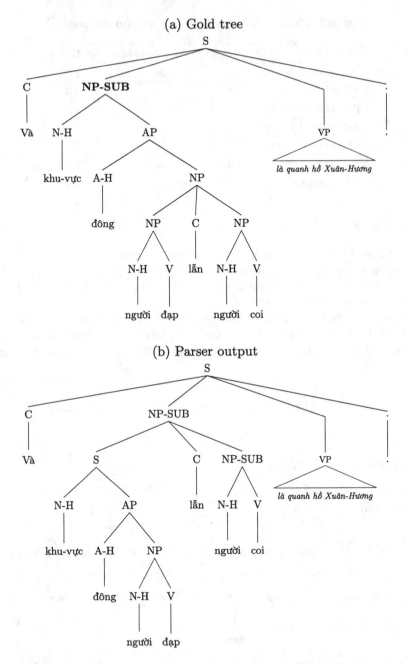

Fig. 4. Example illustrating an ambiguity of NP attachment in Vietnamese

Figure 4 shows a NP attachment error caused by ambiguous construction *"khu-vực đông người đạp lẫn người xem"*. This phrase is predicted for NP that contains 3 subphrases "S C NP" in while that NP contains 2 subphrases "N AP" in gold tree.

Figure 5 shows a PP attachment error caused by ambiguous construction *"của chúng"*. The *"của chúng"* phrase is predicted for PP that belongs to NP-SUB phrase while this phrase belongs to VP in gold tree.

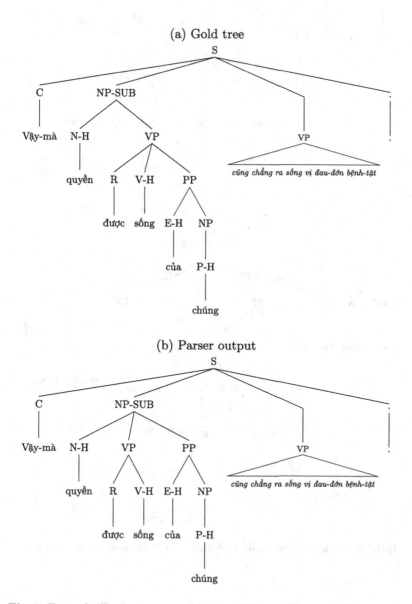

Fig. 5. Example illustrating an ambiguity of PP attachment in Vietnamese

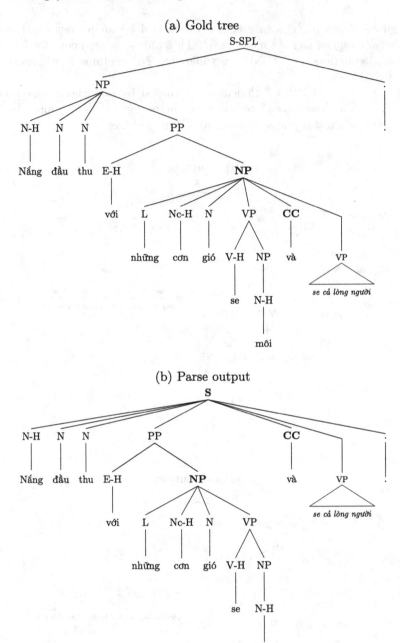

Fig. 6. Example illustrating an ambiguity of CC attachment in Vietnamese

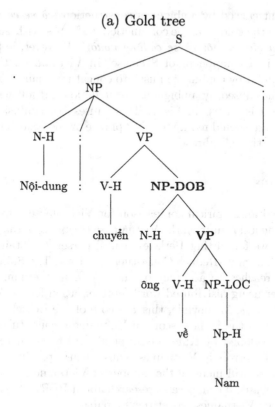

(a) Gold tree

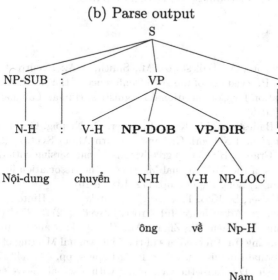

(b) Parse output

Fig. 7. Examples illustrating an ambiguity between coordination and subordinate construction in Vietnamese

CC attachment caused by ambiguous construction *"và se cả lòng người"* is shown in Fig. 6. In the gold tree, the conjunction *"và"* goes with *se cả lòng người* to NP *"những cơn gió se môi và se cả lòng người"*. However, in parser output, this conjunction is a component of S clause. In Vietnamese, this ambiguous construction is caused by conjunction used to concat two clauses or two phrases.

VP attachment caused by ambiguous construction VP following NP phrase and the head verb is shown in Fig. 7. This causes the confusion in deciding whether a VP phrase should modify the NP phrase or the head verb. Meanwhile, it should annotate the NP phrase.

4 Conclusions

We have presented an empirical comparison for Vietnamese constituency parsing. With experimental results on the Vietnamese treebank [9] that consist about 10,000 sentences, we found that Berkeley neural parser [6] obtains significantly higher F_1 score than Shift-Reduce Constituency Parser. The Shift-Reduce Constituency Parser reaches an accuracy of under 72% in Vietnamese. The accuracy of this parser using distributed word representation feature only increased slightly by about 0.2%. Meanwhile, this F_1 score of the Berkeley neural parser is over 80%, which is the highest score in Vietnamese constituency parser. We also use a tool developed by Kummerfeld et al. [19] to evaluate parsers. This tool helps categorize errors in Vietnamese constituency parsing.

In the future, we will increase the number of Vietnamese constituency sentences for Viettreebank. Then, we also research about BERT[3] to enhance further the accuracy of our Vietnamese constituency parser.

References

1. Dyer, C., Kuncoro, A., Ballesteros, M., Smith, N.A.: Recurrent neural network grammars. In: Proceedings of the 2016 Conference of the North American Chapter of the Association for Computational Linguistics: Human Language Technologies, pp. 199–209 (2016)
2. Kuncoro, A., Ballesteros, M., Kong, L., Dyer, C., Neubig, G., Smith, N.A.: What Do Recurrent Neural Network Grammars Learn About Syntax? (2017)
3. Coavoux, M., Crabbé, B.: Neural greedy constituent parsing with dynamic oracles. In: Proceedings of the 54th Annual Meeting of the Association for Computational Linguistics, vol. 1, Long Papers, pp. 172–182 (2016)
4. Vinyals, O., Kaiser, L., Koo, T., Petrov, S., Sutskever, I., Hinton, G.: Grammar as a foreign language. Adv. Neural Inf. Process. Syst. **2**, 2773–2781 (2015)
5. Zhu, M., Zhang, Y., Chen, W., Zhang, M., Zhu, J.: Fast and accurate shift reduce constituent parsing. In: Proceedings of the 51st Annual Meeting of the Association for Computational Linguistics, vol. 1: Long Papers, pp. 434–443 (2013)
6. Kitaev, N., Klein, D.: Constituency parsing with a self-attentive encoder. In: Proceedings of the 56th Annual Meeting of the Association for Computational Linguistics, vol. 1: Long Papers. Melbourne, Australia (2018)

---

[3] Bidirectional Encoder Representations from Transformers.

7. The, N.Q., Le Thanh, H.: Vietnamese syntactic parsing using the Lexicalized Probabilistic Context-free Grammar, pp. 9–10. FAIR Conference, Nha Trang, Vietnam, August 2007

8. Lam, D.B., Le Thanh, H.: Implementing a Vietnamese syntactic parser using HPSG. In: Proceedings of the ICT.rda Conference, Hanoi, Vietnam (2008)

9. Nguyen, P.T., Xuan, L.V., Nguyen, T.M.H., Nguyen, V.H., Le-Hong, P.: Building a large syntactically-annotated corpus of Vietnamese. In: Proceedings of the 3rd Linguistic Annotation Workshop, ACL-IJCNLP, Suntec City, Singapore, pp. 182–185 (2009)

10. Le-Hong, P., Nguyen, T.M.H., Nguyen, P.T., Roussanaly, A.: Automated extraction of tree adjoining grammars from a treebank for VietNamese. In: 10th International Conference on Tree Adjoining Grammars and Related Formalisms. Yale University, New Haven, USA (2010)

11. Le, H.T., Do, L.B., Pham, N.T.: Efficient syntactic parsing with beam search. The 2010 IEEE RIVF Conference, Nov. 01–04, 2010, Hanoi, Vietnam (2010)

12. Le-Hong, P., Roussanaly, A., Nguyen, T.M.H.: A syntactic component for vietnamese language processing. J. Lang. Modell. **3**(1), 145–184 (2015)

13. Nguyen, Q.T., Miyao, Y., Le, H.T.T., Ngan, L.T.N.: Challenges and solutions for consistent annotation of vietnamese treebank. In: Proceedings of the Language Resources and Evaluation Conference (2016)

14. Turian, J., Ratinov, L., Bengio, Y.: Word representations: a simple and general method for semi-supervised learning. In: Proceedings of ACL, pp. 384–394. Uppsala, Sweden (2010)

15. Le-Hong, P., Nguyen, T.-M.-H., Nguyen, T.-L., Ha, M.-L.: Fast dependency parsing using distributed word representations. In: Li, X.-L., et al. (eds.) PAKDD 2015. LNCS (LNAI), vol. 9441, pp. 261–272. Springer, Cham (2015). https://doi.org/10.1007/978-3-319-25660-3_22

16. Phuong, L.H., Thi Minh Huyên, N., Roussanaly, A., Vinh, H.T.: A hybrid approach to word segmentation of vietnamese texts. In: Martín-Vide, C., Otto, F., Fernau, H. (eds.) LATA 2008. LNCS, vol. 5196, pp. 240–249. Springer, Heidelberg (2008). https://doi.org/10.1007/978-3-540-88282-4_23

17. Zhu, M., Zhang, Y., Chen, W., Zhang, M., Zhu, J.: Fast and accurate shift-reduce constituent parsing. In: Proceedings of the 51st Annual Meeting of the Association for Computational Linguistics, pp. 434–443 (2013)

18. Peters, M.E., Neumann, M., Iyyer, M., Gardner, M., Clark, C., Lee, K., Zettlemoyer, L.: Deep contextualized word representations. In: Proceedings of the 2018 Conference of the North American Chapter of the Association for Computational Linguistics: Human Language Technologies (2018)

19. Kummerfeld, J.K., Hall, D., Curran. J.R., Klein, D.: Parser showdown at the wall street corral: an empirical investigation of error types in parser output. In: Proceedings of the 2012 Joint Conference on Empirical Methods in Natural Language Processing and Computational Natural Language Learning, pp. 1048–1059 (2012)

Antonyms-Synonyms Discrimination Based on Exploiting Rich Vietnamese Features

Bui Van Tan[1(✉)], Nguyen Phuong Thai[2], Pham Van Lam[3],
and Dinh Khac Quy[2]

[1] University of Economic and Technical Industries, Hanoi, Vietnam
bvtan@uneti.edu.vn
[2] VNU University of Engineering and Technology, Hanoi, Vietnam
thainp@vnu.edu.vn
[3] Institute of Linguistics, Vietnam Academy of Social Sciences, Hanoi, Vietnam
lampv.il@vass.gov.vn

Abstract. Antonymy and Synonymy are paradigmatic relations which are in the core problems of language. Distinguishing antonyms from synonyms is a key task to achieve high performance in natural language processing systems. Distinguishing between antonyms and synonyms is a hard problem because the co-occurrence distributions of the antonyms or synonyms tend to be highly similar. On the other hand, this issue has been thoroughly studied in English. However, it has not been effectively addressed for Vietnamese. Compared to English, Vietnamese has its own word-level characteristics that indicate the synonymous or antonymous relation. In this paper, we introduce a framework which exploits exhaustively special Vietnamese features to distinguish between antonyms from synonyms. We propose a deep neural network model (ViASNet) that can utilize not only lexico-syntactic information captured from the context of word pairs in a corpus but also its word-level features, and distribution features as well. The experimental results show that the proposed method is effective. Furthermore, our method achieved high performance in comparison to several the state of the art methods.

Keywords: Semantic relation · Antonym synonym distinction · Semantic relation classification

1 Introduction

Antonymy and synonymy are Paradigmatic relations that play an important role in the organization of the mental lexicon knowledge [3,5]. In which, the synonymy is a semantic relation between two words that map to the same meaning [13]. By definition, two words are synonymous each other if they have the same or nearly the same meaning in some or all senses. In contrast, antonymy is defined as the oppositeness between words [9]. Distinguishing between antonymy and

© Springer Nature Singapore Pte Ltd. 2020
L.-M. Nguyen et al. (Eds.): PACLING 2019, CCIS 1215, pp. 374–387, 2020.
https://doi.org/10.1007/978-981-15-6168-9_31

synonymy is important for NLP applications such as Machine Translation [22], Sentiment Analysis [11] and Information Retrieval [8].

Paradigmatic relations such as synonymy, antonymy and hypernymy are notoriously difficult to distinguish because the first-order co-occurrence distributions of the related words tend to be very similar across the relations [22, 25]. An example with regard to the sentence *The laptop/mainframe/computer is fast/slow*. In this sentence, *laptop* and *mainframe* are co-hyponyms, and *computer* a hypernym of them, *fast* and *slow* are antonyms, all of these words occur in identical contexts.

Prior studies on this problem can be classified into three approaches including pattern-based, vector semantics, and unsupervised measure. In pattern-based approaches, Lin et al. [8] present a method to retrieve synonyms from distributionally similar words. The method makes use of two patterns: *either X or Y*, and *from X to Y*. It relies on the hypothesis that *if two words X and Y appear in one of the patterns, they are very unlikely synonymous*. Mohammad et al. [11] indicated that Lin' Patterns have a low coverage for their antonym set. Turney el al. proposes a unified approach to analogies, synonyms, antonyms, and associations. The approach consists of the PairClass algorithm, a feature extraction algorithm, and supervised classification based on a Support Vector Machine. Sabine Schulte im Walde el al. [26] presents a pattern-based approach for distinguishing paradigmatic relations for German words. This work extracted lexico-syntactic patterns between the word pairs and computed pattern frequency vectors, then using a nearest-centroid to distinguish antonyms from synonyms.

In vector semantics approaches, models gain a low-dimension vector space such as Word2Vec, and GloVe, also known as word embedded models. Therein, each word is represented by a real-valued feature vector. An approach based on supervised word embeddings for the task of identifying antonymy is introduced by Ono et al. [20]. They proposed two models to learn word embeddings WE-T and WE-TD. Pham et al. introduced the multitask Lexical Contrast Model (mLCM), an extension of the effective Skip-gram method that optimizes semantic vectors to predict contexts. Nguyen et al. [16] proposed dLCE, an effective model exploiting external lexical resources. The lexical contrast information was integrated into the skip-gram model to learn word embeddings. This model successfully predicted degrees of similarity and identified antonyms and synonyms. According to another way, Mrksic et al. [12] proposed ATTRACT–REPEL model that uses synonymy and antonymy constraints drawn from lexical resources to inject into pre-trained word embedding models.

In unsupervised approaches, distributional measures are proposed to distinguish synonyms from antonyms in an unsupervised manner. Scheible et al. [23] show that adjectival synonyms and antonyms can be distinguished via a word space model by introducing a distributional measure. Another distributional measure, Enrico Santus [4] introduced APAnt model to distinguish antonymy from synonymy. The model is based on an observation that antonyms are often very similiar except in one dimension of meaning. For example [4], both *giant* and *dwarf* refer to a person, with a head, two legs and two feet, but their size is different. APAnt is an adaption of the Average Precision measure. It compares the N most salient contexts of a pair of antonyms and synonyms. This model is based on a hypothesis that synonyms share a number of salient contexts that are significantly higher than the ones shared by antonyms.

Main contributions of this paper are as follows:

– We proposed a deep neural network model exploiting multi-type features to distinguish antonyms from synonyms.
– We introduced a set of word-level patterns which are strong signals to determine the type of paradigmatic relations, and a set of Sino-Vietnamese synonymous word pairs as an essential resource to explore lexical semantic relations.
– We built a Vietnamese dataset for the antonyms-synonyms distinction task which covers a variety of relative circumstances, according to various criteria in Vietnamese.

2   Related Works

Pattern-based methods for inducing semantic relations between pairs of terms (x; y) consider the lexico-syntactic paths that connect the joint occurrences of x and y in a large corpus. A variety of methods have been proposed to distinguish antonyms from other relations [8,22]. These studies proved the effectiveness of the lexico-syntactic patterns. Fundel et al. [6] pointed out the shortest dependency paths (SDP) between two words in their co-occurrence context was shown to be informative for recognizing relations. Deep learning techniques showed good performance in encoding the lexico-syntactic patterns from syntactic parse trees. Several methods show improved performance by using recurrent neural networks (RNN) to process a dependency path edge-by-edge. Xu et al. (2015) apply a separate long short term memory network to capture the indicative information in such paths.

The study closest to our work is the one proposed by [17]. The authors hypothesize that antonymous word pairs co-occur with each other in lexico-syntactic patterns within a sentence more often than synonymous word pairs. This work presentes a pattern-based neural network model (ViASNet) that exploits lexico-syntactic patterns from syntactic parse trees for the task of distinguishing between antonyms and synonyms. The distance between related words along a syntactic path is used as a new feature. In comparison with the ViASNet model, our method has two differences. Firstly, LSTM is also used to capture the lexico-syntactic patterns. However, SDP is not used for the following reasons: (1). Vietnamese dependency parsers achieve much lower accuracy than English [15], (2). A unique feature of the Vietnamese language is that interpretation manners are often used in sentences. In that way, related words can be used to semantically complement for each other. Therefore some words not belonging to SDP can contain useful information. For example, in a co-occurrence context of *lịch_ sự* - *thô_ lỗ* as *"anh ấy là người **lịch_ sự** chứ không **thô_ lỗ**, cục_ cằn, vũ_ phu"* <he is a polite person rather than a rude, grumpy, brute.>, *thô_ lỗ* is a near-synonym of *cục_ cằn* and *vũ_ phu* but them is not in SDP. A bidirectional LSTM is used to encode an unabridged co-occurrence context as a vector denoted the lexico-syntactic patterns. Last but not least, besides advantages, methods based on SDP suffer from its own disadvantage of losing useful information. Can et al. (2019) claims that using the SDP may lead to the omission of useful information (i.e., negation, adverbs, prepositions, etc.). Secondly, our method exploits specific characteristics of Vietnamese captured by word-level patterns.

3   Proposed Method

In this section, we propose a framework that can exploit exhaustively special Vietnamese features to distinguish antonymy from synonymy. We first present Vietnamese word-level contrasted patterns (Sect. 3.1). Section 3.2 describes statistics of the local mutual information score on Vietnamese word pairs. Finally, we present the ViASNet architect which combines soft features and hard features to achieve high performance in this problem (Sect. 3.3). Our framework is demonstrated in Fig. 1.

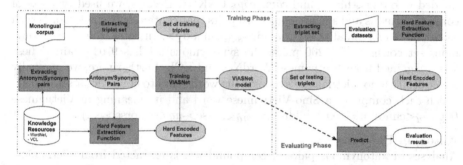

Fig. 1. Overview of the proposed framework.

In our framework, we first extract a set of antonymous/synonymous word pairs from Vietnamese WordNet [19] and Vietnamese Computational Lexicon (VCL)[1]. Then this set is used to extract a set of triplets (triplet set) from a corpus. A triplet is a tuple $<u, v, contextual\ words>$ co-occurred in a sentence. u and v are synonymous/antonymous words. *contextual words* includes the other words in the sentence (excluding u and v). Besides, three features of antonyms and synonyms including local mutual information, Vietnamese word-level patterns, and word similarity score are also extracted as hard encoded features. Both the triplet set and hard encoded features are fed into the ViASNet for training as shown in Fig. 1.

---

[1] https://vlsp.hpda.vn/demo/?page=vcl.

3.1   Vietnamese Word-Level Patterns

The Vietnamese language has special lexical characteristics [7,14,24]. Therefore, along with encoding lexico-syntactic patterns, we also aim to exploit exhaustively these special characteristics of Vietnamese. In the Vietnamese language, there are not only single words but also compound words. For example, compound words of VCL take a proportion of about 70%. Analyses of antonymous/synonymous word pairs show that the semantic relations of syllables are a strong indication of relation classifiability. Typical as if two words contain syllables that form an antonymous word pair then it tends to be an antonymous word pair strongly. For example, since *vui*<happy>-*buồn*<sad> is an antonymous pair then *vui_ vẻ*<merry>-*buồn_ rầu*<moody>is an antonymous pair too. Similarly, since *phụ*<dad>-*cha*<father> and *mẫu*<mom>-*mẹ*<mother> are synonymous word pairs, the pair of their compounds *cha_ mẹ*<parents>-*phụ_ mẫu*<parents> is also a synonymous word pair.

To analyze word-level patterns of pairs of Vietnamese compound words, each word u is split into two components u_1 and u_2 where u_1 is the first syllable and u_2 is the remainder of the word. For example, *dày_ cồm_ cộp*<thick> is divided into *dày*<thick> and *cồm_ cộp*<UNK>[2]. Besides, we used two sets of core antonymous and synonymous word pairs that contain only single words such as *to*<big>-*nhỏ*<small>, *buồn*<sad>-*sầu*<moody>. In total, the core antonymous set consists of 1,200 pairs, the synonymous set has 9,045 pairs. These sets are gained from Vietnamese WordNet and VCL. Furthermore, we manually select 3,003 Sino-Vietnamese synonymous word pairs to construct a set of pairs which each comprises a Sino-Vietnamese word and its meaning in Vietnamese (e.g., *tử*<children>-*con*<children>, *mã*<horse>-*ngựa*<horse>, ...).

Antonym Patterns
Analyses of antonymous pairs of compound words extracted from Vietnamese WordNet and VCL show that *in a pair, components of a word have a cross semantic correlation with components of the remaining word*. As a result, we found 8 word-level patterns which have the oppositeness between words. These

Fig. 2. An illustration of the semantic correlation between components across two words of a synonymous/antonymous pair.

---

[2] UNK denotes to Vietnamese syllables that don't have any corresponding English words.

patterns are indications of the antonymousness. Figure 2 represents a visual illustration of this correlation. In which (A-a) shows that the components of the two antonymous compound words are also antonymous single words. (A-b) show that their second components form to synonymous pairs.

8 patterns of antonymous words:

- $u_{1\_}*$ - $v_{1\_}*$, where u_1 is an antonym of v_1, remaining syllables do not hold any relation: *ác_ độc*<sinister> - *hiền từ*<gentle>.
- $u_1\_u_2$ - $v_1\_v_2$: where u_1 is an antonym of v_1, u_2 is a antonym of v_2: *cao_sang*<topping> - *thấp_ hèn*<lowly>.
- $u_1\_u_2$ - $v_1\_v_2$, where u_1 is an antonym of v_1, u_2 is a synonym of v_2: *hữu_ ích*<helpful> - *vô_ dụng*<futile>.
- u_1_x - v_1_x, where u_1 is an antonym of v_1, remaining syllables are identical: *ác_ tính*<malignant> - *lành_ tính*<benign>.
- $u_{1\_}*$ - v, where u_1 is an antonym of v: *bẩn_ thỉu*<dirty> - *sạch*<clean>.
- u - NW_u, where NW is a word in the negative word set {*vô, phi, bất, chẳng, không*}<all of these words have a meaning corresponding to the word *no* in English. It acts as a prefix of a word and makes the meaning of this word opposite>: *thành_ văn*<explicit> - *bất_ thành_ văn*<implicit>.
- $u_{1\_}*$ - NW_u_1, where NW is a word in the negative word set: *minh_ bạch*<transparent> - *bất_ minh*<ambiguous>.
- $u_1\_u_2$ - $v_1\_v_2$, where u_1 is an synonym of v_1, u_2 is a synonym of v_2: *bôi_ đen*<blacked out> - *tô_ hồng*<dyed pink>.

Synonym Patterns

Similarly to antonymous word pairs, Vietnamese synonymous word pairs can be recognized based on the relation between sub-word components across words. Analyses of synonymous pairs extracted from Vietnamese WordNet and VCL obtain 7 word-level patterns. These patterns are a strong indication of synonyms. Figure 2 represents a visual illustration of this correlation. In which (S-a) shows that components of two synonymous compound words are also synonymous single words and sharing identical components. (S-b) show that two synonym compound words consist of Sino-Vietnamese synonym pairs.

7 patterns of synonymous words:

- x_* - x_*, where two words share the first syllable x, remaining syllables do not hold any relation: *chật_ hẹp*<narrow> - *chật_ chội*<cramped>.
- u - u_*, where one word u is the first syllable of the other word: *dối*<cheat>- *dối_ trá*<scam>.
- u - $*\_u$, where one word u is the second syllable of the other word: *hiếm*<rare>- *khan_ hiếm*<scarce>.
- u_1_x - v_1_x, where u_1 is an synonym of v_1, the other syllables are identical: *kính_ trọng*<revere>- *tôn_ trọng*<respect>.
- $u_{1\_}*$ - $v_{1\_}*$, where u_1 is an synonym of v_1, remaining syllables do not hold any relation: *dối_ trá*<lie>- *lừa_ lọc*<cheat>.

- x_y - y_x, where syllables in a word is a permutation of the other word's: *giảm_ sút*<decrease>- *sút_ giảm*<cutback>.
- u_1-u_2 - v_1-v_2, where u_1 - v_2 and u_2-v_1 are synonyms of each other according to the Sino-Vietnamese meaning, respectively: *thính_ giả*<hearer>- *người_ nghe*<hearer>.

Table 1 shows statistics of word-level patterns in two datasets including ViCon introduced by [18] and ViAS-1000 represented in Sect. 4.

Table 1. The rate of the word-level pattern of the antonymy synonymy in Vietnamese.

Dataset	#Pairs	#Patterns	Percentage (%)
ViCon	1,398	622	52.6
ViAS-1000	1,000	453	54.7
Antonym pairs	24,347	15,126	37.9
Synonym pairs	156,847	88,382	44.7

3.2 Local Mutual Information

Several prior studies have pointed out that the co-occurrence of an antonym pair is more often than expected by chance. Charles and Miller [3] proposed a co-occurrence hypothesis that antonyms relatively frequently co-occur in the same sentence. But this hypothesis is also true for hypernyms, holonyms, meronyms and near-synonyms. Therefore, discriminating antonymous relation from other relation is still difficult. Strong co-occurrence is not a sufficient condition for detecting antonyms, but it is useful. Fellbaum [5] counted the appearance of antonyms in a corpus. Experimental results have shown that the antonym pairs have highly significant numbers of co-occurrences. Computing the Point Mutual Information (PMI) scores of antonyms, synonyms and random pairs from the Princeton WordNet show that the average PMI scores of antonyms were significantly higher than the ones of synonyms and random pairs [11].

We computed the Point Mutual Information and Lexicographers Mutual Information (LMI) [2] of 1000 synonym pairs and 1000 antonym pairs in a Vietnamese corpus (Vcorpus) with more than 500 million tokens. As a result, the average PMI score of antonymous pairs and synonymous pairs are 2.86 and 0.79, respectively. Similarly, the average LMI score of antonymous pairs and synonymous pairs are 72.78 and 13.24, respectively.

3.3 The ViASNet Architecture

In this section, we present a model to distinguish antonyms from synonyms for the Vietnamese (ViASNet). The proposed model makes use of a recurrent neural network with LSTM units to encode the context of word pairs. Figure 3 illustrates the ViASNet model.

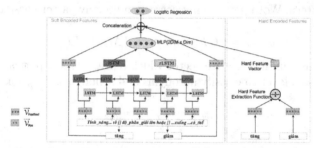

tính_năng tự_động sẽ [tăng] độ_phân_giải lên hoặc [giảm] độ_sáng xuống để hình_ảnh chiếu ra đẹp nhất có_thể
<the automatic function will [increase] the resolution or [decrease] the resolution to get the best image quality possible>

Fig. 3. The architecture of ViASNet for distinguishing between antonyms and synonyms.

Give a triplet $<u, v, contextual\ words>$. Each word is represented by a concatenation vector of fastText embedding and part of speech embedding. Contextual words $w_{1:n}$ is feed into the bidirectional LSTM module and the contextual vector defined as the following vector concatenation:

$$\vec{v}_{biLSTM} = \vec{v}_{lLSTM(w_{1:n})} \oplus \vec{v}_{rLSTM(w_{n:1})} \tag{1}$$

Where l/r represent distinct left-to-right/right-to-left word embeddings of the contextual words. Next, we apply the following non-linear function on the concatenation of the left and right context representations:

$$MLP(\vec{v}_{biLSTM}) = L_2(ReLU(L_1(\vec{v}_{biLSTM}))) \tag{2}$$

Where MLP stands for Multi Layer Perceptron, ReLU is the Rectified Linear Unit activation function, and $L_i(x) = W_i x + b_i$ is a fully connected linear operation. Let denote $\vec{v}_c$ is the representation vector for contextual words, it can be calculated as follows:

$$\vec{v}_c = MLP(\vec{v}_{biLSTM}) \tag{3}$$

Denote $\vec{v}_{soft}$ is the concatenation of $\vec{v}_c$ and representation vectors of $\vec{v}_u$ and $\vec{v}_v$. $\vec{v}_{soft}$ becomes a soft encoded feature:

$$\vec{v}_{soft} = \vec{v}_c \oplus \vec{v}_v \oplus \vec{v}_u \tag{4}$$

Next, we denote the hard encoded feature vector as $\vec{v}_{hard}$. It is a $k-dimensions$ vector which is generated from three values by vector construction function (VCF), including a similarity value calculated by the sigmoid function (SimScore), an LMI score (LMI), and a word-level pattern encoded value (WLP). Antonymy word-level patterns are encoded by value in the range from 0 to 8. In contrary, synonymy word-level patterns are encoded by value in the range from −7 to 0. Where values are different from 0 representing a corresponding word-level pattern. Word pairs do not correspond to any pattern assigned a value of 0.

$$\vec{v}_{hard} = VCF(SimScore, LMI, WLP) \tag{5}$$

finally, $\vec{v}_{triplet}$ is a unified feature vector which represents a triplet is fed into the logistic regression layer to classify antonyms and synonyms. $\vec{v}_{triplet}$ is calculated as follows:

$$\vec{v}_{triplet} = \vec{v}_{hard} \oplus \vec{v}_{soft} \tag{6}$$

4   Vietnamese Dataset Construction

In a prior study, ViCon dataset was introduced by [18]. It is a reliable Vietnamese dataset of lexical contrast pairs for evaluation of models that distinguishes similarity and dissimilarity. The dataset consists of 400 noun pairs, 400 verb pairs, and 600 adjective pairs randomized selected from VCL. Although, this dataset is built elaborately and analyzed meticulously. However, it still has several disadvantages. Firstly, that is the pairs is not covered circumstances of antonymy and synonymy according to various criteria in Vietnamese because the pairs in this dataset are randomly selected. Secondly, the ratio of the number of synonymous/antonymous word pairs through the part of speech (POS) does not correspond to the natural ratio. Last but not least, several pairs contain two words with different POS such as $xuất\_hiện_{verb}$ <appear> - $thiếu_{adjective}$ <absent>, $thanh\_danh_{noub}$<repute> - $ô\_nhục_{adjective}$ <infamous>.

In this section, we introduce a novel Vietnamese Dataset for the antonyms-synonyms distinction task (ViAS-1000 dataset[3]). Antonymous/synonymous pairs are manually selected from Vietnamese WordNet [19] and VCL. Firstly, We extracted all antonymous and synonymous word pairs according to the three part-of-speech categories: noun, verb and adjective. Thereafter, 400 antonymous pairs and 600 synonym pairs are carefully selected with respect to a number of constraints:

- Two words of a pair necessarily have jointly a POS.
- The number of pairs should correspond to the natural ratio according to the part-of-speech. In Vietnamese, the popularity degree of antonymy and synonymy decreases in the order of adjective, verb, noun. Besides, synonyms are also much more popular than antonyms.
- Antonymous/synonymous word pairs cover sufficiently the circumstances of the relation of Vietnamese such as single/compound words, subordinating/coordinating words, reduplicative-words, reverse-words.
- Datasets should contain words in different domains (e.g. sport, emotion, vehicle,...).

---

[3] https://github.com/BuiVanTan2017/ViASNet.

4.1   Selecting Antonymous Word Pairs

Antonymous word pairs are chosen to cover diverse circumstances of the relations as follows:

- Adjective pairs opposite meaning according to the scale.
- Adjective pairs opposite meaning according to the negation.
- Verb pairs opposite meaning according to the negation.
- Single syllable antonymous word pairs.
- Multi-syllable instance noun pairs opposite meaning.
- Multi-syllable abstract noun pairs opposite meaning.
- Subordinate compound adjective pairs opposite meaning.
- Coordinated compound reduplicative-adjective pairs opposite meaning.
- Subordinate compound verb pairs opposite meaning.

4.2   Selecting Synonymous Word Pairs

Synonymous word pairs are chosen to cover diverse circumstances of the relations as follows:

- Single-syllable synonym adjective pairs.
- Multi-syllable synonym adjective pairs.
- Synonym adjective pairs consist of a single-syllable word and a reduplicative multi-syllable word.
- Single-syllable synonym noun pairs.
- Synonym noun pairs consist of a single-syllable word and a reduplicative multi-syllable word.
- Synonym verb pairs consist of a single-syllable word and reduplicative multi-syllable word.

The number of pairs according to the relations and POS including 400 antonymous pairs (Adjective - 250, Verb - 90, Noun - 60), 600 synonymous pairs (Adjective - 250, Verb - 200, Noun - 150).

5   Experiments

5.1   Baseline Models

Word2Vec Model [10]: A well-known word representation method also known word embedding method introduced by Mikolov et al., 2013.

fastText Model [1]: This is a SGNS variant which builds word vectors as the sum of their constituent character n-gram vectors.

GloVe Model [21]: This model learns by constructing a co-occurrence matrix that basically counts how frequently a word appears in contexts introduced by Pennington et al., 2014. While Word2Vec tries to capture co-occurrence one window at a time, GloVe tries to capture the counts of overall statistics.

ATTRACT–REPEL Model [12]: This model is an algorithm for improving the semantic quality of word vectors by injecting constraints extracted from lexical resources.

Distributional Lexical-Contrast Embedding Model- dLCE [16]: This is an effective model which injects lexical contrast information into a pre-trained word embedding model.

5.2   Experimental Settings

Resources for Training

To train embedding models and extract the triplet set, we used a monolingual Vietnamese (Vcorpus[4]). Any words that appear less than 5 times are excluded. We used Python Vietnamese Toolkit[5] for tokenizing and pos-tagging the corpus. Vcorpus includes $28,767,455$ sentences with $567,473,442$ tokens, in which contains $1,603,857$ distinct tokens.

The set of the antonymous/synonymous word pairs are extracted from Vietnamese WordNet and VCL. We excluded from this set any pair in the test set (ViCon, ViAS-1000). As a result, the total number of antonymous and synonymous word pairs are $24,347$ and $156,847$, respectively. To obtain triplets, we pick all of the sentences in the Vcorpus which contains any pair of the antonymous/synonymous set. Note that, two words in a pair must be the same POS. A triplet set extracted that included $3,809,864$ antonymous-triplets and $23,500,420$ synonymous-triplets.

Parameter Settings

In training the ViASNet model, a word represented by a concatenation vector of a 300-dimensions fastText embedded vector and a 5-dimensions POS embedded vector. AdaGrad algorithm is used with the number of the epoch is 20, and the initial learning rate of AdaGrad was set to 0.05. For the baseline models, We used the Multi-layer Perceptron classifier to distinguish relations. The parameter of MLP is set as follows: $hidden\_layer\_sizes = (300, 2)$, $activation = $ 'relu', other network parameters are set by default.

In the experiments, ViASNet is the case the model only uses the hard encoded features. Meanwhile, ViASNet[+] is the case the model more exploiting the hard encoded features such as the similarity score, LMI score, and word-level patterns.

5.3   Experimental Results

We used the F-score (**F1**) as a primary evaluation metric. The F1 is the harmonic mean of precision (**P**) and recall(**R**). Table 2 shows the significant performance of our models in comparison to the baselines.

---

[4] Collected from vietnamese Wikipedia and https://baomoi.com/.
[5] https://github.com/trungtv/pyvi.

Table 2. Performance of the ViASNet model in comparison to the baseline models

Model	ViAS-1000			ViCon		
	P	**R**	**F1**	**P**	**R**	**F1**
GloVe	**0.85**	0.46	0.60	0.77	0.64	0.70
fastText	0.78	0.45	0.57	0.75	0.63	0.68
Word2Vec	0.76	0.48	0.59	0.74	0.63	0.68
dLCE	0.82	0.56	0.67	**0.79**	0.68	0.73
Attract-Repel	0.76	0.46	0.57	0.76	0.68	0.72
ViASNet	0.75	0.70	0.72	0.72	0.80	0.76
ViASNet$^+$	0.79	**0.73**	**0.76**	0.74	**0.82**	**0.78**

The experimental results in Table 2 have shown that baseline methods achieved a high precision score, but a low score of the recall. The ViASNet model has improved significantly on the recall score leading to it gained a high result on the F1-score in comparison to the baseline models. Furthermore, the ViASNet$^+$ model which exploits more the hard encoded features increased performance on both the precision and the recall scores.

6 Conclusion

This paper introduced a deep neural network model exhaustively exploited special Vietnamese features to distinguish between antonyms and synonyms. The proposed model not only utilizes lexico-syntactic information captured from co-occurrence contexts of word pairs in a corpus, but also word-level features, and distribution features as well. The word-level patterns were first introduced as a useful feature to recognize the relations. A set of Sino-Vietnamese synonymous word-pairs was introduced as essential resources to explore lexical semantic relations. Furthermore, the ViAS-1000 dataset was introduced which covered a variety of relative circumstances, according to various criteria in Vietnamese. Our proposed model significantly outperformed five baselines relying on previous work in term F1-score. In the future, we intend to apply the proposed method to discriminate antonymy also from other semantic relations and to automatically extract antonymy-related pairs for the ontologies and lexical resources construction.

Acknowledgments. This paper is a part of project number KHCN-TB.23X/13-18 which is led by Assoc. Prof. Ngo Thanh Quy and funded by Vietnam National University, Hanoi under the Science and Technology Program for the Sustainable Development of Northwest Region.

References

1. Bojanowski, P., Grave, E., Joulin, A., Mikolov, T.: Enriching word vectors with subword information. TACL **5**, 135–146 (2017). http://dblp.uni-trier.de/db/journals/tacl/tacl5.html#BojanowskiGJM17
2. Bordag, S.: A comparison of co-occurrence and similarity measures as simulations of context. In: Gelbukh, A. (ed.) CICLing 2008. LNCS, vol. 4919, pp. 52–63. Springer, Heidelberg (2008). https://doi.org/10.1007/978-3-540-78135-6_5
3. Charles, W.G., Miller, G.A.: Contexts of antonymous adjectives. Appl. Psycholinguist. **10**, 357–375 (1989)
4. Santus, E., Lu, Q., Lenci, A., Huang, C.: Unsupervised antonym-synonym discrimination in vector space. In: The First Italian Conference on Computational Linguistics CLiC-it (2014)
5. Fellbaum, C.: Co-occurrence and antonymy. Int. J. Lexicogr. **8**(4), 281–303 (1995)
6. Fundel, K., Küffner, R., Zimmer, R.: Relex - relation extraction using dependency parse trees. Bioinformatics **23**(3), 365–371 (2007)
7. Lam, P.V.: Study antonyms in Vietnamese. Doctoral thesis (2019)
8. Lin, D., Zhao, S., Qin, L., Zhou, M.: Identifying synonyms among distributionally similar words. In: Proceedings of the 18th International Joint Conference on Artificial Intelligence (IJCAI-2003), pp. 1492–1493 (2003)
9. Lyons, J.: Semantics. Cambridge University Press, Cambridge (1977)
10. Mikolov, T., Chen, K., Corrado, G., Dean, J.: Efficient estimation of word representations in vector space. arxiv:1301.3781 (2013)
11. Mohammad, S., Dorr, B.J., Hirst, G., Turney, P.D.: Computing lexical contrast. Comput. Linguist. **39**(3), 555–590 (2013)
12. Mrksic, N., et al.: Semantic specialisation of distributional word vector spaces using monolingual and cross-lingual constraints. CoRR (2017)
13. Murphy, L.M.: Semantic Relations and the Lexicon. Cambridge University Press, Cambridge (2003)
14. Ngo, T.V., Ha, T.L., Nguyen, P.T., Nguyen, L.M.: Combining advanced methods in Japanese-Vietnamese neural machine translation. In: KSE, pp. 318–322. IEEE (2018)
15. Nguyen, D.Q., Dras, M., Johnson, M.: An empirical study for Vietnamese dependency parsing. In: ALTA, pp. 143–149. ACL (2016)
16. Nguyen, K.A., Walde, S.S., Vu, N.T.: Integrating distributional lexical contrast into word embeddings for antonym-synonym distinction. In: The 54th Annual Meeting of the Association for Computational Linguistics, p. 454 (2016)
17. Nguyen, K.A., Walde, S.S., Vu, N.T.: Distinguishing antonyms and synonyms in a pattern-based neural network. In: EACL, pp. 76–85. Association for Computational Linguistics (2017)
18. Nguyen, K.A., Walde, S.S., Vu, N.T.: Introducing two Vietnamese datasets for evaluating semantic models of (dis-)similarity and relatedness. In: NAACL-HLT, pp. 199–205. Association for Computational Linguistics (2018)
19. Nguyen, P.T., Pham, V.L., Nguyen, H.A., Vu, H.H., Tran, N.A., Truong, T.T.H.: A two-phase approach for building Vietnamese wordnet. In: The 8th Global Wordnet Conference, pp. 259–264 (2015)
20. Ono, M., Miwa, M., Sasaki, Y.: Word embedding-based antonym detection using thesauri and distributional information. In: HLT-NAACL, pp. 984–989 (2015)
21. Pennington, J., Socher, R., Manning, C.D.: Glove: global vectors for word representation. In: EMNLP, vol. 14, pp. 1532–1543 (2014)

22. Roth, M., im Walde, S.S.: Combining word patterns and discourse markers for paradigmatic relation classification. In: Proceedings of the 52nd Annual Meeting of the Association for Computational Linguistics, pp. 524–530. Association for Computational Linguistics (2014)

23. Scheible, S., im Walde, S.S., Springorum, S.: Uncovering distributional differences between synonyms and antonyms in a word space model. In: IJCNLP, pp. 489–497. Asian Federation of Natural Language Processing/ACL (2013)

24. Tan, B.V., Thai, N.P., Lam, P.V.: Construction of a word similarity dataset and evaluation of word similarity techniques for Vietnamese. In: KSE, pp. 65–70. IEEE (2017)

25. Tan, B.V., Thai, N.P., Thuan, N.M.: Enhancing performance of lexical entailment recognition for Vietnamese based on exploiting lexical structure features. In: KSE, pp. 341–346. IEEE (2018)

26. im Walde, S.S., Köper, M.: Pattern-based distinction of paradigmatic relations for German nouns, verbs, adjectives. In: Gurevych, I., Biemann, C., Zesch, T. (eds.) GSCL 2013. LNCS (LNAI), vol. 8105, pp. 184–198. Springer, Heidelberg (2013). https://doi.org/10.1007/978-3-642-40722-2_19

Towards a UMLS-Integratable Vietnamese Medical Terminology

The Quyen Ngo[✉], My Linh Ha, Thi Minh Huyen Nguyen,
Thi Mai Huong Hoang, and Viet Hung Nguyen

VNU University of Science, Hanoi, Vietnam
{ngoquyenbg,huyenntm,hoangthimaihuong_t60}@hus.edu.vn, linhhm@vnu.edu.vn,
hungnv.hus@gmail.com

Abstract. Lexical resources play an essential role in text processing. In this paper, we present our work on the construction of a Vietnamese medical terminology integratable into the UMLS multilingual metathesaurus (*Unified Medical Language System*). The construction of the Vietnamese medical terminology is done by collecting terms from existing lexical sources on one hand, and by extracting terms from Vietnamese medical corpora on the other. In order to draw maximum benefit from the varied sources and corpora that can be collected, we have developed a set of tools adapted to the specificities of each of those resources, based upon proven techniques. This allows us to acquire consequent amounts of good quality mono- and bilingual medical terminology data.

Keywords: UMLS · Vietnamese medical terminology

1 Introduction

For many years, medical text processing has been an active research domain. The exploitation of biomedical data, especially clinical text from electronic medical records (EMR), allows the extraction of much valuable information in the medical field: research on drug interactions, side-effects of drugs, disease progression forecast, disease diagnosis and suggestions for treatment regimens, disease risk prediction, health care to individuals... This has led to the development of numerous software systems for medical text analysis, such as the open source Apache cTAKES (clinical Text Analysis and Knowledge Extraction System[1]) [25], MetaMap[2] [6], MedLEE (Medical Language Extraction and Encoding System) [9], CLAMP (Clinical Language Annotation, Modeling, and Processing) [26], or the open source system I2B2/tranSMART[3].

One of the most important linguistic resources for computer text processing is the lexical base. For the analysis of documents in a specialized domain,

---

[1] http://ctakes.apache.org/.
[2] https://metamap.nlm.nih.gov/.
[3] https://www.i2b2.org/.

© Springer Nature Singapore Pte Ltd. 2020
L.-M. Nguyen et al. (Eds.): PACLING 2019, CCIS 1215, pp. 388–399, 2020.
https://doi.org/10.1007/978-981-15-6168-9_32

the availability of a corresponding terminology or ontology contributes strongly to the performance of a system. In the medical field, different standard ontologies in the form of biomedical terminologies, classifications and vocabularies have been designed, developed and implemented. For example, the International Classification of Disease (ICD, [21]) centralizes different medical concepts such as diseases, injuries, procedures and symptoms; Logical Observation Identifiers Names and Codes (LOINC, [7]) contains standard codes used for determining medical laboratory test observations; Systematized Nomenclature of Medicine - Clinical Terms (SNOMED-CT, [23]) provides code, terminology, synonyms and definition about health and medical areas such as procedures, findings, diseases, micro-organisms, and substances. The most significant resource is the Unified Medical Language System[4] (UMLS). The UMLS project was promoted by the US National Library of Medicine. It is a collection of documents and software that gathers several sources of medical and biomedical vocabularies and standards, allowing many uses such as synchronizing healthcare information, drug names, biomedical terms, *etc.* between various hospital information systems, information retrieval and data mining in the biomedical domain.

The UMLS consists of three main components as described below.

– The first component is Metathesaurus – a multilingual terminology database composed of more than one million concepts acquired from hundreds of different biomedical vocabularies.
– The second one is the semantic network that includes 133 categories of biomedical concepts and 54 relations between these categories.
– The third one is the SPECIALIST Lexicon and Lexical Tools, providing lexical information and tools for biomedical text analysis.

This paper presents our work on the construction of a medical terminology, towards a Vietnamese medical knowledge base like the UMLS sources. In order to ensure the interoperability of our resource with others in a multilingual perspective, we attempt to make our terminology integratable into the UMLS multilingual metathesaurus. Section 2 of this document is dedicated to the presentation of this resource, which constitutes our reference.

Due to its extensiveness and high degree of specialization, building such a resource manually would be a daunting task, requiring the contribution of a large number of experts in various medical fields. We therefore turn to the possibility of building it by automatic acquisition from existing sources, be they less structured lexicons or full-text corpora. In recent years, the issue of automatic terminology acquisition, using statistical and machine-learning techniques, has also received a significant attention, and we present in Sect. 3 an overview of existing works and techniques from which we draw inspiration for our work.

Finally, Sect. 4 presents our work for building a Vietnamese medical lexicon and the first results obtained for this task, before the conclusions and discussions of the future work presented in Sect. 5.

---

[4] https://www.nlm.nih.gov/research/umls/.

2   The Unified Medical Language System

The project of building the Unified Medical Language System [1] (UMLS) started in 1986, under the direction of the National Library of Medicine (NLM), belonging to the National Institutes of Health in the US Department of Health and Human Services. The main purpose of this project is to produce a biomedical metathesaurus and lexical tools, which are essential resources for developing software systems capable of understanding texts in the domains of health and biomedicine. The three main components of UMLS, as mentioned above and shown in Fig. 1, are useful for the construction of software systems for managing and processing biomedical data, as well as for the research in medical informatics. The UMLS helps for efficiently retrieving, integrating and aggregating health data. The UMLS is not designed for a particular application but it is multipurpose. Many tools are developed for customizing the UMLS resource for specific purposes.

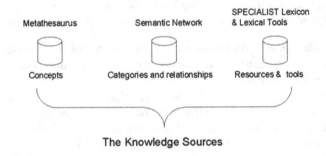

Fig. 1. Three components of the UMLS [1]

More precisely, the three main parts of the UMLS present the following characteristics:

- The Metathesaurus is a multi-purpose and multilingual vocabulary database. It is a large scale biomedical terminology providing biomedical concepts and their associated terms as well as their relationships. The concepts and terms are collected from hundreds of electronic sources such as code sets, dictionaries, thesauri and term lists acquired from clinical documents, health service billing, public health report or medical research. In its latest version (May 7, 2018), the metathesaurus contains:

  - 2,779,711 concepts associated to ∼6,901,000 terms from 207 sources in 20 languages (English, Basque, Czech, Danish, Dutch, French, German, Hebrew, Hungarian, Italian, Japanese, Korean, Latvian, Norwegian, Polish, Portuguese, Russian, Serbo-Croatian, Spanish, Swedish);
  - among these concepts, 2,051,406 are linked to 127 semantic types defined in the Semantic Network.

- Semantic Network: This component allows a consistent categorization of the biomedical and health-related concepts defined in the Metathesaurus and the relations between these concepts.
- SPECIALIST Lexicon and Lexical Tools: The SPECIALIST Lexicon includes orthographic, morphological and syntactic information of a large coverage of general and biomedical English lexical units from a variety of sources. This lexicon serves for the SPECIALIST NLP (*Natural Language Processing*) System which is constituted of three tools: SemRep for extracting semantic predication from biomedical text, MetaMap for discovering instances of UMLS concepts in text and the Lexical Tools for retrieving lexical variants.

While a NLP system for Vietnamese medical text processing should be developed separately due to linguistic specificities, it can for its operation rely upon a multilingual lexical resource like the UMLS metathesaurus. Our objective is to build a Vietnamese medical lexicon integratable into this metathesaurus.

3 Approaches for Term Acquisition

Three main approaches can be considered when building a medical terminology:

- Using existing specialized vocabularies. In the medical field, many existing resources are available: medical dictionary, list of drug names, list of common symptoms or diseases, list of medical ingredients, *etc.* However, these resources are far from exhaustive and can at best form a basis upon which to collect more complete data.
- Extracting terms from corpora. A widely used approach, which we will explore further here, consists in processing domain-specific corpora using statistical and machine learning techniques to extract specialized terms from them.
- Crowdsourcing. For long-term projects, it can be worthwhile to develop a platform that will attract human contributions. While we have not used this approach yet, we introduce it below since it constitutes a promising avenue for our future work.

3.1 Corpus-Based Term Acquisition

As presented in [30] and [14], there have been many studies in term recognition since the 1990s. Currently, we can classify the existing term extraction techniques into the following groups: linguistic methods (e.g. [5,11]), statistical methods (e.g. [3]), hybrid methods (e.g. [10]) and more recently machine learning methods (e.g. [4]).

Statistical and hybrid methods have in common the definition of a numerical "rating" that attemps to reflect the likelihood of a candidate term being an actual domain term. Some notable criteria used for term identification and extraction include:

- TF-IDF (*term frequency - inverse document frequency*): In [24], the authors examine the results of applying TF-IDF to determine what words in a corpus of documents might be more important. The authors use TF-IDF to compute a score for every word of a document, then take the words with high TF-IDF value as terms of documents.
- Word co-occurrence: In [19], the authors introduce an algorithm for extracting keywords from a document without making use of a corpus. First, they extract terms that are frequent in the document. Then, they extract and count the number of co-occurrences of each term and the frequent terms. A term which co-occurs more frequently with a special subset of frequent terms is probably a keyword in that document.
- Distributed representation: This technique consists in building weighted word vectors representing the contexts in which candidate terms appear. In [20], the authors introduce several extensions of distributed representation to improve not only the word vector quality but also the training speed. They then introduce an effective technique for identifying phrases in a text document. They also show that it is possible to learn good vector representations for millions of phrases.
- C-value: The hybrid method presented in [8] combines linguistic features and statistical methods, building a term extraction tool composed of two parts.

- The linguistic part relies on:

  * part-of-speech (POS) information;
  * linguistic filters for excluding tokens that are irrelevant for extraction;
  * a stop-list.

- The statistical part makes use of the C-value measure for computing and ranking the *termhood* of term candidates obtained after the linguistic part.

$$
\text{C-value}(a) = \begin{cases} log_2(|a|)f(a) \\ \qquad a \text{ is not nested} \\ log_2(|a|)(f(a) - \dfrac{1}{P(T_a)} * \sum_{b \in T_a} f(b)) \\ \qquad \text{otherwise} \end{cases}
$$

in which:

  * a is a candidate string
  * $f(.)$ is the measure of occurence frequency of a string in the corpus
  * T_a: is the set of extracted terms that contain a
  * $P(T_a)$ is the number of these candidate terms.

Once computed, only candidates whose C-value is above a certain threshold are kept.

Beside the above approaches for extracting monolingual terms from copora, another group of approaches consists in aligning parallel texts to extract bilingual terms [17,18,29]. Bond [2] chooses to extract bilingual terms from a monolingual corpus.

3.2 Crowdsourcing Approach

The term "Crowdsourcing"[5] was first coined in a Wired Magazine article by Jeff Howe (2006), who presented how businesses were using the Internet to "outsource work to the crowd". Researchers have used this approach in many fields and obtained many remarkable results. Keating and Furberg [13] present a framework of evidence-based methodology that describes concepts, assumptions and practices to assist researchers interested in conducting community research. Paniaqua [22] introduces many types of crowdsourcing, such as problem solving, learning paradigms, open innovation, new product development and so on. In addition, there are several models for crowdsourcing: paid-for market places, games with a purpose, volunteer based platforms. The advantages of crowdsourcing over expert based annotation have been discussed in [12,28].

In the medical field, there are many medical crowdsourced question and answering websites that serve for patients, doctors and society. In [16], the authors propose a system called MKE (*Medical Knowledge Extraction*) that is capable to extract automatically high quality semantic triples of entities from the noisy set of questions and their answers. The system can also estimate expertise for the doctors who give answers on the website. Another example of success in crowdsourcing is the game framework JeuxDeMots [15]. In this game, the players deal with a kind of word game that helps to score terms in French and their relations to obtain a wide and useful lexical-semantic network. As of December 2, 2018, they have gathered an impressive base of 3,328,617 terms with 257,860,581 annotated relations.

In the following section, we present the construction of a medical terminology for the Vietnamese language. We choose to implement various methods for acquiring medical terms and mapping these terms to the UMLS metathesaurus if possible.

4 Construction of a Medical Terminology for Vietnamese

In this section, we first investigate existing resources that can be directly used to collect Vietnamese medical terms. We then propose some corpus-based methods and experiments for Vietnamese medical term extraction on different corpora. For each topic, we also give discussions on the obtained results and future works.

4.1 Terminology Compilation from Existing Vocabularies

For this more direct approach to terminology building, we have been able to extract medical terms from the following sources:

- ICD (International Statistical Classification of Diseases and Related Health Problems): this document provide a coding of 11383 diseases into short codes which constitute a standard for medical research and practice. The current

---

[5] https://en.wikipedia.org/wiki/Crowdsourcing.

version used in Vietnam is the 10th Revision[6]. The ICD helps to insure that communications about patients between medical staff from different countries, areas, or even schools, as well as between different health departments, can be conducted without any ambiguity or risk of misunderstanding.

- Vietnamese Health Ministry list of approved drugs, containing 20,135 drug names. Of those, we have been able based on morphological similarity to map 7,146 drugs to the drug concepts in the UMLS, which contains 87,801 drugs. We could not automatically find the corresponding UMLS concepts for the remaining drugs based on word forms only.
- Vietnamese medical lexicon [27]: this dictionary contains about 4600 terms with concept definition. We have not yet exploited the term definitions in this lexicon, but in the future they could be useful for determining the semantic types of each term.

From the investigation of existing Vietnamese medical vocabularies and the UMLS structures, we can conclude that many things remain to be done to achieve a good and exchangeable medical terminology for Vietnamese:

- term acquisition and term translation in English;
- synonymous term detection;
- term categorization into semantic types and term relation determination.

In the following section, we focus on the problem of acquisition of Vietnamese medical terms and their English translations. The other points will be tackled in future works.

4.2   Corpus-Based Term Extraction

We first present here the corpora that we collected for experimentation, then introduce three different methods that we have developed and experimented for medical terminology extraction.

4.2.1   Medical Corpora Collection

Our first source of medical texts is the Wikipedia corpus[7]: this encyclopedia is composed of 1,187,840 articles in Vietnamese (retrieved in September 2018). A first issue, which we discuss further in the next subsection, is how to extract from this corpus the articles in the medical field.

The second source is websites in Vietnamese specialized in medical topics, and online medical journals. Many medical websites in Vietnamese offer good quality material, for example https://wikimed.vn, https://www.thuocbietduoc. com.vn, https://hellobacsi.com, *etc.* Concerning online medical journals, while the data would be of great interest, it is often technically difficult to access texts in a machine-readableformat. We therefore focus at this first stage of our work on articles from medical websites.

---

[6] http://123.31.27.68/ICD/ICD10.htm.

[7] https://www.wikipedia.org/.

Finally, a few websites offer English-Vietnamese multilingual health information documents, for example the WHO website. In our future work, these documents will be mined to extract aligned bilingual terms.

4.2.2 Monolingual and Bilingual Term Extraction Using the TF-IDF Based Method

In this work, we perform the extraction of monolingual and bilingual medical terms in the following main steps. We employ the medical website corpus as a reference to extract a first kernel of medical keywords that let us identify relevant Wikipedia articles.

1. Collecting medical documents: from medical websites, we can directly acquire medical articles. 43,925 articles have been thus collected.
2. Identifying a kernel set of medical terms: To do that, we make use of the TF-IDF measure for ranking words from articles on medical websites. 6023 words with highest ranks are selected to combine with the set of 4655 terms from the medical dictionary (Sect. 4.1) to get a kernel set of 9987 medical terms.
3. Identifying medical-related articles from Wikipedia: As mentioned in Sect. 4.2, for Wikipedia articles, we have to build an algorithm to identify medical-related articles. Once again, we employ the TF-IDF measure to extract keywords for each Wikipedia articles. We identify as a medical document any article whose keyword set contains at least 4 words in the medical kernel set, and whose number of medical keywords exceeds 10% of the number of keywords in that article.
4. The title of each Wikipedia article is considered as a term in our terminology database, for example the term "**Thuỷ đậu**" (Chicken Pox[8]).
5. In the last step, we extract the English-Vietnamese bilingual medical terms by using the Wikipedia API to extract the title of the corresponding English article (if any) for each selected Vietnamese Wikipedia article.

Thanks to this process, we are first able to add 6023 medical-related keywords to the medical terminology, followed by 1630 medicine-related Wikipedia articles in Vietnamese. Among these, English translation for 1409 terms are found, from which we have identified 1036 English terms present in the UMLS metathesaurus.

In the next subsection, we present the application of C-value algorithm, a hybrid method for extracting Vietnamese medical terms.

4.2.3 Monolingual Term Extraction Using C-Value Algorithm

To develop a C-value algorithm based tool for Vietnamese term extraction, we need to elaborate a set of regular expressions for filtering word phrases which are medical term candidates.

---

[8] https://vi.wikipedia.org/wiki/Thuy_dau.

Table 1. Survey of medical terminology

The term structure	Quantity	Percent
Noun Phrase	1514	64,63%
Verb Phrase	748	31,92%
Adjective Phrase	81	3,45%
Total	*2343*	*100%*

Table 1 show a statistic of term structures according to our small survey on 2343 Vietnamese medical terms. It can be seen that noun phrases make up the largest part of this collection. We have therefore chosen for the work here presented to focus on extracting terms that are noun phrases. A set of rules is written to capture the noun phrase structure in Vietnamese.

A Vietnamese noun phrase is of the form "Pre + Center + Post", "Pre + Center", or "Center + Post". Since the function of a term is to identify a concept clearly and efficiently, terms usually have a concise structure; therefore, almost all terms considered follow the "Center + Post" structure. The Post part is usually made up of adjectives, nouns, verbs or prepositions directly modifying the central noun, without relative pronouns. Due to the difficulty of efficiently distinguishing a noun phrase containing a verb from a full sentence, we restrict ourselves to the following patterns for candidate terms:

- Noun Noun*
- Noun (Adjective—Noun)* Adjective
- Noun Prep (Adjective—Noun)*.

For each candidate thus identified, we compute the C-value as presented in the previous section, before selecting the highest-rated terms for addition into the terminology. The experiment results are shown in Table 2, where we can observe the number of candidates and extracted terms from two medical corpora crawled from two websites hellobacsi.com and yhoctonghop.com after some preprocessing steps (POS tagging, word segmentation, . . .).

Table 2. Term extraction results

Website	# articles	# candidates	# extracted terms
hellobacsi.com	3328	74888	2775
yhoctonghop.com	45	8366	650

The extracted terms are evaluated by 4 persons, reaching an average consensus of 57.86%, with an average of 96.51% of the total terms accepted.

These results are very encouraging for an extended application of this C-value based method to different types of word phrase structures.

4.2.4 Bilingual Term Extraction from Monolingual Medical Corpus

This last extraction technique is based upon the constatation that in Vietnamese medical documents, new or unusual terms often appear accompanied by their English version, for clarification. To take advantage of this feature, we employ a hybrid method taking into account:

- patterns reflecting the expected syntactic structure of a Vietnamese term followed by its English counterpart, typically in brackets,
- the length ratio between the Vietnamese candidate term and supposed translation,
- a measure of mutual information between components of the Vietnamese candidate term,
- several heuristics to weed out undesirable words.

The developed method allows to obtain 626 bilingual terms from our medical corpus. We expect to apply this method on articles from medical journals, which generally show a higher frequency of English term quotes.

5 Conclusions

This paper has shown an overview about the lexical resources in the biomedical field and presented our work toward the construction of a medical terminology for Vietnamese, with the aim of integrating it into the UMLS multilingual metathesaurus. We have combined several approaches to extract information from a variety of sources, developing several techniques for this task:

- Collecting terms from available vocabularies (ICD-10, drug lists, *etc*). Beside monolingual terms, 11383 diseases and 7146 drugs are mapped to English corresponding concepts in the UMLS metathesaurus.
- Extracting terms from articles on medical websites and on Wikipedia, using a TF-IDF based approach. 6023 keywords from medical websites and 1640 terms from Wikipedia are extracted. 1036 Wikipedia terms in Vietnamese are mapped to English concepts in the UMLS.
- Building a tool based on the C-value hybrid approach to extract 3245 noun phrase as medical terms from 3373 online medical articles.
- Implementing a tool for extracting bilingual terms from monolingual corpus, thus obtaining 626 bilingual terms from articles on medical websites.

The experimental results show that we can apply these approaches to get highly reliable terms. We have equally shown that there remain many problems to resolve in the future, as the construction of such a terminology base is a truly ambitious enterprise. In the next steps, we will continue to collect monolingual data and apply the tested methods on all the collected data. Another important task to do is to acquire English Vietnamese bilingual datasets and develop tools for extracting bilingual terms from these datasets.

References

1. Bodenreider, O.: The unified medical language system (UMLS): integrating biomedical terminology. Nucleic Acids Res. **32**, D267–D270 (2004). https://doi.org/10.1093/nar/gkh061
2. Bond, F., Chang, Z., Uchimoto, K.: Extracting bilingual terms from mainly monolingual data. In: 14th Annual Meeting of the Association for Natural Language Processing, Tokyo, Japan (2008)
3. Church, K.W., Hanks, P.: Word association norms, mutual information, and lexicography. Comput. Linguist. **16**(1), 22–29 (1990). https://www.aclweb.org/anthology/J90-1003
4. Conrado, M., Pardo, T., Rezende, S.: A machine learning approach to automatic term extraction using a rich feature set. In: Proceedings of the 2013 NAACL HLT Student Research Workshop, pp. 16–23. Association for Computational Linguistics, Atlanta, Georgia (June 2013). https://www.aclweb.org/anthology/N13-2003
5. Daille, B.: Conceptual structuring through term variations. In: Proceedings of the ACL 2003 Workshop on Multiword Expressions: Analysis, Acquisition, and Treatment, no. 1, pp. 9–16 (2003)
6. Demner-Fushman, D., Rogers, W., Aronson, A.: Metamap lite: an evaluation of a new java implementation of metamap. J. Am. Med. Inform. Assoc.: JAMIA **24**, 841–844 (2017). https://doi.org/10.1093/jamia/ocw177
7. Forrey, A.W., et al.: Logical observation identifier names and codes (LOINC) database: a public use set of codes and names for electronic reporting of clinical laboratory test results. Clin. Chem. **42**(1), 81–90 (1996)
8. Franzi, K., Ananiadou, S.: The C/NC value domain independent method for multiword term extraction. J. Nat. Lang. Process. **6**, 145–180 (1999)
9. Friedman, C.: Towards a comprehensive medical language processing system: methods and issues. In: Proceedings: A Conference of the American Medical Informatics Association/AMIA Annual Fall Symposium. AMIA Fall Symposium, vol. 4, pp. 595–599 (February 1997)
10. Hliaoutakis, A., Zervanou, K., Petrakis, E.: The AMTEx approach in the medical document indexing and retrieval application. Data Knowl. Eng. **68**, 380–392 (2009). https://doi.org/10.1016/j.datak.2008.11.002
11. Justeson, J.S., Katz, S.M.: Technical terminology: some linguistic properties and an algorithm for identification in text. Nat. Lang. Eng. **1**(1), 9–27 (1995). https://doi.org/10.1017/S1351324900000048
12. Karën, F., Gilles, A., Bretonnel, C.K.: Amazon mechanical turk: gold mine or coal mine? Comput. Linguist. **37**(2), 413–420 (2011). https://doi.org/10.1162/COLI_a_00057
13. Keating, M., Furberg, R.D.: A methodological framework for crowdsourcing in research. In: Proceedings of the 2013 Federal Committee on Statistical Methodology (FCSM) Research Conference (2013)
14. Korkontzelos, I., Klapaftis, I.P., Manandhar, S.: Reviewing and evaluating automatic term recognition techniques. In: Nordström, B., Ranta, A. (eds.) GoTAL 2008. LNCS (LNAI), vol. 5221, pp. 248–259. Springer, Heidelberg (2008). https://doi.org/10.1007/978-3-540-85287-2_24
15. Lafourcade, M.: Making people play for lexical acquisition with the JeuxDeMots prototype. In: 7th International Symposium on Natural Language Processing, SNLP 2007, Pattaya, Chonburi, Thailand, p. 7 (December 2007). https://hal-lirmm.ccsd.cnrs.fr/lirmm-00200883

16. Li, Y., et al.: Extracting medical knowledge from crowdsourced question answering website. IEEE Trans. Big Data **6**, 1–1 (2016). https://doi.org/10.1109/TBDATA.2016.2612236

17. Liu, J., Morin, E., Saldarriaga, S.P.: Towards a unified framework for bilingual terminology extraction of single-word and multi-word terms. In: Proceedings of the 27th International Conference on Computational Linguistics (COLING) (2018)

18. Macken, L., Lefever, E., Hoste, V.: TExSIS: bilingual terminology extraction from parallel corpora using chunk-based alignment. Terminology **19**, 1–30 (2013). https://doi.org/10.1075/term.19.1.01mac

19. Matsuo, Y., Ishizuka, M.: Keyword extraction from a single document using word co-occurrence statistical information. Int. J. Artif. Intell. Tools **13**, 157–169 (2003)

20. Mikolov, T., Sutskever, I., Chen, K., Corrado, G., Dean, J.: Distributed representations of words and phrases and their compositionality. In: Proceedings of the 26th International Conference on Neural Information Processing Systems, NIPS 2013, vol. 2, pp. 3111–3119. Curran Associates Inc., USA (2013). http://dl.acm.org/citation.cfm?id=2999792.2999959

21. Organization, W.H.: International Statistical Classification of Diseases and Related Health Problems. Tenth Revision, vol. 2 (2010). https://www.who.int/classifications/icd/ICD10Volume2_en_2010.pdf

22. Paniagua, J., Korzynski, P.: Social Media Crowdsourcing, pp. 1–5. Springer, New York, New York, NY (2017)

23. Patrick, R., Julien, G., Christian, L., Antoine, G.: Automatic medical encoding with SNOMED categories. BMC Med. Inform. Decis. Mak. **8**, S6 (2008)

24. Salton, G., Buckley, C.: Term-weighting approaches in automatic text retrieval. Inf. Process. Manag. **24**(5), 513–523 (1988). https://doi.org/10.1016/0306-4573(88)90021-0

25. Savova, G., et al.: Mayo clinical text analysis and knowledge extraction system (cTAKES): architecture, component evaluation and applications. J. Am. Med. Inform. Assoc.: JAMIA **17**, 507–13 (2010). https://doi.org/10.1136/jamia.2009.001560

26. Soysal, E., et al.: CLAMP - a toolkit for efficiently building customized clinical natural language processing pipelines. J. Am. Med. Inform. Assoc. **25**, ocx132 (2017). https://doi.org/10.1093/jamia/ocx132

27. Trieu, N.Q., Song, P.: Medical Encyclopedia of Vietnam. Medical Publishing House One Member Company Limited, Ha Noi (2011)

28. Wang, A., Hoang, C.D.V., Kan, M.Y.: Perspectives on crowdsourcing annotations for natural language processing. Lang. Resour. Eval. **47**(1), 9–31 (2013)

29. Yang, W., Yan, J., Lepage, Y.: Extraction of bilingual technical terms for Chinese-Japanese patent translation. In: Proceedings of the NAACL Student Research Workshop, pp. 81–87. Association for Computational Linguistics, San Diego (June 2016). https://doi.org/10.18653/v1/N16-2012. https://www.aclweb.org/anthology/N16-2012

30. Zhang, Z., Iria, J., Brewster, C., Ciravegna, F.: A comparative evaluation of term recognition algorithms. In: LREC 2008 (2008). http://www.lrec-conf.org/proceedings/lrec2008/pdf/538_paper.pdf

Vietnamese Word Segmentation with SVM: Ambiguity Reduction and Suffix Capture

Duc-Vu Nguyen[1,3(✉)], Dang Van Thin[1,3], Kiet Van Nguyen[2,3], and Ngan Luu-Thuy Nguyen[2,3]

[1] Multimedia Communications Laboratory, University of Information Technology, Ho Chi Minh City, Vietnam
{vund,thindv}@uit.edu.vn
[2] University of Information Technology, Ho Chi Minh City, Vietnam
{kietnv,ngannlt}@uit.edu.vn
[3] Vietnam National University, Ho Chi Minh City, Vietnam

Abstract. In this paper, we approach Vietnamese word segmentation as a binary classification by using the Support Vector Machine classifier. We inherit features from prior works such as n-gram of syllables, n-gram of syllable types, and checking conjunction of adjacent syllables in the dictionary. We propose two novel ways to feature extraction, one to reduce the overlap ambiguity and the other to increase the ability to predict unknown words containing suffixes. Different from UETsegmenter and RDRsegmenter, two state-of-the-art Vietnamese word segmentation methods, we do not employ the longest matching algorithm as an initial processing step or any post-processing technique. According to experimental results on benchmark Vietnamese datasets, our proposed method obtained a better F_1-score than the prior state-of-the-art methods UETsegmenter, and RDRsegmenter.

Keywords: Vietnamese natural language processing · Word segmentation · Pos tagging

1 Introduction

Word segmentation is an essential task in Vietnamese natural language processing, which has a significant impact on higher processing levels [1,3,8]. Unlike English, white spaces in Vietnamese written text can function as a syllable separator or a word separator. For example, the Vietnamese string "hiện đại hóa đất nước" (modernize$_{hiện\_đại\_hoá}$ country$_{đất\_nước}$), which consists of five syllables, is segmented into "hiện_đại_hoá đất_nước". Underscores denote the white spaces which function as syllable separator, and white spaces are used for word separation. Vietnamese word segmentation can be considered as a binary classification problem with two classes: underscore and white-space [12].

© Springer Nature Singapore Pte Ltd. 2020
L.-M. Nguyen et al. (Eds.): PACLING 2019, CCIS 1215, pp. 400–413, 2020.
https://doi.org/10.1007/978-981-15-6168-9_33

Vietnamese is an isolated language and every Vietnamese word has exactly one form [4]. Vietnamese words are constituted by one or more syllables. According to the statistics reported in [4], and [14], about 16% of Vietnamese words are single-syllable words and 71% are two-syllable words. Single-syllable words account for about 81% of Vietnamese syllables, which means 19% syllables are not meaningful when standing alone. The string "loại hình phạt" (3 syllables) can be segmented as "loại_hình phạt" (type$_{loại\_hình}$ penalize$_{phạt}$) or "loại hình_phạt" (type$_{loại}$ penalty$_{hình\_phạt}$). This phenomenon is called "overlap ambiguity involving three consecutive syllables" by the authors in [4]. All of the above have created challenges in Vietnamese word segmentation [13].

We have an observation that solving overlap ambiguity is essential for the Vietnamese word segmentation task. The authors in [4] proposed the ambiguity resolver, which uses a bi-gram language model. Their proposal has slightly improved the Vietnamese word segmentation result. Additionally, the binary classifier for the Vietnamese word segmentation trained by the authors in [14] still causes overlap ambiguity cases. They used rules based on the dictionary and threshold for the classifier in the post-processing phase to handle overlap ambiguities. Experimental results on the benchmark Vietnamese treebank show that the approach of the authors in [14] outperforms the previous state-of-the-art method of the authors in [4]. Therefore, we decided to inspire the idea from the authors in [14] in handling overlap ambiguities. However, we have assumed how the performance of our method changes when using feature templates to reduce overlap ambiguity cases without post-processing.

From a different point of view, the authors in [7] proposed affixes features as a part of the rich feature set in their Vietnamese POS tagging method. Additionally, the authors in [5] utilized potential affixes to improve the performance of unknown words (accuracy of 80.69% on Vietnamese POS tagging task of Vietnamese treebank [12]). In practice, we can not perform part-of-speech (POS) tagging for unknown words if these unknown words can not be constituted by machine annotated word segmentation. Therefore, we decide to study the impact of affixes on the performance of word segmentation. We approach Vietnamese word segmentation with a uni-directional model in which labels are predicted from left to right of a sentence based on a syllable window. Because those labels from the left hand have been predicted, we can utilize information of suffixes to improve Vietnamese word segmentation.

In this paper, we propose a feature-based method using SVM classifier to solve the Vietnamese word segmentation task. Our method considers Vietnamese word segmentation as a binary classification with two classes: underscore and white-space [14], in which a majority of feature templates are inherited from the research of the authors in [8,14]. Two novel feature templates in our method are to reduce ambiguity cases and capture unknown words containing suffixes. Our proposed method obtained better F_1-score than the previous state-of-the-art methods JVnSegmenter [8], vnTokenizer [4], DongDu [6], UETsegmenter [14], and RDRsegmenter [9] measured on the Vietnamese treebank [12] for Vietnamese word segmentation task. Additionally, we used VnMarMoT [10] on the result of our word segmentation method. On the benchmark Vietnamese treebank [12], we achieved result better F_1-score than previous state-of-the-art result [10] on Vietnamese POS Tagging task when using predicted segmentation instead of gold segmentation.

2   Our Approach

In this section, we first model the word segmentation task. Next, we concentrate on the most critical part of our paper, which is the features extraction phase for the SVM classifier.

2.1   Problem Representation

In the early days of the research on Vietnamese word segmentation, the authors in [1] considered Vietnamese word segmentation as a stochastic transduction problem. They represented the input sentence as an unweighted Finite-State Acceptor (FSA). Recently, the syllable-based and white-space-based representation have been two typical ways of modeling the Vietnamese word segmentation task. The authors in [8] presented the syllable-based representation. In syllable-based representation, three labels B_W, I_W, and O_W are used to indicate syllables that begin a word, syllables inside a word, and syllables outside a word, respectively. Syllables outside a word are punctuation marks such as full stops, commas, question marks, semicolons, and brackets. The authors in [12] presented the white-space-based representation. In this representation, computers are expected to differentiate two types of white space: one appears in between two syllables of the same word, denoted by an underscore; the other separates two different words, denoted by a white space.

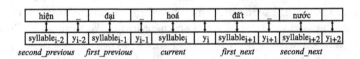

Fig. 1. Example of five-syllable window. In this diagram, the string "hiện đại hóa đất nước" (modernize$_{hiện\_đại\_hoá}$ country$_{đất\_nước}$) consisting of five syllables.

We decided to use white-space-based representation for our Vietnamese word segmentation method because of its clarity. In our approach, we assign underscore or white space labels for each syllable from left to right of the input sentence by utilizing features in the window of five syllables from the current syllable. An example is given in Fig. 1, in which the current syllable is syllable$_i$ ("hoá"), and it needs to be classified. The gold label of syllable$_i$ is y$_i$ (white space). The five-syllable window of the current syllable contains syllable$_{i-2}$ ("hiện"), syllable$_{i-1}$

("đại"), syllable$_i$ ("hoá"), syllable$_{i+1}$ ("đất"), and syllable$_{i+2}$ ("nước"). Additionally, we can utilize previous labels y_{i-1}, y_{i-2} and so on, for feature extraction of the current syllable.

2.2 Feature Extraction

To represent information of each syllable of the input sentence, we use the count vectorization technique. We divide the extracted features into four groups (four-vectors), which are baseline, more-than-four-syllable word, ambiguity reduction, and suffix feature. To obtain only one vector for the current syllable, we concatenated these four vectors.

We would like to introduce some utility operators and functions that we use to present feature templates for Vietnamese word segmentation. Firstly, the f_i symbol represents a function that returns the lowercase-simplified form of syllable$_i$. Secondly, $f_{i:i+k+1}$ returns the concatenation of lowercase-simplified forms of adjacent syllables from syllable$_i$ to syllable$_{i+k}$ with white-space characters between them. For example given five-syllable window in Fig. 1, the value of f_i symbol is "hoá" and value of $f_{i-1:i+2}$ symbol is "đại hoá đất". Besides, we should take syllable types into account for feature extraction. In our research, we inherit from [14] four syllable types: "lower", "upper", "all upper", and "other", which correspond to the following cases: the syllable has all lowercase letters; the syllable has an upper-case initial letter; the syllable has all upper-case letters; and the syllable is a number or other things. In a similar manner as f_i and $f_{i:i+k+1}$, we use t_i and $t_{i:i+k+1}$ symbols for types of syllables. Lastly, range(i, i+k+1) returns the list of integers ranging from i to i+k : (i, i+1, ..., i+k).

2.2.1 Baseline Features

Table 1. Baseline feature templates for word segmentation.

No.	Templates
1	$\{f_j$ for j in **range**(i−2, i+3)$\}$
2	$\{f_{j:j+2}$ for j in **range**(i−2, i+2)$\}$
3	$\{(i−j)$ for j in **range**(i−2, i+2) if **inVNDict**($f_{j:j+2}$)$\}$
4	$\{(i−j)$ for j in **range**(i−2, i+1) if **inVNDict**($f_{j:j+3}$)$\}$
5	$\{(i−j)$ for j in **range**(i−3, i+1) if **inVNDict**($f_{j:j+4}$)$\}$
6	$\{t_{j:j+2}$ for j in **range**(i−2, i+2) if $(t_j \neq$ 'LOWER' and $\neg$**inVNDict**($f_{j:j+2}$))$\}$
7	$\{t_{j:j+3}$ for j in **range**(i−2, i+1) if $(t_j \neq$ 'LOWER' and $\neg$**inVNDict**($f_{j:j+3}$))$\}$
8	$(t_i = t_{i+1} =$ 'LOWER' and $f_i = f_{i+1}$)?
9	$(t_i = t_{i+1} =$ 'UPPER' and **isVNFamilyName**(f_i))?
10	$(t_i = t_{i+1} =$ 'UPPER' and **isVNMiddleName**(f_i))?

Table 1 shows all feature templates of the baseline feature group. We have introduced f_i, $f_{i:i+k+1}$, t_i, $t_{i:i+k+1}$ symbols, and range(i, i+k+1) function in the last paragraph of Subsect. 2.2, for convenience. In Table 1, inVNDict($f_{i:i+k+1}$) returns true if and only if $f_{i:i+k+1}$ is in Vietnamese word dictionary; isVNFamilyName(f_i) returns true if and only if f_i is a Vietnamese family name; isVNMiddleName(f_i) returns true if and only if f_i is a Vietnamese middle name. Notably, we used the Vietnamese words dictionary[1], list of Vietnamese family and middle names from research of the authors in [9].

In this baseline feature group, we inherit two ways of extracting feature with five-syllable window for current syllable from [14], which are the lowercase form of syllables (the first and second templates in Table 1) and syllable types (the sixth and seventh templates in Table 1). We also inherit from [14] the following features: full-reduplicative word (the eighth template), Vietnamese family name (the ninth template), Vietnamese middle name (the tenth template). Additionally, we check if a conjunction of two up to four adjacent syllables in a window of seven syllables exists in the dictionary (the third, fourth, and fifth templates). These feature templates are inherited from the research of the authors in [8] except the fifth template.

2.2.2   More-than-Four-Syllable Word Features

We proposed this feature template based on the research of the authors in [8] to capture the signal of whether the center syllable is a unit of a more-than-four-syllable word. We expect the classifier can predict more-than-four-syllable words although they are rare in Vietnamese.

Table 2. Feature templates for capturing five up to nine syllables words.

No.	Templates
1	$\{(i-j)$ **for j in range**$(i-4, i+1)$ **if inVNDict**$(f_{j:j+5})\}$
2	$\{(i-j)$ **for j in range**$(i-5, i+1)$ **if inVNDict**$(f_{j:j+6})\}$
3	$\{(i-j)$ **for j in range**$(i-6, i+1)$ **if inVNDict**$(f_{j:j+7})\}$
4	$\{(i-j)$ **for j in range**$(i-7, i+1)$ **if inVNDict**$(f_{j:j+8})\}$
5	$\{(i-j)$ **for j in range**$(i-8, i+1)$ **if inVNDict**$(f_{j:j+9})\}$

We recognize that words are containing up to five to nine syllables (we have shown the distribution of unique words according to lengths in Table 4 of Subsect. 3.1). Thus, we only take into account the concatenation of adjacent syllables with length ranging from five to nine. Lastly, we check all concatenations in the dictionary (the first, second, third, fourth, and fifth templates in Table 2).

2.2.3   Ambiguity Reduction Features

We assume that some syllables tend not to combine with other syllables in constituting a two-syllable word. For the convenience of presentation, we call

---

[1] https://github.com/datquocnguyen/RDRsegmenter/blob/master/VnVocab.

the syllable with such a tendency "a separable syllable". We define a separable syllable as a syllable where the number of occurrences a_i of one-syllable words constituted by that syllable is higher than the number of occurrences b_i of more-than-one-syllable words beginning with that syllable.

a) **syllable$_i$ is a separable syllable:**

syllable$_i$	?	syllable$_{i+1}$	?	syllable$_{i+2}$	?	syllable$_{i+3}$	?	syllable$_{i+4}$
current		*first_next*		*second_next*		*third_next*		*fourth_next*

b) **syllable$_{i-1:i+2}$ can be a word:**

syllable$_{i-1}$	_	syllable$_i$	?	syllable$_{i+1}$	?	syllable$_{i+2}$	?	syllable$_{i+3}$
first_previous		*current*		*first_next*		*second_next*		*third_next*

c) **syllable$_{i-2:i+2}$ can be a word:**

syllable$_{i-2}$	_	syllable$_{i-1}$	_	syllable$_i$	?	syllable$_{i+1}$	?	syllable$_{i+2}$
second_previous		*first_previous*		*current*		*first_next*		*second_next*

d) **syllable$_{i-3:i+2}$ can be a word:**

syllable$_{i-3}$	_	syllable$_{i-2}$	_	syllable$_{i-1}$	_	syllable$_i$	?	syllable$_{i+1}$
third_previous		*second_previous*		*first_previous*		*current*		*first_next*

Fig. 2. Four situations were used in designing ambiguity reduction feature templates.

However, we do not consider a syllable as a separable syllable if $a_i + b_i$ is not higher than the average of $a_j + b_j$ of all possible separable syllables because of we want to get rid of an uncertain separable syllable. In Vietnamese, there are some conspicuous separable syllables such as "những" (these), "nhưng" (but), "cũng" (also), "đây" (here), and "với" (with). The syllable "văn" (literature) is a non-separable syllable. For example, syllable "văn" usually is the first syllable of many two-syllable words such as "văn_ bản" (document), "văn_ hoá" (culture), "văn_sĩ" (writer), and "văn_ kiện" (documentation).

Table 3. Feature templates in case of a current syllable is a separable syllable, and the first previous label is SPACE.

No.	Templates
1	{**inVNDict**($f_{j:j+2}$) **for j in range**(i, i+4)}
2	{**inVNDict**($f_{j:j+3}$) **for j in range**(i, i+3)}
3	{**inVNDict**($f_{j:j+4}$) **for j in range**(i, i+2)}
4	{**inVNDict**($f_{j:j+5}$) **for j in range**(i, i+1)}

The noticeable difference between our method from research of [14] is that we do not use post-processing for dealing with overlap ambiguities. We proposed a novel way of feature extraction, in which we used boolean variables to record signals of overlap ambiguity cases. In case of the current syllable is a separable syllable and the first-previous label is SPACE (as we can see in Fig. 2), we check the concatenations of lowercase-simplified forms of adjacent syllables in Vietnamese dictionary: $\{f_{i:i+2}, f_{i+1:i+3}, f_{i+2:i+4}, f_{i+3:i+5}\}$ (the first template in Table 3); $\{f_{i:i+3}, f_{i+1:i+4}, f_{i+2:i+5}\}$ (the second template in Table 3), $\{f_{i:i+4}, f_{i+1:i+5}\}$ (the third template in Table 3); $\{f_{i:i+5}\}$ (the fourth template in Table 3. In other words, we check all combinations of every two, three, four, and five adjacent syllables in a five-syllable window (as we can see in Fig. 2) in Vietnamese dictionary. This manipulation records all signals of overlap ambiguity cases, which are considered as features. We perform the same manipulation in case of syllable$_{i-1:i+2}$, syllable$_{i-2:i+2}$, and syllable$_{i-3:i+2}$ can be a word (described in Fig. 2).

2.2.4   Suffix Features

In Vietnamese, suffixes are tail-affixes (syllables or one-syllable words) that are placed after a word to create larger words [11]. In our research, we obtain potential suffixes by statistics instead of linguistic knowledge. To obtain potential suffixes, we counted the number of occurrences of the last lower syllables in an out-of-vocabulary three-syllable or four-syllable words. However, we do not consider a syllable as a suffix if its number of occurrences is not higher than the average number of occurrences of all possible suffixes because we want to get rid of uncertain suffixes.

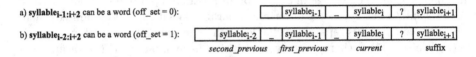

a) **syllable$_{i-1:i+2}$** can be a word (off_set = 0):

b) **syllable$_{i-2:i+2}$** can be a word (off_set = 1):

Fig. 3. Diagram of suffix case describes whether we choose "underscore" or "space" for current syllable.

We design suffix features with the expectation that the classifier can predict three-syllable or four-syllable words more accurate in case of next syllable is a suffix (as we can see in Fig. 3). In other words, we want the classifier to pay special attention to the case where the next syllable is a suffix. In case the next syllable is a suffix, we derive current lowercase-simplified forms of conjunction of adjacent syllables $f_{i-1-off\_set:i+1}$ (the value of "off_set" follows Fig. 3) as a feature for classifier. The next syllable f_{i+1} is also treated as a feature. Finally, we derive left and right contexts of the current suffix which are $f_{i-2-off\_set}$, $f_{i-3-off\_set}$, f_{i+2}, and f_{i+3} as features. For example, we assume that in the training set we have the string "xây_dựng cơ_sở vật_chất theo hướng **hiện_đại_hoá**, hoàn_thành việc xoá lớp_học tạm_bợ" (build facilities towards modernization, finish eradicating unsettled classrooms) and in the test set there is the string "xây_dựng nhà dân theo hướng **kiên_cố_hoá** để phòng_chống lụt_bão" (build residential houses following solidified methods to protect against storms and floods). We also assume that in this example "**kiên_cố_hoá**" (solidified) is out-of-vocabulary. The syllable "**hoá**" is a suffix in Vietnamese. In this case, the classifier can not predict the word "**kiên_cố_hoá**" in the test set because it is out-of-vocabulary. However, if we leverage the context of this suffix when training, we may predict the word "**kiên_ cố_ hoá**". Because it has the same the left context, uni-gram "hướng" and bi-gram "theo_hướng", as the word "**hiện_ đại_ hoá**" (modernized).

3   Experiment and Result

3.1   Corpora

In our research, we compared the performance of our Vietnamese word segmentation method with published results of other well-known state-of-the-art approaches. Additionally, we studied the impact of our word segmentation method on the performance of the POS tagging task. For these purposes, we evaluated our methods on the VLSP 2013 WordSeg and VLSP 2013 POSTag corpus[2], which was released for competition. Both of the two corpora are provided for research or educational purpose by the national project on Vietnamese language and speech processing VLSP[3]. The training dataset of VLSP 2013 WordSeg consists of 75,389 manually word-segmented sentences (approximately 23 words per sentence on average), which is part of Vietnamese treebank corpora [12]. The test dataset of VLSP 2013 WordSeg consists of 2,120 sentences (approximately 31 words per sentence). The training dataset of VLSP 2013 POSTag consists of 26,999 manually word-segmented sentences (about 22.5 words per sentence on average), which was collected from two sources of the national VLSP project [12] and the Vietnam Lexicography Center[4]. The test dataset of VLSP 2013 POSTag consists of 2,120 sentences. Specially, we also experimented with the Vietnamese word segmentation corpus, which was provided by the authors in [8]. In this paper, we temporarily call this corpus "VNWordSeg"[5]. VNWordSeg consists of 7,807 manually word-segmented sentences (about 19 words per sentence on average), which was divided into 5 folds for later research [8].

Table 4 shows the distribution of unique words according to the number of syllables in a word in VNWordSeg, Training dataset of VLSP 2013 POSTag, and

Table 4. Distribution of unique words according to number of syllables in a word (%).

Corpus	Number of syllables in a word					
	1	2	3	4	5–9	>9
VNWordSeg	38.21	53.59	07.57	00.52	00.11	00.00
Training dataset of VLSP 2013 POSTag	31.66	58.51	07.33	02.03	00.45	00.02
Training dataset of VLSP 2013 WordSeg	36.49	48.92	11.54	02.63	00.41	00.01

---

[2] http://vlsp.org.vn/vlsp2013/eval/ws-pos.
[3] http://vlsp.org.vn.
[4] https://www.vietlex.com.
[5] https://www.jaist.ac.jp/~hieuxuan/vnwordseg/data.

Training dataset of VLSP 2013 WordSeg. The majority of the three datasets are one- and two- syllables words. More-than-four-syllable words are rare in the three datasets. However, words containing from five to nine syllables account for the notable small ratios (0.11%, 0.45%, and 0.41% in VNWordSeg, Training dataset of VLSP 2013 POSTag, and Training dataset of VLSP 2013 WordSeg, respectively). For more detail, there are 136, 305, and 321 separable syllables (described in Subsect. 2.2.3) in VNWordSeg, Training dataset of VLSP 2013 POSTag, and Training of VLSP 2013 WordSeg, respectively.

3.2 Experimental Setup

Vietnamese word segmentation has to solve the large-scale classification problem [8]. Therefore, we decided to use the Linear Support Vector Classification (LinearSVC) [15] as a tool for SVM classifier implementation. The LinearSVC on Python 3 programming language was based on LIBLINEAR written on C programming language [2]. By using LinearSVC, we tuned only one parameter, which is the penalty parameter C of the error term in the SVM classifier. We chose the best value of C based on the main evaluation metric F_1 score by using gird search experiments, in which value of C can be 0.001, 0.01, 0.1, 1, 10, or 100.

3.3 Feature Selection Results

To explore the impacts of feature groups on the performance, we conducted feature selection experiments with all combinations of features on three datasets VNWordSeg, Training dataset of VLSP 2013 POSTag, and Training dataset of VLSP 2013 WordSeg. We denoted "base", "long", "sep", and "sfx" for baseline, more-than-four-syllable word, ambiguity reduction, and suffixes feature groups, respectively.

Table 5 presents feature selection results with all combinations of feature groups. More-than-four-syllable word features have impacts on the Training dataset of VLSP 2013 POSTag (0.02+%) slightly, and Training dataset of VLSP 2013 WordSeg (0.03+%) in comparison with the baseline groups. The ambiguity reduction features have the most substantial impacts on VNWordSeg (0.08+%). We can also observe that the suffixes features, which have minimal impacts on three corpora (according to our experiments, there are 2, 4, and 3 suffixes on VNWordSeg, Training dataset of VLSP 2013 POSTag, and Training dataset of VLSP 2013 WordSeg, respectively).

Table 5. Our word segmentation results using 5-fold cross-validation with all combinations of features (%). We also re-trained UETsegmenter [14] and RDRsegmenter [9] methods with the same training datasets and testing datasets with the aim of reference.

Prior methods/features	Corpus					
	VNWordSeg		Training set of VLSP 2013 POSTag		Training set of VLSP 2013 WordSeg	
	C	F_1-score	C	F_1-score	C	F_1-score
UETsegmenter [14]	-	92.0986	-	97.9820	-	98.7954
RDRsegmenter [9]	-	93.7811	-	98.3069	-	99.0726
base	1.0	94.4866	0.1	98.5080	0.1	99.2630
base + long	1.0	94.4858	0.1	98.5371	0.1	99.2762
base + sep	1.0	94.5686	0.1	98.5647	0.1	99.2963
base + sfx	1.0	94.4881	0.1	98.5104	0.1	99.2669
base + long + sep	1.0	94.5686	0.1	98.5848	0.1	99.3024
base + long + sfx	1.0	94.4910	0.1	98.5434	0.1	99.2811
base + sep + sfx	1.0	**94.5752**	0.1	98.5666	0.1	99.2979
base + long + sep + sfx	1.0	94.5743	0.1	**98.5870**	0.1	**99.3032**

3.4 Main Results

Table 6 compares the Vietnamese word segmentation results of our method with results published in previous research works, using the same training and test datasets. Table 6 shows that our method achieved the highest precision, recall, and F_1-score. Our method obtains 0.29+% higher F_1-score than RDRsegmenter [9], which is the recent state-of-the-art approach. It should be noted that the results of vnTokenizer [4], JVnSegmenter [8] and DongDu [6] were reported by the authors in [14].

Table 6. Word segmentation results on test dataset of VLSP 2013 WordSeg (%).

Method	Precision	Recall	F_1-score
vnTokenizer [4]	96.98	97.69	97.33
JVnSegmenter-Maxent [8]	96.60	97.40	97.00
JVnSegmenter-CRFs [8]	96.63	97.49	97.06
DongDu [6]	96.35	97.46	96.90
UETsegmenter [14]	97.51	98.23	97.87
RDRsegmenter [9]	97.46	98.35	97.90
Our WordSeg {all features}	**97.81**	**98.57**	**98.19**

Table 7 shows the Vietnamese word segmentation 5-fold cross-validation results of our method with results published in previous research on the VNWordSeg corpus. Method of the authors in [17] had been holding the highest F_1-score on VNWordSeg. However, our method obtains the highest recall score on the VNWordSeg corpus.

Table 7. Word segmentation results using 5-fold cross-validation on VNWordSeg corpus (%).

Method	Precision	Recall	F_1-score
Method of the authors in [8]	94.00	94.45	94.23
Method of the authors in [17]	**96.71**	93.89	**95.30**
Our WordSeg {base + sep + sfx}	94.24	**94.92**	94.58

3.5 Analyses

In order to analyze the word segmentation results in more detail, we computed F_1 score according to number of syllables in a word and three and four syllables words containing suffixes. Additionally, we also re-trained UETsegmenter [14] with the Vietnamese words dictionary of RDRsegmenter [9] and vice versa. As we can see in Table 8, our method obtains higher F_1 scores than UET-Segmener [14], and RDRsegmenter [9] on one and two syllables words (1 & 2). On three-syllable words (3^a), RDRsegmenter [9] achieves the highest F_1 score.

Table 8. Word segmentation results (F_1 score) on **test dataset of VLSP 2013 WordSeg** according to number of syllables in a word (%). For convenience, we denote three and four syllables unknown words containing suffixes by 3^b and 4^b (unknown words are detected by checking in the Vietnamese words dictionary of RDRsegmenter [9]). And conversely, we use 3^a and 4^a, indicating three and four syllables words which are not 3^b or 4^b. Notably, we temporarily use **UETws, RDRws**, and **UITws** as abbreviations for **UETsegmenter** [14], **RDRsegmenter** [9], and **our word segmentation method using all features**. We also provide proportions of words (%) in parentheses.

Vietnamese dictionary resource	Method	Number of syllables in a word							Total
		1 (57.75)	2 (40.42)	3^a (00.74)	3^b (00.13)	4^a (00.68)	4^b (00.05)	5–9 (00.22)	
UETws [14]	UETws [14]	98.46	**97.97**	79.96	**89.74**	78.62	100.00	21.30	97.87
	RDRws [9]	98.37	97.68	85.41	89.03	74.23	100.00	23.60	97.74
	UITws	**98.59**	97.96	**85.77**	**89.74**	77.26	100.00	**34.02**	**98.01**
RDRws [9]	UETws [14]	98.47	97.90	80.40	0.00	**79.51**	**26.32**	34.97	97.79
	RDRws [9]	98.57	97.85	**86.30**	79.19	75.74	0.00	23.60	97.90
	UITws	**98.82**	**98.14**	85.23	**80.20**	78.60	0.00	**46.83**	**98.19**

On four-syllable words (4^a), UETsegmenter [14] achieves the highest F_1 score. Notably, UETsegmenter [14] used another Vietnamese words dictionary[6] which contains all 7 three-and-four-syllable unknown words that they predict correctly. Besides, UETSegmener [14] can not predict three syllables words containing suffixes (3^b) when training with the Vietnamese words dictionary of RDRsegmenter [9]. Therefore, we can conclude that RDRsegmenter [9] and our word segmentation method have not solved unknown words containing suffixes badly (3^b). Lastly, different from the result of UETsegmenter [14] on three-syllable words (3^a) and RDRsegmenter [9] on four-syllable words (4^a), our result on three-syllable and words four-syllable words are not left far away by the highest result.

Lastly, Table 9 shows POS tagging performance on the test dataset of VLSP 2013 POSTag with the predicted word segmentation. We re-trained the UETsegmenter tool on VLSP 2013 POSTag. Our Vietnamese word segmentation method has helped VnMarMot [10] of increase in performance on VLSP 2013 POSTag with 0.3+% improvement of F_1 score by comparing with (VnMarMoT [10] using RDRsegmenter [9]) approach.

Table 9. POS Tagging performance with predicted word segmentation on test dataset of VLSP 2013 POSTag (%).

Method	F_1-score	
	WordSeg	POSTag
RDRPOSTagger with RDRsegmenter [10]	97.75	93.39
(BiLSTM-CRF + CNN-char) with RDRsegmenter [10]	97.75	93.55
VnMarMoT with RDRsegmenter [10]	97.75	93.96
VnMarMoT [10] with Our WordSeg {all features}	**98.06**	**94.27**

4 Conclusion and Future Work

In this paper, we propose a novel feature-based method using the SVM classifier for Vietnamese word segmentation. Overlap ambiguity and unknown words containing suffixes phenomena are real challenges in Vietnamese word segmentation. We prove that our proposed features, ambiguity reduction and suffix-capturing features, help to improve the performance of word segmentation. Experiments on the benchmark Vietnamese datasets show that our method obtains a higher F_1-score score than state-of-the-art approaches. Finally, according to the experimental results, our Vietnamese word segmentation method has a positive impact on Vietnamese POS tagging. However, the greatest weakness of our ambiguity reduction and suffix features is that we do not care about parts-of-speech information. Therefore, we are planning to refer to the ambiguity solving method of the authors in [16] for our further research. Our code is open-source and available at https://github.com/ngannlt/UITws-v1.

---

[6] https://github.com/phongnt570/UETsegmenter/blob/master/dictionary.

Acknowledgment. This research is funded by University of Information Technology-Vietnam National University HoChiMinh City under grant number D1-2019-16.

References

1. Dinh, D., Hoang, K., Nguyen, V.T.: Vietnamese word segmentation. In: Proceedings of the Sixth Natural Language Processing Pacific Rim Symposium, pp. 749–756 (2001)
2. Fan, R.E., Chang, K.W., Hsieh, C.J., Wang, X.R., Lin, C.J.: LIBLINEAR: a library for large linear classification. J. Mach. Learn. Res. **9**, 1871–1874 (2008)
3. Ha, L.A.: A method for word segmentation in Vietnamese. In: Proceedings of the Corpus Linguistics 2003 Conference, pp. 282–287 (2003)
4. Hông Phuong, L., Thi Minh Huyên, N., Roussanaly, A., Vinh, H.T.: A hybrid approach to word segmentation of Vietnamese texts. In: Martín-Vide, C., Otto, F., Fernau, H. (eds.) LATA 2008. LNCS, vol. 5196, pp. 240–249. Springer, Heidelberg (2008). https://doi.org/10.1007/978-3-540-88282-4_23
5. Le, H.P., Roussanaly, A., Nguyen, T.M.H., Rossignol, M.: An empirical study of maximum entropy approach for part-of-speech tagging of Vietnamese texts. In: Traitement Automatique des Langues Naturelles - TALN 2010, p. 12. ATALA (Association pour le Traitement Automatique des Langues), Montréal, Canada (2010)
6. Luu, T.A., Yamamoto, K.: Ứng dụng phu'o'ng pháp Pointwise vào bài toán tách tù'cho tieng Viet (2012). http://www.vietlex.com/xu-li-ngon-ngu/117-Ung_dung_phuong_phap_Pointwise_vao_bai_toan_tach_tu_cho_tieng_Viet
7. Nghiem, M., Dinh, D., Nguyen, M.: Improving Vietnamese POS tagging by integrating a rich feature set and support vector machines. In: 2008 IEEE International Conference on Research, Innovation and Vision for the Future in Computing and Communication Technologies, pp. 128–133 (2008)
8. Nguyen, C.T., Nguyen, T.K., Phan, X.H., Nguyen, L.M., Ha, Q.T.: Vietnamese word segmentation with CRFs and SVMs: an investigation. In: The 20th Pacific Asia Conference on Language. Information and Computation: Proceedings of the Conference, pp. 215–222. Tsinghua University Press, Huazhong Normal University, Wuhan (2006)
9. Nguyen, D.Q., Nguyen, D.Q., Vu, T., Dras, M., Johnson, M.: A fast and accurate Vietnamese word segmenter. In: Proceedings of the 11th International Conference on Language Resources and Evaluation (LREC 2018), pp. 2582–2587 (2018)
10. Nguyen, D.Q., Vu, T., Nguyen, D.Q., Dras, M., Johnson, M.: From word segmentation to POS tagging for Vietnamese. In: Proceedings of the Australasian Language Technology Association Workshop 2017, pp. 108–113. Brisbane (2017)
11. Nguyen, D.H.: Vietnamese. London Oriental and African Language Library. John Benjamins, Amsterdam (1997)
12. Nguyen, P.T., Vu, X.L., Nguyen, T.M.H., Nguyen, V.H., Le, H.P.: Building a large syntactically-annotated corpus of Vietnamese. In: Proceedings of the Third Linguistic Annotation Workshop, ACL-IJCNLP 2009, pp. 182–185. Association for Computational Linguistics (2009)
13. Nguyen, Q.T., Nguyen, N.L., Miyao, Y.: Comparing different criteria for Vietnamese word segmentation. In: Proceedings of the 3rd Workshop on South and Southeast Asian Natural Language Processing, pp. 53–68. The COLING 2012 Organizing Committee, Mumbai (2012)

14. Nguyen, T.P., Le, A.C.: A hybrid approach to Vietnamese word segmentation. In: 2016 IEEE RIVF International Conference on Computing Communication Technologies, Research, Innovation, and Vision for the Future (RIVF), pp. 114–119 (2016)
15. Pedregosa, F., et al.: Scikit-learn: machine learning in python. J. Mach. Learn. Res. **12**, 2825–2830 (2011)
16. Pham, D.D., Tran, G.B., Pham, S.B.: A hybrid approach to Vietnamese word segmentation using part of speech tags. In: 2009 International Conference on Knowledge and Systems Engineering, pp. 154–161 (2009)
17. Tran, O.T., Le, C.A., Ha, T.Q.: Improving Vietnamese word segmentation and POS tagging using MEM with various kinds of resources. J. Nat. Lang. Process. **17**(3), 3_41–3_60 (2010). https://doi.org/10.5715/jnlp.17.3_41

An Assessment of Substitute Words in the Context of Academic Writing Proposed by Pre-trained and Specific Word Embedding Models

Chooi Ling Goh[1]([⊠]) and Yves Lepage[2]

[1] The University of Kitakyushu, Kitakyushu, Fukuoka, Japan
goh@kitakyu-u.ac.jp
[2] Waseda University, Kitakyushu, Fukuoka, Japan
yves.lepage@waseda.jp

Abstract. Researchers who are non-native speakers of English always face some problems when composing scientific articles in this language. Most of the time, it is due to lack of vocabulary or knowledge of alternate ways of expression. In this paper, we suggest to use word embeddings to look for substitute words used for academic writing in a specific domain. Word embeddings may not only contain semantically similar words but also other words with similar word vectors, that could be better expressions. A word embedding model trained on a collection of academic articles in a specific domain might suggest similar expressions that comply to that writing style and are suited to that domain. Our experiment results show that a word embedding model trained on the NLP domain is able to propose possible substitutes that could be used to replace the target words in a certain context.

Keywords: Word embedding · Word similarity · Dictionary lookup · Synonym · Academic writing

1 Introduction

Many researchers face problems in composing scientific articles. For non-native speakers of English, the problem becomes more severe. They can use machine translation systems to translate from their mother tongues to English, but most of the time, the translation output quality is not satisfactory, or does not comply with the academic writing style. A bilingual dictionary or a thesaurus may be used to search for suitable expressions, when only simple words come across the mind. However, not all expressions suggested comply to the academic writing style. Moreover, it is difficult to know whether the alternate ways of expression are in style and in lexicon.

In this paper, we suggest to use word embeddings to search for substitute words in the context of academic writing. Word embeddings have been applied

© Springer Nature Singapore Pte Ltd. 2020
L.-M. Nguyen et al. (Eds.): PACLING 2019, CCIS 1215, pp. 414–427, 2020.
https://doi.org/10.1007/978-981-15-6168-9_34

to many natural language processing tasks such as information retrieval, sentiment analysis, question answering and document classification. As opposed to a thesaurus, which usually provides only semantically similar words or expressions, word embeddings may not only show semantically similar words but also other words with similar word vectors, that could be even better. Furthermore, if we train an embedding model on only a collection of academic articles, then, possibly, similar expressions which comply to that writing style and which are suited to the domain might be suggested. This can be very helpful for non-native speakers of English to guide them to write articles in style and in lexicon. As an example, a less proficient person may know the easy word "*but*", but word vectors may propose "*however*" or "*although*" as alternative words. Similarly, a target word like "*show*" may be replaced with more sophisticated words like "*reveal*" or "*depict*".

One may suggest to use the English lexical database WordNet [11] for finding the synonyms while writing an article. Certainly, it is a good idea to consult a well defined semantic network for this purpose, but we are wondering whether word embeddings would propose some different expressions. It has been reported that the number of human-judged synonyms extracted from word embeddings is about twice the number given by WordNet in a survey [6] for extracting synonyms to be used in the machine translation evaluation metric METEOR [3]. Therefore, there is a high possibility that word embeddings could give better suggestion of word proposal in academic writing.

The goal of this paper is to compare the word similarity results for a word embedding model built on a specific domain with some other general large pre-trained models. We use the ACL Anthology Reference Corpus[1] (ACL-ARC hereafter) in the natural language processing (NLP) domain as our specific domain. ACL Anthology is a digital archive of research papers in the premium conferences in NLP and the English language quality of the papers is reputed. We would like to know whether a specific model trained on specific domain could give equivalent or better word similarity than large pre-trained models.

2   Specific Word Embedding Model Trained on ACL-ARC

As mentioned above, we used ACL-ARC to build a specific word embedding model. ACL-ARC is a subset of ACL Anthology[2]. The corpus consists of the publications about computational linguistics and natural language processing from selected conferences and journals since 1979 until 2015. It consists of 22,878 articles.

We used the gensim[3] implementation of Word2Vec to build our model. We trained our model with the continuous bag-of-words (CBoW) model as this model has been shown to be the optimal choice for building better models for English [6,9]. The parameter settings are as follows.

---

[1] https://acl-arc.comp.nus.edu.sg/.
[2] https://aclanthology.coli.uni-saarland.de/.
[3] https://radimrehurek.com/gensim/.

- Dimensionality of the word vectors: size = 300
- Distance between the current word with the predicted word: window = 5
- Minimum count of word occurrence: min_count = 5

As pre-processing, we extracted the texts from the XML output generated by the commercial optical character recognition (OCR) software, Nuance Omnipage. The front pages from the conferences are excluded. We also excluded the section references in the papers. However, there still exists some noise or uncleaned texts. Most of the noise is coming from conference names, mathematical equations, figures and tables. All the texts are lowercased, and words containing numbers, symbols or punctuations are removed.

Table 1 shows some statistics on the corpus used for building our word embedding model. From 88 million tokens, we built a model containing 66k word vectors.

Table 1. Statistics on the word embedding model built on ACL-ARC

# of articles used	21,636
# of tokens	88,006,598
# of distinct word	578,960
# of word vectors (include references)	77,311
# of word vectors (exclude references)	66,453

3   Large Pre-trained Models

There exist three standard models for word embeddings at the moment[4]: Word2vec [9], GloVe [13] and fastText [2]. These models allow us to compute the semantic similarity between two words, so as to find the most similar words given a target word. The ability to obtain word vectors for out-of-vocabulary words is featured in fastText [2] by capturing subword information. While Word2vec [9] is limited to a vector space locally, GloVe [13] also considers word co-occurrence globally.

We will use large pre-trained models available from the three methods above to compare with our specific word embedding model trained on ACL-ARC. All the models, including the specific model, are trained with 300 dimensions which has been proven to deliver optimal performance [7,9].

- Word2vec[5]: trained on GoogleNews, GoogleNews-vectors-negative300.bin.gz, 3 billion tokens, 3 million word vectors.

---

[4] We leave aside the more recent ELMo [14] that is based on deep context, and BERT [4] that uses masked language model.
[5] https://code.google.com/archive/p/word2vec/.

– GloVe[6]: trained on Wikipedia 2014 + Gigaword 5, glove.6B.zip (300d), 6 billion tokens, 400 thousand word vectors.
– fastText[7]: trained on Wikipedia 2017 + UMBC webbase corpus + statmt.org news dataset, wiki-news-300d-1M.vec.zip [10], 16 billion tokens, 1 million word vectors.

The GoogleNews model[8] contains compound words (e.g. *"ANTARA_News_PRNewswire_AsiaNet"* and *"eerily_similar"*), whereas in other models, no compound word is found. Besides, GoogleNews and fastText models have more uncleaned items, like erroneous spelling, than other models (e.g. *"baed"*, *"similiar"*, *"infomation"*). Furthermore, these models are case-sensitive, e.g. *"show"* and *"Show"*, both exist in the models. A preliminary experiment has shown that these noise words appeared in higher ranking of word similarity, it is therefore better to remove them from the models.

Hence, in order to have a fair comparison, we further filter the large pretrained models, so that they only contain words that are found in the ACL-ARC word vectors. A large number of word vectors are removed by this filtering. The number of word vectors left is shown in Table 2. After filtering, all the words in all models are in lowercase, and no compound words or erroneous words left.

Table 2. Statistics on the word embedding models after filtering

	ACL-ARC	GoogleNews	GloVe	fastText
Training size	88M	3B	6B	16B
Before filtering	66,453	3M	400k	1M
After filtering	66,453	24,912	33,231	32,569

4   Experiments

We chose 12 highly frequent words from ACL-ARC which look like producing more choice of substitute words to be the target words. We extracted similar words using the four models presented above. The 12 target words used for evaluation are shown below. Table 3 shows the synonyms taken from the WordNet lemmas.

> *with, by, each, using, results, some, however, methods, see, very, thus, shows*

---

[6] https://nlp.stanford.edu/projects/glove/.
[7] https://fasttext.cc/docs/en/english-vectors.html.
[8] For simplicity, the four models are referred as ACL-ARC, GoogleNews, GloVe and fastText hereafter.

Table 3. 12 frequent words selected from ACL-ARC used for evaluation. Right column shows the synonyms taken from the WordNet lemmas.

target word	Synonyms from WordNet
with	[NOT FOUND]
by	aside, away, past
each	apiece, for_each_one, from_each_one, to_each_one
using	apply, employ, expend, exploitation, habituate, practice, use, utilise, utilize, victimisation, victimization
results	answer, consequence, effect, ensue, event, final_result, issue, lead, leave, outcome, resolution, result, resultant, resultant_role, solution, solvent, termination, upshot
some	about, approximately, around, close_to, just_about, more_or_less, or_so, roughly
however	all_the_same, even_so, nevertheless, nonetheless, notwithstanding, still, withal, yet
methods	method, method_acting
see	ascertain, assure, attend, catch, check, come_across, consider, construe, control, date, determine, discover, encounter, ensure, envision, escort, examine, experience, fancy, figure, find, find_out, get_a_line, get_wind, get_word, go_out, go_steady, go_through, hear, image, insure, interpret, learn, look, meet, pick_up, picture, project, realise, realize, reckon, regard, run_across, run_into, see_to_it, take_care, take_in, understand, view, visit, visualise, visualize, watch, witness
very	identical, rattling, real, really, selfsame
thus	frankincense, gum_olibanum, hence, olibanum, so, thence, therefore, thusly
shows	appearance, bear_witness, demo, demonstrate, depict, designate, display, establish, evidence, evince, exhibit, express, indicate, picture, point, present, prove, read, record, register, render, shew, show, show_up, testify, usher

For each word, we extracted the 10 nearest neighbor words based on cosine similarity from each word embedding model (see Sects. 2 and 3). According to [1], word embedding performance is affected by various factors such as corpus size, length of individual texts or existence of specific content. [1] suggest to measure the stability of the performance by testing on multiple bootstrap samples. On the other hand, word vectors are concatenated in [8] in order to combine two vectors so as to obtain better performance in some extrinsic tasks. In our approach, we did not combine word embeddings, but only combined similar words extracted from the four models (i.e. JointModel) using the heuristic below.

1. For each proposed word, sum up the cosine similarity values from all models.
2. Order the list by number of occurrences and total cosine similarity in descending order.
3. Consider only the 10 highest ranking words in that order.

In order to evaluate the performance of each model, we conducted two experiments:

- evaluation on an extrinsic task using the machine translation output, and
- an intrinsic evaluation by human judgement.

4.1 Evaluation Using Machine Translation

For each target word, we collected 10 sentence pairs from English-French translation pairs in Linguee[9]. We tried not to collect sentences that are too long or too short. Too long sentences may not produce satisfactory machine translation results and too short sentences may not provide enough context. In average, the length is about 17 words per sentence. We chose the English-French language pair as the translation pair because it exhibits better machine translation results currently. For 12 words, we collected 120 sentence pairs in total. We then used the deepL[10] translator to translate from English to French. Since deepL is trained on top of Linguee, we also translated the sentences using Google Translate[11] for comparison. We evaluate the translation results using the BLEU metric [12]. Higher BLEU scores are obtained by translations that are closer to the target reference translations, which imply better translations. The translation performance for deepL and Google Translate are shown at the top part of Table 4. It shows that deepL delivers better translation results than Google Translate. In the following experiments, we consequently used only deepL for translation.

Table 4. Machine translation results using BLEU scores

	BLEU score
deepL	45.09
Google translate	39.48
Word embedding model	
ACL-ARC	39.05
GoogleNews	39.61
GloVe	38.69
fastText	38.65
JointModel	40.02

---

[9] https://www.linguee.com/.
[10] https://www.deepl.com/translator.
[11] https://translate.google.com/.

For each target word in each model, we replaced the corresponding 10 sentences with the 10 candidate words, i.e., each candidate word is replaced in 10 sentences, therefore, we generated 100 sentences per target word. In total for 12 target words, we have 1,200 sentences per model. We translated these sentences using deepL and calculated the BLEU scores. The bottom part of Table 4 shows the translation results. As for individual models, GoogleNews proposes the best candidates and fastText is the worst. By combining all the models into a JointModel, better candidates are proposed.

4.2   Human Judgement

We also evaluate candidate word proposals obtained by human judgement. For each target word, if a proposed candidate word can be used to replace the original word, by any form of rephrasing, then it is considered as a possible substitute (1 point), or else it is not (0 point). The substitute word must also conform to morphological features, i.e., it should exhibit the correct word form according to the tense, number, etc and it should be semantically similar. Basically, different word forms of the same lemma are not necessarily substitutable, e.g. "*using*" and "*used*". We enquire how many possible substitutes are proposed by each model.

Table 5. Results by human judgement

Model	Total	Avg/person	Avg/word
Before inconsistency correction			
ACL-ARC	227	32.43	2.70
GoogleNews	223	31.86	2.65
GloVe	125	17.86	1.49
fastText	244	34.86	2.90
JointModel	251	35.86	2.99
After inconsistency correction			
ACL-ARC	237	33.86	2.82
GoogleNews	240	34.29	2.86
GloVe	131	18.71	1.56
fastText	259	37.00	3.08
JointModel	**263**	**37.57**	**3.13**

We asked seven postgraduate students who are conducting research in the NLP domain to evaluate the word candidates. These students are non-native speakers of English: two are French, one is Thai, and four others are Chinese. These students have experience on writing at least one conference paper or their thesis in English.

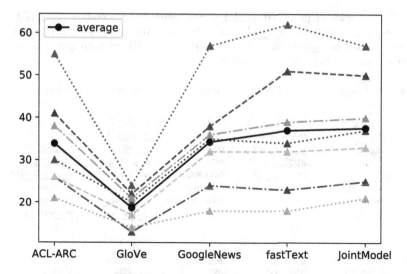

Fig. 1. Comparison of results of evaluators. Each dotted line shows a different evaluator. The black solid line shows the average. The lines between the points are just for better identification of evaluators.

As usually observed with human evaluators, there exists some inconsistency, where evaluators have chosen a word in a model, but have not chosen the same word in another model. For example, from the target word *"very"*, one of the evaluators had chosen the word *"remarkably"* in the JointModel, but did not choose it in the specific ACL-ARC model and the fastText model. This phenomenon happened for almost all the evaluators. We corrected this kind of mistakes made by the evaluators by double checking the selections, and adding the corresponding selections to the possible substitute word lists for all models.

Table 5 shows the results for human judgement. The top part shows the original results made by the evaluators without correction and the bottom part is the version corrected for inconsistency. In average, each word has about two to three possible substitute proposals. Although the specific ACL-ACR model is much smaller than the large pre-trained models, it gives comparable results for finding similar words. Figure 1 shows the annotations for each evaluator[12]. In general, the JointModel gives the best word proposals and the GloVe model has the least suitable candidates. All the evaluators show similar tendency to all the models. The human judgement has a fixed-marginal Fleiss's kappa [5] value of 0.49, which is considered as moderate agreement among the evaluators.

4.3   Discussion

Tables 6 and 7 show some examples of proposed similar words. Double underline shows words selected by at least four evaluators and single underline by at least

---

[12] The line graph is used just to better identify the evaluators.

Table 6. Examples of proposed similar words sorted by descending order of cosine similarity. Double underline (<u>word</u>) shows words selected by at least four evaluators and single underline (<u>word</u>) by at least one evaluator. Words with gray background shows mutual agreement by all evaluators.

Target word	ACL-ARC	GoogleNews	GloVe	fastText	JointModel
using	employing, utilizing, applying, exploiting, uses, via, used, relying, employs, utilizes	use, utilizing, used, uses, incorporating, applying, employing, utilize, utilized, utilizes	used, use, uses, method, methods, instead, techniques, technique, example, can	utilizing, employing, utilising, applying, incorporating, use, constructing, combining, creating, substituting	use, utilizing, used, uses, employing, applying, incorporating, utilizes, utilising, constructing
however	moreover, furthermore, although, but, nevertheless, unfortunately, therefore, because, though, nonetheless	though, although, nevertheless, nonetheless, that, meanwhile, also, but, only, not	although, though, but, that, not, nevertheless, because, only, fact, also	although, though, nevertheless, but, therefore, nonetheless, indeed, unfortunately, consequently, yet	although, though, but, nevertheless, nonetheless, that, not, only, also, therefore
methods	techniques, approaches, algorithms, methodologies, strategies, method, metrics, mechanisms, technique, schemes	techniques, method, methodologies, technique, tactics, methodology, strategies, approaches, mechanisms, procedures	techniques, method, technique, using, methodologies, methodology, use, tactics, procedures, tools	techniques, method, methodologies, strategies, approaches, methodology, procedures, technique, ways, tactics	techniques, method, methodologies, technique, approaches, methodology, strategies, tactics, procedures, mechanisms

one evaluator. Words with gray background shows mutual agreement by all evaluators. There is not much mutual agreement among the evaluators. This could be caused by the different English proficiency levels among them. However, in general, many of the proposed words look good, and conform to the lexicon used in scientific articles. Moreover, word embeddings also propose words that are not in their WordNet synsets (referred to Table 3). For example, the target word *"very"* has the lemmas *"identical, rattling, real, really, selfsame"* in its WordNet synset, but our models suggested that *"quite, fairly, extremely, pretty, remarkably, highly"* etc. are possible substitutions. We conclude that it is possible to use word embedding models to find appropriate substitutions in the context of academic writing.

Table 7. Continue from Table 6.

target word	ACL-ARC	GoogleNews	GloVe	fastText	JointModel
some	many, numerous, certain, several, borderline, lots, conflicting, themselves, exceptional, all	many, few, lot, several, plenty, those, lots, these, all, little	many, few, those, other, have, more, several, others, these, even	many, several, these, certain, all, most, various, other, those, few	many, several, few, these, those, all, other, certain, lots, have
very	quite, fairly, extremely, relatively, remarkably, too, comparatively, overly, reasonably, sufficiently	extremely, quite, pretty, fairly, particularly, really, so, obviously, too, especially	extremely, quite, so, pretty, too, really, well, always, especially, but	extremely, pretty, quite, remarkably, highly, relatively, fairly, too, most, somewhat	extremely, quite, too, pretty, fairly, so, really, remarkably, relatively, especially
shows	demonstrates, illustrates, showing, reveals, depicts, summarizes, displays, indicates, suggests, show	show, shown, showed, showing, indicates, reveals, demonstrates, suggests, illustrates, confirms	show, shown, showing, showed, seen, tv, television, featured, appears, recent	show, showing, indicates, shown, demonstrates, showed, illustrates, reveals, displays, depicts	show, showing, shown, demonstrates, showed, illustrates, reveals, indicates, depicts, displays

Based on the evaluation in Table 5, the specific ACL-ARC model provides a slightly lower number of possible substitutes compared to other models (except the GloVe model). But, it exhibits a larger variety of proposals which conform to the academic writing style. The human evaluation is very much dependent on the English proficiency level of the evaluators. Some of the proposed words seem to be too difficult for them to judge, especially in the absence of any context. This experiment was useful to help us in the design of a writing aid tool: we understood that just proposing a list of possible substitutes is not enough. We shall provide writers with usage samples of the possible substitutes.

Tables 8 and 9 show two examples of translations after substituting the proposed words to the target word. Table 8 shows the target word "*using*" with the substitute words proposed by the specific ACL-ARC model; Table 9 shows the target word "*however*" with the substitute words proposed by the Joint-Model model. The substitute sources with double underline are words selected by at least four evaluators and single underline by at least one evaluator. The translation outputs show that even we use the words with different form for the same lemma, as "*uses*" and "*used*", to replace the target word "*using*", it will be

Table 8. An example for translation of *"using"* with the substitute words proposed by the ACL-ARC model.

Source	Reference
Sometimes you may want to create a window running a program directly, without **using** a shell first	Parfois vous voulez créer une fenêtre executant directement un programme, sans **passer** par l'invite de commande
Translation by deepL	Parfois, vous pouvez vouloir créer une fenêtre exécutant un programme directement, sans **utiliser** un shell au préalable
Translation by Google Translate	Parfois, vous souhaiterez peut-être créer une fenêtre exécutant un programme directement, sans **utiliser** d'abord un shell

Substitution source	Translation by deepL
Sometimes you may want to create a window running a program directly, without **employing** a shell first	Parfois, vous pouvez vouloir créer une fenêtre exécutant un programme directement, sans **utiliser** un shell au préalable
Sometimes you may want to create a window running a program directly, without **utilizing** a shell first	Parfois, vous pouvez vouloir créer une fenêtre exécutant un programme directement, sans **utiliser** d'abord un shell
Sometimes you may want to create a window running a program directly, without **applying** a shell first	Parfois, vous pouvez vouloir créer une fenêtre exécutant un programme directement, sans **appliquer** d'interpréteur de commandes au préalable
Sometimes you may want to create a window running a program directly, without **exploiting** a shell first	Parfois, vous pouvez vouloir créer une fenêtre exécutant un programme directement, sans **exploiter** d'abord un shell
Sometimes you may want to create a window running a program directly, without **uses** a shell first	Parfois, vous pouvez vouloir créer une fenêtre exécutant un programme directement, sans **utiliser** d'interpréteur de commandes au préalable
Sometimes you may want to create a window running a program directly, without **via** a shell first	Parfois, vous pouvez vouloir créer une fenêtre exécutant un programme directement, sans **passer** par un shell d'abord
Sometimes you may want to create a window running a program directly, without **used** a shell first	Parfois, vous pouvez vouloir créer une fenêtre exécutant un programme directement, sans **utiliser** d'interpréteur de commandes au préalable
Sometimes you may want to create a window running a program directly, without **relying** a shell first	Parfois, vous pouvez vouloir créer une fenêtre exécutant un programme directement, sans **avoir besoin de faire appel à** un shell au préalable
Sometimes you may want to create a window running a program directly, without **employs** a shell first	Parfois, vous pouvez vouloir créer une fenêtre exécutant un programme directement, sans **utiliser** un shell au préalable
Sometimes you may want to create a window running a program directly, without **utilizes** a shell first	Parfois, vous pouvez vouloir créer une fenêtre exécutant un programme directement, sans **utiliser** d'abord un shell

Table 9. An example for translation of *"however"* with the substitute words proposed by the JointModel model.

Source	Reference
It is not **however** the sole tool, nor in the end is it the most important one	Ce n'est **cependant** pas le seul, ni le plus important au bout du compte
Translation by deepL	**Mais** ce n'est pas le seul outil, ni en fin de compte le plus important
Translation by Google Translate	Ce n'est **cependant** pas le seul outil, ni à la fin le plus important
Substitution source	Translation by deepL
It is not **although** the sole tool, nor in the end is it the most important one	Ce n'est pas le seul outil, ni en fin de compte le plus important
It is not **though** the sole tool, nor in the end is it the most important one	Ce n'est pas le seul outil, ni en fin de compte le plus important
It is not **but** the sole tool, nor in the end is it the most important one	Ce n'est pas **seulement** le seul outil, ni en fin de compte le plus important
It is not **nevertheless** the sole tool, nor in the end is it the most important one	Ce n'est **cependant** pas le seul outil, ni en fin de compte le plus important
It is not **nonetheless** the sole tool, nor in the end is it the most important one	**Il n'en reste pas moins que** ce n'est pas le seul outil, ni le plus important en fin de compte
It is not **that** the sole tool, nor in the end is it the most important one	Ce n'est pas **que** le seul outil, ni en fin de compte le plus important
It is not **not** the sole tool, nor in the end is it the most important one	Ce n'est pas le seul outil, ni en fin de compte le plus important
It is not **only** the sole tool, nor in the end is it the most important one	Ce n'est pas **seulement** le seul outil, ni en fin de compte le plus important
It is not **also** the sole tool, nor in the end is it the most important one	Ce n'est pas **non plus** le seul outil, ni en fin de compte le plus important
It is not **therefore** the sole tool, nor in the end is it the most important one	Ce n'est **donc** pas le seul outil, ni en fin de compte le plus important

translated into the same French word *"utiliser"*. However, the right contexts are different. For the target word *"however"*, some substituted words are omitted in the translations, as in the case of *"although"* and *"though"*. For this kind of conjunction word, it is difficult to replace the target word directly, but one needs to rewrite the whole sentence in order to keep it semantically similar. Hence, it is also difficult to judge by translation output.

5   Conclusion

The purpose of this paper was to inspect the use of various word embedding models, in order to look for substitute words for a certain target word in the context of academic writing in place of dictionary lookup. Our experiment focused on proposing words for articles in the natural language processing domain, using the

ACL-ARC as a corpus for training a specific word embedding model. We limited the word vectors of the large pre-trained models to the vocabulary found in the specific ACL-ARC model for a fair comparison. Compared to large pre-trained models, the specific model proposed more words conform to the academic writing style. By combining the proposed words from all the models into a JointModel model, we further improved the word proposals.

We conclude that word embeddings are useful for suggesting substitute words for writing academic articles. They can help a non-native speaker of English to transform a low level proficiency text into proper academic style writing.

In the future, we will explore into suggesting different expressions, not only at the word level, but also at the phrase, sentence or even paragraph level. We also want to enforce functional similarity using substitute vectors [7,15], so that proposed words are conform not only by semantic similarity, but also by morphological similarity. Finally, some words may lead to more lexical choice than other words, which points at varying the number of proposed substitutes. We may apply relative cosine similarity as suggested by [6], and decide on a threshold so as to suggest a relevant variable number of word proposals.

Acknowledgment. This work was supported by JSPS KAKENHI Grant Number JP18K11446 .

References

1. Antoniak, M., Mimno, D.: Evaluating the stability of embedding-based word similarities. Trans. Assoc. Comput. Linguist. **6**, 107–119 (2018)
2. Bojanowski, P., Grave, E., Joulin, A., Mikolov, T.: Enriching word vectors with subword information. Trans. Assoc. Comput. Linguist. **5**(1), 135–146 (2017)
3. Denkowski, M., Lavie, A.: Meteor universal: language specific translation evaluation for any target language. In: Proceedings of the EACL 2014 Workshop on Statistical Machine Translation, pp. 376–380 (2014)
4. Devlin, J., Chang, M.W., Lee, K., Toutanova, K.: BERT: pre-training of deep bidirectional transformers for language understanding. In: Proceedings of NAACL-HLT, pp. 4171–4186 (June 2019)
5. Fleiss, J.L.: Measuring nominal scale agreement among many raters. Psychol. Bull. **76**(5), 378–382 (1971)
6. Leeuwenberg, A., Vela, M., Dehdari, J., Genabith, J.: A minimally supervised approach for synonym extraction with word embeddings. Prague Bull. Math. Linguist. **105**, 111–142 (2016)
7. Melamud, O., Dagan, I., Goldberger, J.: Modeling word meaning in context with substitute vectors. In: Proceedings of the NAACL, pp. 472–482 (2015)
8. Melamud, O., McClosky, D., Patwardhan, S., Bansal, M.: The role of context types and dimensionality in learning word embeddings. In: Proceedings of the NAACL-HLT, pp. 1030–1040 (June 2016)
9. Mikolov, T., Chen, K., Corrado, G.S., Dean, J.: Efficient estimation of word representations in vector space. In: Proceedings of Workshop at ICLR (2013)
10. Mikolov, T., Grave, E., Bojanowski, P., Puhrsch, C., Joulin, A.: Advances in pre-training distributed word representations. In: Proceedings of LREC (2018)

11. Miller, G.A.: WordNet: a lexical database for English. Commun. ACM **38**(11), 39–41 (1995)
12. Papineni, K., Roukos, S., Ward, T., Zhu, W.J.: BLEU: a method for automatic evaluation of machine translation. In: Proceedings of ACL, pp. 311–318 (July 2002)
13. Pennington, J., Socher, R., Manning, C.D.: GloVe: global vectors for word representation. In: Proceedings of EMNLP, pp. 1532–1543 (2014)
14. Peters, M.E., et al.: Deep contextualized word representations. In: Proceedings of NAACL-HLT, pp. 2227–2237 (June 2018)
15. Yatbaz, M.A., Sert, E., Yuret, D.: Learning syntactic categories using paradigmatic representations of word context. In: Proceedings of the EMNLP-CoNLL, pp. 940–951 (July 2012)

Effective Approach to Joint Training of POS Tagging and Dependency Parsing Models

Xuan-Dung Doan[1(✉)], Tu-Anh Tran[1], and Le-Minh Nguyen[2]

[1] Viettel Cyberspace Center, Viettel Group, Hanoi, Vietnam
{dungdx4,anhtt182}@viettel.com.vn
[2] Japan Advanced Institute of Science and Technology, Ishikawa 923-1292, Japan
nguyenml@jaist.ac.jp

Abstract. We propose a joint model for POS tagging and dependency parsing. Our model consists of a BiLSTM-CNN-CRF-based POS tagger [26] and a Deep Biaffine Attention-based dependency parser [24]. A combined objective function is used to jointly train both models. Experiment results show very competitive performance on several languages of the Universal Dependencies (UD) v2.2 Treebanks [11].

Keywords: Dependency parsing · Biaffine Attention · Joint training

1 Introduction

Dependency parsing is the task of identifying binary grammatical relations between pairs of words in a sentence. Two common methods of dependency parsing are: transition-based [12,18], and graph-based [10,19] parsing. Recently, there has been a surge in the use of deep learning approaches to dependency parsing [3–5,9,24,25], which help alleviate the need for hand-crafted features, take advantage of the vast amount of raw data through word embeddings, and achieve state-of-the-art performances. Some popular techniques used are Long Short-Term Memory networks (LSTM) [22], and attention mechanism [8,23].

Among the deep learning approaches, joint multi-task models are gaining traction [6,7,14,28], due to the intuition that information sharing among related tasks during training can help the model generalize better on each single task [16,21]. In a joint multi-task model, the tasks are either structured in a hierarchical order, where one task takes as its input the output of another task, or learned concurrently through multiple prediction heads in the last layer.

In this work, we demonstrate a joint training paradigm of a part-of-speech (POS) tagger and a dependency parser. We use a BiLSTM-CNN-CRF model [26] for our POS tagger, a graph-based Biaffine Attention model [24] for our dependency parser, and jointly train both models through a combined objective function. We evaluate our model on the UD v2.2 Treebanks [11]. Our model achieves better parsing scores than the Biaffine Attention model [24] and the joint model from [6,7] on some languages, and very competitive results in general.

© Springer Nature Singapore Pte Ltd. 2020
L.-M. Nguyen et al. (Eds.): PACLING 2019, CCIS 1215, pp. 428–435, 2020.
https://doi.org/10.1007/978-981-15-6168-9_35

2   Related Works

Several works have attempted to use the multitask learning paradigm to simultaneously solve several NLP tasks within a single model. With regards to dependency parsing, [28] is one of the earlier works to jointly learn POS tagging and dependency tagging for the Chinese language. The authors use learned POS tags as features for graph-based dependency parsers, and also to decide the spanning context in the decoding phase. For deep learning-related methods, [14] uses a cascaded model to jointly learn five different tasks, achieving state-of-the-art or competitive performances in all individual tasks. In their model, outputs from the POS tagging component are fed into the higher-level chunking component, whose outputs are in turn fed into the dependency parsing component.

Recently, Nguyen et al. [6] propose a joint multi-task model where a POS tagger and a graph-based dependency parser are learned hierarchically. Their model uses bidirectional LSTMs to extract features that are then shared with both components. They are able to achieve state-of-the-art results in both POS tagging and dependency parsing on 19 languages of the UD v1.2 treebanks. After that, they [7] improve their model by extending the BIST graph-based dependency parser [9], by incorporating a POS tagger using a similar hierarchical structure.

3   Methodology

In this section, we describe each component of the joint model and the combined objective function. In our model, an input sentence of n words $w = w_1, w_2, ..., w_n$ is fed to each of the two component networks to learn separate token embeddings. We outline the learning process below.

3.1   Graph-Based Dependency Parsing

For graph-based dependency parsing, we pose the task as a structured prediction problem [1, 19]:

$$predict(w) = \underset{y \in \mathcal{Y}(w)}{argmax}\, score_{global}(w, y) \tag{1}$$

$$score_{global}(w, y) = \sum_{part \in y} score_{local}(w, part) \tag{2}$$

Given a sequence of word vectors $w_{1:n}$, we find the highest-scoring parse tree y among the possible valid parses $\mathcal{Y}(w)$ over w. The global score is generally computed by summing over local scores of each part of the graph.

3.2   Encoder

Our encoder is based on the bidirectional LSTM-CNN architecture (BiLSTM-CNN) [26]. The CNN encodes character-level information of each word into a

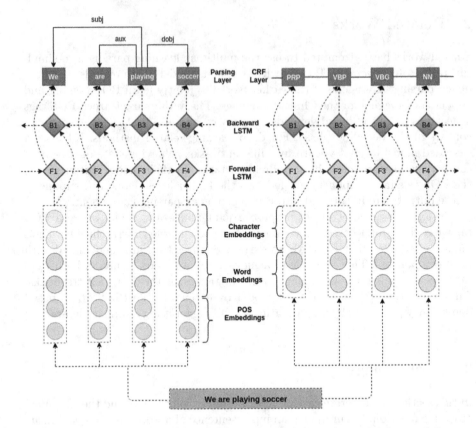

Fig. 1. Illustration of our joint POS tagging and dependency parsing model.

vector representation, which is then concatenated with its corresponding word embedding and POS tag embedding before being fed to a BiLSTM layer. The BiLSTM is used to extract context features of each word. For the task of dependency parsing, we also add POS tag embeddings to further enrich each word's representation. Finally, the encoder outputs a sequence of hidden states s_i (Fig. 1).

3.3   Biaffine Attention Mechanism

We incorporate the Biaffine attention mechanism described in [24] into our dependency parser. The task is posed as a classification problem, where given a dependent word, the goal is to predict the head word (or the incoming arc). Formally, let s_i and h_t be the BiLSTM output states for the dependent word and a candidate head word respectively, the score for the arc between s_i and h_t is calculated as:

$$e_i^t = h_t^T W s_i + U^T h_t + V^T s_i + b \tag{3}$$

where W, U, V are weight matrices, and b is the bias vector. An additional multilayer perceptron (MLP) is added after the BiLSTM layer to reduce dimensionality and overfitting.

Similarly, the dependency label classifier also uses a biaffine function to score each label, given the head word vector h_t and child vector s_i as inputs. Again, we use MLPs to transform h_t and s_i before feeding them into the classifier.

3.4 Part-of-Speech Tagging

As proposed by [26], we use the BiLSTM-CNN-CRF architecture for our POS tagging model. The hidden states output by the encoder described in section B are fed through a Conditional Random Field (CRF) layer [2,13] to predict the POS tags.

3.5 Objective Function of the Joint Training Model

We implement joint training of both dependency parsing and POS tagging models by optimizing a combined objective function of three individual losses:

$$\mathcal{L} = \mathcal{L}_{POS} + \mathcal{L}_{arc} + \mathcal{L}_{rel} \tag{4}$$

where $\mathcal{L}_{POS}$, $\mathcal{L}_{arc}$, $\mathcal{L}_{rel}$ are the POS tagging loss, the arc labeling loss, and the relation labeling loss respectively.

4 Experiments

4.1 Setup

We evaluate our system on 14 treebanks from the UD v2.2 Treebanks [11]. The statistics of the treebanks are summarized in [29].

We adopt similar hyper-parameters as [24,25] and [26]. Pre-trained embeddings are not used in any experiments.

Evaluation Metrics: parsing performance is measured using two metrics: unlabeled attachment score (UAS), labeled attachment score (LAS).

4.2 Main Results

For a fair comparison of parsing performance, we compare our results with the graph-based Deep Biaffine (BiAF) parser [24], which achieved state-of-the-art results on many languages. This model refers to supervised models RNN-Graph in [29]. We also compare our model with the jPTDP parser [6,7]. Table 1 lists our evaluation results on the test sets.

Our model achieves state-of-the-art UAS and LAS on eight out of fourteen languages. Our parser obtains significantly better results than jPTDP. On other languages, the performance of our parser is competitive with BiAF.

Table 1. Results (UAS%/LAS%) on the test sets

Lang	Our model	BiAF	jPTDP
da	83.73/80.15	**87.16/84.23**	82.17/78.88
en	89.41/87.59	**90.44/88.31**	87.55/84.71
et	**86.90/84.06**	86.76/83.28	85.45/82.13
hi	**95.76/93.30**	95.63/92.93	93.25/89.83
hr	89.08/83.28	**89.66/83.81**	88.74/83.62
id	84.90/78.01	**87.19/82.60**	84.64/77.71
ja	93.71/91.03	89.06/78.74	**94.21/92.02**
la	**83.42/78.87**	81.05/76.33	74.95/69.76
nl	**91.79/89.06**	90.59/87.52	-
no	92.55/90.52	**94.52/92.88**	-
ro	**90.64/84.92**	90.07/84.50	88.74/83.54
sk	**90.29/86.81**	90.19/86.38	85.88/81.89
vi	**75.54/72.21**	74.91/71.39[a]	67.72/58.27
zh	**86.11/83.63**	73.62/67.67	82.50/77.51

[a]We use the BiAF model in our experiments

5 Error Analysis

Following [15, 20], we analyze our model's parsing performance on the English language and the Vietnamese language with respect to:

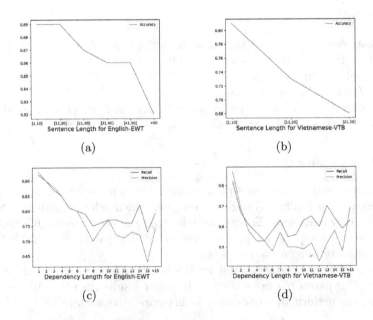

(a)

(b)

(c)

(d)

Fig. 2. Parsing performance of our model

- Sentence length: the number of words in a sentence.
- Dependency length: the distance between the head word and the dependent word in absolute value.

Figure 2(a) and 2(b) show the parsing accuracy of our model with respect to sentence lengths on the English language and the Vietnamese language. Our model tends to achieve better parsing accuracy on shorter sentences, which require fewer parsing decisions.

Similarly, Fig. 2(c) and 2(d) measure the precision and recall with respect to dependency lengths on the English language and the Vietnamese language. Once again, our model is better at identifying closer dependencies.

6 Conclusion

We present a joint POS tagging and graph-based dependency parsing model. Our model shows strong results on the Universal Dependencies v2.2 treebanks [11], achieving better parsing scores than the similar model on several languages. In the future, we plan to expand our experiments to more datasets, and assess the viability of using shared embeddings and shared LSTM representations between the Biaffine Attention and BiLSTM-CNN-CRF models. We also look to improve our model by using self-attention approaches as proposed by [29–31]. Our implementation is available at: http://github.com/dungdx34/dependency_joint_postag.

References

1. Taskar, B., Chatalbashev, V., Koller, D., Guestrin, C.: Learning structured prediction models: a large margin approach. In: Proceedings of the Twenty-Second International Conference on Machine Learning (ICML 2005), Bonn, Germany, August 7–11, 2005, pp. 896–903 (2005)
2. Sutton, C., McCallum, A.: An introduction to conditional random fields for relational learning (2006)
3. Dyer, C., Ballesteros, M., Ling, W., Matthews, A., Smith, N.A.: Transition-based dependency parsing with stack long short-term memory. In Proceedings of ACL-2015, Long Papers, vol. 1, pp. 334–343, Beijing (2015)
4. Fernández-González, D., Gómez-Rodríguez, C.: Left-to-right dependency parsing with pointer networks. In: Proceedings of the: Annual Conference of the North American Chapter of the Association for Computational Linguistics (NAACL-HLT 2019), p. 2019, Minneapolis (2019)
5. Chen, D., Manning, C.: A fast and accurate dependency parser using neural networks. In: Proceedings of EMNLP-2014, Doha, Qatar, pp. 740–750 (2014)
6. Nguyen, D.Q., Dras, M., Johnson, M.: A novel neural network model for joint pos tagging and graph-based dependency parsing. In: Proceedings of the CoNLL 2017 Shared Task: Multilingual Parsing from Raw Text to Universal Dependencies (CoNLL), pp. 134–142 (2017)

7. Nguyen, D.Q., Verspoor, K.: An improved neural network model for joint POS tagging and dependency parsing. In: Proceedings of the CoNLL 2018 Shared Task: Multilingual Parsing from Raw Text to Universal Dependencies (CoNLL), pp. 81–91 (2018)
8. Bahdanau, D., Cho, K., Bengio, Y.: Neural machine translation by jointly learning to align and translate. In: Proceedings of ICLR-2015 (2015)
9. Kiperwasser, E., Goldberg, Y.: Simple and accurate dependency parsing using bidirectional lstm feature representations. Trans. Assoc. Comput. Linguist. **4**, 313–327 (2016)
10. Eisner, J.M.: Three new probabilistic models for dependency parsing: an exploration. In Proceedings of COLING, pp. 340–345 (1996)
11. Nivre, J., Abrams, M., et al.: Universal dependencies 2.2 (2018). http://hdl.handle.net/11234/12837
12. Nivre, J.: An efficient algorithm for projective dependency parsing. In: Proceedings of the 8th International Workshop on Parsing Technologies (IWPT), pp. 149–160 (2003)
13. Lafferty, J., McCallum, A., Pereira, F.C.N.: Conditional random fields: probabilistic models for segmenting and labeling sequence data. In: Proceedings of ICML-2001, vol. 951, pp. 282–289 (2001)
14. Hashimoto, K., Xiong, C., Tsuruoka, Y., Socher, R.: A joint many-task model: growing a neural network for multiple NLP tasks. In: The 2017 Conference on Empirical Methods in Natural Language Processing (EMNLP 2017) (2017)
15. Van Nguyen, K., Nguyen, N.L.T.: Error analysis for vietnamese dependency parsing. In: The 7th International Conference on Knowledge and System Engineering (KSE), Hochiminh, Vietnam, vol. 10 (2015)
16. Caruana, R.: Multitask learning. Mach. Learn. **28**(1), 41–75 (1997)
17. Collobert, R., Weston, J., Bottou, L., Karlen, M., Kavukcuoglu, K., Kuksa, P.: Natural language processing (almost) from scratch. J. Mach. Learn. Res. **12**, 2493–2537 (2011)
18. McDonald, R., Pereira, F.: Online learning of approximate dependency parsing algorithms. In: Proceedings of EACL, pp. 81–88 (2006)
19. McDonald, R., Crammer, K., Pereira, F.: Online large-margin training of dependency parsers. In: Proceedings of ACL, pp. 91–98 (2005)
20. McDonald, R., Nivre, J.: Analyzing and integrating dependency parsers. Comput. Linguist. **37**(1), 197–230 (2011)
21. Ruder, S.: An overview of multi-task learning in deep neural networks. arXiv preprint arXiv:1706.05098 (2017)
22. Hochreiter, S., Schmidhuber, J.: Long short-term memory. Neural Comput. **9**(8), 1735–1780 (1997)
23. Luong, T., Pham, H., Manning, C.D.: Effective approaches to attention-based neural machine translation. In: Proceedings of EMNLP-2015, Lisbon, Portugal, pp. 1412–1421 (2015)
24. Dozat, T., Manning, C.D.: Deep biaffine attention for neural dependency parsing. In: Proceedings of ICLR-2017, Long Papers, Toulon, France, vol. 1 (2017)
25. Ma, X., Hu, Z., Liu, J., Peng, N., Neubig, G., Hovy, E.H.: Stack-pointer networks for dependency parsing. In: Proceedings of the 56th Annual Meeting of the Association for Computational Linguistics, ACL 2018, Melbourne, Australia, July 15–20, 2018, Long Papers, vol. 1, pp. 1403–1414 (2018)
26. Ma, X., Hovy, E.: End-to-end sequence labeling via bi-directional LSTM-CNNs-CRF. In: Proceedings of the 54th Annual Meeting of the Association for Computational Linguistics (ACL 2016), Berlin, Germany, pp. 1064–1074 (August 2016)

27. LeCun, Y., et al.: Backpropagation applied to handwritten zip code recognition. Neural Comput. **1**, 541–551 (1989)
28. Li, Z., Zhang, M., Che, W., Liu, T., Chen, W., Li, H.: Joint models for Chinese POS tagging and dependency parsing. In: Proceedings of the 2011 Conference on Empirical Methods in Natural Language Processing (EMNLP-2011), Edinburgh, Scotland, UK, July 2011, pp. 1180–1191 (2011)
29. Ahmad, W.U., Zhang, Z., Ma, X., Hovy, E., Chang, K.-W., Peng, N.: On difficulties of cross-lingual transfer with order differences: a case study on dependency parsing. In: NAACL (2019)
30. Che, W., Liu, Y., Wang, Y., Zheng, B., Liu, T.: Towards better UD parsing: deep contextualized word embeddings, ensemble, and treebank concatenation. In: Proceedings of the CoNLL 2018 Shared Task: Multilingual Parsing from Raw Text to Universal Dependencies, pp. 55–64 (2018)
31. Wang, W., Chang, B., Mansur, M.: Improved dependency parsing using implicit word connections learned from unlabeled data. In: Proceedings of the 2018 Conference on Empirical Methods in Natural Language Processing, pp. 2857–2863 (2018)

Information Extraction

Towards Computing Inferences from English News Headlines

Elizabeth Jasmi George[✉][iD] and Radhika Mamidi[iD]

LTRC, International Institute of Information Technology, Hyderabad,
Hyderabad 500 032, India
elizabeth.george@research.iiit.ac.in, radhika.mamidi@iiit.ac.in
https://www.iiit.ac.in/

Abstract. Newspapers are a popular form of written discourse, read by many people, thanks to the novelty of the information provided by the news content in it. A headline is the most widely read part of any newspaper due to its appearance in a bigger font and sometimes in colour print. In this paper, we suggest and implement a method for computing inferences from English news headlines, excluding the information from the context in which the headlines appear. This method attempts to generate the possible assumptions a reader formulates in mind upon reading a recent headline. The generated inferences could be useful for assessing the impact of the news headline on readers, including children. The understandability of the current state of social affairs depends significantly on the assimilation of the headlines. As the inferences that are independent of the context depend mainly on the syntax of the headline, dependency trees of headlines are used in this approach, to find the syntactic structure of the headlines and to compute inferences out of them. Considering the headline as the entry point to a piece of news and a source rich information about the news, we explored a headline's potential of giving an idea about the current state of affairs, leveraging the syntax structure.

Keywords: Computing inferences · Presuppositions · Conventional implicatures · Pragmatics · News discourse · News headline

1 Introduction

The headline of a news report appears at the top of the news report and is often printed in a bigger font and some times in bright colour. The marketability of a news story depends to a great extent on the ability of the headline to attract readers. A headline generally tries to summarise the content of the news story, with a firm intention of communicating the context to the reader. Headlines also try to attract the attention of the newsreaders, prompting them to read on through the news story. Headline functions as a number of speech acts. It urges,

Supported by organization Advainet Solutions Private Limited.

warns, and informs the reader [10]. This work views headline as a potential source of rich information capable of generating multiple inferences relevant to the current social state, making it worthy of adding to the general knowledge.

This work was done as a part of building a system for children to learn about current affairs in a simpler way. In this work, we consider headline as a standalone unit of discourse, without any context or supporting background information and compute the inferences that arise from the headline alone. Our experiment attempts to compute inferences based on syntactic triggers. Headlines are self-contained utterances in which the news editor is the speaker, and the newsreader is the listener. Interpretations of an utterance can consist of inferences such as presuppositions, entailments, conventional implicatures and conversational implicatures. This paper focuses on inferences, in particular presuppositions and conventional implicatures, which are independent of context and omit conversational implicature, which requires context information to formulate. The number of triggers used in this experiment is limited, and the results include negatives in some cases.

1.1　Inferences

Levinson [12] identifies the order of addition of inferences to the context of an utterance as (i) the entailments of the uttered sentence S (ii) the clausal conversational implicatures of S (iii) the scalar conversational implicatures of S (iv) the presuppositions of S. Presupposition is used to describe any kind of background assumption against which an action, theory, expression or utterance makes sense or is rational. Conventional implicatures are non-truth-conditional inferences that are not derived from superordinate pragmatic principles like the maxims [9] but are simply attached by convention to particular lexical items or expressions.

Fromkin et al. [5] state presuppositions as implicit assumptions about the world, required to make an utterance meaningful or appropriate. Conventional implicatures are associated with specific words and result in additional conveyed meanings when those words are used [17]. Presuppositions are denoted by '>>', and conventional implicatures are denoted by '≈'. For an utterance, "*The king of France is wise.*" there can be a presupposition that >> There is a present king of France [12]. For an utterance, "*Amelia is a toddler, but she is quiet.*" there can be a conventional implicature that ≈ Toddlers are not usually quiet [2].

As an example, upon reading a headline, '*Schaeuble says British were 'deceived' in Brexit campaign*', a reader may make the following inferences. (i) Schaeuble exists. (ii) Schaeuble said something. (iii) Schaeuble believes that the British were 'deceived' in the Brexit campaign. (iv) The Brexit campaign happened. (v) Brexit can have campaign. (vi) The British government was deceived in the Brexit campaign. (vii) The British citizens were deceived in the Brexit campaign. The inferences (vi) and (vii), which are conversational implicatures, need more contextual information along with the headline under consideration to support them. So generating inferences such as (vi) and (vii) is not attempted in this work, and we try to generate inferences similar to those stated from (i) to (v).

1.2 Related Work

Pekar [24] proposed a method for automatic discovery of pairs of verbs related by entailment, such as *X buy Y => X own Y* and *appoint X as Y => X becomes Y*. Learning verb inference rules by Hila Weisman et al. [25] introduce linguistically motivated indicators that are specific to verbs and may signal the semantic relation between verb pairs and present a supervised classification model for detecting lexical entailment between verbs.

Cianflone et al. [1] have introduced the novel task of predicting adverbial presupposition triggers, and this paper explores the scope of computing presupposition statements from the syntax structure provided by dependency trees of news headlines. The approach used in that paper uses deep learning, while this paper demonstrates a rule-based approach. The Recognizing Textual Entailment (RTE) challenge [18] dataset consisted of *text(t)-hypothesis(h)* pairs with the task of judging for each pair, whether *t* entails *h*. In this work, we attempt to generate hypotheses for news headlines rather than judging whether a hypothesis is correct. Burger and Ferro [19] attempted to generate a large corpus of textual entailment pairs from the lead paragraph and headline of a news article. To the best of our knowledge, ours is the first work towards computing presuppositions and conventional implicatures from English news headlines.

1.3 Linguistic Definitions and Characteristics of Headlines

Dor [3] defines headlines as the negotiators between stories and readers and identifies their four functions of summarising, highlighting, attracting, and selecting. The headline, together with the lead or the opening paragraph summarises a news story [4]. Gattani [7] identifies three broad macro headline functions. (i) The informative headline, which gives a good idea about the topic of the news story. (ii) The indicative headline, which addresses what happened in the news story. (iii) Eye catcher headline, which does not inform about the content of the news story but is designed to entice people to read the story. The higher the mental effort required for processing a headline, the less relevant it becomes [3]. While reading a headline, the reader should be able to construct assumptions, either based on what can be perceived in their immediate environment or based on assumptions already stored in their memory. The relevance of a headline is directly proportional to the amount of contextual effects and inversely proportional to the cognitive processing effort required to recover these effects [15].

Headlines are characterized by the density of the information present in them, and they have the syntactic characteristics of telegraphic speech. They may contain bold expressions, polarisation, exaggerations, and provocative wording [11]. While processing headlines, more information should be expected from a shorter span of words. The grammatical rules for proper English sentences would be frequently violated either for filling more information in the short space available or for promoting the curiosity of the reader. News headlines use a unique language called 'block language', a name first coined by Straumann [8]. Block language has a structure different from the usual clause or sentence structure, but it often conveys a complete message. This language usually consists of lexical items lower than sentences.

1.4   Relevance of This Work

This work computes inferences from headlines. The inferences generated can be fed to a learning system that grades the impact created by the headline, based on sensitivity, child-Friendliness, clarity, and various other parameters as required. It is advantageous to evaluate the impact because an ordinary reader naturally reads through the headlines in the newspaper before starting to read the whole news articles. The understandability of the headline contributes towards the ease of understanding of the news story that follows it.

2   Data

The dataset used in this work comprises around 350 headlines collected manually from different news websites [21–23] about four popular events which appeared continuously in news reports for a period of a few months. The topics selected for including in the dataset are 'Brexit', 'disputes over the South China sea', 'Syrian refugee crisis', and 'Pyeongchang winter olympics'. In the dataset, the headlines were arranged in chronological order to facilitate their use in studying the gradual evolution of the headlines, assuming that the reader has already read the previous headlines for the same news item. The timestamp associated with headlines in the dataset is not used in the present work. However, it might be useful for future developments to evaluate how headlines evolve as the news on that topic progresses in the course of time and how readers understand them based on their awareness of the previous headlines on the same topic.

2.1   Format of the Data

The data used as input for computing inferences using our rule-Based system are in the format: Headline [source: News source Timestamp]. A subset of the same dataset is used for collecting human inferences for evaluation purposes. Some Examples of headlines used as data is given below.

> *Schaeuble says British were "deceived" in Brexit campaign* [source: Reuters June 23, 2017, 07:18 PM IST]
> *Olympics: IOC will not exclude Asian cities from 2026 Games bid* [source: Reuters February 06, 2018, 02:58PM IST]

3 Proposed Method

In this work, it is assumed that only the headline is available to the reader for understanding the topic of the news and that the reader is entirely ignorant of the previous happenings under the same topic of news. The inferences of headlines are computed based on some logical conclusions attained, rooted in certain grammatical relations present in the headline. Rusu et al. [16] suggest subject-predicate-object triplet extraction from sentences that motivated this work. In the case of a news headline, the participants are the composer of the headline, who is the speaker, and the ordinary person reading the headline, who is the addressee. For computing inferences, we begin with the extraction of nouns and verbs.

In the algorithm, we start with dependency parsing the headline, thus obtaining the verbs occurring in the headline with their dependencies. We get the headline tagged with POS tagger from Stanford and then extract the list of nouns and list of verbs in the headline. The verbs are also lemmatized to get the base form of the verbs present in the headline. The lemmatized form is used when a different form of the verb other than the tense form in which it appears in the headline, is required for a changed tense form in the computed inferences. A few rule-Based approaches are implemented to get inferences from the headline. Stanford openIE [6] gives inferences that are directly stated in the headline. The headline "How the company kept out 'subversives'" gives the inference "company kept out 'subversives'" by openIE [6]. More inferences assumed from the syntactic structure of the headline are generated by the rule-Based system.

3.1 Extracting the Dependencies

The Stanford dependencies are binary grammatical relations held between a 'governor' and a 'dependent' [14]. The dependency tree is a singly rooted directed acyclic graph with an intermediate level of granularity of dependency relations, motivated by the needs of practical applications [26]. Dependency tree representations are considered as a right balance of complexity and expressivity for use in NLP related tasks [27]. The dependencies obtained from the Stanford CoreNLP dependency parser [13] are generated as a dependency tree that contains dependencies as tuples like those in the examples given below for the headline '*Rescue rules by Bank of England will divide Britain*'.

Each tuple consists of a dependency type *dep*, a *governorGloss*, and a *dependentGloss*. The current representation by Stanford contains approximately 50 grammatical relations. *governorGloss* is the word which is the governor in the grammatical relation, and *dependentGloss* is the word which is the dependent in the grammatical relation. We iterate through these tuples to find interrelation between words, which give rise to inferences.

(i) {'dep': 'case', 'governor': 6, 'governorGloss': 'England', 'dependent': 5, 'dependentGloss': 'of'}

(ii) {'dep': 'aux', 'governor': 8, 'governorGloss': 'divide', 'dependent': 7, 'dependent Gloss': 'will'}

(iii) {'dep': 'dobj', 'governor': 8, 'governorGloss': 'divide', 'dependent': 9, 'dependent Gloss': 'Britain'}

3.2   Rule-Based System for Inference Generation

In this work, we use a rule-Based system that comprises rules based on commonly occurring syntactic patterns. These patterns are modeled as inference triggers. Inference generation logic for an associated inference trigger is configured as a rule. Multiple iterations are performed on the dependency relations to generate inferences. Node JS tense conjugator [20] is used to find the required tense form of the verb to be attached in the computed inferences.

Since this work demonstrates the use of syntax structures to generate inferences using only a few triggers in the scope of inference triggers, the addition of more known triggers like iterative—anymore, return, another time, to come back, restore, repeat, etc. change of state verbs—*stopped, began, continued, start, finish, carry on, cease, leave, enter, come, go, arrive,* etc. factive verbs—*regrets, aware, realize, know, be sorry that, be proud that, be indifferent that, be glad that, be sad that,* etc. verbs of judging—*accuse, criticize, blame, apologize, forgive, condemn, impeach,* etc. which humans are better at making inferences upon should be included for more accurate results. The rules could be expanded with a string comparison of the verb under consideration with these triggers mentioned earlier. The current set of inference triggers and rules used in computing inferences from headlines are elaborated in the following sections. The set of rules can be extended with more patterns to improve the quality of the generated inferences.

Presence of a Future Tense Verb. The presence of a future tense verb in the headline could suggest that we can infer that the event described by the noun is yet to happen. If dependent is 'aux' (auxiliary) and 'dependentGloss' is the string 'will', then iterate once again through the dependencies to find a dependent 'dobj' (direct object) which is the noun phrase which is the (accusative) object of the verb where the 'governorGloss' of both dependency relations match. The dependency tree for a headline containing a future tense verb is given in Fig. 1.

For example, "Russian state television will not broadcast Olympics without national team." can have an inference >> "Olympics is not yet broadcast".

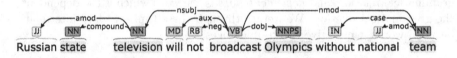

Fig. 1. Dependency tree for the headline "Russian state television will not broadcast Olympics without national team".

Algorithm 1. Computing Inferences Based on the Presence of a Future
Tense in a Headline.

Result: Inferences from headlines
Consider VD as the set of verbs in the headline with their dependencies,
    obtained from parser ;
for *each dependency tuple D in VD* **do**
    | **if** *'dep' of D is= 'aux' and 'dependentGloss' of D is = 'will'* **then**
    | | **for** *each dependency tuple ND in VD* **do**
    | | | **if** *'dep' of ND is = 'dobj' and 'governorGloss' of ND is =*
    | | | *'governorGloss' of D* **then**
    | | | | output 'dependentGloss' of ND;
    | | | | output "is not yet";
    | | | | output past tense of ('governorGloss' of D) ;
    | | | **end**
    | | **end**
    | **end**
end

Presence of Nominal Modifier 'of'. If there is a nominal modifier 'of' then, it
could suggest that we can infer that the dependent 'has' governor. For example,
"Governor of New Jersey meets PM Narendra Modi." can have an inference
>> "New Jersey has a Governor". The dependency tree for a headline with a
nominal modifier 'of' is given in Fig. 2.

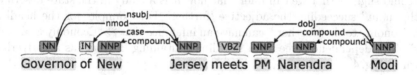

Fig. 2. Dependency tree for the headline "Governor of New Jersey meets PM Narendra
Modi".

Presence of the Conjunction 'but'. The presence of the conjunction 'but'
could suggest that we can infer that the subject was expected to undergo 'negation' of that which is mentioned in the part of the headline after the conjunction
'but'. The dependency tree for a headline with a conjunction 'but' is given in
Fig. 3. For example, "Flybe to close Isle of Man base in 2020 but flights will continue" can have an inference >> "Closing Isle of Man base was expecting flights
not continuing".

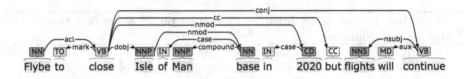

Flybe to   close   Isle of Man   base in   2020 but flights will continue

Fig. 3. Dependency tree for the headline "Flybe to close Isle of Man base in 2020 but flights will continue".

Algorithm 2. Computing Inferences Based on the Presence of Conjunction 'but' in a Headline.

Result: Inferences from headlines

Consider VD as the set of verbs in the headline with their dependencies, obtained from parser ;

for *each dependency tuple D in VD* **do**

    if *'dep' of D is = 'conj:but'* **then**

        output "being ";

        output 'governorGloss' of D;

        output " was [not] expecting ";

        output Gerund of ('dependentGloss' of D) ;

    end

end

Presence of 'further' as an Adjective. The presence of 'further' as an adjective could suggest that we can infer that now it is already in the state described by the 'noun' succeeding the adjective 'further'. For example, for the headline "UK economy to slow further." can have an inference that >> "Economy is already slow". The dependency tree for a headline containing 'further' as an adverb is shown in Fig. 4.

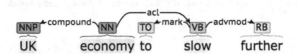

UK   economy to   slow   further

Fig. 4. Dependency tree for the headline "UK economy to slow further".

Presence of a 'noun compound'. The presence of noun compounds such as 'Brexit campaign' could suggest that we may infer that 'Brexit' that is the first part 'N1' of the noun compound can be/can have a 'campaign', that is the second part 'N2' of the noun compound. The problem of computing semantic relation of the nouns N1 and N2 in the noun compound is not dealt with in this experiment. Only common sense assimilation that "N1 can be N2" or "N1 can have N2" is generated. For example, "Russia's Olympic ban strengthens Putin's reelection hand." can have an inference >> "Olympic can be/can have ban". The dependency tree for a headline containing a noun compound is shown in Fig. 5.

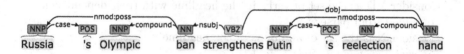

Fig. 5. Dependency tree for the headline "Russia's Olympic ban strengthens Putin's reelection hand".

Presence of a 'verb' in Past Tense. If the 'verb' is in the past tense in a headline, it could suggest that we can infer that the event has already happened. For example, "How women won the right to vote in 1918" can have an inference >> "women won the right to vote". The dependency tree for a headline with a verb in the past tense is shown in Fig. 6.

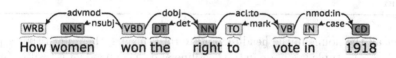

Fig. 6. Dependency tree for the headline "How women won the right to vote in 1918".

Presence of 'again' in a Clause with a Verb. The presence of 'again' as an adverbial modifier in a clause with a verb could suggest that we can infer that the event described by the noun has already happened. For example the headline "Norway regulator again rejects "Donut" fish farm volume plan." can have an inference >> "Norway regulator has rejected "Donut" fish farm volume plan before". The dependency trees for a headlines with 'again' as adverbial modifier is given in Fig. 7.

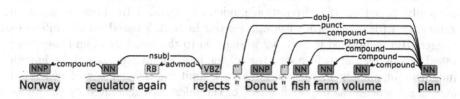

Fig. 7. Dependency tree for the headline "Norway regulator again rejects "Donut" fish farm volume plan".

Algorithm 3. Computing Inferences of a Headline Which has Presence of 'again' in a Clause with a Verb.

Result: Inferences from headlines

Consider VD as the set of verbs in the headline with their dependencies, obtained from parser ;

Consider N is the set of all nouns in the headline ;

for *each dependency tuple D in VD* **do**

    if *'dep' of D is = 'advmod' and 'dependentGloss' of D is 'again' or any noun in N is 'dependentGloss' with 'dep' of D = 'nsubj'* **then**

        for *each dependency tuple ND in VD* **do**

            if *'dep' of ND is = 'nsubj' and 'governorGloss' of ND is = 'governorGloss' of D* **then**

                output 'dependentGloss' of ND;

                output past tense of ('governorGloss' of D) "before";

            end

        end

    end

end

Get all the nouns in the headline and iterate through them until the 'dependentGloss' of a tuple is a noun in the headline and the dependent is 'nsubj' (nominal subject) that is a noun phrase which is the syntactic subject of a clause or if dependency relation is 'advmod' (adverb modifier). Then if the 'governorGloss' is 'again', follow from the inner 'for-loop' of Algorithm 3.

4 Results and Discussion

The unavailability of annotated inferences makes the comparison and evaluations difficult for this task. The inferences generated with the system are compared with manually annotated inferences for 100 randomly collected headlines. Annotators are two research scholars doing research in Linguistics and fluent in English. They did the annotation of the subset of the dataset for evaluation manually, based on the annotation guidelines provided to them. Annotation guidelines with explanatory examples given in Sect. 5 based on the inference triggers [12] mentioned in Sect. 3.2 were given to the annotators, and they were asked to look for the surface structure of the headline in general and use human judgment in making inferences. No upper limit on the number of generated human inferences was imposed. 11.8% of the inferences generated by the annotators were of the existential types, such as those beginning with a clause like "*there exists*". The inference triggers other than the existential ones are occurring less in headlines compared to normal discourse, due to the peculiarity of block language used.

Table 1. Accuracy and generated percentage of inferences computed.

Inference trigger	Percentage of accurate inferences	Percentage of inaccurate inferences	Percentage of missing inferences
But	69.3	0	30.7
Again	82.7	8.3	9
Further	94	6	0
Future tense	93	3	4
Noun compound	54.4	40.2	5.4

Table 2. Comparison of manually annotated inferences with computed inferences for a headline.

Headline	Manually annotated inferences	Computed inferences	Correct inferences	Incorrect results
IOC extends North Korea deadline for Pyeongchang games	1. IOC has power to extend deadline 2. North Korea has deadline 3. Deadline can be extended 4. There exists North Korea 5. There exists Pyeonchang games	1. Korea can have deadline 2. Pyeongchang has games 3. Games has deadline	40%	0%
Olympics: Medals at Winter Olympics through years	1. There exists Winter Olympics 2. Olympics has medals 3. Olympics had been happening through years 4. There exists medals in years	1. Winter can have olympics 2. Olympics has medals 3. Years had medals	75%	0%
Schaeuble Says British were "deceived" in Brexit campaign	1. Schaeuble exists 2. Schaeuble believes that the British were "deceived" in Brexit campaign 3. Brexit can have campaign 4. Schaeuble said something 5. Schaeuble believes that the British were 'deceived' in Brexit campaign 6. Brexit campaign happened	1. Schaeuble Says British were "deceived" 2. Brexit can be/can have campaign 3. Campaign has deceived	16.7%	33%

The percentages of computed inferences for some inference triggers used in this experiment are given in Table 1. For a headline 'Britain takes step towards Brexit with repeal bill' our system generates the following inferences (i) Britain takes step (ii) Britain takes step towards Brexit (iii) Britain takes step with repeal bill (iv) repeal can be/can have bill (v) Brexit has step.

Table 2 shows the comparison results of manually annotated inferences with the computed inferences for the three headlines in the first column and gives the percentage of correct computed inferences and percentage of incorrect results out of the computed inferences for those headlines. For example for the last headline in Table 2, "*Schaeuble Says British were 'deceived' in Brexit campaign*" only one of the manually annotated inferences—"Brexit can be/can have campaign" is computed by our rule-based system thus making the percentage of correct computed inferences to be 16.7%. Out of the three computed inferences, "campaign has deceived" is wrong. Therefore the percentage of incorrect results in the computed inferences is 33%.

5   Annotation Guidelines

5.1   Purpose of Annotation

This Annotation task targets to provide the possible presuppositions for a news headline. Presuppositions can be any background assumption against which the headline makes sense or is rational. A '>>' symbol denotes presuppositions. A sentence and its negative counterpart share the same set of presuppositions.

5.2   Guidelines for Annotating Presuppositions

For annotating, look for presupposition triggers, which are the linguistic items that are particular words or some aspects of the surface structure of the headline in general [12], which generates presuppositions.

Presupposition Triggers. The following are some presupposition triggers with examples.

Definite Descriptions. Example- Hunterston B: Pictures show cracks in Ayrshire nuclear reactor. >> There exist cracks in Ayrshire nuclear reactor.

Factive Verbs. Factive verbs like regrets, aware, realize, know, be sorry that, be proud that, be indifferent that, be glad that, be sad that, etc. Example- Corbyn 'regrets' Labour MPs' resignations. >> Labour MPs resigned.

Implicative Verbs. Implicative verbs like manage, remember, bother, get, dare, care, venture, condescend, happen, be careful, have the misfortune, have the sense, take the time, take the trouble, take the opportunity, etc. Example- How Russia Managed to Destroy Saudi Arabia? >> Russia destroyed Saudi Arabia.

Change of State Verbs. Change of state verbs such as stopped, began, continued, start, finish, carry on, cease, leave, enter, come, go and arrive. Example- China has stopped stockpiling metals. >> China had been stockpiling metals.

Iterative. Iterative words such as again, anymore, return, another time, to come back, restore, repeat and for the nth time. Example- HTC in talks with Micromax, Lava and Karbonn to return to Indian market. >> Micromax, Lava and Karbonn had been in the Indian market previously.

Verbs of Judging. Verbs of judging such as accuse, criticize, blame, apologize, forgive, condemn and impeach. Example- Amnesty criticizes Hungary over treatment of migrants. >> Amity thinks that Hungary was treating migrants badly.

Temporal Clauses. Temporal clauses such as before, while, after, when, during and whenever. Example- Britons were endlessly lied to during Brexit campaign >> There was a Brexit campaign.

Cleft Sentences. Cleft sentences like (i) What he wanted to buy was a Fiat, (ii) It is John for whom we are looking, (iii) All we want is peace etc. Example- It is Julian for whom we are looking >> We are looking for someone.

Implicit Clefts with Stressed Constituents. Implicit clefts with stressed constituents like capital letters, or bold type, or underlined type can give rise to presuppositions.

Comparisons and Contrasts. Comparisons and contrasts such as too, back and in return can give rise to presuppositions. Example- Russia is a better negotiator than Italy >> Italy is a negotiator.

Future Tense Verb. The presence of a future tense verb in the headline can create a presupposition that the event described in the noun has not happened yet. Example- Russian state television will not broadcast Olympics without national team. >> Olympics is not yet broadcast by Russian state television.

The Conjunction 'but'. Suggest a Contrast. The Conjunction 'but' Suggest a Contrast. Example- Olympics-It's ready but will they come? >> Being ready was expecting them to come.

Gender-Specific Statements. Example- New Zealand Prime Minister Jacinda Ardern gives birth to first child. >> Jacinda Ardern is a female.

Since the headlines use tricky language to attract readers, human intuition while listing the presuppositions is required. Presuppositions should be expressed as simple sentences in English.

6   Conclusions and Future Work

In this work, we considered headline as a stand-alone unit of text without attaching any information from the context in which it appeared in a news report. Based on the observation that the presence of certain words and tense conditions can trigger inferences from a headline, we tried to generate inferences based on a set of rules, formulated based on certain grammatical relations present in the headline. In the future, the rule set could be expanded to include more observations and complex rules to compute more inferences. These inferences can be used to measure the impact and sensitivity of a headline mainly for checking the appropriateness when used in a platform designed for children. This experiment was more of an attempt towards computing inferences from the headline, and the results are not complete due to the limited proportion of rules implemented compared to the large list of cases generating presuppositions and conventional implicatures. This approach of applying logic on the syntactic structure to generate inferences stands different from alternative methods using deep learning techniques because of the lesser data, time, and compute requirement.

Acknowledgements. We would like to thank Dr. Monojit Choudhury, Microsoft Research-Bangalore, for suggesting this topic of research as part of the Computational Socio-pragmatics course he taught at IIIT-H. We would also like to thank all the anonymous reviewers for carefully reading through a previous version of this document and for offering valuable suggestions for improvement.

References

1. Cianflone, A., Feng, Y., Kabbara, J., Cheung, J.C.K.: Let's do it "again": a first computational approach to detecting adverbial presupposition triggers. In: Proceedings of the 56th Annual Meeting of the Association for Computational Linguistics, Volume 1: Long Papers (2018). https://aclweb.org/anthology/P18-1256. Accessed 4 Sept 2019
2. Potts, C.: Into the conventional-implicature dimension (2006). https://doi.org/10.1111/j.1747-9991.2007.00089.x
3. Dor, D.: On newspaper headlines as relevance optimizers. J. Pragmat. **35**, 695–721 (2003). https://doi.org/10.1016/s0378-2166(02)00134-0
4. van Dijk, T.A.: News as Discourse (1990)
5. Fromkin, V., Rodman, R., Hyams, N.: An introduction to Language, 8th edn. Thomson/Wadsworth, Boston (2007)
6. Angeli, G. Premkumar, M.J., Manning, C.D.: Leveraging linguistic structure for open domain information extraction. In: Proceedings of the Association of Computational Linguistics (ACL) (2015). https://doi.org/10.3115/v1/p15-1034
7. Gattani, A.: Automated natural language headline generation using discriminative machine learning models. (2007). Simon Fraser University Homepage. https://summit.sfu.ca/item/2546. Accessed 2 Sept 2019
8. Straumann, H.: Newspaper Headlines: A Study in Linguistic Method. G. Allen & Unwin, Limited, London (1935). https://trove.nla.gov.au. Accessed 2 Sept 2019

9. Paul Grice, H.: Logic and conversation. In: Cole, P., Morgan, J.L. (eds.) Speech Acts, pp. 41–58. Academic Press, New York (1975). https://doi.org/10.1057/9780230005853_5

10. Iarovici, E., Amel, R.: The strategy of the headline. Semiotica **77**–4, 441–459 (1989). https://doi.org/10.1515/semi.1989.77.4.441

11. Kronrod, A., Engel, O.: Accessibility theory and referring expressions in newspaper headlines. J. Pragmat. **33**, 683–699 (2001). https://doi.org/10.1016/s0378-2166(00)00013-8

12. Levinson, S.C.: Pragmatics. Cambridge University Press, Cambridge (1983). https://doi.org/10.1017/CBO9780511813313

13. Manning, C.D., Surdeanu, M., Bauer, J., Finkel, J., Bethard, S.J., McClosky, D.: The Stanford CoreNLP natural language processing toolkit. In: Proceedings of the 52nd Annual Meeting of the Association for Computational Linguistics: System Demonstrations, pp. 55–60 (2014). https://doi.org/10.3115/v1/p14-5010

14. Marneffe, M.-C., Manning, C.D.: Stanford typed dependencies manual (2008). Revised for the Stanford Parser v. 3.7.0 (2016). Stanford NLP group homepage https://nlp.stanford.edu/. Accessed 2 Sept 2019

15. Pilkington, A.: Poetic Effects: A Relevance Theory Perspective. John Benjamins, Amsterdam (2000). https://doi.org/10.1075/pbns.75

16. Rusu, D., Dali, L., Fortuna, B., Grobelnik, M., Mladenic, D.: Triplet extraction from sentences. In: Proceedings of the 10th International Multi-Conference Information Society-IS, pp. 8–12 (2007). SemanticScholar homepage https://pdfs.semanticscholar.org/

17. Yule, G.: Pragmatics. Oxford University Press, Oxford (1996). https://doi.org/10.1017/CBO9780511757754.011

18. Dagan, I., Glickman, O., Magnini, B.: The PASCAL recognising textual entailment challenge. In: Quiñonero-Candela, J., Dagan, I., Magnini, B., d'Alché-Buc, F. (eds.) MLCW 2005. LNCS (LNAI), vol. 3944, pp. 177–190. Springer, Heidelberg (2006). https://doi.org/10.1007/11736790_9

19. Burger, J., Ferro, L.: Generating an entailment corpus from news headlines. In: Proceedings of the ACL Workshop on Empirical Modeling of Semantic Equivalence and Entailment, Ann Arbor, Michigan, pp. 49–54. Association for Computational Linguistics, June 2005. https://doi.org/10.3115/1631862.1631871

20. Spencer Kelly and many contributors: Compromise-modest natural-language processing in JavaScript. https://www.npmjs.com/package/compromise. Accessed 2 Sept 2019

21. Reuters Homepage. https://in.reuters.com/. Accessed 2 Sept 2019

22. The Hindu Homepage. https://www.thehindu.com/. Accessed 2 Sept 2019

23. BBC Homepage. https://www.bbc.com/. Accessed 2 Sept 2019

24. Pekar, V.: Discovery of event entailment knowledge from text corpora. Comput. Speech Lang. **22**, 1–16 (2008)

25. Weisman, H., Berant, J., Szpektor, I., Dagan, I.: Learning verb inference rules from linguistically-motivated evidence. In: EMNLP-CoNLL (2012)

26. de Marneffe, M.-C., MacCartney, B., Manning, C.D.: Generating typed dependency parses from phrase structure parses. LREC (2006)

27. de Marneffe, M.-C., Nivre, J.: Dependency grammar. Annu. Rev. Linguist. **5**, 197–218 (2019)

Extraction of Food Product and Shop Names from Blog Articles Using Named Entity Recognition

Ryuya Ikeda[1] and Kazuaki Ando[2]([✉])

[1] Graduate School of Engineering, Kagawa University, Kagawa, Japan
s18g454@stu.kagawa-u.ac.jp
[2] Faculty of Engineering and Design, Kagawa University, Kagawa, Japan
ando@eng.kagawa-u.ac.jp

Abstract. Named Entity Recognition (NER) is an important research topic in natural language processing, and has been widely studied for a long time. Recently, NER models using neural networks have been proposed, and have achieved the high performance for formal texts written in English such as the dataset of CoNLL2003. We are currently developing a system that provides useful information about food souvenirs that can be purchased only at the particular location or area (called the local limited food souvenirs). Therefore, it is important to apply an efficient NER method to extract food souvenirs and shop names as named entities from noisy user-generated texts written in Japanese such as texts of blog, Q&A, and online review sites. However, most of the existing NER methods using neural networks have not been evaluated with noisy user-generated texts or texts in languages other than English. In this paper, we propose a Conditional Random Fields-based model and compare the performance of the proposed CRF model and the existing state-of-the-art models using neural networks on a dataset constructed by blog texts written in Japanese. From the experimental results, it turned out that a simple neural network model with part-of-speech embeddings for NER has shown the best performance for named entities that were not included in the training data, and the proposed CRF model has shown the best precision in our task.

Keywords: Japanese named entity recognition · Information extraction · Blog article · Food souvenir · Conditional Random Field · Neural network

1 Introduction

Named Entity Recognition (NER) is one of the fundamental research problems in natural language processing. NER systems have been widely researched for a long time. In recent years, NER models using neural networks have achieved high performance [7,11,13]. Most of the models which achieved the state-of-the-art performance were evaluated with formal texts written in English such as

© Springer Nature Singapore Pte Ltd. 2020
L.-M. Nguyen et al. (Eds.): PACLING 2019, CCIS 1215, pp. 454–468, 2020.
https://doi.org/10.1007/978-981-15-6168-9_37

the dataset of CoNLL2003. It has hardly been verified whether these models can achieve the high performance for non-formal, noisy texts and/or texts in languages other than English.

We are currently developing a system that presents useful information about local limited food souvenirs, in order to provide support for choosing food souvenirs [8]. We need an efficient NER method to extract food souvenirs and shop names as named entities on noisy user-generated texts. Common entities such as names of person, location, and organization often include clues that can distinguish the entity classes in the notation of the entity itself. Names of the local limited food souvenirs are novel, emerging and rare entities, and have more variety of the notation. Therefore, it is difficult to detect and extract them compared to NER of common entities.

In this paper, we propose a Conditional Random Fields (CRF) based model, and verify the performance of the CRF model and the existing state-of-the-art models [7,11,13,15] on noisy user-generated texts written in Japanese by comparative experiments for recognizing names of food products and shops as named entities.

2 Related Work

2.1 Named Entity Recognition

Conventional NER systems with machine learning algorithms use hand-crafted features such as part-of-speech (POS) tags and the notations. In these systems, Support Vector Machine (SVM) or CRF was applied to sequence labeling tasks [9,12,18,21].

In recent years, neural network models (neural models) which do not depend on hand-crafted features achieved the high performance in the NER task. Huang et al. proposed Bidirectional-LSTM-CRF (BiLSTM-CRF) model combining Bidirectional-LSTM (BiLSTM) with CRF, and achieved high performance in NER tasks [7]. This model gives word representations obtained by inputting word embeddings to BiLSTM as inputs of CRF instead of using hand-crafted features. Many of the neural models for NER are based on the BiLSTM-CRF model. Lample et al. proposed a model that obtains subword information by giving characters in a word to BiLSTM [11]. Ma and Hovy proposed a Bidirectional-LSTM-CNN-CRF (BiLSTM-CNN-CRF) model that obtains subword information by using CNN instead of BiLSTM [13]. In both models, subword information and word embeddings are concatenated as an input of BiLSTM in BiLSTM-CRF. Recently, models using neural language model have achieved state-of-the-art performance in NER tasks [4].

2.2 Japanese NER

As for Japanese NER, several conventional models have been proposed. Sassano and Kurohashi proposed a SVM-based model using structural information such as cache features or coreference relations as hand-crafted features [21]. Iwakura proposed a boosting-based model using rules and word information extracted from unlabeled data [9].

In recent years, neural models for Japanese NER have been proposed. Misawa et al. implied that BiLSTM-CNN-CRF proposed by Ma and Hovy is not effective for Japanese NER, and proposed a Character-Bidirectional-LSTM-CRF (Char-BiLSTM-CRF) model [15]. This model which predicts entity labels for each character has achieved the state-of-the-art performance in Japanese NER. Mai et al. proposed a neural model for Japanese fine-grained named entities [14]. This model has improved the performance of Japanese fine-grained NER by using dictionary information based on Wikipedia texts and embeddings of entity categories.

2.3   NER in Noisy User-Generated Texts

Many NER models are evaluated using formal texts such as news articles (e.g. CoNLL2003 dataset). NER in noisy user-generated texts have received attention in recent years. Ritter et al. proposed a NER method for tweets [20]. Aguilar et al. proposed a method using a multitask learning and achieved state-of-the-art F-measure for the WNUT2017 dataset based on multiple social media domains [3].

Typically, the performance of NER in noisy user-generated texts tends to be lower because such text includes various expressions, vocabulary, and spelling errors. For example, the state-of-the-art NER method which was proposed by Akbik et al. [4] achieved the F-measure of 93.09% for the CoNLL2003 dataset. On the other hand, a method which was proposed by Aguilar et al. [3] achieved the F-measure of 45.55% for the WNUT2017 dataset.

3   Extraction of Food Product and Shop Names

We are currently developing a system that presents useful information about local limited food souvenirs, in order to provide support for choosing food souvenirs in Japan [8,16]. Therefore, we need an efficient NER method to extract names of food souvenirs and shops as named entities on noisy user-generated texts such as texts of blog and Q&A sites.

In this paper, we propose a CRF model for NER and choose the suitable method for our task by comparing the performance of the existing state-of-the-art models using neural networks.

3.1   Construction of Training Data

We explain the procedure to construct training data for NER. The procedure consists of two steps:

Collection of Sentences. Sentences about food souvenirs are extracted from blog articles. Each sentence is tokenized using Japanese morphological analysis.

Annotation. We annotate each token (word) of the results of morphological analysis. We annotate names of all food products with "PRO" labels, because there is a possibility that food products will be souvenirs in the future, even if the products are not written as food souvenirs in the sentence. Shop names selling food souvenirs are annotated with "SHO" labels. Others are annotated with "O" labels.

We use the BIOLU labeling scheme [19] : "B" for the first token of a entity; "I" for a token inside an entity; "O" for outside an entity; "L" for the last token of an entity; and "U" for the unit length entity (e.g. entity consisting of one word).

3.2   CRF Model

We have conducted the preliminary experiments to consider effective features of our CRF model [8]. The following features were obtained to contribute the performance improvement of the CRF model:

- Token string
- POS
- Subcategory of POS (SubPOS)
- Character types
- Set a flag to word(s) in brackets (inBracket).

Character types have seven types: number; lower case; upper case; Hiragana; Katakana; Kanji; and others. If a token string includes some character types, all character types which are included in the token string are used for the feature of the token (e.g., "買う (purchase)" includes Kanji and Hiragana, therefore the features of this token are "Kanji" and "Hiragana"). We use "inBracket" because names of food products and shops are written in brackets frequently. The CRF model uses these features for experiments in this paper. A window size is set to two, namely, the feature of each token includes the features of two previous tokens and next two tokens, and the current token. If the current token is the first of the sentence, the features of the current token are the features for the next two tokens and the current token.

3.3   Neural Models

We use four neural models for NER: BiLSTM-CRF (Huang model) [7]; BiLSTM-CRF (Lample model) [11]; BiLSTM-CNN-CRF [13]; and Char-BiLSTM-CRF [15] models. First three models achieved the high performance for NER, and are often used in current NER systems. The last model also achieved the state-of-the-art performance for Japanese NER. The Lample and BiLSTM-CNN-CRF models are subword information to BiLSTM layer in the Huang model. Sub-word information are obtained by inputting embeddings of characters which are included in the word to BiLSTM or CNN.

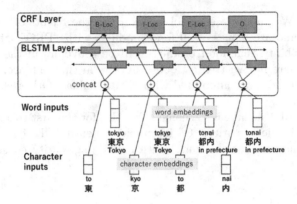

Fig. 1. A neural model proposed by Misawa et al. (sited from [15])

Figure 1 shows Char-BiLSTM-CRF model. This model consists of BiLSTM and CRF layers and predicts a label for every character independently. Word boundaries are not clear in Japanese, however, this model deals with boundary conflicts by predicting a label for every character.

4   Experiment

We evaluate the NER performance of the proposed CRF models and the state-of-the-art neural models by comparative experiments using blog texts in Japanese. We use MeCab [10] for Japanese morphological analysis and CRFsuite [17] for CRF in the experiments. A dictionary used for MeCab is IPADIC, and hyper parameters of CRFsuite is the default value. We implement neural models using AllenNLP toolkit [6].

4.1   Experimental Data

We collected names of food souvenirs from OMIYA! [1], and searched the category of "Sweets and Dessert" in Yahoo! Japan Blogs [2] by the names of food souvenirs. As the result of search, 680 articles including names of food souvenirs were obtained. We annotated named entity labels for 13,970 sentences in blog articles. There are many captions for images of souvenirs and sentences consisting only of a souvenir name or shop name in blog articles. Therefore, we excluded sentences according to the following two conditions from the annotated data in this experiment, in order to evaluate the performance of NER correctly.

- A sentence consisting only of nouns and postpositional particles.
- A sentence that does not include verbs, adjectives or auxiliary verbs.

Finally, 9,483 sentences are used as experimental data. The number of product names is 1,418 and the number of shop names is 750 in the data. The number of unique product names is 939 and the number of unique shop names is 476. Figure 2 shows the distribution of the length of sentences in the experimental data. The length of them was measured by the number of words included in a sentence or a named entity. From Fig. 2, we can see that many sentences in the experimental data consist of 10 to 30 words. Figure 3 shows the distribution of the length of sentences including a named entity in the experimental data. The left and right side graphs in Fig. 3 show the distribution of the length of sentences including a product entity or a shop entity, respectively. There is no great difference between these distributions on the experimental data. Figure 4 shows the distribution of the length of named entities in the experimental data. These distributions in Fig. 4 have also no great difference.

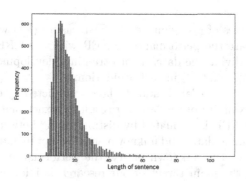

Fig. 2. Distribution of the length of sentences in the experimental data

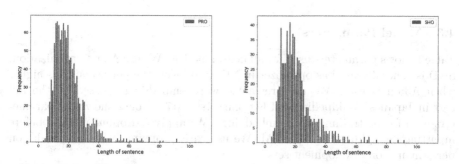

Fig. 3. Distribution of the length of sentences by named entities in the experimental data

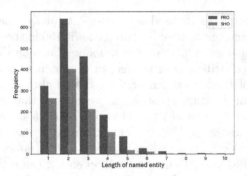

Fig. 4. Distribution of the length of named entities in the experimental data

4.2 Evaluation

We measure the values of precision, recall, F1-measure (F1) with 10-fold-cross-validation to compare the performance of NER with the CRF model and four neural models. We divide the dataset into training, development, and test sets at a ratio of 90%, 5%, and 5% in each validation.

We compare the correct labels and the labels predicted by each NER model, and determine the validity of the labeling result by exact match. In particular, the performance of NER is evaluated by distinguishing between known entities (included in the training data) and unknown entities (not included in the training data). In the case of the unknown entity, the performance of NER tends to be lower [5]. Since the main target of our research is information about local limited food souvenirs, it is important to extract unknown entities effectively. The average ratio of unknown product and shop names in each test set is about 40%.

4.3 Model Parameters

Table 1 shows parameters used for all neural models. We use Adam optimization. SGD is generally used as optimizer in NER, however, the performance is higher when Adam is used. We use pretrained word embeddings[1] based on the full text in Japanese Wikipedia as of February 1st 2017. Table 2 shows the parameters used for pretraining word embeddings. We apply a dropout method before inputting embeddings to BiLSTM. We use early stopping based on validation performance on development sets.

Next, we show the parameters of neural models using character embeddings. The latter half of Table 1 shows the parameters of the BiLSTM-CNN-CRF and the Lample models. We set the dimension of character embeddings to 50 and the hidden size of CNN or BiLSTM layer to 50, because the number of character types in Japanese is larger than that of English. By referring to [13] character

---

[1] http://www.cl.ecei.tohoku.ac.jp/~m-suzuki/jawiki_vector/.

Table 1. Parameters used for all neural models

Common parameters	
Dimension of word BiLSTM hidden layer	128
Number of units of word BiLSTM	1
Maximum number of epoch	50
Batch size	32
Learning rate	0.001
Dropout rate	0.5
Gradient clipping	5.0
Optimizer	Adam
Early stopping patience	20
Parameters of BiLSTM-CNN-CRF model	
Number of CNN filter	50
Window size of CNN	2
Parameters of Lample model	
Hidden size of character BiLSTM	50
Number of units of character BiLSTM	1

Table 2. Parameters used for pretraining word embeddings

Model architecture	cbow
Embedding size	200
Window size	5
Number of negative samples	5
Downsample setting for frequent words	0.001

embeddings are initialized with uniform samples from $[-\sqrt{\frac{3}{dim}}, \sqrt{\frac{3}{dim}}]$, where the *dim* represents the dimension of character embeddings. The character embeddings are updated during training of neural models. We set the window size of CNN to two, because the length of Japanese words tends to be shorter than that of English. The hyper parameter of Char-BiLSTM-CRF is only the dimension of character embeddings. The number of the dimension of character embeddings is set to 50 as with other models.

4.4   Results

Table 3 and Table 4 show the evaluation results of the CRF and neural models. The best performance in all models is highlighted by boldface type.

First, we focus on the result for "all" entities in two tables. It turns out that the f1-measure of the Huang model is the highest in both the results for PRO and SHO entities. Next we focus on the result of "unknown" entities. It can be seen that the BiLSTM-CNN-CRF model has achieved the best f1-measure for

Table 3. Experimental results for PRO

Model	All			Unknown			Known		
	Precision	Recall	F1-measure	Precision	Recall	F1-measure	Precision	Recall	F1-measure
CRF	**76.51**	61.03	67.61	**59.44**	49.39	53.51	**88.51**	68.74	77.03
Huang model	70.65	74.09	**72.19**	56.01	62.33	58.55	85.77	**85.22**	**85.42**
BiLSTM-CNN-CRF	70.03	73.67	71.67	56.02	**62.99**	**59.30**	85.84	83.73	84.50
Lample model	68.93	73.55	70.95	53.69	62.34	57.13	86.30	84.61	85.18
Char-BiLSTM-CRF	67.24	**74.43**	70.07	48.65	60.20	52.61	84.12	83.81	83.80

Table 4. Experimental results for SHO

Model	All			Unknown			Known		
	Precision	Recall	F1-measure	Precision	Recall	F1-measure	Precision	Recall	F1-measure
CRF	**83.45**	56.13	66.64	**61.10**	34.57	41.84	94.49	69.70	79.84
Huang model	73.08	73.60	**73.12**	53.03	**61.47**	**56.14**	**94.78**	84.12	88.87
BiLSTM-CNN-CRF	72.23	71.06	71.33	49.37	53.05	50.96	93.04	86.35	89.03
Lample model	74.90	70.28	72.39	52.42	50.36	50.93	94.22	**86.89**	**90.19**
Char-BiLSTM-CRF	71.07	**73.97**	72.10	44.30	55.09	47.79	92.44	85.27	88.57

unknown PRO entities, and the Huang model has achieved the best f1-measure for unknown SHO entities. Although the CRF model has achieved the best precision for "all" and "unknown" entities, the recall of the CRF model was lower than that of the neural models. In particular, the recall of the CRF was about 27 points lower than the best recall of the neural models for unknown SHO entities. The precision of the Char-BiLSTM-CRF model was also lower than that of the other neural models. The Char-BiLSTM-CRF had achieved the state-of-the-art performance for Japanese NER [15], however, this model was not effective for our task.

4.5   Additional Experiments and Results

We verify the effective method to improve the performance of recognizing unknown entities by adding following features to the neural models.

– POS embeddings
– Pretrained word embeddings using blog texts.

Aguilar et al. reported that the performance for the WNUT2017 dataset was improved by POS embeddings [3]. The POS feature also greatly contributed to the improvement of the performance of NER of our CRF model. Therefore, we expect further improvement by adding POS embeddings to neural models.

We prepare POS embeddings and SubPOS embeddings separately, and initialize them in the same way as character embeddings. We use information of POS and SubPOS which are obtained from the results of morphological analysis in the same way as the CRF model. POS and SubPOS embeddings are also updated during training of neural models. The dimension of POS embeddings is set to 10, which has obtained the high performance in our preliminary experiments.

Table 5. Experimental results for PRO by neural models with POS embeddings

Model	all			unknown		
	precision	recall	f1-measure	precision	recall	f1-measure
CRF	**76.51**	61.03	67.61	**59.44**	49.39	53.51
Huang model	71.45(70.65)	**77.90**(74.09)	**74.46**(72.19)	57.60(56.01)	**68.25**(62.33)	**62.18**(58.55)
BiLSTM-CNN-CRF	67.50(70.65)	73.03(73.67)	69.01(71.67)	53.24(56.02)	59.59(62.99)	55.34(59.30)
Lample model	71.14(68.93)	74.49(73.55)	72.45(70.95)	57.59(53.69)	63.94(62.34)	59.66(57.13)
Char-BiLSTM-CRF	68.78(67.24)	73.63(74.43)	70.55(70.07)	51.19(48.65)	60.52(60.20)	54.48(52.61)

Table 6. Experimental results for SHO by neural models with POS embeddings

Model	all			unknown		
	precision	recall	f1-measure	precision	recall	f1-measure
CRF	**83.45**	56.13	66.64	**61.10**	34.57	41.84
Huang model	75.85(73.08)	**75.61**(73.60)	**75.61**(73.12)	56.30(53.03)	60.67(**61.47**)	**58.06**(56.14)
BiLSTM-CNN-CRF	71.99(72.23)	67.27(71.06)	69.11(71.33)	47.42(49.37)	47.71(53.05)	46.74(50.96)
Lample model	74.11(74.90)	68.58(70.28)	70.94(72.39)	51.80(52.42)	48.01(50.36)	49.39(50.93)
Char-BiLSTM-CRF	73.17(71.07)	74.61(73.97)	73.58(72.10)	45.74(44.30)	55.81(55.09)	49.58(47.49)

Table 7. Experimental results by different pretrained word embeddings

Model	all					
	PRO			SHO		
	Wiki	Blog	Blog + Wiki	Wiki	Blog	Blog + Wiki
Huang model	72.19	71.49	**72.39**	73.12	72.19	**73.47**
BiLSTM-CNN-CRF	71.67	71.96	**72.55**	71.33	71.68	**72.97**
Lample model	70.95	70.90	**73.93**	**72.39**	71.49	71.80
Char-BiLSTM-CRF	70.07	68.69	**70.74**	72.10	72.52	**74.61**
Model	unknown					
	PRO			SHO		
	Wiki	Blog	Blog + Wiki	Wiki	Blog	Blog + Wiki
Huang model	58.55	**58.65**	56.69	**56.14**	53.23	52.49
BiLSTM-CNN-CRF	**59.30**	58.55	56.94	50.96	**53.21**	49.95
Lample model	57.13	56.21	**59.44**	50.93	**53.29**	50.20
Char-BiLSTM-CRF	**52.61**	49.52	52.10	47.49	46.87	**48.59**

Table 5 and Table 6 show the evaluation results of the CRF model and the neural models with POS embeddings. In these tables, the values in each parentheses indicate the result of the neural model without POS embeddings and SubPOS embeddings. From Table 5 and Table 6, it is confirmed that the performance of the Huang model was improved. The f1-measure of the Huang model for unknown PRO entities has been improved by two points or more. The f1-measure for unknown SHO entities has been also improved by about two points. However, the performance of the BiLSTM-CNN-CRF model and the Lample model have decreased overall by adding POS embeddings.

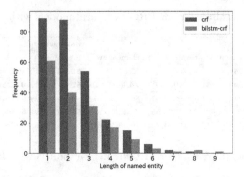

Fig. 5. Distribution of the frequency of extraction errors in product entities by the number of words

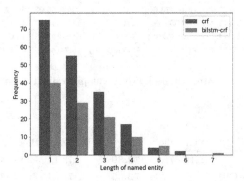

Fig. 6. Distribution of the frequency of extraction errors in shop entities by the number of words

Next, we conduct the experiments to confirm the effectiveness of text domains used for generating word embeddings. In the previous experiments, the pretrained word embeddings based on Japanese Wikipedia was used. In this experiment, we use full texts and comments in blog articles obtained from Yahoo! Japan Blog [2] to pretrain word embeddings. We extracted blog articles including "土産 (souvenir)" in the set of articles from January 1st, 2005 to December 31st, 2014, and the following two pretrained embeddings.

- Pretrained embeddings by blog texts (Blog)
- Pretrained embeddings by texts of blog articles and Japanese Wikipedia as of May 27th, 2019 (Blog + Wiki).

Table 2 shows the parameters used for pretraining the embeddings. Table 7 shows the f1-measure of neural models with each of pretrained word embeddings. As shown in Table 7, we could not find the significant improvement of the NER performance from the results by the neural models with each of pretrained embeddings.

5 Discussion

5.1 Quantitative Evaluation

Figure 5 and Fig. 6 show the distribution of the number of errors in product and shop entities extracted by the CRF model and the neural model in 10-fold-cross-validation. The neural model in Fig. 5 and 6 is the Huang model with POS embeddings and the pretrained word embeddings based on Japanese Wikipedia.

From Fig. 5 and Fig. 6, it can be seen that the CRF model has more errors in extracting short entities compared with the neural model. We consider that the neural model is effective for short entities, because it can use all words as clues during labeling a sequence using the Bidirectional LSTM.

Next, we analyze extraction errors. We classify extraction errors into the following 4 types in analysis of errors.

- Labeling a named entity as "O" (NoLabel)
- Labeling non named entity (MisLabel)
- Mistake of range for labeling (MisRange)
- Labeling a wrong entity class (MisClass).

Figure 7 shows the rate of extraction errors of the CRF model and the neural model. The neural model in Fig. 7 is the same as the neural model in Fig. 5 and 6. From Fig. 7, we can see that the CRF model has many errors in NoLabel. On the other hand, errors in NoLabel of the neural model are lower than that of the CRF model. Moreover, the neural model has more errors in MisLabel and MisRange than the CRF model. From these results, we found that the CRF model and the neural model differ in the tendency of extraction errors, and there are types of sentences and named entities suitable for each model.

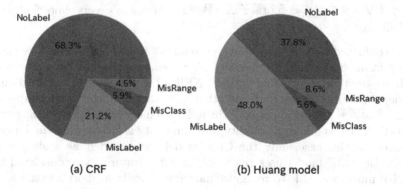

Fig. 7. Rate of extraction error

5.2 Error Analysis

We analyze the named entities recognized only by the neural models and only by the CRF model respectively.

The followings show examples of named entities recognized only by the neural models and sentences including the named entities.

- Product name: 山川 (Yamakawa), 若草 (Wakakusa)
  桂月堂さんの「山川」と「若草」と「薄小倉」を購入致しました。((I) purchased "Yamakawa", "Wakakusa" and "Usuogura" of Keigetsudo.)
- Product name: カマンベールチーズフロランタン (Camember Chees Florentin)
  千葉県マザー牧場「カマンベールチーズフロランタン」チーズ味がしっかりとした焼き菓子。
  (Chiba Mother Farm "Camembert Cheese Florentin" Baked sweets with rich cheese taste.)

We confirmed that the neural models can recognize product names by learning information about long-distance words such as "購入する (purchase)" or "菓子 (sweets)". For example, "山川 (Yamakawa)", "若草 (Wakakusa)" and "薄小倉 (Usuogura)" are food product names in the first example sentence, the CRF model only recognize "薄小倉 (Usuogura)" as a souvenir name. The CRF model can not use information in the word "購入する (purchase)" for labeling "山川 (Yamakawa)" and "若草 (Wakakusa)". In contrast, there are cases where neural models can use information in some words far from the named entity.

Next, the followings show examples of named entities recognized only by the CRF model and sentences including the named entities.

- Shop name: 丸玉製菓 (Marutamaseika)
  丸玉製菓が1番美味しいです。(Marutamaseika is the most delicious.)
- Shop name: 藤屋 (Fujiya)
  バイオリンタウンにある和菓子店「藤屋」。(Japanese sweets shop "Fujiya" in Violin Town.)

We confirmed that the neural models tend not to be able to recognize named entities from short sentences. Usually, short sentences do not have clue words which are used for recognizing entities. NER often fails even if a named entity includes clue words, characters, and other information. In the first example sentence, "丸玉製菓 (Marutamaseika)" includes a clue word "製菓 (seika; confectionery)". There is a high possibility that the entity is a shop name in Japanese sentences. In this case, only the CRF model recognized it as a shop name. Because the CRF model uses the word itself for features, we considered that the CRF model was able to recognize named entities from short sentences.

Next, the followings show example sentences including named entities labeled only by the Huang model.

- PRO entity: みそまん (Misoman)
  さすが有名な「みそまん」だけあって美味しいですね。(As expected, famous "Misoman" is delicious.)

- SHO entity: うさぎや (Usagiya)
  出かけたのは日本橋のうさぎやです。(I went to Usagiya in Nihonbashi.)
- PRO entity: 東京ばな奈 (Tokyo banana)
  期間限定ということで東京ばな奈と一緒に迷わず購入。(As it is a limited time offer, I bought it with Tokyo Banana without hesitation.)

We confirmed that the neural models with character embeddings tend not to be able to recognize named entities including Hiragana (the Japanese syllabary) characters. Compared with Kanji or Katakana, Hiragana is rarely included in the product and shop entities. Therefore, we consider that Hiragana in named entities is possible to become noise for learning and recognizing.

We also confirmed that the neural models with character embeddings tend not to be able to recognize named entities due to spelling inconsistencies related to character classes. In the third example sentence, "東京ばな奈" is the exact product entity. There is a spelling inconsistency such as "東京ばなな" and "東京バナナ" in the training data. This kind of spelling inconsistencies tend to appear in user-generated contents written in Japanese frequently. The existing state-of-the-art NER models with character embeddings could not recognize such entities.

From whole of experiments, we can say that the Huang model is effective for extracting product and shop entities from blog texts.

6   Conclusion

In this paper, we have proposed the method to extract names of food products and shops from blog texts by NER to collect information about food souvenirs from noisy user-generated texts in Japanese. By comparing the extraction performance of the proposed model with the state-of-the-art NER models, we confirmed that the effectiveness of the BiLSTM-CRF model proposed by Huang et al. has achieved the best performance for unknown and known entities. We also confirmed that the proposed CRF model has achieved the best precision.

In the future, we will improve the performance of extracting unknown named entities. Then, we will propose a method to extract other information (e.g., price, review) and a method to identify whether it is local limited food souvenirs. Finally, we will collect information about local limited food souvenirs from web sites using proposed method.

Acknowledgment. This work was partially supported by JSPS KAKENHI Grant Number JP19K12271.

References

1. OMIYA!. https://omiyadata.com/jp/
2. Yahoo! Japan Blog (At 2019–12-15, The service was end). https://blogs.yahoo.co.jp/

3. Aguilar, G., López Monroy, A.P., González, F., Solorio, T.: Modeling noisiness to recognize named entities using multitask neural networks on social media. In: Proceedings of the NAACL (2018)
4. Akbik, A., Blythe, D., Vollgraf, R.: Contextual string embeddings for sequence labeling. In: Proceedings of the COLING (2018)
5. Augenstein, I., Derczynski, L., Bontcheva, K.: Generalisation in named entity recognition: a quantitative analysis. Comput. Speech Lang. **44**, 61–83 (2017)
6. Gardner, M., et al.: AllenNLP: a deep semantic natural language processing platform. In: Proceedings of Workshop for NLP Open Source Software (NLP-OSS) (2018)
7. Huang, Z., Xu, W., Yu, K.: Bidirectional LSTM-CRF models for sequence tagging. arXiv: 1508.01991 (2015)
8. Ikeda, R., Ando, K.: Extraction of product and shop names of souvenirs from blog articles using conditional random fields. In: Proceedings of the Fourth International Conference on Electronics and Software Science (2018)
9. Iwakura, T.: A named entity recognition method using rules acquired from unlabeled data. In: Proceedings of the RANLP (2011)
10. Kudo, T., Yamamoto, K., Matsumoto, Y.: Applying conditional random fields to Japanese morphological analysis. In: Proceedings of the EMNLP (2004)
11. Lample, G., Ballesteros, M., Subramanian, S., Kawakami, K., Dyer, C.: Neural architectures for named entity recognition. In: Proceedings of the NAACL (2016)
12. Li, Y., Bontcheva, K., Cunningham, H.: SVM based learning system for information extraction. In: Winkler, J., Niranjan, M., Lawrence, N. (eds.) DSMML 2004. LNCS (LNAI), vol. 3635, pp. 319–339. Springer, Heidelberg (2005). https://doi.org/10.1007/11559887_19
13. Ma, X., Hovy, E.: End-to-end sequence labeling via bi-directional LSTM-CNNS-CRF. In: Proceedings of the ACL (2016)
14. Mai, K., et al.: An empirical study on fine-grained named entity recognition. In: Proceedings of the COLING (2018)
15. Misawa, S., Taniguchi, M., Miura, Y., Ohkuma, T.: Character-based bidirectional LSTM-CRF with words and characters for Japanese named entity recognition. In: Proceedings of the First Workshop on Subword and Character Level Models in NLP (2017)
16. Nagao, N., Ando, K.: Extraction of product names for constructing a database of souvenir information. In: Proceedings of the Fifth International Conference on Informatics and Applications (2016)
17. Okazaki, N.: CRFsuite: a fast implementation of conditional random fields (CRFS) (2007). http://www.chokkan.org/software/crfsuite/
18. Passos, A., Kumar, V., McCallum, A.: Lexicon infused phrase embeddings for named entity resolution. In: Proceedings of the CoNLL (2014)
19. Ratinov, L., Roth, D.: Design challenges and misconceptions in named entity recognition. In: Proceedings of the CoNLL (2009)
20. Ritter, A., Clark, S., Mausam, Etzioni, O.: Named entity recognition in tweets: an experimental study. In: Proceedings of the EMNLP (2011)
21. Sasano, R., Kurohashi, S.: Japanese named entity recognition using structural natural language processing. In: Proceedings of the IJCNLP (2008)

Transfer Learning for Information Extraction with Limited Data

Minh-Tien Nguyen[1,2](✉), Viet-Anh Phan[3], Le Thai Linh[1], Nguyen Hong Son[1], Le Tien Dung[1], Miku Hirano[1], and Hajime Hotta[1]

[1] Cinnamon Lab, 10th Floor, Geleximco Building, 36 Hoang Cau, Dong Da District, Hanoi, Vietnam
{linhlt,levi,nathan,miku,hajime}@cinnamon.is
[2] Hung Yen University of Technology and Education, Hung Yen, Vietnam
tiennm@utehy.edu.vn
[3] Le Quy Don Technical University, Hanoi, Vietnam
anhpv@mta.edu.vn

Abstract. This paper presents a practical approach to fine-grained information extraction. Through plenty of authors' experiences in practically applying information extraction to business process automation, there can be found a couple of fundamental technical challenges: (i) the availability of labeled data is usually limited and (ii) highly detailed classification is required. The main idea of our proposal is to leverage the concept of transfer learning, which is to reuse the pre-trained model of deep neural networks, with a combination of common statistical classifiers to determine the class of each extracted term. To do that, we first exploit BERT to deal with the limitation of training data in real scenarios, then stack BERT with Convolutional Neural Networks to learn hidden representation for classification. To validate our approach, we applied our model to an actual case of document processing using a public data of competitive bids for development projects in Japan. We used 100 documents for training and testing and confirmed that the model enables to extract fine-grained named entities with a detailed level of information preciseness specialized in the targeted business process, such as a department name of application receivers.

Keywords: Information extraction · Transfer learning · BERT

1 Introduction

The recent growth of AI technologies is dramatic, as many applications have already been deployed to real business cases [11]. One of the most significant topics in the business scene is the conversion of unstructured text into structured one because it is stated as an entrance to the digital transformation [6,18]. It has been studied by many proposed methods [1]. In business services, the extraction of specific information matters such as organization names, personal names,

© Springer Nature Singapore Pte Ltd. 2020
L.-M. Nguyen et al. (Eds.): PACLING 2019, CCIS 1215, pp. 469–482, 2020.
https://doi.org/10.1007/978-981-15-6168-9_38

addresses, or date plays an important role to facilitate document processing systems. Therefore, named entity recognition (NER) is one of the key technologies of natural language processing (NLP) which help AI software contributes to the commercial applications [5].

NER has been received attention due to its important role in many NLP applications [3,7,9]. Conventional approaches use dictionary-based [5], which requires a huge amount of time to set up to keep it up-to-date. Recent approaches have been utilized machine learning to reduce the efforts of dictionary building and maintenance [9]. Majority of the researches have successfully done the simple tasks of NER, yet those have focused on relatively-easy tasks, e.g. extracting names of person or organizations; hence it is not always straightforward to apply the proposed algorithms to real cases due to two gaps. The first gap is about the amount of data. Document processing typically focuses on narrow, specific topics rather than general, wide ones, where only a small amount of annotated data is available. Let's take our case as an example. We only received 100 bidding documents for both training and testing. It challenges neural-network-based natural architectures, which usually require more than 10,000 annotated training examples. For instance, CoNLL 2003 provided 20,000 annotated words for NER [2]. For actual cases, in a narrow domain, annotating the such number of training data is a time-consuming and non-trivial task. For the second gap, the identification of information types cannot be merely the categories [4,10,14]. The usual scope of NERs is to extract the categories of information such as organization names; however, in many cases, there can be two types of organizations, such as payee and payer in invoices. We believe that this level of detailed understanding is the key to the practical use of NER.

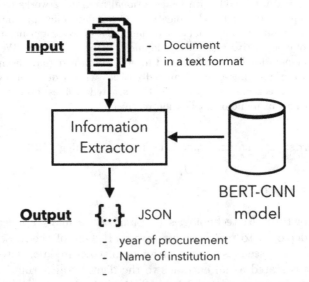

Fig. 1. The overview of the system.

In this research, we aim to identify a particular type of named entities from documents with the limited amount of training data. The proposed model utilizes the idea of transfer learning to fine-grained NER tasks. As shown in Fig. 1, the information extractor employs BERT [2] for addressing the small number of training data (the first gap) and then stacks Convolutional Neural Network (CNN) with multilayer perceptron (MLP) to extract detail information, e.g. the year of procurement and the name of institution (the second gap). Outputs are stored in the JSON format for other tasks of document processing in our system. Our contributions are two-fold:

- We propose a practical model of fine-grained NER which employs BERT [2] as an element of transfer learning [12]. By employing BERT and retraining the model with a small amount of data, our model achieves significant improvements compared to strong methods in extracting fine-grained NER. To the best our knowledge, we are the first study for the extraction of fine-grained NER for Japanese bidding documents.
- We present an application to the actual case of competitive bids for development projects in Japan using a public data set and confirm that the accuracy of the proposed model can be acceptably high enough for practical use.

We applied our model to a real scenario of extracting information from bidding documents. Statistical analyses show that our model with small training data achieves improvements in term of F-score compared to strong baselines.

The rest of the paper is organized as follows: Sect. 2 presents the relevant research. Our proposed model is described in Sect. 3. Settings and evaluation metric are shown in Sect. 4. We show the results, discussion, and error analysis in Sects. 5 and 6. Section 7 concludes our investigations.

2 Related Work

Conventional methods utilize dictionaries [16] or take advantage of machine learning for named entity extraction. The dictionary-based method usually uses a pre-defined dictionary of entities to match tokens in documents. This method can achieve high accuracy, but it is time-consuming and labor-expensive to prepare the dictionary. In contrast, the machine learning method exploits features to train a classifier which can distinguish entities. This method has been shown efficiency for NER. Recently, the success of deep learning attracts researchers to apply this technique to information extraction. A recent study employs Long-Short Term Memory (LSTM) with a Conditional Random Field (CRF) to classify the contextual expressions [9]. More precisely, the authors used LSTM to learn the hidden representation of data and then stacked CRF for classification. This method showed promising results. In practice, several research projects focus on the nested named entities and have a great progress so far [3,7].

For NER, high-level concepts such as people, places, organizations usually need to extract. However, for practical applications, categories must be in a more detailed level [8]. Here, fine-grained entity type classification was proposed, especially in the field of question answering, information retrieval [10,14], or the automatic generation of ontology [4]. However, the main challenge of fine-grained NER is the amount of training data required to train the classifier. To tackle this problem, transfer learning [12] is an appropriate solution. It leverages pre-trained models trained by a large amount of out-domain data to build a new model with a small number of training data in a new domain. Transfer learning is efficient because we daily have faced with limited training data. Let's take our scenario as an example, we need to extract values for 24 tags in long Japanese documents while the number of the training documents is only 78. Thus, transfer learning is one of the most efficient technique in such scenario [15,17].

One of the highlights of transfer learning is the recent open source named Bidirectional Encoder Representation from Transformers (BERT) [2]. BERT is a form of transfer learning and has achieved state-of-the-art results on 11 NLP tasks, including the very competitive Stanford Question Answering Dataset (SQuAD). In this work, we develop a model based on BERT for our business task, which extracts information in long Japanese documents.

3   Proposed Approach

This section first introduces the problem and then describes our proposed model for information extraction from very long texts of Japanese bidding documents with limited training data.

(2)   調達物件の特質等　入札説明書及び仕様書による。

(3)   供給期間　平成30年10月1日午前0時から平成31年9月30日
午後12時まで

(4)   供給場所　宮崎県小林総合庁舎

(5)   入札方法　(1)の調達物件について入札を実施する。入札金額

Fig. 2. A part of bidding document. Clause 3 (two first lines) mean *"supply period from 04/01/2019 to 03/31/2020 (MM/dd/YYYY)"*.

3.1   Task Definition

This research focuses on an actual case of document processing with a public actual data of competitive bids for development projects in Japan. The task is to extract values for fields (tags) from bidding documents of an electric power company. A bidding document is a long document which has 3 sections 1) specifications, 2) invitation to bid, 3) instructions to bid.

Table 1. Information need to be extracted from bidding contract documents. Content type of name can be the name of an entity such as a company, department, or person.

No	Tag name	Type
1	Year of procurement	year
2	Prefecture	text
3	Tittle of bidding	text
4	Name of institution	text
5	Address for demand	address
6	Start date of procurement	date
7	End date of procurement	date
8	Contract value	number
9	Amount of value	number
10	Class of reserved value	number
11	Amount of reserved value	number
12	Public announcement date	date
13	Deadline for delivery the specification	date
14	Deadline for questionnaire	date
15	Deadline for applying qualification	date
16	Deadline for biding	date
17	Opening application date	date
18	PIC of inquiry of question	name
19	TEL&FAX of inquiry of questions	Tel/Fax
20	Address for submission of application of qualification	address
21	Department & PIC for submission of application of qualification	name
22	Address of submitting of application of bidding	address
23	Department & PIC for submission of application of bidding	name
24	Place of opening bid	address

Table 1 shows target information to be extracted. As observed, many tags may have the same value or data types, e.g. deadlines for the questionnaire (tag

14) and bidding (tag 16). Hence, locating the value of a specific tag is challenging and needs to understand the context in the document.

Figure 2 shows a part of a bidding document with three clauses 3, 4, 5 of the specification section. The yellow highlighted texts are the values for tag 6 (start date of procurement), and tag 7 (end date of procurement). Another difficulty is that a tag is not represented by specific terms. Therefore, to retrieve the values correctly, the learning model needs to understand the document structures and meaning of the text. For example, given a paragraph like Fig. 3, where the yellow line is the value of tags 20, 22, and the green line is the value of tags 18, 21, 23, 24, the model needs to decide which is the tag of the yellow and green lines correctly. To do that, we introduce a model which is the combination of BERT and CNN as follows.

(4)　封筒（長形3号）に入れ、入札書に押印した印鑑と同じもので封印し、表側に申請する工事（業務）名、開札日及び「入札書在中」の旨を明記し、「親展」と記載するとともに、裏側の左下部に入札参加者名を記載のうえ、次の宛先へ送付してください。（別紙「郵便入札の方法」参照）

〒860−8601（専用郵便番号のため、住所は省略して差し支えない。）
熊本市中央区手取本町1番1号
熊本市総務局契約監理部工事契約課

Fig. 3. Example of a paragraph in a document. (Color figure online)

3.2　Proposed Model

As mentioned, our model takes advantage of a pre-trained model (BERT), combined with CNN to learn the local context of each document for classification. Figure 4 shows the overall architecture of the proposed model. The model has three main components: (i) the input vector representations of input tokens, (ii) BERT for learning hidden vectors for every token from the input tag and the document, and (iii) a convolution layer for capturing the local context and a softmax to predict the value location. The rest of this section will describe all parts of the model.

Input Representation. Each input data includes the tag and the bidding document. The tag is treated as a single sequence, and the document is split into segments with a length of 384 tokens. Each token vector representation is determined based on the embeddings of token, sequence, and position. To differentiate among the input sequences, a special token (SEP) is inserted between them. For example, as showed in Fig. 4, the tag i has N_i tokens, in which $token_1$ corresponds to embedding E_1^i. The document has k sequences separated by token [SEP]. Most of the specification of embeddings follows the original paper [2].

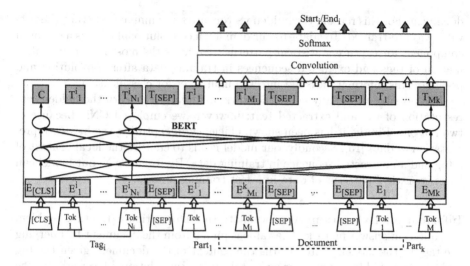

Fig. 4. The BERT-CNN model for extracting the value of a tag.

BERT. BERT is a multi-layered bidirectional Transformer encoder, which allows our model to represent the context of a word by considering its neighbors [2]. This is unlike unidirectional models that learn the contextual representation for each word using the words in one side (left or right). For example, considering the word "bank" in two sentences "I went to the bank to deposit some money" and "I went to the bank of the river", the representations of "bank" are identical for a left-context unidirectional model, but they are distinguished with the use of BERT. This characteristic is compatible with our problem where many tags have homogeneous values. Let's take a look to Table 1, given an address, it may be the address for demand (tag 5), submitting the application of qualification (tag 20), submitting for of bid (tag 22), or place of the opening bid (tag 24). Therefore, in our problem, the context aspect is critical to determine the tag that the address belongs to.

In our research, a pre-trained BERT was employed for two reasons. Firstly, BERT has shown state-of-the-art performance on many NLP tasks ranging over single/pair sentence classification, question answering, and sentence tagging. We, therefore, take advantage of BERT for our task of information extraction. Secondly, a pre-trained model is an appropriate solution to tackle our scenario of lacking training data. More precisely, we received only 100 documents for training and testing from our client. This is unpractical to train the whole network from scratch. To tackle this problem, we decided to use BERT as a type of transfer learning to fine-tune our model on bidding documents.

Convolutional Layer. A convolutional layer was stacked on the top of BERT to learn local information of its input. Applying BERT is to take advantage of data in other domain, and produces hidden vectors from the tag and the bidding document. For this reason, the hidden vectors are for common texts that the network was pre-trained. To adapt to our domain, a convolutional layer is a essential component for capture statistical patterns. It retains the most important information of tags and extracted sequences in training data after convolution and pooling. This information is fed into the final layer for classification.

It is possible to use any neural network architecture to learn the hidden representation of tag and extracted text; however, we employed CNN because of two reasons. Firstly, it has been shown efficiency in capturing the hidden representation of data [15]. Secondly, our model needs to capture the local context of a tag and an extracted sequence in training data. By using CNN, our model can learn statistical patterns of data in a fast training process.

Information Extraction. We formulate the information extraction problem as a question answering task. The value is pulled from the document by querying the tag. In the model, BERT learns the context of the document given the tag and produces hidden vectors for every token; the convolutional layer adjusts the vectors towards to our domain. Finally, a softmax layer is used to predict the location of the value. Each token is predicted to one of three outcomes including the start/end positions, and irrelevant to the tag. The extracted value is gathered based on the start to end positions with the highest probabilities.

Training. The training process of our model includes two stages: (i) pre-training and (ii) fine-tuning. For the first stage, the pre-trained weights of BERT were reused, while the weights of the rest layers were randomly initialized. The BERT was trained with a large text corpus of Japanese collected from Wikipedia. The training task is to predict whether a sentence is the next or just a random of other sentence [2]. The setting is the same as the original paper [2].

The whole network was fine-tuned in 20 epochs with our training data by using cross-entropy. We used multilingual BERT-based model trained for 102 languages on a huge amount of texts from Wikipedia. The model has 12 layers, a hidden layer of 768 neurons, 12 heads and 110M parameters. Convolution uses 768 filters with the window size = 3.

4   Settings and Evaluation Metric

4.1   Dataset

The dataset contains 100 bidding documents, in which 78 documents were used for training and 22 documents for testing. From Tables 1 and 2, we can observe that the extraction extracts many similar types of short values in a very long document, which has an average of 616 sentences.

Table 2. Statistics of the dataset.

Statistics	Mean	Std.
#training samples	78	–
#testing samples	22	–
#characters/sample	22,537	17,191
#sentences/sample	616	368

4.2 Baselines

To verify the efficiency of our approach, we compare our model to five baselines as follows.

- **BERT:** is the basic model which obtains state-of-the-art performance on many natural language processing tasks, including QA [2]. We directly applied BERT for QA on our testing data, without any additional training.
- **BERT+CRF:** stacks CRF on BERT for prediction. This is because CRF is a conventional method for information extraction and NER. This method was trained on training data and then applied to test data.
- **BERT+LSTM+CRF:** This model uses LSTM to capture the hidden representation of sequences and employs CRF for classification. This is a variation of LSTM-CRF for NER [9]. We also tried with BiLSTM but its results are not good to report.
- **n-grams+MLP+regex:** was trained with n-gram features (n in $[1,4]$). The MLP was used to predict the paragraph containing the values of tags. The regular expression was finally applied to extract the values.
- **Glove+CNN+BiLSTM+CRF:** We used a deep neural network including layers of convolution, bidirectional Long Short Term Memory (BiLSTM), and conditional random fields (CRFs) to automatically extract tags of input texts. For token embedding, we use Glove [13].

4.3 Evaluation Metric

We used F-score (F-1) to evaluate the performance of our model as well as the baselines. We matched extracted outputs to ground-truth data to compute precision, recall, and F-score. The final F-score was computed on all tags.

5 Results and Discussion

Table 3 reports the comparison according to F1-score. On average, our method outperforms others notably. This is because our model exploits the efficiency of BERT trained on a large amount of data; therefore, it can potentially capture the hidden representation of data. By stacking convolution and retraining the model with 78 training documents, our model has the ability to adapt to a new domain. As a result, the model can correctly extract information on bidding documents and can improve the performance over BERT-QA of 4.55%. The improvements come from two possible reasons: (i) we employ BERT to tackle the limited number of training data and (ii) we take advantage of CNN for fine-tuning to capture local context. This confirms our assumption in Sect. 1 that we can utilize transfer learning for information extraction in a narrow domain with limited data.

Table 3. Comparison of methods according F1-score

Method	Avg. of F1
BERT (QA)	0.8607
BERT+CRF	0.1773
BERT+LSTM+CRF	0.3817
BERT+CNN+MLP (Our model)	**0.9062**
n-grams+MLP+Regex	0.8523
CNN+BiLSTM+CRF+MLP	0.6766

The BERT-QA model is the second best. It is understandable that this model also uses BERT as a pre-trained model to extract hidden representation. However, it needs to be adapted to a new domain. It shows the efficiency of retraining the model on a new domain. Even with a small amount of data (78 training documents), our model can improve the F-score. Interestingly, BERT+CRF and BERT+LSTM+CRF do not show improvements. For BERT-CRF, it may lack the patterns learned from training data because it directly uses outputs from BERT for CRF to do classification. For BERT-LSTM-CRF, the possible reason is bidding documents are very long (Table 2); therefore, long-term dependencies may affect these models. The model using MLP with n-grams features achieves competitive results. This is because this model uses two steps to extraction candidates. The first step uses MLP to detect whether a sentence contains extracted information or not. The second step uses regular expressions to extract candidates. However, in some cases, the expressions cannot cover the patterns of text (please refer to Sect. 6). An interesting point is that the model CNN+BiLSTM+MLP outputs low scores, in which its performance is lower than n-grams+MLP of 17.57%. As mentioned above, training deep neural networks from scratch with limited data is difficult to convergence because we need to optimize a large number of parameters. The feature-based method requires less training data but it easily suffers from the definition of regular expressions.

We observed F-score on each tag to analyze how each model works. To do that, we matched extracted text to reference and compute scores, which are plotted in Fig. 5. We did not plot the results of BERT+CRF, BERT+LSTM+CRF, and CNN due to its low scores. As can be seen, our model works stably on all tags, and usually achieves the best F1 scores in comparison with BERT-QA and

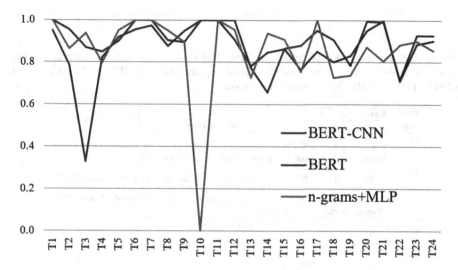

Fig. 5. F1-scores per tag.

n-grams-MLP. This again supports results in Table 3 where our model is the best. It should be noted that the feature-based method obtains 0% of F1 for tag 10 (classification of the reserved values related to the case). The reason is that n-grams features and the regular expressions may not capture well the patterns of these values (tag 10). This shows that the n-grams+MLP+regex method can be affected by the definition of regular expressions. As the result, it could not find any values for tag 10 on the test set. By contrast, our model trained on BERT can exploit rich hidden representation from large pre-trained data and with 78 training documents of bids, the model can correctly extract tag 10.

6   Error Analysis

In this section, we investigate how our model works on real data of bids. We observed the output of our model and the baselines. Table 4 shows some predicted values for three tags 21, 22, 23 (Table 1) in documents of two contracts. The first part shows a case that our model do not work and the second part presents a case that our model extracts correct sequences.

After observing, we found that for the feature-based method, it may suffer from the out of vocabulary problem because regular expressions were created by on keywords defined by humans. Although the classifier can predict correctly the sentence containing the value, the regular expression fails to locate the information. Taking the contract 0553 as an example, since the address of the company is not in the training data, the regular expression only captures the number part. Unlikely, the deep models can predict both the text and number parts in this case. However, the model trained from scratch usually produces uncompleted text, especially with the long-text values. This comes from the lack of training

Table 4. Some examples of model outputs. The correct answers and correct predictions are highlighted in blue. We did not show outputs of BERT, BERT+CRF and BERT+LSTM+CRF due to space limitation.

File	Sample A
Tag	Correct answer
21	熊谷市東区役所東町支所1階総務民生課 (Kumagaya City Higashi Ward Office Higashimachi Branch 1F General Affairs and Welfare Division).
22	〒23583東町郵便局留 (Higashi-machi Post Office Station, postal code: 23583).
23	熊谷市役所東区役所東町支社宛 (Kumagaya City Hall Higashi Ward Office Higashi-machi Branch Office).
	n-gram MLP
21	熊谷市東区役所東町支所1階総務民生課
22	〒23583東町郵便局留
23	熊谷市役所東区役所東町支社宛
	CNN+BiLSTM
21	熊谷市東区東町東45番地熊谷市東区役所東町支所宛
22	〒23-585東町郵便局留熊谷市役所東区役所東町支所宛
23	〒23-585東町郵便局留熊谷市役所東区役所東町支所宛
	Our model
21	東町支所総務民生課
22	熊谷市東区東町東45番地
23	東町支所総務民生課

File	Sample B
Tag	Correct answer
21	新潟県会計管理局物品管理調達課物品調達担当 (Niigata Prefecture Accounting Administration Department Goods Management Procurement Section Goods Procurement).
22	〒11-878新潟市小路2丁目1番地10 (postal code 11-878 Niigata, Niigata, Koji, 2 Chome―1―10).
23	県立新潟病院総務課整備担当 (Prefectural Niigata Hospital General Affairs Division Maintenance Section).
	n-gram MLP
21	(232)6118
22	目1番地10郵便番号11-878電話番号10
23	(19)618
	CNN+BiLSTM
21	目1番地10郵便番号11
22	新潟市小路2丁
23	新潟市小路2丁
	Our model
21	新潟県会計管理局物品管理調達課物品調達担当
22	〒11-878新潟市小路2丁目1番地10
23	新潟県会計管理局物品管理調達課物品調達担当

data that is one of the essential keys to allow the convergence of a deep model. In contrast, the pre-trained model can predict the whole content.

We also found that the feature-based methods can predict short values like date times, and numbers efficiently. In this case, the classifier finds the sentence using features and the regular expression can locate the information easily. For these types of fields, the performance of the feature-based and the deep models is similar. This is shown in Fig. 5.

7   Conclusion

We presented a transfer learning method for information extraction with limited data. In the experiment, we used 100 documents for the training and validation, and confirmed that our method achieves high accuracy of fine-grained classification with the limited number of annotated data. We believe that the result implies a positive aspect of the concept of transfer learning for real information extraction scenarios.

A possible direction of our research is to investigate more final layers for classification. We also encourage to add heuristic features generated from humans and change the structure of our model by replacing CNN by more deeper structures, e.g. CNN-BiLSTM.

Acknowledgement. We would like to thank to NLP and publication board members of Cinnamon Lab for useful discussions and insightful comments on earlier drafts. We also thank to anonymous reviewers for their comments for improving our paper.

References

1. Deng, L., Liu, Y.: Deep Learning in Natural Language Processing. Amazon Digital Services LLC (2018)
2. Devlin, J., Chang, M.W., Lee, K., Toutanova, K.: Bert: pre-training of deep bidirectional transformers for language understanding. arXiv preprint arXiv:1810.04805 (2018)
3. Finkel, J.R., Manning, C.D.: Nested named entity recognition. In: Proceedings of the 2009 Conference on Empirical Methods in Natural Language Processing: Volume 1, vol. 1, pp. 141–150. Association for Computational Linguistics (2009)
4. Fleischman, M., Hovy, E.: Fine grained classification of named entities. In: Proceedings of the 19th International Conference on Computational linguistics-Volume 1, pp. 1–7. Association for Computational Linguistics (2002)
5. Gokhan Tur, R.D.M.: Spoken Language Understanding: Systems for Extracting Semantic Information from Speech. Wiley, New York (2011)
6. Herbert, L.: Digital Transformation: Build Your Organization's Future for the Innovation Age. Bloomsbury Business, London (2017)
7. Ju, M., Miwa, M., Ananiadou, S.: A neural layered model for nested named entity recognition. In: Proceedings of the 2018 Conference of the North American Chapter of the Association for Computational Linguistics: Human Language Technologies, Volume 1 (Long Papers), vol. 1, pp. 1446–1459 (2018)

8. Andrew, J.J., Tannier, X.: Automatic extraction of entities and relation from legal documents. In: Proceedings of the Seventh Named Entities Workshop, pp. 1–8 (2018)

9. Lample, G., Ballesteros, M., Subramanian, S., Kawakami, K., Dyer, C.: Neural architectures for named entity recognition. arXiv preprint arXiv:1603.01360 (2016)

10. Lee, C., et al.: Fine-grained named entity recognition using conditional random fields for question answering. In: Ng, H.T., Leong, M.-K., Kan, M.-Y., Ji, D. (eds.) AIRS 2006. LNCS, vol. 4182, pp. 581–587. Springer, Heidelberg (2006). https:// doi.org/10.1007/11880592_49

11. McKinsey Global Institute: A future that works: automation, employment and productivity (2017). https://www.mckinsey.com/~/media/mckinsey/featured %20insights/Digital%20Disruption/Harnessing%20automation%20for%20a %20future%20that%20works/MGI-A-future-that-works-Executive-summary. ashx

12. Pan, S.J., Yang, Q.: A survey on transfer learning. IEEE Trans. Knowl. Data Eng. **22**(10), 1345–1359 (2010). https://doi.org/10.1109/TKDE.2009.191

13. Pennington, J., Socher, R., Manning, C.: Glove: global vectors for word representation. In: Proceedings of the 2014 Conference on Empirical Methods in Natural Language Processing (EMNLP), pp. 1532–1543 (2014)

14. Shimaoka, S., Stenetorp, P., Inui, K., Riedel, S.: An attentive neural architecture for fine-grained entity type classification. arXiv preprint arXiv:1604.05525 (2016)

15. Shin, H.C., et al.: Deep convolutional neural networks for computer-aided detection: CNN architectures, dataset characteristics and transfer learning. IEEE Tran. Med. Imaging **35**(5), 1285–1298 (2016)

16. Watanabe, Y., Asahara, M., Matsumoto, Y.: A graph-based approach to named entity categorization in Wikipedia using conditional random fields. In: Proceedings of the 2007 Joint Conference on Empirical Methods in Natural Language Processing and Computational Natural Language Learning (EMNLP-CoNLL) (2007)

17. Weiss, K., Khoshgoftaar, T.M., Wang, D.D.: A survey of transfer learning. J. Big Data **3**(1), 1–40 (2016). https://doi.org/10.1186/s40537-016-0043-6

18. Inmon, W.H., Nesavich, A.: Tapping into Unstructured Data: Integrating Unstructured Data and Textual Analytics into Business Intelligence. Prentice Hall, Upper Saddle River (2007)

Self-deprecating Humor Detection:
A Machine Learning Approach

Ashraf Kamal[1] and Muhammad Abulaish[2(✉)]

[1] Jamia Millia Islamia, (A Central University), New Delhi, India
ashrafkamal.mca@gmail.com
[2] South Asian University, New Delhi, India
abulaish@sau.ac.in

Abstract. Humor is one of the figurative language categories, and it is mainly used in human communication to express emotions and sentiments. Due to the complex structure of humorous texts, automatic humor detection is a challenging task. The detection becomes more challenging when we consider *self-deprecating humor*, which is a special category of humor in which users generally criticize and put themselves down. Interestingly, in recent years *self-deprecating humor* has been seen as a new business marketing trend, such as brand endorsement, advertisement, and content marketing. In this paper, we propose a novel *self-deprecating humor* detection approach using machine learning technique with an aim to enhance *self-deprecating humor* based marketing strategies. We have identified 16 new features related to three different feature categories – *self-deprecating pattern*, *exaggeration*, and *word-embedding*, and considered 11 humor-centric features from baseline works, and trained random forest classifier for detecting self-deprecating humor in Twitter. The proposed approach is evaluated over Twitter and two baseline datasets, and it performs significantly better in terms of standard information retrieval metrics.

Keywords: Social network analysis · Humor detection · Self-deprecating humor · Machine learning

1 Introduction

Humor is a special category of figurative language to express emotions and sentiments using laughter, jokes, and amusement in human communication. Its computational detection is an interesting, challenging and important task due to its complex characteristics and non-literal expression [1]. It can be useful for many applications, such as opinion mining and sentiment analysis, e-education, trends discovery, and e-advertisement [2]. *Self-deprecating humor* is a special category of humor in which users put themselves down. The online Urban dictionary[1] (last accessed on 07-Sep-19) defines *self-deprecating humor* as a *"humor where*

---

[1] https://goo.gl/JoSaak.

© Springer Nature Singapore Pte Ltd. 2020
L.-M. Nguyen et al. (Eds.): PACLING 2019, CCIS 1215, pp. 483–494, 2020.
https://doi.org/10.1007/978-981-15-6168-9_39

you put yourself down. Sometimes funny, but sometimes overused fishing for compliments or a signal of low self-esteem". For example, consider a humorous text *"out of all the things I lost I miss my mind the most"*, in which the phrase *I miss my mind the most* is used for self-deprecating.

Recently, *self-deprecating humor* is seen as a new marketing trend and many brands considered this category of humor for self-promotion through deprecating and disparaging themselves by accepting their own flaws in a humorous fashion without losing the brand value. Self-deprecating style of endorsement creates a sense of community by making humanized a relationship with the customer[2]. It is mainly used for brands[3], content marketing[4], and advertisement[5] (last accessed on 07-Sep-19). Interestingly, it is also seen in celebrities interviews[6] and politician speech[7] (last accessed on 07-Sep-19).

In this paper, we propose a novel approach for *self-deprecating humor* detection with an aim to enhance *self-deprecating humor* based marketing strategies. The proposed approach follows a layered design in which first layer uses a semi-automated process to identify candidate *self-around* instances from the dataset, and second layer performs feature extraction and classifier learning for self-deprecating humor detection. The proposed approach identifies 16 new features based on *self-deprecating pattern, exaggeration,* and *word-embedding,* and uses 11 humor-centric features from baseline works to train random forest classifier for detecting self-deprecating humor in Twitter. It is evaluated over three datasets and performs significantly better in terms of standard information retrieval metrics.

The remainder of this paper is organized as follows. Section 2 presents a brief review of the state-of-the-art techniques and approaches for computational humor detection. It also highlights how our proposed approach differs from existing state-of-the-art techniques. Section 3 presents the functional details of our proposed approach. Section 4 explains the formulation of various newly identified and existing features. Section 5 presents the statistics of the datasets, including the data crawling and pre-processing processes. Section 6 presents our experimental setting and evaluation results. Finally, Sect. 7 concludes the paper with future research directions.

2   Related Work

Automatic humor recognition is considered as a classification problem and the main task is to judge whether a textual message is humorous or non-humorous [1]. Mihalcea and Strapparava [4] classified humorous or non-humorous text in One-Liners English jokes, news sentences, BNC corpus, and proverb lists

---

[2] https://goo.gl/dfyXTr.

[3] https://goo.gl/2bMYqZ.

[4] https://goo.gl/H3W3n7.

[5] https://bit.ly/2Woblev.

[6] https://goo.gl/uDyvwf.

[7] https://goo.gl/KN1jHi.

datasets. Reyes et al. [5] considered humor generation using supervised machine learning techniques. Zhang and Liu [2] and Raz [6] applied humor detection in Twitter. Yang et al. [3] considered a random forest classifier to identify humor using 10-fold cross validation technique. They considered features, such as incongruity, ambiguity, interpersonal effect, and phonetic style. They also introduced humor anchors in the form of words or phrases that play a role in recognizing humor.

Zhang et al. [7] introduced features like contextual knowledge, affective polarity, and subjectivity for humor recognition. Liu et al. [1] proposed a sentiment association for humor recognition in discourse relations. Beukel and Aroyo [8] introduced homonym as an indicative feature for humor recognition. Chen and Soo [11] applied a deep learning approach for humor recognition using a convolutional neural network. Ortega-Bueno et al. [9] proposed an attention-based recurrent neural network for humor detection in Spanish language and Ermilov et al. [10] applied the supervised approach in the Russian language for humor recognition. Gultchin et al. [12] applied word embedding using the Euclidean vector for humor recognition.

It can be seen from the above discussion that none of the existing works aims to identify *self-deprecating humor*, in which users generally undervalue, disparage, or deprecate themselves using humorous words or phrases.

3   Proposed Approach

This section presents the functional details of the proposed self-deprecating humor detection technique. As stated earlier, the proposed approach follows a layered design in which first layer aims to identify candidate self-around instances from the dataset, and second layer focuses on feature extraction and classifier learning for self-deprecating humor detection. Further details about both layers are presented in the following sub-sections.

3.1   Layer-1

After an in-depth analysis of the datasets, it is noticed that in many instances (messages) generally users refer themselves. For example, consider the message "I love to stay at home on Sunday." We consider all such instances as *self-around*. Moreover, many *self-around* instances are found to be *self-deprecating humor*, in which users deprecate, undervalue or disparage themselves using humorous words or phrases.

The first layer applies a semi-automated process to identify all candidate *self-around* instances from the datasets. Using this filtration process, only candidate *self-around* instances are retained for further processing and rest of the instances are discarded to enhance the efficacy of the *self-deprecating humor* detection process by the second layer.

Algorithm 1 presents the pseudo codes of the filtration process mentioned above. Initially, Spacy[8] tagger is applied to generate tokens for each instances

---

[8] https://spacy.io/.

of the dataset. In an instance, if any of the token is first person singular (plural) personal pronoun, such as '*i*', '*am*', '*me*', '*my*', '*mine*', '*myself*', '*we*', '*are*', '*us*', or '*our*' then it is added to the *self-around explicit* file, otherwise it is added to the *self-around implicit* file. In Algorithm 1, steps 3–9 show *self-around explicit* file creation process, whereas step 10 shows the *self-around implicit* file creation process.

The *self-around implicit* file is analyzed manually to identify candidate self-around instances. After manual analysis, both *self-around explicit* and *self-around implicit* files are merged and considered as the candidate *self-around* dataset, which is passed to the second layer for feature extraction and classification model learning.

Algorithm 1: Self-Around Instance Detection

    Input   : A file F containing pre-processed instances
    Output: Files F_{exp} and F_{imp} containing self-around explicit and implicit instances, respectively

1 **foreach** *instance in F* **do**
2     $k \leftarrow$ spacy (instance) ;   `/* tokenization */`
3     $n \leftarrow$ no-of-token(k)
4   **for** $i \leftarrow 1$ *to* n **do**
5   **if** $k[i].token \in \{i, am, me, my, mine, myself, we, are, us, our\}$ **then**
6   append(F_{exp}, k) ;   `/* k is self-around explicit instance */`
7   goto step 1;
8   **endif**
9   **endfor**
10   append(F_{imp}, k) ;   `/* k is self-around implicit instance */`
11 **endforeach**

3.2 Layer-2

The second layer (i.e., layer-2) considers candidate self-around dataset as an input for feature extraction and classifier learning. A total number of 16 new features from three different feature categories are identified, out of which nine are based on *self-deprecating patterns*, three are *exaggeration*, and four are based on *word embeddings*. In addition, 11 features from one of the baseline works by Yang et al. [3] are used in this study. Further details about these features are presented in the following section. Finally, layer-2 learns one of the popular machine learning techniques, random forest classifier, for detecting *self-deprecating humor*.

4 Feature Extraction

This section presents the details about existing baseline features and newly identified features for detecting *self-deprecating humor*.

4.1   The Features Followed from Baseline

We have considered baseline features from Yang et al. [3]. A total number of 11 semantic structure associated humor-centric features are taken from 3 feature categories, such as *ambiguity, interpersonal effect,* and *phonetic style.* These semantic structures reflect important information about an instance to be a humorous.

4.2   Newly Proposed Features

Apart from baseline features, we have identified a total number of 16 features from three feature categories – *self-deprecating pattern, exaggeration,* and *word-embedding.* A detailed discussion about these feature categories is given in the following paragraphs.

Self-Deprecating Pattern (SDP)

The *self-deprecating pattern* features is inspired from Abulaish and Kamal [14] work in which they proposed *self-deprecating sarcasm* detection task. Since *sarcasm* is considered as the aggressive form of *humor* [15], we have identified a similar but more reliable nine self-deprecating pattern-based features related to humor. These features mainly target self-deprecating patterns in an instance of the dataset. The patterns are either based on the relative order of the Part-of-Speech (POS) tags and tokens or frequency count of the tokens in an instance. All features in this category are binary in nature and they are briefly described in the following paragraphs.

$\mathcal{F}_1$ **(Interjection Followed by '*i*' or '*we*').** In an instance, if an interjection (UH-tagged word) is found and an immediate next token is either 'i' or 'we' (i.e., $UH \rightarrow i(we)$), then the value of this feature is considered as 1, otherwise 0. For example, "*Wow we love going to school after fever!*.", is a self-deprecating humor using this pattern.

$\mathcal{F}_2$ **(Common Self-deprecating Pattern).** In an instance, if any of the following common self-deprecating pattern matches ($P1$ to $P5$) is found, then the value of this feature is set to 1, otherwise 0.

- $P1$: If the token '*it*' is followed by a question word (WRB-tagged word), i.e., $it \rightarrow WRB$.
- $P2$: If a question word (WRB-tagged word) is followed by a personal pronoun (PRP-tagged word), i.e., $WRB \rightarrow PRP$.
- $P3$: If the token '*i*' is followed by the word 'love', i.e., $i \rightarrow love$. During analysis, 'love' is the most frequently used word found in self-deprecating instances.
- $P4$: If an adjective (JJ-tagged word) is followed by an adverb (RB-tagged word), i.e., $JJ \rightarrow RB$.
- $P5$: If an adverb (RB-tagged word) is followed by an adjective (JJ-tagged word), i.e., $RB \rightarrow JJ$.

$\mathcal{F}_3$ **(Token '*i*' or '*we*' Followed by Negative Modal Verb).** In an instance, if either a token '*i*' or '*we*' is found and an immediate next tag is a modal verb (MD) and then followed by token '*not*' (i.e., $i(we) \rightarrow MD \rightarrow not$), then the value of this feature is set to 1, otherwise 0. For example, "*I can not delay to go office anymore for more work.*", is a self-deprecating humor using this pattern.

$\mathcal{F}_4$ **(Token '*i*' or '*we*' Followed by Verb).** In an instance, if either a token '*i*' or '*we*' is found and an immediate next tag is verb (non-3rd person singular present) (i.e., $i(we) \rightarrow VBP$), then the value of this feature is considered as 1, otherwise 0. For example, "*We love deadlines at work, especially on Christmas eve.*", is a self-deprecating humor using this pattern.

$\mathcal{F}_5$ **(Token '*i*' or '*we*' Followed by Past Tense Verb).** In an instance, if either a token '*i*' or '*we*' is found and an immediate next tag is past tense verb (i.e., $i(we) \rightarrow VBD$), then the value of this feature is considered as 1, otherwise 0. For example, "*I used to be a doctor but i did not have the patients.*", is a self-deprecating humor using this pattern.

$\mathcal{F}_6$ **(Conjunction Followed by Token '*i*' or '*we*').** In an instance, if a conjunction tag is found and an immediate next token is either '*i*' or '*we*' (i.e., $CC \rightarrow i(we)$), then the value of this feature is considered as 1, otherwise 0. For example, "*We have got some powdered water but we do not know what to add.*", is a self-deprecating humor using this pattern.

$\mathcal{F}_7$ **(Question Word Followed by Token '*i*' or '*we*').** In an instance, if a question word tag is found and an immediate next token is either '*i*' or '*we*' (i.e., $WRB \rightarrow i(we)$), then the value of this feature is considered as 1, otherwise 0. For example, "*when i am not in my right mind my left mind gets pretty crowded.*", is a self-deprecating humor using this pattern.

$\mathcal{F}_8$ **(Token '*i*' Count).** In an instance, if the token '*i*' occurs at least two, then the value of this feature is considered as 1, otherwise 0. For example, "*i am not cheap but i am on special this week.*", is a self-deprecating humor using this pattern.

$\mathcal{F}_9$ **(Token '*my*' Count).** In an instance, if the token '*my*' occurs at least two, then the value of this feature is considered as 1, otherwise 0. For example, "*my weight is perfect for my height which varies.*", is a self-deprecating humor using this pattern.

Exaggeration (EXA)

Exaggeration plays an important role to create over emphasis, over-statement in a humorous instance. The exaggeration contains an intensifier in the form of adverb, adjective, and interjection in an instance, and its frequency count is considered for self-deprecating humor detection. All features in this category are also *binary* and their values are either 0 or 1. A total number of three *exaggeration* features are identified to capture the frequency count of adverbs, adjectives, and interjections.

$\mathcal{F}_{10}$ **(Interjection Count).** In an instance, if an interjection tag 'UH' occurs at least twice, then the value of this feature is assigned as 1, otherwise 0. For example, "*oh wow! I am complete without holidays.*", is a self-deprecating humor using this pattern.

$\mathcal{F}_{11}$ **(Adjective Count).** In an instance, if an adjective tag 'JJ' occurs at least twice, then the value of this feature is assigned as 1, otherwise 0. For example, "*It is so great I am late as always !!1 perfect Monday.*", is a self-deprecating humor using this pattern.

$\mathcal{F}_{12}$ **(Adverb Count).** In an instance, if an adverb tag 'RB' occurs at least twice, then the value of this feature is assigned as 1, otherwise 0. For example, "*I am not a liar he just arranges the truth in his favor.*", is a self-deprecating humor using this pattern.

Word Embedding (WE)

Humor relies on a certain type of opposition or contradiction. These features are taken from Joshi et al. [13] in which authors considered semantic similarity or discordance works as a clue to handle context incongruity in an instance. A total number of four word embedding-based features are considered with an aim to capture those self-deprecating humorous instances in which sentiment-bearing words are absent, but semantic similarity or discordance between words are present. The word2vec[9] approach is implemented for the word embedding-based features. We trained the word2vec model over all three datasets and used Eq. 1 to calculate *Cosine* similarity between all word-pairs (W_m, W_n) in an instance. Finally, dissimilarity values between word-pairs are determined using Eq. 2.

$$Cosine(W_m, W_n) = \frac{W_m \cdot W_n}{|W_m| \cdot |W_n|} \tag{1}$$

$$Dissimilarity(W_m, W_n) = 1 - Cosine(W_m, W_n) \tag{2}$$

$\mathcal{F}_{13}$ **(Maximum Score of Most Similar Word Pair).** The value of this feature is considered as the highest score of the most similar word-pair among all possible word-pairs of an instance.

$\mathcal{F}_{14}$ **(Minimum Score of Most Similar Word Pair).** The value of this feature is considered as the lowest score of the most similar word-pair among all possible word-pairs of an instance.

$\mathcal{F}_{15}$ **(Maximum Score of Most Dissimilar Word Pair).** The value of this feature is considered as the highest score of the most dissimilar word-pair among all possible word-pairs of an instance.

$\mathcal{F}_{16}$ **(Minimum Score of Most Dissimilar Word Pair).** The value of this feature is considered as the lowest score of the most dissimilar word-pair among all possible word-pairs of an instance.

---

[9] https://code.google.com/archive/p/word2vec/.

5   Datasets

In this section, we discuss all datasets, including baselines, used for the empirical evaluation of the proposed self-deprecating humor detection technique.

- Baseline datasets: Baseline datasets are collected from two sources. First, *Pun of the Day* dataset from Yang et al. [3]. Pun belongs to a wordplay in which similar sounding words or words with multiple meanings indicate humorous effects. Second, 16000 *One-Liners* dataset is taken from Mihalcea and Strapparava [4].
- Twitter dataset: We have crawled the Twitter dataset using REST API in Python 2.7 during May 2019. We considered a hashtag-based approach to curate the Twitter dataset. Using hashtags, users self-label their Twitter instances (i.e., tweets), such as #humor and #not. Humorous instances are collected using #humor, #fun, and #love hashtags, whereas non-humorous instances are collected using the #not and #hate hashtags.

After crawling the Twitter dataset, we have applied a number of pre-processing tasks to obtain the fine-tuned dataset for the experimental evaluation of the proposed approach. The Twitter-specific pre-processing module consists of various data cleaning steps, such as removal of URLs, @mentions, retweets, and hashtags from the Tweets. In addition, other data cleaning steps, such as removal of numbers, duplicate instances, dots, ampersands, extra white spaces, hexa-characters, quotes, emoticons, and upper-case letters to lower- case conversion are applied on both baseline and Twitter datasets.

Table 1 presents the statistics of the dataset. Table 2 presents the statistics of the retained instances after filtration through the semi-automated process of layer-1.

Table 1. Statistics of the datasets

Datasets	#Humor	#Non-humor	Total (#instances)
16000 One-Liners	16000	16002	32002
Pun of the Day	2423	2403	4826
Twitter	10000	10000	20000
Total (#instances)	28423	28405	56828

6   Experiment Setup and Results

This section presents the experimental setup and evaluation results of our proposed approach. It also presents a comparative analysis of the proposed approach with one of the existing state-of-the-art approaches. All experimental tasks were evaluated on a machine with a configuration of 3.40 GHz Intel-Xeon processor, 16 GB RAM, and Windows 8.1 Pro (64-bit) operating system. Most of the functioning modules are implemented in Python 2.7. However, Random Forest (RF)

Table 2. Retained instances after filtration in layer-1

Datasets	#Humor	#Non-humor	Total (#instances)
16000 One-Liners	4520	1414	5934
Pun of the Day	618	728	1346
Twitter	2910	3405	6315
Total (#instances)	8048	5547	13595

classifier is implemented in WEKA tool kit 3.8 and evaluated through 10-fold cross validation using standard information retrieval metrics discussed in the following sub-sections.

6.1 Evaluation Metrics

Standard information retrieval metrics such as *precision, recall,* and *f-score* are used for the experimental evaluation of the proposed approach. *Precision* measures the correctness, whereas *recall* measures the completeness of a classifier. *F-score* is calculated as the harmonic mean of the *precision* and *recall* values. Furthermore, these metrics are calculated using True Positives (*TP*), False Positives (*FP*), and False Negatives (*FN*). *TP* is defined as the number of correctly retrieved *self-deprecating humor* instances, whereas *FP* is defined as the number of *non-self-deprecating humor* instances that are identified as *self-deprecating humor*. Finally, *FN* is defined as the number of *self-deprecating humor* instances that are missed. Formally, *precision, recall,* and *f-score* are defined in Eqs. 3, 4, and 5 respectively.

$$Precision~(P) = \frac{TP}{TP + FP} \tag{3}$$

$$Recall~(R) = \frac{TP}{TP + FN} \tag{4}$$

$$F\text{-}score~(F) = \frac{2 \times P \times R}{P + R} \tag{5}$$

6.2 Evaluation Results and Comparative Analysis

In this section, we discuss evaluation results and comparative analysis of our proposed self-deprecating humor detection approach. Table 3 presents the performance evaluation results using RF classifier with 10-fold cross validation on the datasets presented in Table 2. Similarly, Figs. 1, 2 and 3 present the visualization of the comparative analysis of the newly proposed features and baseline features in terms of *precision, recall,* and *f-score* on *16000 One-Liners, Pun of the Day,* and *Twitter* datasets, respectively.

Table 3. Performance evaluation results using RF classifier with 10-fold cross validation on datasets presented in table 2

Features ↓	Datasets →	16000 One-Liners			Pun of the Day			Twitter		
		P	R	F	P	R	F	P	R	F
Yang et al. [3]	HCF	0.78	0.88	0.83	0.52	0.48	0.50	0.56	0.60	0.56
Newly proposed features	SDP	0.76	0.98	0.85	0.72	0.50	0.60	0.53	0.63	0.58
	EXA	0.76	0.97	0.86	0.65	0.62	0.61	0.52	0.74	0.62
	WE	0.78	0.88	0.83	0.61	0.62	0.62	0.55	0.54	0.55
Yang et al. [3] + Newly proposed features	Hybrid (HCF + SDP + EXA + WE)	0.79	0.95	0.87	0.71	0.72	0.72	0.64	0.58	0.61

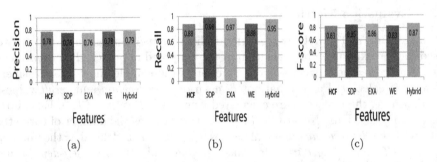

Fig. 1. Comparative analysis of the HCF (Yang et al. [3]) and newly proposed features over 16000 One-Liners dataset (a) Precision (b) Recall (c) F-score

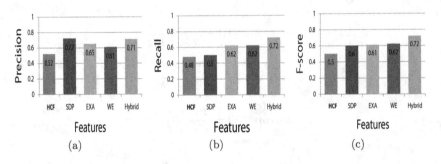

Fig. 2. Comparative analysis of the HCF (Yang et al. [3]) and newly proposed features over Pun of the Day dataset (a) Precision (b) Recall (c) F-score

To the best of our knowledge, *self-deprecating humor* detection is a new problem and no study has been reported yet in the existing literatures. Therefore, comparative evaluation of our proposed approach is not possible. However, Yang et al. [3] have considered humor-centric-features (HCF) for humor detection. Therefore, we have analyzed the effectiveness of their features for

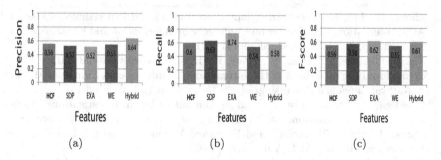

Fig. 3. Comparative analysis of the HCF (Yang et al. [3]) and newly proposed features over Twitter dataset (a) Precision (b) Recall (c) F-score

self-deprecating humor detection. We have also analyzed the combined effect of Yang et al. [3] features and our newly proposed features towards self-deprecating humor detection.

Table 3 presents evaluation results using the HCF, newly proposed features (i.e., SDP, EXA, and WE), and their combination. It can be observed from this table that our newly proposed features show significantly better classification results. The highest *precision* of 0.78, 0.72, 0.55 are obtained over *One Liner*, *Pun of the Day*, and *Twitter* datasets using newly proposed features. Similarly, it can be observed that combining HCF and newly proposed features resulted in improved *f-score* over two datasets.

7 Conclusion and Future Work

In this paper, we have considered a new problem of *self-deprecating humor* detection and modeled it as a binary classification task. We have identified a list of 16 new features that are clubbed with existing humor-centric features to learn classification models. We have adopted a layered-design approach in which the first layer filters out all those instances that are not self-around. The main intent behind this work is to enhance those systems that target self-deprecating humor-based marketing strategies. Detecting new *self-deprecating* discriminating features and experimental evaluation of the proposed approach over large datasets seems one of the future directions of research.

Acknowledgment. This publication is an outcome of the R&D work undertaken project under the Visvesvaraya PhD Scheme of Ministry of Electronics & Information Technology, Government of India, being implemented by Digital India Corporation.

References

1. Liu, L., Zhang, D., Song, W.: Modeling sentiment association in discourse for humor recognition. In: Proceedings of the 56th Association for Computational Linguistics (Short Papers), pp. 586–591. ACL, Melbourne (2018)

2. Zhang, R., Liu, N.: Recognizing humor on Twitter. In: Proceedings of the 23rd ACM International Conference on Conference on Information and Knowledge Management, pp. 889–898. ACM, Shanghai (2014)
3. Yang, D., Lavie, A., Dyer, C., Hovy, E.: Humor recognition and humor anchor extraction. In: Proceedings of the International Conference on Empirical Methods in Natural Language Processing, pp. 2367–2376. ACL, Lisbon (2015)
4. Mihalcea, R., Strapparava, C.: Making computers laugh: investigations in automatic humor recognition. In: Proceedings of the International Conference on Human Language Technology and Empirical Methods in Natural Language Processing, pp. 531–538. ACL, Vancouver (2005)
5. Reyes, A., Rosso, P., Buscaldi, D.: From humor recognition to irony detection: the figurative language of social media. Data Knowl. Eng. **74**, 1–12 (2012)
6. Yishay, R.: Automatic humor classification on Twitter. In: Proceedings of the Conference of the North American Chapter of the Association for Computational Linguistics: Human Language Technologies, Student Research Workshop, pp. 66–70. ACL, Montréal (2012)
7. Zhang, D., Song, W., Liu, L., Du, C., Zhao, X.: Investigations in automatic humor recognition. In: 10th International Symposium on Computational Intelligence and Design, pp. 272–275. IEEE, Hangzhou (2017)
8. Van den Beukel, S., Aroyo, L.: Homonym detection for humor recognition in short text. In: Proceedings of the 9th Workshop on Computational Approaches to Subjectivity, Sentiment and Social Media Analysis, pp. 286–291. ACL, Brussels (2018)
9. Ortega-Bueno, R., Muniz-Cuza, C.E., Pagola, J.E.M., Rosso, P.: UO UPV: deep linguistic humor detection in Spanish social media. In: Proceedings of the 3rd Workshop on Evaluation of Human Language Technologies for Iberian Languages, pp. 204–213. Seville (2018)
10. Ermilov, A., Murashkina, N., Goryacheva, V., Braslavski, P.: Stierlitz meets SVM: humor detection in Russian. In: Ustalov, D., Filchenkov, A., Pivovarova, L., Žižka, J. (eds.) AINL 2018. CCIS, vol. 930, pp. 178–184. Springer, Cham (2018). https://doi.org/10.1007/978-3-030-01204-5_17
11. Chen, P.-Y., Soo, V.W.: Humor recognition using deep learning. In: Proceedings of the 16th Conference on the North American Chapter of the Association for Computational Linguistics: Human Language Technologies (Short Papers), vol. 2, pp. 113–117. ACL, New Orleans (2018)
12. Gultchin, L., Patterson, G., Baym, N., Swinger, N., Kalai, A.T.: Humor in word embeddings: cockamamie gobbledegook for nincompoops. In: Proceedings of the 36th International Conference on Machine Learning, pp. 2474–2483. PMLR, Long Beach (2019)
13. Joshi, A., Tripathi, V., Patel, K., Bhattacharyya., Carman, M.: Are word embedding-based features useful for sarcasm detection?. In: Proceedings of the Conference on Empirical Methods in Natural Language Processing, pp. 1006–1011. ACL, Austin (2016)
14. Abulaish, M., Kamal, A.: Self-deprecating sarcasm detection: an amalgamation of rule-based and machine learning approach. In: Proceedings of the IEEE/WIC/ACM International Conference on Web Intelligence, pp. 574–579. IEEE, Santiago (2018)
15. Stieger, S., Formann, A.K., Burger, C.: Humor styles and their relationship to explicit and implicit self-esteem. Personal. Individ. Differ. **50**(5), 747–750 (2011)

Grammar Error and Plagiarism Detection

Deep Learning Approach for Vietnamese Consonant Misspell Correction

Ha Thanh Nguyen[✉], Tran Binh Dang, and Le Minh Nguyen

Japan Advanced Institute of Science and Technology, Nomi, Japan
{nguyenhathanh,s1710457,nguyenml}@jaist.ac.jp

Abstract. Vietnamese words are combinations of consonants, vowels, and diacritics. Previous studies on Vietnamese spelling correction often focused on mistyped errors. Misspelled errors are more common and difficult to detect. Based on our literature review, there is no direct study to address this issue. A misspelled Vietnamese word can become another word does exist in the vocabulary but make the sentence a different meaning or meaningless. While mistyped errors are typographical errors, misspelled errors may appear in any type of text including typed documents and handwritten text. Compared to mistyped errors, misspelled errors are harder to detect, especially by people who type it out. This error comes from the wrong understanding about the spelling of the word. For that reason, checking a sentence with a vocabulary filter does not guarantee that the sentence is spelled correctly. Checking Vietnamese spelling errors is a difficult problem. There have been many articles trying to solve this problem with different approaches but they have their own limitations. In this paper, we propose a deep learning approach focusing on consonant misspell errors with superior accuracy compared to the existing methods. The accuracy of our model makes a significant gap compared to the current state-of-the-art model.

Keywords: Vietnamese consonant misspell correction · Misspell direction encoding · Deep learning

1 Introduction

1.1 Vietnamese Language

The Vietnamese language is an isolating language belongs to the Austroasiatic languages class, which is a synthetic covering 168 languages in southern Asia. There are two categories as Munda and Mon-Khmer. The Vietnamese language belongs to the Mon-Khmer category. It is a language in which almost every word consists of a single morpheme, differ from a multi-morpheme language like English. In English, the word "come" may have different morphemes such as "comes", "came", "coming". In Vietnamese, two different morphemes represent two different words and mostly convey two different meaning.

© Springer Nature Singapore Pte Ltd. 2020
L.-M. Nguyen et al. (Eds.): PACLING 2019, CCIS 1215, pp. 497–504, 2020.
https://doi.org/10.1007/978-981-15-6168-9_40

In practice of using Vietnamese, the errors in documents are sometimes very challenging to find out. A word, which is in the vocabulary and very similar to the correct one, can be the misspell token. Vietnam has differences in pronunciation in words in localities, which leads to spelling errors when people write Vietnamese. In addition, Some Vietnamese words are borrowed and recorded from ancient Chinese, which leads to difficulties in determining the correct spelling of words.

In the Vietnamese language, some consonant sounds are created by only one letter, other consonant sounds are two letter digraph and others are written with more than one letter or digraph. In daily life, people in different areas in Vietnam use different consonant in the same word. Nevertheless, they use the same spelling in the written language.

Table 1. Examples of consonant misspelled errors in Vietnamese

Misspelling type	Original and misspelled sentences	Meaning
ch-tr	Tôi đắp **chăn** cho cô ấy	I covered her with a **blanket**
	Tôi đắp **trăn** cho cô ấy	I covered her with a **python**
s-x	**Sương** còn đọng trên lá	**Dew** remains on leaves
	Xương còn đọng trên lá	**Born** remains on leaves
d-r-gi	Cô ấy rất thích ăn **dâu**	She loves eating **strawberry**
	Cô ấy rất thích ăn **râu**	She loves eating **beard**
n-l	Thuyền của tôi đang **neo** trong bến	My boat **is anchored** in the dock
	Thuyền của tôi đang **leo** trong bến	My boat **is climbing** in the dock

1.2 Mistyped Errors and Misspelled Errors

To understand the practicality and novelty of this study, it is necessary to understand the difference between mistyped errors and misspelled errors. Mistyped errors are errors related to typing. Previous studies often focus on this type of error. However, mistyped errors can be detected if the typist reviews the text carefully without the support of the computer. Compared to mistyped errors, misspelled errors are harder to detect, especially by people who type it out. This error comes from the wrong understanding of the spelling of the word.

Most of the mistyped errors are out of vocabulary. This error can be easily detected by a vocabulary-based filter or reviewer. Misspelled error are errors that the writer himself didn't know it was an error. In addition, if it is not indicated that this is an error, the writer will continue to use this wrong word in future documents.

In practice, this problem is due to the complex origin of some difficult words in Vietnamese and the difference in pronunciation of different regions. Based on "Vietnamese spelling dictionary",[1] the common consonant misspellings included: **ch-tr, s-x, d-r-gi** and **l-n**.

---

[1] Binh Mai, Lam Ngoc, "Từ điển chính tả Tiếng Việt" [Vietnamese spelling dictionary], Hong Duc publisher.

In many cases, the misspelled words are still in the vocabulary but make sentences meaningless or wrong meaning. Table 1 shows some of the examples of consonant misspelled errors.

2 Related Works

The dictionary-based approach is the simplest method for spell checking. The first, syllables that do not appear in the dictionary are identified as errors. The second, a candidate set is built based on similarity for each syllable. If there is any candidate that combining with surrounding syllable create a word in the dictionary, the current syllable is identified as an error. Obviously, this method has a limit in detecting the error by single syllable and the error syllable that exists in the vocabulary.

Nguyen Duc Hai and Nguyen Pham Hanh Nhi [1] applied sentence parsing to a Vietnamese spell checking model. The input sentence is parsed by the Earley algorithm, if the sentence has a spelling error, it will not be parsed correctly. The logic behind this model is that the syllables that make sentence cannot be parsed can be defined as spelling errors. However, this model has three limitations. Firstly, the complexity of the Earley algorithm is costly. With n syllables in the input sentence, the complexity of the algorithm is $O(n^3)$ leading to difficulties in handling long documents. Secondly, the Vietnamese language has approximately 3000 rules, collecting and using all of the rules is a huge challenge. Finally, there are ambiguities in Vietnamese. As a result, there could be a sentence that is misspelled but still able to be parsed.

Nguyen Hong Vu [4] developed a normalization method for Vietnamese tweets based on a language model with dictionaries and Vietnamese vocabulary structures. The tweet with spelling errors will be detected based on the similarity of the words. After detecting the errors, the system fixes them by the link of vowels and consonants and evaluate by the language model. In this model, the author improved based on Dice model and SRILM (a language model).

In 2014, Nguyen Thi Xuan Huong et al. [3] applied large N-gram for spell checking system. This model is created based on a statistical approach. In order to deal with the spare data problem, the authors used a large corpus for training. The large corpus is collected from many text resources in various topics to reduce the number of unknown words. It also helps to increase the number of combinations of syllables, enabling to exploit more relations among syllables and their distribution. The system suggests a syllable as an error based on a candidate set. The system creates candidates based on syllable structure corresponding to a typing error, tones error, consonant error, and region error.

To decrease the candidate set's size, a dictionary is used. The candidate words which do not appear in the dictionary will be removed. The best candidate is chosen statistically by an N-gram model with the window size is 5. Based on the frequency of unigram, bigram and trigram of a candidate in a context, the N-gram score is calculated. System's suggestions are based on this score. This approach is currently state-of-the-art for Vietnamese spell checking problem with accuracy approximate 94% F-score on their experimental data.

3 Model Construction

3.1 Dataset Preparation

Unlike the method of using rules, machine learning models, especially deep learning, need to be trained with data. So we need a corpus containing correct spelled and misspelled sentences. Assuming that the official published documents on online newspapers and other informational sites are spelled correctly, we built the data set with the following procedure.

Raw data is crawled from online Vietnamese websites including newspaper and other informational sites. Then, for each original sentence, we create their derivative by modifying confusing consonant words, **ch-tr**, **s-x**, **d-r-gi** and **l-n**. Candidates in the final data set include the original sentence and derivatives including all of the consonant confusion possibilities. Adding the original sentence into the data set ensures that the model actually distinguishes the correct spelling cases and misspellings rather than simply relying on a list of confusing consonants.

3.2 Misspell Direction Encoding

Correcting misspelling is to find the error and correct it. For binary errors, if the error is found, it means that we can turn it into the correct word by replacing the current consonant with the remaining consonant in the pair. For example, if the model that discovered the word "chanh" (lemon) was misspelled, we can easily correct the error by replacing "ch" with "tr" to get the word "tranh" (painting).

In contrast, with the "d-r-gi" confusion set, the problem is not that simple. Finding the wrong word does not mean that it can be turned into the right word. In this case, each wrong position has two consonant changing options. Suppose that the model detected "giêm giúa" is a wrong word, in this case, "diêm dúa", "diêm rúa", "diêm giúa", "riêm dúa", "riêm rúa", "riêm giúa", "giêm dúa", "giêm rúa" are possible options for changing the word but only "diêm dúa" is correct.

From the above observations, we designed a misspell direction encoding mechanism. Using this encoding, with an input sequence, the model needs to predict the corresponding coding sequence, based on that, the sentence could be corrected. The encoding is shown in Table 2.

3.3 Deep Learning Model

A deep neural network is obtained and trained for predicting the code sequences of given input sentences. Sentences are first tokenized into a list of words, then converted into a list of embedding vectors. The vocabulary contains all words in the corpus and their misspelled derivatives, the embedding dimension is 100. Other tokens, that are not in the vocabulary, are represented as a random vector standing for *unknown word*.

Table 2. Misspell direction encoding

Original consonant	Direction code		
	−1	0	1
ch	-	ch	tr
tr	-	tr	ch
s	-	s	x
x	-	x	s
l	-	l	n
n	-	n	l
d	gi	d	r
gi	r	gi	d
r	d	r	gi

In order to train the model by the examples in batches, we need to make the examples the same length. All sentences were pruned and padded into 20 words long. In the training phase, we use Adam optimizer with Cross Entropy Loss to train the neural network with bidirectional stacked LSTM [5] architecture. The parameters are shown in Table 3.

Table 3. Parameters

Parameters	Value
Embedding dimension	100
Hidden dimension	32
Number of hidden layers	2
Batch size	512
Learning rate	0.001
Drop out rate	0.4

LSTM architecture enables the model to capture the signal of a sequence of inputs. Using gated mechanism, LTSM is able to focus only on important parts of the sequence to avoid vanishing gradient problem. Information in LSTM is repeatedly calculated through different time steps. As a result, this architecture is able to capture the context of the whole sentence. The computation in an LSTM cell can be formulated as below, in which, i, h, f and o represent for input node, hidden node, forget gate and output gate respectively:

$$i_t = \sigma(x_t U^i + h_{t-1} W^i)$$
$$f_t = \sigma(x_t U^f + h_{t-1} W^f)$$
$$o_t = \sigma(x_t U^o + h_{t-1} W^o)$$
$$\tilde{C} = tanh(x_t U^g + h_{t-1} W^g)$$
$$C = \sigma(f_t * C_{t-1} + i_t * \tilde{C}_t)$$
$$h_t = tanh(C_t) * o_t$$

3.4 Experimental Results

To evaluate the effectiveness of the method, we train and test the model with the data prepared as described in Sect. 3.1. The training data set contains 210,000 sentences and the test data set contains 21,000 other sentences. These data sets have no duplicate sentences and contain all the consonant errors as described.

During training, we record the loss value, accuracy on position and accuracy on the whole sentence. The model reachs the minimum loss value at around $10th$ epoch.

Table 4. Examples of outputs of Nguyen et al.'s model and our model

Input sentence	Correct sentence	Nguyen et al. model output	Our model output
Lúc đó, bả ngán tôi **nắm** vì tôi ốm nhom, **nại** nhìn **như** thằng bụi đời.	**Lúc** đó, bả ngán tôi **lắm** vì tôi ốm nhom, **lại** nhìn **như** thằng bụi đời.	**Núc** đó, bả **quán** tôi **lắm** vì tôi ốm nhom, **nại** nhìn **dư** thằng bụi đời.	**Lúc** đó, bả ngán tôi **lắm** vì tôi ốm nhom, **lại** nhìn **như** thằng bụi đời.
Chường của **tráu** bạn **lào** cũng được **tra** mẹ đưa **se** hơi đi học.	**Trường** của **cháu** bạn **nào** cũng được **cha** mẹ đưa **xe** hơi đi học.	**Chường** của **tráu** bạn **lào** cũng được **cha** mẹ đưa **xe** hơi đi học.	**Trường** của **cháu** bạn **nào** cũng được **cha** mẹ đưa **xe** hơi đi học.
Cúm A/H5N1 ở người phụ thuộc vào **rịch** cúm ở **da** cầm.	Cúm A/H5N1 ở người phụ thuộc vào **dịch** cúm ở **gia** cầm.	Cúm Cúm A/H5N1 ở người phụ thuộc vào **dịch** cúm ở **nha** cầm.	Cúm A/H5N1 ở người phụ thuộc vào **dịch** cúm ở **gia** cầm.
Phó Thủ tướng, Bộ **trưởng** Ngoại giao Phạm **Da** Khiêm **trả** lời phỏng vấn.	Phó Thủ tướng, Bộ **trưởng** Ngoại giao Phạm **Gia** Khiêm **trả** lời phỏng vấn.	Phó Thủ tướng, Bộ **trưởng** Ngoại giao Phạm **Gia** Khiêm **trả** lời phỏng vấn.	Phó Thủ tướng, Bộ **trưởng** Ngoại giao Phạm **Ra** Khiêm **trả** lời phỏng vấn.

To compare our model with Nguyen Thi Xuan Huong et al.'s current state-of-the-art model in Vietnamese spelling error, we implement the model and test with our data. There is a big gap when our model reaches 96.94% accuracy on positions and 90.56% accuracy on sentences, the model of Nguyen et al. only achieving these two measurements are 88.45% and 72.18% respectively. The comparison between the two models is shown in Fig. 1.

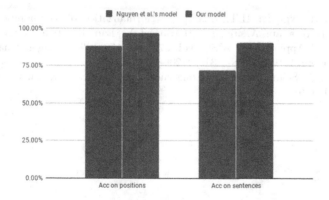

Fig. 1. Comparison between Nguyen et al.'s model and our model

Next, we examine the errors of the two models. Both models are statistically based, so they fail with rare words. However, the model of Nguyen et al. poorly effective in fixing errors. There are cases that this model detected an error but turned it into another error, and there are also situations that this model turned the correct word to the wrong word. Our model is designed to focus on consonant errors, thus avoiding such problems. Table 4 shows some examples of output from the two models.

4 Conclusion

Unlike previous studies focusing on mistyped errors in Vietnamese, in this study we directly solve the misspelled errors on consonants. We have introduced a deep learning approach to solving the problem. The main idea of this approach is to use the misspell direction encoding to identify and correct error positions. The deep learning network architecture used in the proposed method is bidirectional stacked LSTM. In terms of misspelled error correction, our model has superior accuracy compared to current methods in Vietnamese spell checking. The results of this study are highly applicable to practical applications. In future work, we will analyze and resolve the Vietnamese misspellings for vowels and diacritics.

References

1. Hai, N.D., Nhi, N.P.H.: Syntactic parser in Vietnamese sentences and its application in spell checking. University of Science Ho Chi Minh City (1999)
2. Duy, N.T.N., Dien, D.: An approach in Vietnamese spell checking. University of Science Ho Chi Minh City (2004)
3. Thi Xuan Huong, N., Dang, T.-T., Nguyen, T.-T., Le, A.-C.: Using large N-gram for Vietnamese spell checking. In: Nguyen, V.-H., Le, A.-C., Huynh, V.-N. (eds.) Knowledge and Systems Engineering. AISC, vol. 326, pp. 617–627. Springer, Cham (2015). https://doi.org/10.1007/978-3-319-11680-8_49

4. Nguyen, V.H., Nguyen, H.T., Snasel, V.: Normalization of Vietnamese tweets on Twitter. In: Abraham, A., Jiang, X.H., Snášel, V., Pan, J.-S. (eds.) Intelligent Data Analysis and Applications. AISC, vol. 370, pp. 179–189. Springer, Cham (2015). https://doi.org/10.1007/978-3-319-21206-7_16
5. Hochreiter, S., Schmidhuber, J.: Long short-term memory. Neural Comput. 9(8), 1735–1780 (1997)

Grammatical Error Correction for Vietnamese Using Machine Translation

Nghia Luan Pham[1,3]([✉]), Tien Ha Nguyen[2], and Van Vinh Nguyen[3]

[1] Hai Phong University, Haiphong, Vietnam
luanpn@dhhp.edu.vn
[2] VNU University of Science, Hanoi, Vietnam
tienhapt@gmail.com
[3] VNU University of Engineering and Technology, Hanoi, Vietnam
vinhnv@vnu.edu.vn

Abstract. Correction of Vietnamese grammatical errors plays an important role in Natural Language Processing. In this paper, we propose a new method using Machine Translation. We consider the grammatical error correction problem like machine translation problem with source language as grammatical wrong text and target language as grammatical right texts, respectively. Additionally, we carry out pre-processing step with grammatical wrong text using spelling checker such as MS Word spelling tool before using Machine translation model.

Our experiments based on the state-of-the-art Machine Translation systems combining with pre-processing step. Experimental results achieved 84.32 BLEU score with Vietnamese grammatical error correct based on SMT architecture and 88.71 BLEU score system based on NMT architecture, which indicates that our method achieves promising results.

Keywords: Vietnamese grammatical error correction · Statistical Machine Translation · Neural Machine Translation

1 Introduction

Nowadays, correction of grammatical errors is an active research topic, this topic based on Machine Translation has been applied to English, but there is not any research which uses Machine Translation for Vietnamese.

Vietnamese is not easy to learn, even both Vietnamese people and Vietnamese learners usually make grammatical errors in the text. There are several types of error, such as spelling mistakes, using wrong words. A Vietnamese grammatical error correction (GEC) system will have the benefit for Vietnamese and Vietnamese learners. Also, the GEC models can be applied to Natural Language Processing systems. The difference in our method is that we apply the model to Vietnamese, which is much harder than English. As the increasing number of information, we have a chance to access to the valuable source of knowledge about potential customers. Information extraction from Vietnamese online text, however, is a critical natural language understanding. This is the most challenge.

© Springer Nature Singapore Pte Ltd. 2020
L.-M. Nguyen et al. (Eds.): PACLING 2019, CCIS 1215, pp. 505–512, 2020.
https://doi.org/10.1007/978-981-15-6168-9_41

We propose a new method for Vietnamese grammatical error correction. It is useful for a non-native Vietnamese learner and for a native speaker. Our presentation is structured: Sect. 2 summarizes the related work. Section 3 described our method. Section 4 presents the experiments. Finally, conclusions are presented in Sect. 5.

2 Related Work

As we mentioned above, the correction of grammatical errors is an active research topic. Therefore, many studies have been published. In this section, we present some approaches to correct grammatical errors in recent years.

In [7], Courtney Napoles and Chris Callison-Burch presented an investigation about components of a statistical machine translation pipeline then authors customized for grammatical error correction. They showed that extending the translation grammar with generated rules for spelling correction can improve the Max-Match metric score by as much as 20%.

In [1], Kai-Fu proposed an approach to grammatical error correction using neural machine translation for Chinese. Their staged approach includes: first they remove the surface errors. Then they built the grammatical error correction system using neural machine translation.

In [2], authors proposed the method that combines two popular approaches *(SMT and NMT)* to build a system for automated grammatical error correction. This combination system gains new results on the CoNLL-2014 and JFLEG benchmarks.

The methods above are most related to our method, but our method is different from these methods as some points:

1. We carry out pre-processing step using spelling checker with the Vietnamese input text, then put it in the machine translation system to correct remaining grammatical errors.
2. We also solve grammatical errors correction in Vietnamese language using Machine Translation. According to our understanding, this is the research that applying Machine Translation for Vietnamese grammatical errors correction, the first time.

3 Our Method

We treat the Vietnamese grammar detection and correction problem like machine translation problem, so this task, we propose a method using machine translation. In particular, wrong grammar and right grammar texts are considered like source and target language respectively. Machine translation model detect and correct grammar errors.

3.1 Machine Translation

Phrase-based Statistical Machine Translation: The input texts are segmented into a number of sequences of words or phrases. Each phrase in the source sentence is translated into the target language. The translation model is built on the noisy channel model [4]. This model uses Bayes rules to reformulate translation probabilities to translate a foreign sentence f into e. The best translation for a foreign sentence f is the Eq. 1:

$$e = arg \max_e p(e)p(e|f) \tag{1}$$

The above equation consists of two main components: the language model p(e) and the translation model p(e∥f). Monolingual data in the target side is used for training language model and parallel data is used for training translation model, parameters are estimated from parallel data, the best output sentence e for the input sentence f according to the equation

$$e = arg \max_e p(e|f) = arg \max_e \sum_{m=1}^{M} \lambda_m h_m(e, f) \tag{2}$$

where h_m is a feature function such as language model, translation model and λ_m corresponds to a feature weight.

Neural Machine Transaltion: Given a sentence in source side $x = (x_1, ..., x_m)$ and its corresponding sentence in target side $y = (y_1, ..., y_n)$. In paper, we use the attentional NMT architecture proposed by [5]. In their work, the encoder, which is a bidirectional recurrent neural network, reads the source sentence and generates a sequence of source representations $h = (h_1, ..., h_m)$. The decoder is another recurrent neural network, produces the target sentence at a time. The log conditional probability thus can be decomposed as follows:

$$\log p(y|x) = \sum_{i=1}^{n} \log p(y_t|y_{<t}, x) \tag{3}$$

where $y_{<t} = (y_1, ..., y_{t-1})$. As described in Eq. 4, the conditional distribution of $p(y_t|y_{<t}, x)$ like a function of the previously predicted output y_{t-1}, the hidden state of the decoder s_t, and the context vector c_t.

$$p(y_t|y_{<t}, x) \propto \exp\{g(y_{t-1}, s_t, c_t)\} \tag{4}$$

The context vector c_t is used to determine the relevant part of the source sentence to predict y_t. It is computed as the weighted sum of source representations $h_1, ..., h_m$. Each weight α_{ti} for h_i implies the probability of the target symbol y_t being aligned to the source symbol x_i:

$$c_t = \sum_{i=1}^{m} \alpha_{ti} h_i \tag{5}$$

Given a parallel data of size N, the parameter θ of NMT model is trained to maximize the probabilities for all sentence pairs $\{(x^n, y^n)\}_{n=1}^{N}$:

$$\theta^* = arg\max_{\theta} \sum_{n=1}^{N} \log p(y^n|x^n) \tag{6}$$

where θ^* is the optimal parameter.

3.2 Our Method for Vietnamese Grammatical Error Correction

Each language has its own characteristics, and so is Vietnamese. To correct Vietnamese grammatical errors, we must recognize as much error types as possible. Generally, the grammatical error types in Vietnamese can be divided into two groups, as below:

Errors in Sentence Structure: These errors include errors such as sentence components missing, overlapping sentence components and sentences components wrongly ordering.

- *Missing sentence component:* there is a lot of shortened sentences which have only component subject or predicate, thus it makes the sentence meaning ambiguous.
- *Overlapping sentence component:* These errors are often caused by learner's unclear ideas or their limited language ability.
- *The sentence components are in the wrong order:* Unlike English, in Vietnamese, the order of components in a sentence is very important. When we make this kind of error, it makes the sentence meaningless or ambiguous.

Errors in Punctuation: Punctuation in the text is very important because it defines the grammatical structure and expresses the meaning of the sentence. Therefore, errors in punctuation can negatively affect the learners' purpose, which can lead to serious misunderstandings.

The main idea of this paper is correction grammatical errors be considered like translation problem, so the input text in the source language as Vietnamese grammatical wrong and output text is Vietnamese grammatical right as the target language. To solve this problem, we proposed a new method which is described in Fig. 1.

A key advantage of the machine translation is that errors are learned from parallel data automatically. To evaluate the effect of our method, we conduct experiments on the state-of-the-art Machine Translation systems: Statistical Machine Translation (SMT) and Neural Machine Translation (NMT).

4   Experiments

4.1 Dataset

We first collect 317,596 Vietnamese sentences from news sites like dantri.com.vn; vnexpress.net and then cleaning and make grammatical error types from the to

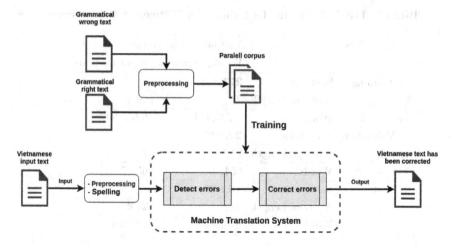

Fig. 1. Illustration for our method. A parallel corpus is collected from grammatical wrong text and grammatical right text, this parallel corpus is used to build a Vietnamese GEC system using Machine Translation (SMT - NMT)

build about 271,822 parallel sentence pairs for training, 29,895 sentences pairs for validation, and 15,879 sentences pairs for the test. The Table 1 is the data statistics for training our Vietnamese GEC systems.

4.2 Settings

We used Moses[1] and OpenNMT[2] [3] to training our Vietnamese GEC systems.

The NMT system is trained Long Short-Term Memory (LSTM) network [5], we use 2-layer, 500 hidden units on the encoder/decoder and the general attention type of Thang Luong [6].

To evaluate the quality of our Vietnamese GEC, we use the BLEU score that standard metric to evaluate the quality of translation systems.

4.3 Results and Discussions

We trained two Vietnamese grammatical error correction systems based on SMT and NMT with the same parallel corpus, they are called Vietnamese GEC_SMT and Vietnamese GEC_NMT. We evaluate the quality of these two systems with two types of input text:

- **None-Spelling:** Vietnamese input text is pre-processed, do not carry out the spelling check step *(Vietnamese GEC_SMT and NMT)*;
- **Spelling:** Vietnamese input text is pre-processed and carry out the spelling check step *(Spell+Vietnamese GEC_SMT and NMT)*.

---

[1] http://statmt.org/moses/.
[2] https://github.com/OpenNMT/OpenNMT-py.

Table 1. The data statistics for training our Vietnamese GEC systems.

Data Sets		Vietnamese language	
		Wrong grammar	Right grammar
Training	Sentences	**271,822**	
	Average Length	21.1	20.8
	Words	5,735,444	5,653,897
Validation	Sentences	**29,895**	
	Average Length	21.9	21.8
	Words	654,700	651,711
Test	Sentences	**15,879**	
	Average Length	21.8	21.6
	Words	346,162	342,986

We measured by BLEU score with the same data set for test, experimental results are described as in the Fig. 2.

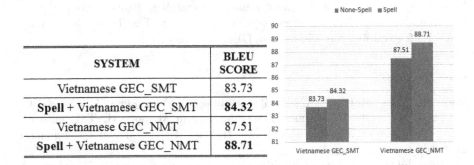

SYSTEM	BLEU SCORE
Vietnamese GEC_SMT	83.73
Spell + Vietnamese GEC_SMT	**84.32**
Vietnamese GEC_NMT	87.51
Spell + Vietnamese GEC_NMT	**88.71**

Fig. 2. The BLEU score: Vietnamese GEC_SMT vs Vietnamese GEC_NMT

In the Fig. 2 show experiemental results of the Vietnamese Grammatical error correction systems, the BLEU score achieved **83.73** points for the Vietnamese GEC_SMT system and **87.51** points for the Vietnamese GEC_NMT system. If the input text is pre-processed and spelling correction before appling Machine Translation models, our systems get better results: the BLEU score achieved **84.32** points for the Spell+Vietnamese GEC_SMT system and **88.71** points for the Spell+Vietnamese GEC_NMT system.

The Fig. 3 shows some example outputs of our systems. From these results, it shows that the NMT system is better than SMT system in Vietnamese grammatical error correction. Both the Vietnamese GEC_SMT system and the Vietnamese GEC_NMT system are restricted in correcting errors that sentence is lacked of characters, rhythm, etc.

The Vietnamese GEC_NMT system correct unk errors *(errors that it unknown)* are not good, but it can correct grammatical errors well. We could get better results when we carry out pre-processing step with the input text using spelling checker tool before using Machine Translation model.

Input sentence	Reference sentence	Output sentence of system			
		Vietnamese GEC_SMT	Vietnamese GEC_NMT	Spell+Vietnamese GEC_SMT	Spell+Vietnamese GEC_NMT
Lập doanh trại ở làng dgần nhăất	Lập doanh trại ở làng gần nhất	Lập doanh trại ở làng dgần nhất	Lập doanh trại ở làng gần nhất	Lập doanh trại ở làng gần nhất .	Lập doanh trại ở làng gần nhất .
rMặt nđó sưng lên vì ăng đau.	Mặt nó sưng lên vì răng đau .	Mặt sưng lên vì nó tăng đau	Mặt nó sưng lên vì đau đau .	Mặt sưng lên vì nó răng đau .	Mặt nó sưng lên vì răng đau .
tôi vừa nhìn . thảy được , một hình người trong bóng tôi .	Tôi vừa nhìn thảy được một hình người trong bóng tôi .	Tôi vừa nhìn thảy được một hình người trong bóng tôi .	Tôi vừa nhìn thảy được một hình người trong bóng tôi .	Tôi vừa nhìn thảy được một hình người trong bóng tôi .	Tôi vừa nhìn thảy được một hình người trong bóng tôi .
Nhìn bao quát vù uê từ trên đỉnh đồ.	Nhìn bao quát vùng quê từ trên đỉnh đồi .	Nhìn bao quát quê từ trên đỉnh đồi .	Nhìn bao quát vù quê từ trên đỉnh đồ.	Nhìn bao quát vùng quê từ trên đỉnh đồi .	Nhìn bao quát vùng quê từ trên đỉnh đồi .
Có s traảo chính trị giữa haai .	Có sự trao đổi từ chính trị giữa hai nước .	Có sự traảo chính trị giữa hai .	Có lẽ chính trị giữa hai .	Có sự trao chính trị giữa hai .	Có sự chính trị giữa hai .
Qua bản báo cao cho ta thảy được thực trạng ô nhiễm môi trường hiện nay.	Bản báo cáo cho ta thảy được thực trạng ô nhiễm môi trường hiện nay.	Bản báo Qua cao cho ta thảy được thực hiện trạng ô nhiễm môi trường nay .	Bản báo cao cho ta thảy được thực trạng ô nhiễm môi trường hiện nay .	Qua bản báo cáo cho ta thảy được thực hiện trạng ô nhiễm môi trường nay .	Bản báo cáo cho ta thảy được thực trạng ô nhiễm môi trường hiện nay .

Fig. 3. Some outputs of Vietnamese grammatical error correction systems

5   Conclustion and Future Work

In this paper, we presented a new method for Vietnamese grammatical errors correction. We have investigated the effectiveness of models trained with SMT model and NMT model *(the state-of-the-art MT now)* when we applied to solve this GEC problem for Vietnamese. The experimental results show that the quality of grammatical errors correction is promising and could apply this method in real-world.

In the future, we will focus on improving quality. First, we can use the bigger amount of data to train our GEC system, bigger training data is, the more accurate model is. Second, we will use a hybrid SMT and NMT system for GEC system. Finally, we also will focus on collecting and analyzing data, as long as creating more quality data to improve the system.

Acknowledgments. This work is funded by the project: Building a machine translation system to support translation of documents between Vietnamese and Japanese to help managers and businesses in Hanoi approach Japanese market, under grant number TC.02-2016-03 and the project of VNU University of Engineering and Technology, Hanoi, Vietnam.

References

1. Fu, K., Huang, J., Duan, Y.: Youdao's winning solution to the NLPCC-2018 task 2 challenge: a neural machine translation approach to Chinese grammatical error correction. In: Inproceedings (2018)
2. Grundkiewicz, R., Junczys-Dowmunt, M.: Near human-level performance in grammatical error correction with hybrid machine translation. In: Proceedings of the 2018 Conference of the North American Chapter of the Association for Computational Linguistics: Human Language Technologies, Volume 2 (Short Papers) (2018)
3. Kim, Y. Deng, Y., Senellart, J., Klein, G., Rush, A.M.: OpenNMT: open-source toolkit for neural machine translation. arXiv preprint arXiv:1701.02810 (2017)
4. Koehn, P.: Statistical Machine Translation. Cambridge University Press, Cambridge (2010)
5. Pham, H., Luong, M.-T., Manning, C.D.: Effective approaches to attention-based neural machine translation. arXiv preprint arXiv:1508.04025 (2015)
6. Pham, H., Luong, M.-T., Manning, C.D.: Effective approaches to attention-based neural machine translation. In: Proceedings of EMNLP (2015)
7. Napoles, C., Callison-Burch, C.: Systematically adapting machine translation for grammatical error correction. In: Proceedings of the 12th Workshop on Innovative Use of NLP for Building Educational Applications, Copenhagen, Denmark, 8 September 2017, pp. 345–356. Association for Computational Linguistics (2017)

Developing a Framework
for a Thai Plagiarism Corpus

Santipong Thaiprayoon[✉], Pornpimon Palingoon[✉],
Kanokorn Trakultaweekoon[✉], Supon Klaithin[✉],
Choochart Haruechaiyasak[✉], Alisa Kongthon[✉],
Sumonmas Thatpitakkul[✉], and Sawit Kasuriya[✉]

National Electronics and Computer Technology Center, Khlong Luang,
Pathumthani, Thailand
{santipong.tha, pornpimon.pal, kanokorn.tra, supon.kla,
choochart.har, alisa.kon, sumonmas.tha,
sawit.kas}@nectec.or.th

Abstract. One problem of building a Thai plagiarism corpus is the unavailability of the corpus with real examples of plagiarized texts. To solve the problem, we present a new design and construction of a Thai plagiarism corpus, called TPLAC-2019, to evaluate the plagiarism detection algorithms for Thai. The process of Thai plagiarism corpus creation consists of two methods: 1) simulated plagiarism method, and 2) artificial plagiarism method. For the simulated plagiarism method, we provided a Thai plagiarism tagging tool called PlaTool and a Thai plagiarism guideline for assisting human annotators to plagiarize the text passages. As for artificial plagiarism method, plagiarized documents are automatically generated by a machine. Besides, a new method to automatically create plagiarized text passages is proposed in the artificial plagiarism method. The objective of this proposed method is to automatically create plagiarized text passages that resemble human language. To evaluate the performance of machine-generated Thai plagiarized text passages, we prepared the test sets which are generated from the baseline and the proposed methods. The experiments are set up to compare the readability of human-readable texts in plagiarized documents between two different methods. The experimental results show that the proposed method helps improve the readability of human-readable texts which is increased up to 40%.

Keywords: Thai plagiarism corpus · Simulated plagiarism · Artificial plagiarism · Text obfuscation

1 Introduction

In recent years, plagiarism is a critical issue that attracts a lot of attention in academic and educational communities [1–4]. With the ability to easily access content on online sources such as digital libraries, websites and others, some students often plagiarize their assignments by copying and modifying texts obtained from the online sources without proper acknowledgments. As a result, the number of plagiarisms has increased dramatically in higher education institutions. In the last two decades, a myriad of

L.-M. Nguyen et al. (Eds.): PACLING 2019, CCIS 1215, pp. 513–522, 2020.
https://doi.org/10.1007/978-981-15-6168-9_42

plagiarism detection algorithms and tools have been developed to deal with this issue. Previous research studies on plagiarism detection algorithms [5] revealed that two approaches are generally used to detect plagiarism: intrinsic and external approaches. In the intrinsic approach, plagiarized texts are detected within a suspicious document that is inconsistent in writing style without comparing it with the source documents. The external approach, on the other hand, adopts different techniques to find plagiarized passages in a suspicious document by comparing them against the source documents. Both approaches are difficult and challenging due to a variety of plagiarism cases such as copying, rewording (obfuscating) and paraphrasing. In this paper, we focus on the external plagiarism approach that depends on external source documents. Therefore, a standardized evaluation corpus containing various cases of plagiarized texts is necessary for developing and evaluating algorithms to detect different cases of plagiarism.

Much research has been done on methodologies involving plagiarism in corpus construction in various languages such as Arabic, Urdu, Persian and English [6–12] In general, plagiarism corpus construction can be divided into three major methods: (1) real cases of plagiarism (real plagiarized documents from someone else's work), (2) simulated plagiarism (manually created plagiarized documents by humans to simulate plagiarism cases), and (3) artificial plagiarism (automatically generated plagiarized documents by machine). However, the construction of the corpus containing real plagiarism cases is quite expensive and difficult due to legal and ethical issues. Simulated and artificial plagiarism methods, on the other hand, are often employed to create the plagiarism corpus in any languages. Simulated plagiarism method is more realistic in terms of simulating the real behavior of plagiarists and can create plagiarized text passages that are not much different from the real cases of plagiarism while artificial plagiarism method can generate many plagiarized text passages rapidly. Although both methods are widely used in creating plagiarism corpora, there are some drawbacks. Constructing a simulated plagiarism corpus is laborious and time-consuming because there are many steps to create a plagiarized document. Moreover, simulated plagiarism construction usually lacks annotation guidelines and tools to help human annotators to imitate plagiarizing the text passages. As for artificial plagiarism, text passages generated with this method is still human-unreadable because the text passages are generated randomly by machine without analyzing the semantic sentence.

To construct a Thai plagiarism corpus, in this paper we use both simulated and artificial plagiarism models. However, as previously mentioned, the lack of Thai annotation guidelines makes it hard to implement the simulated plagiarism. Another problem concerns artificial plagiarism is the complexity of Thai language. Since Thai is an unsegmented language and complex morphology, it is difficult to identify word and sentence boundaries in order to automatically generate artificial plagiarism cases. To overcome the problems, we proposed a framework for constructing the Thai plagiarism corpus containing simulated and artificial cases of plagiarized documents called TPLAC-2019 corpus. This corpus is based on plagiarism cases and linguistic mechanisms in Thai plagiarism which are used to evaluate the plagiarism detection algorithms. We use Thai Wikipedia articles as the main sources for plagiarizing text passages. In the simulated plagiarism method, we create suspicious documents manually using simple four cases of Thai plagiarism, namely, copy and paste, substitution, insertion, and deletion. These cases are performed to obfuscate text passages from a source document to become plagiarized texts in a suspicious document. In the artificial plagiarism method, the plagiarized text passages of aforementioned plagiarism cases are created by a machine and automatically inserted into the suspicious document. A new method is also developed to automatically create the plagiarized text passages that are naturally close to human language in Thai. An evaluation method is suggested to measure the quality of TPLAC-2019 corpus by considering human-readable text passages in terms of automatic text passages generation by machine.

Our main contributions in this paper are to propose a framework for constructing a Thai plagiarism corpus, containing four plagiarism cases provided in our annotation scheme as a guideline for annotators. A new method to automatically generate plagiarized texts that are close to the natural language usage in Thai will be proposed.

2 Related Corpora

Research in plagiarism pays a lot of attention to the construction of plagiarism corpora in different languages. This section presents some existing plagiarism corpora both Thai and other languages.

2.1 Existing Plagiarism Corpora

Most of the research studies aim to create resources that contain artificial and simulated examples of plagiarism. We discuss two such resources, (1) The corpus of plagiarized short answers [1] (simulated plagiarism), and (2) the PAN-PC Corpora [6, 7, 13] (simulated and artificial plagiarism). The short answer corpus [1] contains plagiarized and non-plagiarized texts in English language. The corpus is manually created by human consisting of 100 documents of length between 200–300 words. The documents are created with four levels of reuse, namely, near copy, light revision, heavy revision, and non-plagiarism. The corpus has 5 source documents which are used to create 57 plagiarized and 38 non-plagiarized documents. The PAN-PC corpora have been developed and matured over the years, and contain documents from Project Gutenberg. In these corpora, the plagiarized documents contain either artificial, simulated or both cases of plagiarism. A number of modification strategies were applied to create different levels of obfuscation.

2.2 Thai Plagiarism Corpus

The ATPC [2] corpus is an academic plagiarism corpus, a corpus that collects simulated academic plagiarism texts in Thai. The corpus uses two main methods: manually created by participants and automatically generated by a program. This corpus consists of two main types of texts: plagiarized and non-plagiarized texts. Plagiarized texts are categorized into four types based on the degree of linguistic mechanisms used in plagiarism.

3 Thai Plagiarism Corpus Construction

In this section, we describe the framework of Thai plagiarism corpus construction which consists of two main methods. Each method of plagiarism document construction will be given in more detail in the subsequent sections.

3.1 The Framework of Thai Plagiarism Corpus Construction

The proposed framework of Thai plagiarism corpus construction is comprised of simulated and artificial plagiarism methods. This framework focuses on simple four cases of Thai plagiarism. For simulated plagiarism method, we aim to design and develop a tool for creating the plagiarized documents that imitate the plagiarism scenario from plagiarists. For artificial plagiarism method, we aim to create the plagiarized documents that resemble human language. The overall methods are shown (see Fig. 1). It indicates that the same steps of the construction are repeated whether the methods are simulated or artificial plagiarism.

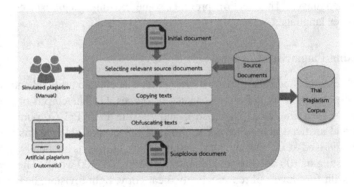

Fig. 1. A framework of Thai plagiarism corpus construction

For the simulated plagiarism method, suspicious documents containing simulated plagiarism cases are manually created by human annotators based on plagiarism cases. These plagiarism cases are adopted to create plagiarized text passages and inserted them into an initial document. After the process of plagiarizing text passages is finished, the initial document will be converted to the suspicious document and imported to a database. To help human annotators make the plagiarized text passages including Thai plagiarism cases, we provide a Thai plagiarism tagging tool called PlaTool and a Thai plagiarism guideline for assisting human annotators to plagiarize the text passages from Thai Wikipedia articles and web page sources. In short, the simulated method consists of three steps: (1) selecting relevant source documents (2) copying texts, and (3) obfuscating texts.

For the artificial plagiarism method, we propose a new method to automatically create plagiarized text passages that resemble human language. First, suspicious documents containing artificial cases of plagiarism are automatically generated by a machine. This method simulates plagiarized text passages by automatically selecting a text passage that meets particular conditions in an article from Thai Wikipedia articles. These conditions will be explained in Sect. 3.3. This method copies and obfuscates the selected text passage with one type of plagiarism cases. After that, they are inserted into the initial document and saved as the suspicious document which in turn is imported to the database. The suspicious documents generated by the simulated and artificial plagiarism methods are written in the standard XML format. This collection of all suspicious documents in the database is the Thai plagiarism corpus.

3.2 The Creation of Simulated Thai Plagiarism Documents

The preliminary creation of simulated Thai plagiarism documents consists of the initial and source documents. The initial documents are prepared for being plagiarized documents. The source documents are a collection of all source documents that are prepared for searching articles corresponding to given keyword title in the initial document. To create the initial and source documents, we collected 110,000 articles from Thai Wikipedia and web pages. These articles are divided into two groups, namely, 109,000 articles for source documents and 1,000 articles for the initial documents selected by linguists from several domains such as science, sports, and travel. These articles are then imported to PlaTool for the human annotators to manually create the plagiarized text passages easily (see Fig. 2).

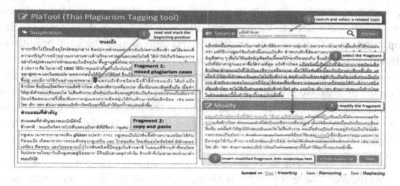

Fig. 2. PlaTool (Thai plagiarism tagging tool)

From Fig. 2, an initial document is first displayed in the left column. The process of annotation consists of many steps. A human annotator starts reading the article and marking the position of text where he or she needs to put the plagiarized text passage in the initial document (see number 1 in Fig. 2). Then, the annotator searches and selects a related article and then the annotator copies some parts of the text in the related article approximately 150–250 characters, called text passage, from the source documents (see number 2 and 3 in Fig. 2). Next, the annotator intentionally and freely obfuscates the text passage by using only one in four plagiarism case or mix them together (see number 4 in Fig. 2). After completing the obfuscations of the text passages, the annotator inserts the obfuscated text passage into the initial document (see number 5 in Fig. 2). After plagiarizing has finished, the initial document becomes suspicious document and then the suspicious document is imported to the database.

3.3　The Creation of Artificial Thai Plagiarism Documents

To create the artificial Thai plagiarism documents, we collected 110,000 articles from Thai Wikipedia and web pages. Out of 110,000 articles, we selected 1,000 vital and featured articles which are chosen by editors in Thai Wikipedia to be the initial documents. The remaining articles of 109,000 articles are the source documents. The method of creating the suspicious documents by machine consists of three main steps: 1) selecting relevant source documents, 2) copying texts, and 3) obfuscating on texts.

1) **Selecting relevant source documents:** This step applies keyword titles appearing in an initial document to perform the relevant source documents selection. The keyword title is words or phrases that can give a text definition and link to another article in the source documents. There are two conditions in finding the keyword titles. First, the words or phrases occurring in the initial document must be the keyword title. For the second condition, the next word of the keyword title must be the transition words such as ดังนั้น (therefore, so, consequently, thus), โดยทั่วไป (generally), อย่างไรก็ตาม (however), หลังจากนั้น (after), นอกจากนี้ (besides, in addition, moreover, furthermore), and แม้ว่า (although, though, even though). Due to the problem of sentence boundaries in Thai, we adopt these transition words to indicate the end of the sentence. When the words or phrases in the initial document meet the conditions, the article related to the keyword title from the source documents is selected.

2) **Copying texts:** This step is to extract the text definition of the keyword title in the selected article from the source documents. The length of the text definition is between 150–250 characters. To solve the problems of incomplete breaking the text definition in Thai, we use whitespace to break the text definition between two texts.

3) **Obfuscating texts:** This step aims to obfuscate the text passage by using one of the four plagiarism cases but keeping the original meaning. The detail of obfuscating in each case is described below.

- *Copy and paste*: The text passage is copied without text obfuscation.
- Substitution: The text passage is obfuscated by using a few words with the same or nearly the same meaning such as ผู้หญิง (woman) and สตรี (lady), ดอกไม้ (flower) and บุปผา (blossom), พ่อ (dad, daddy), and บิดา (father).
- *Insertion*: The text passage is obfuscated by adding a few words with no effects on text meaning such as ก็ (then, also), ไว้ (keep, hold), อาจ (may, might), and ได้ (able to).
- *Deletion*: The text passage is obfuscated by deleting unimportant words such as ก็ (then, also), ขณะที่ (while), อาจ (may, might), ได้ (able to), โดยที่ (whereas), อีก (again), นี้ (this, that), ยัง (still), จะ (will, would), and ซึ่ง (that, which).

The words adopted in each plagiarism case are a result of analysis on language patterns in Thai. In case of words insertion and deletion, we chose some linguistic categories of function words such as auxiliary verbs and particles and content words such as adverbs that have no effects on text meaning whether they are inserted or deleted in the text passage. As for words substitution, we use a synonym for obfuscating the text passages in our work. Finally, the text passage including plagiarism cases is inserted right after the keyword title in the initial document. Then, the initial document is saved as the suspicious document and is inserted into the database.

4 An Example of Thai Plagiarism Cases

This section shows an example of plagiarism cases obfuscated by both simulated and artificial methods. The example case of copy and paste is shown in Table 1.

Table 1. The example case of copy and paste

Plagiarism case	Simulated plagiarism method	Artificial plagiarism method
Copy and paste	**[In Thai]** บ้านพิษณุโลกเคยใช้เป็นฉากของภาพยนตร์เรื่อง "บ้านทรายทอง" <u>ซึ่งถูกสร้างจากนักประพันธ์ผู้มีชื่อเสียง</u> นับว่าเป็นภาพยนตร์ที่ประสบความสำเร็จอย่างยิ่งในยุคสมัยนั้น **[Translation]** The Phitsanulok House was used for the scene of the movie titled "Baan Sai Thong", <u>which was created by a famous novelist.</u> It was very famous in that era. **[Description]** As illustrated in the underlined text, the underlined text is selected and copied manually from the article in source documents without any modifications.	**[In Thai]** นอกจากนี้คาเฟอีนเป็นสารกระตุ้นระบบประสาทส่วนกลาง <u>ซึ่งเป็นส่วนที่มีโครงสร้างที่ใหญ่ที่สุดของระบบประสาทและกระบวนการเผาผลาญอาหารในร่างกาย</u> **[Translation]** In addition, caffeine is a stimulant to the central nervous system <u>which is the part that contains the biggest structure of the nervous system</u> and the metabolic process in the human body. **[Description]** As illustrated in the underlined text, the keyword title in the text is "ระบบประสาทส่วนกลาง/the central nervous system". This keyword links to its article in the source documents. The text definition in the source documents is then copied and inserted automatically without any modifications.

We can conclude that the cases of copy and paste in both simulated and artificial plagiarism methods are to copy exact text passage from the article in the source documents without any modifications. As for the substitution cases, the synonym, which is adopted to replace the word in the text, is a vivid example of replacing a word by keeping the same meaning. While the insertion and deletion cases use the function words to obfuscate without changing the principle meaning of the text passage.

5   The General Statistics of Thai Plagiarism Corpus

Our corpus consists of source documents and suspicious documents. The number of source documents is 110,000 articles, 109,200 Thai Wikipedia articles, and 800 web pages articles. The suspicious documents are divided into two groups; first is 1,000 documents for human annotators and second is 1,000 documents for machine simulation. The statistics of plagiarism cases created from our framework is shown in Table 2.

Table 2. The statistics of plagiarism cases created from our framework

Plagiarism cases	Number of plagiarism cases	
	Simulated plagiarism method	Artificial plagiarism method
Copy and paste	2,010	250
Substitution	1,863	250
Insertion	1,947	250
Deletion	1,990	250
Total	7,810	1,000

From Table 1, it is noteworthy that the number of plagiarism cases created by the simulated plagiarism method is greater than the artificial plagiarism method because a human annotator can create many plagiarism cases of text passages in one suspicious document. Meanwhile, the number of plagiarism cases created by the artificial plagiarism method is equal in every plagiarism case because this method is set the parameters to create only one case of text passage in one suspicious document. However, the number of plagiarism cases is a non-significant variable for evaluating plagiarism detection algorithms.

6   The Experimental Results and Discussions

To evaluate the readability of machine-generated text passages, we prepared two test sets, which generated automatically by the artificial plagiarism method. The first test set contains 100 plagiarized text passages randomly created by following the obfuscation strategy of PAN-PC-2009 that consists of shuffling, removing, inserting, and replacing operation. This operation is considered to be the baseline method. The second test set also contains 100 plagiarized text passages created by the proposed method. Each of the plagiarized text passages in the test sets includes only one case of plagiarism. We then compared both methods in terms of the human-readable texts in plagiarized documents. As for evaluation, we asked five linguists to approve both test sets in terms of human-readable texts. After this process, we counted the votes of all linguists to identify which text passages give a better human-readable text in both methods. The evaluation results are summarized in Table 3.

Table 3. The evaluation results

Methods	Human-readable texts
Baseline method	7
Proposed method	47

From Table 3, our proposed method yields a better performance of generating the human-readable texts in plagiarized documents than the baseline method. The baseline method achieved 7% average readability of plagiarized text passages and the proposed method achieved up to 47% average readability of plagiarized text passages. Hence, the proposed method improves 40% average readability of plagiarized text passages from the baseline method. Our method yields a better result than the baseline method because the proposed method is able to select the relevant article and place the text definition at the end of the sentence correctly. Besides, the proposed method applies the function words and synonyms to obfuscate the text passages. One reason we adopt these words is that these words still keep the text meaning and sentence structure whether these words are inserted, deleted or replaced in the text passage. From the error analysis, we found that some plagiarized text passages are ambiguous texts in terms of text meaning and text naturalness due to the problem of Thai sentence segmentation. As mentioned in the previous section, there are no explicit markers to identify sentence boundaries in Thai written texts, therefore, using whitespace to segment the text definition could obtain incomplete sentences.

7 Conclusions and Future Works

We have presented a framework for constructing the Thai plagiarism corpus containing simulated and artificial cases of plagiarized documents called TPLAC-2019 corpus. This corpus is based on plagiarism cases and linguistic mechanisms in Thai. The process of corpus creation consists of two methods: (1) simulated plagiarism method, and (2) artificial plagiarism method. For the simulated plagiarism method, we provided a Thai plagiarism tagging tool called PlaTool and a Thai plagiarism guideline for assisting human annotators to plagiarize the text passages. As for artificial plagiarism method, plagiarized documents are automatically generated by a machine. A new method to automatically create plagiarized text passages is also proposed in the artificial plagiarism method. The objective of the proposed method is to automatically create plagiarized text passages that resemble human language. To evaluate the performance of machine-generated Thai plagiarized text passages, we prepared the test sets which are generated from the baseline and the proposed methods. The experiments are set up to compare the readability of human-readable texts in plagiarized documents between two different methods. The experimental results show that the proposed method helps improve the readability of human-readable texts. The baseline method achieved 7% average readability of plagiarized text passages and the proposed method achieved up to 47% average readability of plagiarized text passages. For the proposed method, the average readability of human-readable texts is increased by 40%. In the future to improve our proposed method, we plan to apply syntactic analysis and Thai sentence segmentation algorithm to divide the given text passages into meaningful sentences. This algorithm will help improve the performance in part of accurately selecting the meaningful sentences that result in the creation of unambiguous texts and naturalness of texts. Finally, we will encourage the Thai plagiarism corpus to be available as the standard corpus for research and development of Thai plagiarism detection algorithms.

References

1. Clough, P., Stevenson, M.: Developing a corpus of plagiarised short answers. Lang. Resour. Eval. **45**(1), 5–24 (2011)
2. Taerungruang, S., Aroonmanakun, W.: Constructing an academic Thai plagiarism corpus for benchmarking plagiarism detection systems. J. Lang. Stud. **18**(3), 186–202 (2018)
3. Miranda-Jiménez, S., Stamatatos, E.: Automatic generation of summary obfuscation corpus for plagiarism detection. J. Appl. Sci. **14**(3), 99–112 (2017)
4. Juričić, V., Štefanec, V., Bosanac, S.: Multilingual plagiarism detection corpus. In: 35th International Convention MIPRO, pp. 1310–1314. IEEE, Croatia (2012)
5. Barrón-Cedeño, A., Potthast, M., Rosso, P., Stein, B., Eiselt, A.: Corpus and evaluation measures for automatic plagiarism detection. In: The Seventh Conference on International Language Resources and Evaluation, Malta (2010)
6. Potthast, M., Stein, B., Barrón-Cedeño, A., Rosso, P.: An evaluation framework for plagiarism detection. In: 23rd International Conference on Computational Linguistics, pp. 997–1005. Association for Computational Linguistics, China (2010)
7. Potthast, M., Stein, B., Eiselt, A., Barrón-Cedeño, A., Rosso, P.: Overview of the 1st international competition on plagiarism detection. In: SEPLN 2009 Workshop on Uncovering Plagiarism, Authorship, and Social Software Misuse (PAN 09), pp. 1–9 (2009)
8. Mohtaj, S., Asghari, H., Zarrabi, V.: Developing monolingual English corpus for plagiarism detection using human annotated paraphrase corpus. In: Working Notes of CLEF 2015 (2015)
9. Siddiqui, M.A., Khan, I.H., Jambi, K.M., Elhaj, S.O., Bagais, A.: Developing an Arabic plagiarism detection corpus. In: The International Conference on Computer Science, Engineering and Information Technology (CSEIT-2014), Australia, pp. 261–269 (2014)
10. Sharjeel, M., Rayson, P., Muhammad, R., Nawab, A.: UPPC-Urdu paraphrase plagiarism corpus. In: 10th International Conference on Language Resources and Evaluation Conference (LREC), pp. 1832–1836. Lancaster University (2016)
11. Barrón-Cedeño, A., Vila, M., Marti, M.A., Rosso, P.: Plagiarism meets paraphrasing: Insights for the next generation in automatic plagiarism detection. Comput. Linguist. **39**(4), 917–947 (2013)
12. Clough, P., Gaizauskas, R., Piao, S.S., Wilks, Y., METER: MEasuring TExt Reuse. In: 40th Annual Meeting of the Association for Computational Linguistics, pp. 152–159. Association for Computational Linguistics, Pennsylvania (2002)
13. Potthast, M., Barrón-Cedeño, A., Eiselt, A., Stein, B., Rosso, P.: Overview of the 2nd international competition on plagiarism detection. In: Notebook Papers of CLEF 2010 LABs and Workshops (2010)

Author Index

Abulaish, Muhammad 483
Ahamed Khan, M. K. A. 334
Ando, Kazuaki 454
Andrew, Judith Jeyafreeda 191
Araki, Kenji 181

Babieno, Mateusz 181
Bai, Jing 145
Bollegala, Danushka 56, 67, 123, 218, 280

Cao, Rui 145
Chen, Guan-Yuan 155
Chen, Wenye 56
Coenen, Frans 280

Dang, Tran Binh 497
Dang, Vu 231
Dias, Gaël 191
Doan, Xuan-Dung 428
Dung, Le Tien 469
Duong-Trung, Nghia 134

Ehara, Yo 88

Ferrari, Stéphane 191

George, Elizabeth Jasmi 439
Giguet, Emmanuel 191
Goh, Chooi Ling 414

Ha, My Linh 388
Hakami, Huda 56, 67
Haruechaiyasak, Choochart 513
Hashiguchi, Tomoya 245
Hayashi, Yoshihiko 43
Hirano, Miku 469
Ho, Vong Anh 319
Hoang, Thi Mai Huong 388
Hotta, Hajime 469

Ikeda, Ryuya 454

Kamal, Ashraf 483
Kasuriya, Sawit 513
Kelleher, John D. 206
Kieu, Thanh-Binh 3
Klaithin, Supon 513
Ko, Daiki 79
Kobayashi, Tetsunori 43
Komachi, Mamoru 99, 112, 347
Kongthon, Alisa 513
Koreeda, Yuta 15
Kotani, Katsunori 295
Kurosawa, Michiki 99

Lam, Pham Van 374
Le-Hong, Phuong 257, 360
Lepage, Yves 414
Linh, Le Thai 469
Luong, Tho Chi 257

Ma, Wen 145
Mamidi, Radhika 439
Mandya, Angrosh 280
Maurel, Fabrice 191
Minh, Tuan Luu 28
Miyano, Tomoya 307

Nagata, Ryo 245
Ngo, The Quyen 388
Nguyen, Anh Hoang-Tu 169
Nguyen, Chi-Ngon 134
Nguyen, Danh Hoang 319
Nguyen, Duc-Vu 169, 319, 400
Nguyen, Duong Huynh-Cong 319
Nguyen, Ha Thanh 497
Nguyen, Kiet Van 319
Nguyen, Le Minh 497
Nguyen, Le-Minh 428
Nguyen, Minh-Tien 469
Nguyen, Ngan Luu-Thuy 169, 319, 400
Nguyen, Thi Minh Huyen 388
Nguyen, Tien Ha 505
Nguyen, Van Vinh 505

Nguyen, Viet Hung 388
Nguyen-Thi, Luong 360

O'Neill, James 123
Ogawa, Tetsuji 43
Omori, Hikaru 347
Ozaki, Hiroaki 15

Palingoon, Pornpimon 513
Parsons, Simon 218
Pham, Linh Thi-Van 319
Pham, Nghia Luan 505
Pham, Son Bao 3
Phan, Viet-Anh 469
Phan, Xuan-Hieu 3
Piccardi, Massimo 3

Quach, Luyl-Da 134
Quoc, Viet Nguyen 28
Quy, Dinh Khac 374

Rajendran, Pavithra 218
Ross, Robert J. 206
Rzepka, Rafal 181

Sadoun, Driss 245
Saito, Hiroaki 307
Sembokuya, Tomonobu 267

Shimada, Kazutaka 267
Shinnou, Hiroyuki 145
Son, Nguyen Hong 469
Soo, Von-Wun 155

Takahashi, Keigo 347
Takeuchi, Koichi 79
Tan, Bui Van 374
Tanaka, Hirotaka 145
Thai, Nguyen Phuong 374
Thaiprayoon, Santipong 513
Thanh, Huong Le 28
Thatpitakkul, Sumonmas 513
Trakultaweekoon, Kanokorn 513
Tran, Duc Chung 334
Tran, Tu-Anh 428
Tsuyuki, Hiroaki 43

Van Nguyen, Kiet 169, 400
Van Thin, Dang 169, 400
Vu, Anh-Van 134

Wang, Fei 206
Wang, Mengru 15

Yamagishi, Hayahide 112
Yanai, Kohsuke 15
Yoshimi, Takehiko 295

Printed in the United States
By Bookmasters